# A TEXTBOOK OF MODERN TOXICOLOGY

## THIRD EDITION

# A TEXTBOOK OF MODERN TOXICOLOGY

**THIRD EDITION**

Edited by

Ernest Hodgson
Department of Environmental
and Biochemical Toxicology
North Carolina State University

WILEY-
INTERSCIENCE

A JOHN WILEY & SONS, INC., PUBLICATION

*Library of Congress Cataloging-in-Publication Data:*

Hodgson, Ernest, 1932–
   A textbook of modern toxicology / Ernest Hodgson.—3rd ed.
     p.   cm.
Includes bibliographical references and index.
   ISBN 0-471-26508-X
    1. Toxicology. I. Title.

   RA1211.H62 2004
   615.9—dc22                                          2003017524

# CONTENTS

**III  Toxicant Processing In vivo**                                         **75**

**6  Absorption and Distribution of Toxicants**                             **77**
*Ronald E. Baynes and Ernest Hodgson*

**18  Respiratory Toxicity                                                         317**

*Ernest Hodgson, Patricia E. Levi, and James C. Bonner*

**19  Immunotoxicity                                                               327**

*MaryJane K. Selgrade*

# PREFACE

There are some excellent general reference works in toxicology, including *Casarett and Doull's Toxicology, 6th, edition*, edited by Klaassen; a 13-volume *Comprehensive Toxicology*, edited by Sipes, Gandolfi, and McQueen; as well as many specialized monographs on particular topics. However, the scarcity of textbooks designed for teacher and student to use in the classroom setting that impelled us to produce the first and second editions of this work is still apparent. With the retirement of Dr. Levi, a mainstay of the first two editions, and the continuing expansion of the subject matter, it seemed appropriate to invite others to contribute their expertise to the third edition. All of the authors are, or have been, involved in teaching a course in general toxicology at North Carolina State University and thus have insights into the actual teaching process as well as the subject matter of their areas of specialization.

At North Carolina State University, we continue to teach a course in general toxicology that is open to graduate students and undergraduate upperclassmen. In addition, in collaboration with Toxicology Communications, Inc., of Raleigh, North Carolina, we present an accelerated short course at the same level. Our experience leads us to believe that this text is suitable, in the junior or senior year, for undergraduate students with some background in chemistry, biochemistry, and animal physiology. For graduate students it is intended to lay the foundation for subsequent specialized courses in toxicology, such as those in biochemical and molecular toxicology, environmental toxicology, chemical carcinogenesis, and risk assessment.

We share the view that an introductory text must present all of the necessary fundamental information to fulfill this purpose, but in as uncomplicated a manner as possible. To enhance readability, references have been omitted from the text, although further reading is recommended at the end of each chapter.

Clearly, the amount of material, and the detail with which some of it is presented, is more than is needed for the average general toxicology course. This, however, will permit each instructor to select and emphasize those areas that they feel need particular emphasis. The obvious biochemical bias of some chapters is not accidental, rather it is based on the philosophy that progress in toxicology continues to depend on further understanding of the fundamental basis of toxic action at the cellular and molecular levels. The depth of coverage of each topic represents that chapter author's judgment of the amount of material appropriate to the beginning level as compared to that appropriate to a more advanced course.

Thanks to all of the authors and to the students and faculty of the Department of Environmental and Molecular Toxicology at North Carolina State University and to Carolyn McNeill for much word processing. Particular thanks to Bob Esposito of John Wiley and Sons, not least for his patience with missed deadlines and subsequent excuses.

ERNEST HODGSON
*Raleigh, North Carolina*

# CONTRIBUTORS

**Baynes, Ronald E.,** Cutaneous Pharmacology and Toxicology Center, College of Veterinary Medicine, North Carolina State University, Raleigh, NC

**Blake, Bonita L.,** Department of Pharmacology and Neuroscience Center, University of North Carolina at Chapel Hill, Chapel Hill, NC

**Bonner, James C.,** National Institute of Environmental Health Sciences, Research Triangle Park, NC

**Branch, Stacy,** Department of Environmental and Molecular Toxicology, North Carolina State University, Raleigh, NC

**Cope, W. Gregory,** Department of Environmental and Molecular Toxicology, North Carolina State University, Raleigh, NC

**Cunny, Helen,** Bayer Crop Science, Research Triangle Park, NC

**Hodgson, Ernest,** Department of Environmental and Molecular Toxicology, North Carolina State University, Raleigh, NC

**LeBlanc, Gerald A.,** Department of Environmental and Molecular Toxicology, North Carolina State University, Raleigh, NC

**Leidy, Ross B.,** Department of Environmental and Molecular Toxicology, North Carolina State University, Raleigh, NC

**Levi, Patricia E.,** Department of Environmental and Molecular Toxicology, North Carolina State University, Raleigh, NC

**Meyer, Sharon A.,** Department of Toxicology, University of Louisiana, Monroe, LA

**Rose, Randy L.,** Department of Environmental and Molecular Toxicology, North Carolina State University, Raleigh, NC

**Selgrade, MaryJane K.,** United States Environmental Protection Agency, Research Triangle Park, NC

**Shea, Damian,** Department of Environmental and Molecular Toxicology, North Carolina State University, Raleigh, NC

**Smart, Robert C.,** Department of Environmental and Molecular Toxicology, North Carolina State University, Raleigh, NC

# PART I

# INTRODUCTION

■■■■■■ CHAPTER 1

# Introduction to Toxicology

ERNEST HODGSON

## 1.1 DEFINITION AND SCOPE, RELATIONSHIP TO OTHER SCIENCES, AND HISTORY

### 1.1.1 Definition and Scope

Toxicology can be defined as that branch of science that deals with poisons, and a poison can be defined as any substance that causes a harmful effect when administered, either by accident or design, to a living organism. By convention, toxicology also includes the study of harmful effects caused by physical phenomena, such as radiation of various kinds and noise. In practice, however, many complications exist beyond these simple definitions, both in bringing more precise meaning to what constitutes a poison and to the measurement of toxic effects. Broader definitions of toxicology, such as "the study of the detection, occurrence, properties, effects, and regulation of toxic substances," although more descriptive, do not resolve the difficulties. Toxicity itself can rarely, if ever, be defined as a single molecular event but is, rather, a cascade of events starting with exposure, proceeding through distribution and metabolism, and ending with interaction with cellular macromolecules (usually DNA or protein) and the expression of a toxic end point. This sequence may be mitigated by excretion and repair. It is to the complications, and to the science behind them and their resolution, that this textbook is dedicated, particularly to the *how* and *why* certain substances cause disruptions in biologic systems that result in toxic effects. Taken together, these difficulties and their resolution circumscribe the perimeter of the science of toxicology.

The study of toxicology serves society in many ways, not only to protect humans and the environment from the deleterious effects of toxicants but also to facilitate the development of more selective toxicants such as anticancer and other clinical drugs and pesticides.

Poison is a quantitative concept, almost any substance being harmful at some doses but, at the same time, being without harmful effect at some lower dose. Between these two limits there is a range of possible effects, from subtle long-term chronic toxicity to immediate lethality. Vinyl chloride may be taken as an example. It is a potent hepatotoxicant at high doses, a carcinogen with a long latent period at lower

*A Textbook of Modern Toxicology, Third Edition,* edited by Ernest Hodgson
ISBN 0-471-26508-X  Copyright © 2004 John Wiley & Sons, Inc.

doses, and apparently without effect at very low doses. Clinical drugs are even more poignant examples because, although therapeutic and highly beneficial at some doses, they are not without deleterious side effects and may be lethal at higher doses. Aspirin (acetylsalicylic acid), for example, is a relatively safe drug at recommended doses and is taken by millions of people worldwide. At the same time, chronic use can cause deleterious effects on the gastric mucosa, and it is fatal at a dose of about 0.2 to 0.5 g/kg. Approximately 15% of reported accidental deaths from poisoning in children result from ingestion of salicylates, particularly aspirin.

The importance of dose is well illustrated by metals that are essential in the diet but are toxic at higher doses. Thus iron, copper, magnesium, cobalt, manganese, and zinc can be present in the diet at too low a level (deficiency), at an appropriate level (maintenance), or at too high a level (toxic). The question of dose-response relationships is fundamental to toxicology (see Section 1.2).

The definition of a poison, or toxicant, also involves a qualitative biological aspect because a compound, toxic to one species or genetic strain, may be relatively harmless to another. For example, carbon tetrachloride, a potent hepatotoxicant in many species, is relatively harmless to the chicken. Certain strains of rabbit can eat *Belladonna* with impunity while others cannot. Compounds may be toxic under some circumstances but not others or, perhaps, toxic in combination with another compound but nontoxic alone. The methylenedioxyphenyl insecticide synergists, such as piperonyl butoxide, are of low toxicity to both insects and mammals when administered alone but are, by virtue of their ability to inhibit xenobiotic-metabolizing enzymes, capable of causing dramatic increases in the toxicity of other compounds.

The measurement of toxicity is also complex. Toxicity may be acute or chronic, and may vary from one organ to another as well as with age, genetics, gender, diet, physiological condition, or the health status of the organism. As opposed to experimental animals, which are highly inbred, genetic variation is a most important factor in human toxicity since the human population is highly outbred and shows extensive genetic variation. Even the simplest measure of toxicity, the LD50 (the dose required to kill 50% of a population under stated conditions) is highly dependent on the extent to which the above variables are controlled. LD50 values, as a result, vary markedly from one laboratory to another.

Exposure of humans and other organisms to toxicants may result from many activities: intentional ingestion, occupational exposure, environmental exposure, as well as accidental and intentional (suicidal or homicidal) poisoning. The toxicity of a particular compound may vary with the portal of entry into the body, whether through the alimentary canal, the lungs, or the skin. Experimental methods of administration such as injection may also give highly variable results; thus the toxicity from intravenous (IV), intraperitoneal (IP), intramuscular (IM), or subcutaneous (SC) injection of a given compound may be quite different. Toxicity may vary as much as tenfold with the route of administration. Following exposure there are multiple possible routes of metabolism, both detoxifying and activating, and multiple possible toxic endpoints (Figure 1.1).

Attempts to define the scope of toxicology, including that which follows, must take into account that the various subdisciplines are not mutually exclusive and are frequently interdependent. Due to overlapping of mechanisms as well as use and chemical classes of toxicants, clear division into subjects of equal extent or importance is not possible.

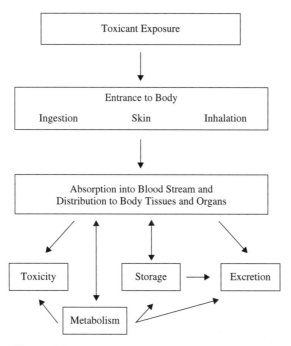

**Figure 1.1**  Fate and effect of toxicants in the body.

Many specialized terms are used in the various subdisciplines of toxicology as illustrated in the *Dictionary of Toxicology*, 2nd edition (Hodgson et al., 1998). However, some terms are of particular importance to toxicology in general; they are defined in the glossary to be found at the end of this volume.

A.  Modes of Toxic Action. This includes the consideration, at the fundamental level of organ, cell and molecular function, of all events leading to toxicity in vivo: uptake, distribution, metabolism, mode of action, and excretion. The term mechanism of toxic action is now more generally used to describe an important molecular event in the cascade of events leading from exposure to toxicity, such as the inhibition of acetylcholinesterase in the toxicity of organophosphorus and carbamate insecticides. Important aspects include the following:

1.  *Biochemical and molecular toxicology* consider events at the biochemical and molecular levels, including enzymes that metabolize xenobiotics, generation of reactive intermediates, interaction of xenobiotics or their metabolites with macromolecules, gene expression in metabolism and modes of action, and signaling pathways in toxic action.

2.  *Behavioral toxicology* deals with the effects of toxicants on animal and human behavior, which is the final integrated expression of nervous function in the intact animal. This involves both the peripheral and central nervous systems, as well as effects mediated by other organ systems, such as the endocrine glands.

3.  *Nutritional toxicology* deals with the effects of diet on the expression of toxicity and with the mechanisms of these effects.

4. *Carcinogenesis* includes the chemical, biochemical, and molecular events that lead to the large number of effects on cell growth collectively known as cancer.

5. *Teratogenesis* includes the chemical, biochemical, and molecular events that lead to deleterious effects on development.

6. *Mutagenesis* is concerned with toxic effects on the genetic material and the inheritance of these effects.

7. *Organ toxicity* considers effects at the level of organ function (neurotoxicity, hepatotoxicity, nephrotoxicity, etc.).

B. Measurement of Toxicants and Toxicity. These important aspects deal primarily with analytical chemistry, bioassay, and applied mathematics; they are designed to provide the methodology to answer certain critically important questions. Is the substance likely to be toxic? What is its chemical identify? How much of it is present? How can we assay its toxic effect, and what is the minimum level at which this toxic effect can be detected? A number of important fields are included:

1. *Analytical toxicology* is a branch of analytical chemistry concerned with the identification and assay of toxic chemicals and their metabolites in biological and environmental materials.

2. *Toxicity testing* involves the use of living systems to estimate toxic effects. It covers the gamut from short-term tests for genotoxicity such as the Ames test and cell culture techniques to the use of intact animals for a variety of tests from acute toxicity to lifetime chronic toxicity. Although the term "bioassay" is used properly only to describe the use of a living organism to quantitate the amount of a particular toxicant present, it is frequently used to describe any in vivo toxicity test.

3. *Toxicologic pathology* is the branch of pathology that deals with the effects of toxic agents manifested as changes in subcellular, cellular, tissue, or organ morphology.

4. *Structure-activity* studies are concerned with the relationship between the chemical and physical properties of a chemical and toxicity and, particularly, the use of such relationships as predictors of toxicity.

5. *Biomathematics and statistics* relate to many areas of toxicology. They deal with data analysis, the determination of significance, and the formulation of risk estimates and predictive models.

6. *Epidemiology* as it applies to toxicology, is of great importance as it deals with the relationship between chemical exposure and human disease in actual populations rather than in experimental settings.

C. Applied Toxicology. This includes the various aspects of toxicology as they apply in the field or the development of new methodology or new selective toxicants for early application in the field setting.

1. *Clinical toxicology* is the diagnosis and treatment of human poisoning.

2. *Veterinary toxicology* is the diagnosis and treatment of poisoning in animals other than humans, particularly livestock and companion animals, but not excluding feral species. Other important concerns of veterinary toxicology are the possible

transmission of toxins to the human population in meat, fish, milk, and other foodstuffs and the care and ethical treatment of experimental animals.

3. *Forensic toxicology* concerns the medicolegal aspects, including detection of poisons in clinical and other samples.

4. *Environmental toxicology* is concerned with the movement of toxicants and their metabolites and degradation products in the environment and in food chains and with the effect of such contaminants on individuals and, especially, populations. Because of the large number of industrial chemicals and possibilities for exposure, as well as the mosaic of overlapping laws that govern such exposure, this area of applied toxicology is well developed.

5. *Industrial toxicology* is a specific area of environmental toxicology that deals with the work environment and constitutes a significant part of *industrial hygiene*.

D. Chemical Use Classes. This includes the toxicology aspects of the development of new chemicals for commercial use. In some of these use classes, toxicity, at least to some organisms, is a desirable trait; in others, it is an undesirable side effect. Use classes are not composed entirely of synthetic chemicals; many natural products are isolated and used for commercial and other purposes and must be subjected to the same toxicity testing as that required for synthetic chemicals. Examples of such natural products include the insecticide, pyrethrin, the clinical drug, digitalis, and the drug of abuse, cocaine.

1. *Agricultural chemicals* include many compounds, such as insecticides, herbicides, fungicides, and rodenticides, in which toxicity to the target organism is a desired quality whereas toxicity to "nontarget species" is to be avoided. Development of such selectively toxic chemicals is one of the applied roles of comparative toxicology.

2. *Clinical drugs* are properly the province of pharmaceutical chemistry and pharmacology. However, toxic side effects and testing for them clearly fall within the science of toxicology.

3. *Drugs of abuse* are chemicals taken for psychological or other effects and may cause dependence and toxicity. Many of these are illegal, but some are of clinical significance when used correctly.

4. *Food additives* are of concern to toxicologists only when they are toxic or being tested for possible toxicity.

5. *Industrial chemicals* are so numerous that testing them for toxicity or controlling exposure to those known to be toxic is a large area of toxicological activity.

6. *Naturally occurring substances* include many phytotoxins, mycotoxins, and minerals, all occurring in the environment. The recently expanded and now extensive use of herbal remedies and dietary supplements has become a cause of concern for toxicologists and regulators. Not only is their efficacy frequently dubious, but their potential toxicity is largely unknown.

7. *Combustion products* are not properly a use class but are a large and important class of toxicants, generated primarily from fuels and other industrial chemicals.

E. Regulatory Toxicology These aspects, concerned with the formulation of laws, and regulations authorized by laws, are intended to minimize the effect of toxic chemicals on human health and the environment.

1. *Legal aspects* are the formulation of laws and regulations and their enforcement. In the United States, enforcement falls under such government agencies as the Environmental Protection Agency (EPA), the Food and Drug Administration (FDA), and the Occupational Safety and Health Administration (OSHA). Similar government agencies exist in many other countries.

2. *Risk assessment* is the definition of risks, potential risks, and the risk-benefit equations necessary for the regulation of toxic substances. Risk assessment is logically followed by *risk communication* and *risk management*.

### 1.1.2 Relationship to Other Sciences

Toxicology is highly eclectic science and human activity drawing from, and contributing to, a broad spectrum of other sciences and human activities. At one end of the spectrum are those sciences that contribute their methods and philosophical concepts to serve the needs of toxicologists, either in research or in the application of toxicology to human affairs. At the other end of the spectrum are those sciences to which toxicology contributes.

In the first group chemistry, biochemistry, pathology, physiology, epidemiology, immunology, ecology, and biomathematics have long been important while molecular biology has, in the last two or three decades, contributed to dramatic advances in toxicology.

In the group of sciences to which toxicology contributes significantly are such aspects of medicine as forensic medicine, clinical toxicology, pharmacy and pharmacology, public health, and industrial hygiene. Toxicology also contributes in an important way to veterinary medicine, and to such aspects of agriculture as the development and safe use of agricultural chemicals. The contributions of toxicology to environmental studies has become increasingly important in recent years.

Clearly, toxicology is preeminently an applied science, dedicated to the enhancement of the quality of life and the protection of the environment. It is also much more. Frequently the perturbation of normal life processes by toxic chemicals enables us to learn more about the life processes themselves. The use of dinitrophenol and other uncoupling agents to study oxidative phosphorylation and the use of $\alpha$-amanitin to study RNA polymerases are but two of many examples. The field of toxicology has expanded enormously in recent decades, both in numbers of toxicologists and in accumulated knowledge. This expansion has brought a change from a primarily descriptive science to one which utilizes an extensive range of methodology to study the mechanisms involved in toxic events.

### 1.1.3 A Brief History of Toxicology

Much of the early history of toxicology has been lost and in much that has survived toxicology is of almost incidental importance in manuscripts dealing primarily with medicine. Some, however, deal more specifically with toxic action or with the use of poisons for judicial execution, suicide or political assassination. Regardless of the paucity of the early record, and given the need for people to avoid toxic animals and plants, toxicology must rank as one of the oldest practical sciences.

The Egyptian papyrus, *Ebers*, dating from about 1500 BC, must rank as the earliest surviving pharmacopeia, and the surviving medical works of Hippocrates, Aristotle, and Theophrastus published during the period 400 to 250 BC all include some mention of poisons. The early Greek poet Nicander treats, in two poetic works, animal toxins (*Therica*) and antidotes to plant and animal toxins (*Alexipharmica*). The earliest surviving attempt to classify plants according to their toxic and therapeutic effects is that of Dioscorides, a Greek employed by the Roman emperor Nero about AD 50.

There appear to have been few advances in either medicine or toxicology between the time of Galen (AD 131–200) and Paracelsus (1493–1541). It was the latter who, despite frequent confusion between fact and mysticism, laid the groundwork for the later development of modern toxicology by recognizing the importance of the dose-response relationship. His famous statement—"All substances are poisons; there is none that is not a poison. The right dose differentiates a poison and a remedy"—succinctly summarizes that concept. His belief in the value of experimentation was also a break with earlier tradition.

There were some important developments during the eighteenth century. Probably the best known is the publication of Ramazini's *Diseases of Workers* in 1700, which led to his recognition as the father of occupational medicine. The correlation between the occupation of chimney sweeps and scrotal cancer by Percival Pott in 1775 is almost as well known, although it was foreshadowed by Hill's correlation of nasal cancer and snuff use in 1761.

Orfila, a Spaniard working at the University of Paris in the early nineteenth century, is generally regarded as the father of modern toxicology. He clearly identified toxicology as a separate science and, in 1815, published the first book devoted exclusively to toxicology. An English translation in 1817, was entitled *A General System of Toxicology or, A Treatise on Poisons, Found in the Mineral, Vegetable and Animal Kingdoms, Considered in Their Relations with Physiology, Pathology and Medical Jurisprudence.* Workers of the late nineteenth century who produced treatises on toxicology include Christian, Kobert, and Lewin. The recognition of the site of action of curare by Claude Bernard (1813–1878) began the modern study of the mechanisms of toxic action. Since then, advances have been numerous—too numerous to list in detail. They have increased our knowledge of the chemistry of poisons, the treatment of poisoning, the analysis of toxicants and toxicity, modes of toxic action and detoxication processes, as well as specific molecular events in the poisoning process.

With the publication of her controversial book, *The Silent Spring*, in 1962, Rachel Carson became an important influence in initiating the modern era of environmental toxicology. Her book emphasized stopping the widespread, indiscriminate use of pesticides and other chemicals and advocated use patterns based on sound ecology. Although sometimes inaccurate and with arguments often based on frankly anecdotal evidence, her book is often credited as the catalyst leading to the establishment of the US Environmental Protection Agency and she is regarded, by many, as the mother of the environmental movement.

It is clear, however, that since the 1960s toxicology has entered a phase of rapid development and has changed from a science that was largely descriptive to one in which the importance of mechanisms of toxic action is generally recognized. Since the 1970s, with increased emphasis on the use of the techniques of molecular biology, the pace of change has increased even further, and significant advances have been made in many areas, including chemical carcinogenesis and xenobiotic metabolism, among many others.

## 1.2   DOSE-RESPONSE RELATIONSHIPS

As mentioned previously, toxicity is a relative event that depends not only on the toxic properties of the chemical and the dose administered but also on individual and interspecific variation in the metabolic processing of the chemical. The first recognition of the relationship between the dose of a compound and the response elicited has been attributed to Paracelsus (see Section 1.1.3). It is noteworthy that his statement includes not only that all substances can be toxic at some dose but that "the right dose differentiates a poison from a remedy," a concept that is the basis for pharmaceutical therapy.

A typical dose-response curve is shown in Figure 1.2, in which the percentage of organisms or systems responding to a chemical is plotted against the dose. For many chemicals and effects there will be a dose below which no effect or response is observed. This is known as the *threshold dose*. This concept is of significance because it implies that a *no observed effect level* (NOEL) can be determined and that this value can be used to determine the safe intake for food additives and contaminants such as pesticides. Although this is generally accepted for most types of chemicals and toxic effects, for chemical carcinogens acting by a genotoxic mechanism the shape of the curve is controversial and for regulatory purposes their effect is assumed to be a no-threshold phenomenon. Dose-response relationships are discussed in more detail in Chapter 21 on toxicity testing.

## 1.3   SOURCES OF TOXIC COMPOUNDS

Given the enormous number of toxicants, it is difficult to classify them chemically, either by function or by mode of action, since many of them would fall into several classes. Some are natural products, many are synthetic organic chemicals of use to society, while others are by-products of industrial processes and waste disposal. It is useful, however, to categorize them according to the expected routes of exposure or according to their uses.

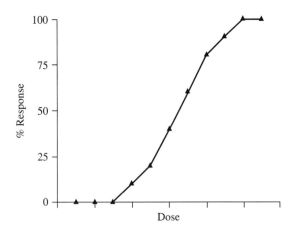

**Figure 1.2**   A typical dose-response curve.

### 1.3.1 Exposure Classes

Exposure classes include toxicants in food, air, water, and soil as well as toxicants characteristic of domestic and occupational Settings. Toxicant exposure classes are described in detail in Chapter 4.

### 1.3.2 Use Classes

Use classes include drugs of abuse, therapeutic drugs, agricultural chemicals, food additives and contaminants, metals, solvents, combustion products, cosmetics, and toxins. Some of these, such as combustion products, are the products of use processes rather than being use classes. All of these groups of chemicals are discussed in detail in Chapter 5.

## 1.4 MOVEMENT OF TOXICANTS IN THE ENVIRONMENT

Chemicals released into the environment rarely remain in the form, or at the location, of release. For example, agricultural chemicals used as sprays may drift from the point of application as air contaminants or enter runoff water as water contaminants. Many of these chemicals are susceptible to fungal or bacterial degradation and are rapidly detoxified, frequently being broken down to products that can enter the carbon, nitrogen, and oxygen cycles. Other agricultural chemicals, particularly halogenated organic compounds, are recalcitrant to a greater or lesser degree to metabolism by microorganisms and persist in soil and water as contaminants; they may enter biologic food chains and move to higher trophic levels or persist in processed crops as food contaminants. This same scenario is applicable to any toxicant released into the environment for a specific use or as a result of industrial processes, combustion, and so on. Chemicals released into the environment are also susceptible to chemical degradation, a process often stimulated by ultraviolet light.

Although most transport between inanimate phases of the environment results in wider dissemination, at the same time dilution of the toxicant in question and transfer among living creatures may result in increased concentration or bioaccumulation. Lipid soluble toxicants are readily taken up by organisms following exposure in air, water, or soil. Unless rapidly metabolized, they persist in the tissues long enough to be transferred to the next trophic level. At each level the lipophilic toxicant tends to be retained while the bulk of the food is digested, utilized, and excreted, thus increasing the toxicant concentration. At some point in the chain, the toxicant can become deleterious, particularly if the organism at that level is more susceptible than those at the level preceding it. Thus the eggshell thinning in certain raptorial birds was almost certainly due to the uptake of DDT and DDE and their particular susceptibility to this type of toxicity. Simplified food chains are shown in Figure 1.3.

It is clear that such transport can occur through both aquatic and terrestrial food chains, although in the former, higher members of the chains, such as fish, can accumulate large amounts of toxicants directly from the medium. This accumulation occurs because of the large area of gill filaments, their intimate contact with the water and the high flow rate of water over them. Given these characteristics and a toxicant with a high partition coefficient between lipid membranes and water, considerable uptake is inevitable.

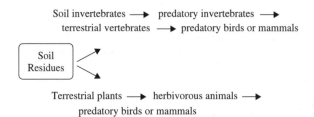

**Figure 1.3** Examples of simplified food chains.

These and all other environmental aspects of toxicology are discussed in Part VII.

## SUGGESTED READING

Hodgson, E., and R. C. Smart, eds. *Introduction to Biochemical Toxicology*, 3rd ed. New York: Wiley, 2001.

Hodgson, E., R. B. Mailman, and J. E. Chambers, eds. *Dictionary of Toxicology*, 2nd ed. London: Macmillan, 1998.

Klaassen, C. D. ed. *Casarett and Doull's Toxicology: The Basic Science of Poisons*, 6th ed. New York: McGraw-Hill, 2001.

Timbrell, J. A. *Principles of Biochemical Toxicology*, 3rd ed. London: Taylor and Francis, 2000.

Wexler, P. *Information Resources in Toxicology*, 3rd ed. San Diego: Academic Press, 2000.

# CHAPTER 2

# Introduction to Biochemical and Molecular Methods in Toxicology

ERNEST HODGSON, GERALD A. LEBLANC, SHARON A. MEYER,
and ROBERT C. SMART

## 2.1 INTRODUCTION

This chapter is not designed to summarize biochemical methods long used in toxicology such as colorimetric and radiometric methods for the investigation of xenobiotic metabolism, either in vivo or in vitro, but rather to give a brief summary of the methods of molecular and cellular biology that have become, more recently, of critical importance in toxicological research. The chapter owes much to Chapters 2 through 4 of the third edition of *Introduction to Biochemical Toxicology* (see Suggested Reading), and the reader is referred to these chapters for additional information.

## 2.2 CELL CULTURE TECHNIQUES

While scientists have had the ability to culture many unicellular organisms for some time, recent advances in the culture of cells from multicellular organisms have played a pivotal role in recent advances in toxicology. Cells can be isolated and either maintained in a viable state for enough time to conduct informative experiments or, in some cases, propagated in culture. The advantages of cultured cells are that they can provide living systems for the investigation of toxicity that are simplified relative to the intact organism and they can be used as replacements for whole animal toxicity testing if the toxic end point can be validated. Human cells play an important role in the extrapolation of toxic effects, discovered in experimental animals, to humans. Cultured cells, from humans or other mammals, are utilized in many of the molecular methods mentioned below. There are, however, limitations in the use of cellular methods. It has not been possible to culture many cell types, and of those that have been cultured, the loss of differentiated cell function is a common problem. Extrapolation of findings to the intact animal is often problematical and the use of undefined media constituents such as serum, often essential for cell viability, may have unwanted or undefined effects on cell function and toxicant bioavailability.

*A Textbook of Modern Toxicology, Third Edition,* edited by Ernest Hodgson
ISBN 0-471-26508-X  Copyright © 2004 John Wiley & Sons, Inc.

Studies have been carried out on cells isolated from tissues and maintained in suspension culture or on cells that have formed monolayers.

### 2.2.1 Suspension Cell Culture

Circulating blood cells or cells easily obtained by lavage such as peritoneal and alveolar macrophages can normally survive in suspension culture when provided with a suitable nutrient medium. Cells from organized solid organs or tissues must be separated from the tissue and, if possible, separated into cell types, before being suspended in such a medium.

Cell association within organs depends on protein complex formation, which in turn is $Ca^{2+}$ dependent. Consequently dissociation media generally contain a proteolytic enzyme and the $Ca^{2+}$ chelator EDTA. There are a number of methods available to separate cell types from the mixture of dispersed cells, the commonest being centrifugation without a density gradient, wherein cells are separated by size, or centrifugation through a density gradient wherein cells are separated on the basis of their buoyant density.

Cells in suspension may be maintained for a limited period of time in defined media or for longer periods in nutrient, but less well-defined, media. In either case these cultures are often used for studies of xenobiotic metabolism.

### 2.2.2 Monolayer Cell Culture

Proliferation of most cells in culture requires attachment to a substrate and occurs until limited by cell-to-cell contact, resulting in the formation of a cellular monolayer. The substrate provided for attachment is usually polystyrene modified to carry a charge. The medium for continued maintenance and growth contains salts and glucose, usually with a bicarbonate buffer. Because of the bicarbonate buffering system these cultures are maintained in a $5-10\%$ $CO_2$ atmosphere in a temperature and humidity controlled incubator. Many cells require serum for optimal growth, inducing considerable variability into the experimental system. Since the factors provided by serum are numerous and complex, defined serum substitutes are not always successful. The factors provided by serum include proteins such as growth factors, insulin and transferrin (to provide available iron), small organic molecules such as ethanolamine, and pyruvate and inorganic ions, such as selenium.

### 2.2.3 Indicators of Toxicity in Cultured Cells

Routine observation of cultured is usually carried out by phase contrast microscopy, utilizing the inverted phase contrast microscope. More recently, more detailed observations have become possible utilizing fluorescent tags and inverted fluorescent microscopes. Fluorescent tags currently in use permit the assessment of oxidant status and mitochondrial function as well as the intracellular concentration of sulfhydryl groups, $Ca^{2+}$, $H^+$, $Na^+$, and $K^+$.

Toxicity to cultured cells may be the result either of inadequacies in the culture or the toxicity effects of the chemical being investigated. Short-term toxicity is usually

**Table 2.1  Examples of Application of Cell Lines Retaining Differentiated Properties in the Study of Toxic Effects***

| Cell Line | Source | Differentiated Cell Type | Toxicant | Measured End Point |
|---|---|---|---|---|
| N1E-115 | Mouse neuroblastoma | Cholinergic neuron | Lead | Blockage of voltage-dependent $Ca^{2+}$ channels |
| | | | Pyrethroid insecticide | Prolonged open time for voltage-dependent $Na^+$ channels |
| PC12 | Rat pheochromocytoma (adrenal medullary tumor) | Adrenergic neuron | Tricresyl phosphate (organophosphate) | Inhibition of neurofilament assembly and axonal growth |
| SK-N-S11 | Human neuroblastoma | Neuron | $N_2O$ (anesthetic) | Depressed cholinergic $Ca^{2+}$ signaling |
| Hepa-1 | Mouse hepatoma | Hepatocyte | 2,3,7,8-tetrachloro-dibenzodioxane (TCCD) | Induction of CYP1A1 and 1B1 |
| H114 E | Rat hepatoma | Hepatocyte | Polychlorinated biphenyls (PCBs) | Induction of CYP1A1 |
| HepG2 | Human hepatoblastoma | Hepatocyte | Cyclophosphamide (antineoplastic) | Cytochrome P450-dependent genotoxicity |
| 3T3-L1 | Mouse embryo fibroblasts | Adipocytes | TCDD | Inhibition of glucose transport and lipoprotein lipase |
| Y1 | Mouse adrenocortical tumor | Adrenocortical cell | Methyl sulfone metabolites of DDT and PCBs | Inhibition of corticosterone synthesis by competitive inhibition of cytochrome P450 |
| LLC-PK1 | Pig kidney | Renal tubule epithelial cell | Cadmium | Cytotoxicity, apoptosis |
| MDCK | Dog kidney | Renal tubule epithelial cell | Organic mercury compounds | Cytotoxicity, transepithelial leakiness |

*Source*: E. Hodgson and R. C. Smart, eds., *An Introduction to Modern Toxicology*, 3rd ed. New York: Wiley, 2001.

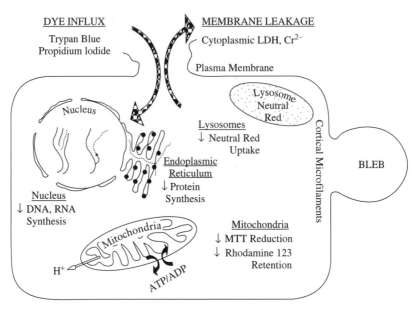

**Figure 2.1**   Idealized diagram of a cell to illustrate parameters often used to measure cytotoxicity and the corresponding affected subcellular organelle. (From *An Introduction to Biochemical Toxicology*, 3rd ed., E. Hodgson and R. C. Smart, eds., Wiley, 2001.)

evaluated by examination of end points that indicate effects on cellular organelles such as leakage of cell constituents into the medium, uptake of dyes into the cell and the formation of surface "blebs." This is illustrated in Figure 2.1.

Longer term assessments of cell toxicity are highly dependent on the relevant toxic end point. They may include measurement of growth competence, apoptosis, and/or necrosis, incorporation of radioactive precursors into essential cellular constituents such as RNA, DNA, and protein and specialized cellular functions. Some examples of the use of cultured cell lines in the study of toxicity effects are shown in Table 2.1.

## 2.3   MOLECULAR TECHNIQUES

Recombinant DNA techniques, including molecular cloning, have provided recent dramatic advances in many areas of both fundamental and applied biology, toxicology not excepted. Responses to toxicants may involve changes in gene expression and the new microarray techniques enable the simultaneous examination of the level of expression of many genes. The completion of the Human Genome Project will permit toxic effects in humans to be investigated and will facilitate extrapolation from experimental animals. The human genome will also provide the essential genetic background information for studies of polymorphisms in xenobiotic-metabolizing and other enzymes. Such polymorphisms have already been shown to be very important in individual sensitivity to clinical drugs and in the definition of populations and/or individuals at increased risk from particular toxicants. Chemically induced mutations, particularly in oncogenes and tumor-suppressor genes are important in chemical carcinogenesis. The

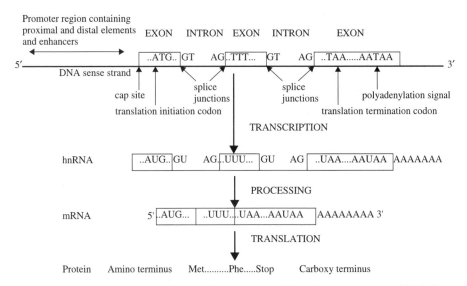

**Figure 2.2** Transcription, mRNA processing, and translation. DNA sense strand is designated by bold lines, hnRNA and mRNA by thinner lines. Exons are shown as rectangles and introns as the intervening spaces between exons. (From *An Introduction to Biochemical Toxicology*, 3rd edition, E. Hodgson and R. C. Smart, eds., Wiley, 2001.)

ability to develop "knockout" animals lacking a particular gene and transgenic animals with an additional transgene is also proving important in toxicological studies.

Gene structure and any of the processes involved in DNA expression including transcription, mRNA processing and translation and protein synthesis (Figure 2.2) can all be examined by molecular techniques. In toxicology this may include toxic effects on these processes or the role of the processes in the mechanism of toxic action.

### 2.3.1  Molecular Cloning

The basic principle of molecular cloning is the insertion of a DNA segment into a suitable vector. The vector is an autonomously replicating DNA molecule and the inserted DNA segment may be as large as a gene or a small as a few nucleotides. The vector containing the DNA is inserted into a cell such as yeast, where it can be replicated many times, and either the DNA or the expressed protein subsequently isolated (Figure 2.3).

### 2.3.2  cDNA and Genomic Libraries

cDNA or genomic libraries are collections of DNA fragments incorporated into a recombinant vector and transformed into an appropriate host cell. In the case of cDNA libraries, the cDNAs complementary to all of the mRNAs in the tissue or cell sample are synthesized in a procedure using reverse transcriptase, before incorporation into the vector. With genomic DNA libraries the genomic DNA is digested, before cloning into the vector, with a restriction enzyme to produce an overlapping set of DNA fragments of some 12 to 20 kb.

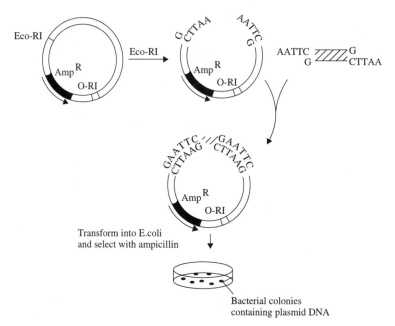

**Figure 2.3**   Molecular cloning using a plasmid vector. (From *An Introduction to Biochemical Toxicology*, 3rd ed., E. Hodgson and R. C. Smart, eds., Wiley,  2001.)

These libraries are used in many screening procedures and many transgenic proteins now routinely available were obtained by their use. Although in some applications the use of cDNA and genomic libraries has been superceded by other methods, particularly those based on PCR, they are still used to advantage in many applications.

### 2.3.3   Northern and Southern Blot Analyses

Northern analysis is usually used to identify and quantitate specific mRNAs in a sample. Southern analysis is used to determine whether or not a gene of interest is present as well as its copy number. Other uses for Southern analysis include identifying restriction fragment length polymorphisms and changes in heterozygosity.

In both Southern and Northern analyses restriction-digested DNA fragments, mRNA, and polyA mRNA are separated by size when electrophoresed on agarose gel. The separated molecules are transferred, by electroblotting or capillary blotting, on to a nylon or nitrocellulose membrane. The immobilized RNA or DNA is reacted with a radiolabeled, chemiluminescent, or fluorescent probe that is complementary to the DNA/RNA of interest, unbound probe is washed off, and the membrane exposed, in the case of radioactive probes, to radioautographic film to visualize the sample of interest.

### 2.3.4   Polymerase Chain Reaction (PCR)

PCR is a powerful technique that can, starting with amounts of DNA as small as those found in single cells, amplify the DNA until large amounts are available for many different kinds of research. Twenty to 40 cycles of can provide up to $10^5$ times the original DNA sample.

It is necessary to know as much of the sequence of the DNA of interest as possible in order to construct appropriate primers. These primers are complementary to the sequence at each end of the DNA sequence to be amplified. The DNA is incubated in a thermal cycler with thermostable DNA polymerase, all four dNTP, and the primers. The incubation temperature is raised to separate the DNA strands, lowered to permit annealing of the primers to the complementary regions of the DNA and then raised to permit the polymerase to synthesize DNA. This cycle is then repeated up to 40 times. The PCR technique has been used for many types of toxicological investigation including; uncovering polymorphisms in xenobiotic-metabolizing enzymes, isolating genes from cDNA and genomic libraries and for mutational analysis, to name only a few.

### 2.3.5   Evaluation of Gene Expression, Regulation, and Function

The methods used for the evaluation of regulation of gene expression are too numerous to be described in detail here. They include Northern analysis to determine levels of a particular mRNA, nuclear run on to determine whether an increase in mRNA is due to an increase in the rate of transcription, and promoter deletion analysis to identify specific elements in the promoter region responsible for the control of expression. Of much current interest is the use of microarrays that permit the study of the expression of hundreds to thousands of genes at the same time. Reverse transcriptase–polymerase chain reaction and RNase protection assay techniques are used to amplify and quantitate mRNAs, while the electrophoretic mobility shift assay is used to measure binding of a transcription factor to its specific DNA consensus sequence.

Gene function in cultured cells can be investigated by expression of the gene product in a suitable expression system or, in vivo, by the creation of transgenic mice, either knockout mice in which the gene in question has been functionally deleted or mice into which a transgene has been introduced.

A general, but more detailed and specific, account of these methods may be found in Smart (2001; see Suggested Reading).

## 2.4   IMMUNOCHEMICAL TECHNIQUES

Most of the recently developed methods for the detection, characterization, and quantitation of proteins are immunoassays based on the fact that proteins are antigens, compounds that can be recognized by an antibody. It is also true that by combining small molecules (haptens) with a larger carrier molecule such as a protein, these methods can be extended to small molecules of interest since antibodies can be produced that recognize epitopes (specific sites on the antigen recognized by the antibody) that include the hapten.

The antibodies used may be polyclonal or monoclonal, each with characteristics fitting them for use in particular immunochemical methods. Injection of a mammal with a foreign protein (immunogen) gives rise to an immune reaction that includes the generation of antibodies from B lymphocytes. Each B lymphocyte gives rise to only a single antibody type that recognizes a single epitope on the antigen. However, since these antibodies are derived from many different B lymphocytes the mixture of antibodies can recognize and bind to many different epitopes on the antigen. This mixture of antibodies can be isolated from the serum of the treated animal and is known, collectively, as *polyclonal antibodies*. However, if individual B lymphocytes from a treated animal can be isolated and cultured, because they are of a single clonal origin, they will produce a specific *monoclonal antibody* that recognizes only a single epitope on the antigen (Figure 2.4). Because of the multiple sites for binding polyclonal antibodies are highly reactive. They are also relatively easy to produce. Monoclonal antibodies, although more difficult to produce, are, on the other hand, more specific. The advantages and disadvantages of each must be considered to determine which is the antibody of choice for a particular application. The most important immunochemical methods include the following:

*Immunolocalization* is a technique for identifying the presence of a protein within the cell, its relative abundance and its subcellular localization. After suitable preparation of the cells, they are treated with an antibody (the primary antibody) that binds to the protein of interest. An antibody that binds to the primary antibody (the secondary antibody) is then allowed to bind and form an antigen—primary antibody—secondary antibody complex. The detection system generally consists of the formation of a colored insoluble product of an enzymatic reaction, the enzyme, such as alkaline phosphatase or horseradish peroxidase, being covalently linked to the secondary antibody.

*Immunoaffinity purification* involves the use of antibodies, bound to an insoluble matrix, for chromatography. The advantage of this method is that it is highly specific, often permitting purification in a single step. *Immunoprecipitation* is a variant of immunoaffinity purification and is a means to remove a protein from a complex mixture in a highly specific manner.

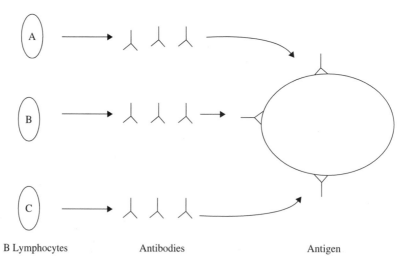

B Lymphocytes                  Antibodies                  Antigen

**Figure 2.4**   The generation of antibodies of several clonal origins (polyclonal antibodies) with antibodies from each clonal origin (monoclonal antibodies A, B and C) recognizing a distinct epitope on the antigen. (From *An Introduction to Biochemical Toxicology*, 3rd ed., E. Hodgson and R. C. Smart, eds., Wiley, 2001.)

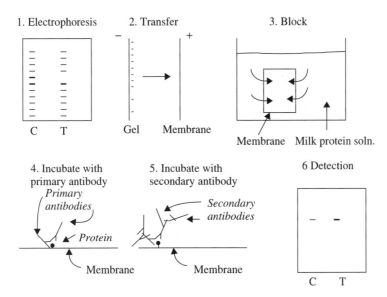

1. Electrophoresis    2. Transfer    3. Block

C    T    Gel    Membrane    Membrane    Milk protein soln.

4. Incubate with primary antibody    5. Incubate with secondary antibody    6 Detection

Primary antibodies    Secondary antibodies

Protein

Membrane    Membrane    C    T

**Figure 2.5**  Diagrammatic representation of the use of immunoblotting to assess relative levels of a P450 protein following treatment of rats with a PCB. C = hepatic microsomal proteins from a control, untreated rat; T = hepatic microsomal proteins from a rat treated with PCBs. (From *An Introduction to Biochemical Toxicology*, 3rd ed., E. Hodgson and R. C. Smart, eds., Wiley, 2001.)

*Western blotting*  is a widely used technique in which antibodies are used to detect proteins following electrophoresis, generally SDS polyacrylamide gel electrophoresis that permits the separation of proteins on the basis of their molecular weights (Figure 2.5). Western blotting can be used to determine the presence and relative amount of a particular protein in a biological sample as well as its molecular weight.

*Radioimmunoassay*  (RIA) is a very sensitive method used to measure minute quantities of an antigen. Since this method is most often used to measure drugs, toxicants, and other xenobiotics, the antigen used to produce the antibody is the small molecule (hapten) linked covalently to a protein. Among the techniques used in the actual measurement, the antigen capture method, in which the competition between radiolabeled antigen and the unlabeled antigen in the sample, is the most common.

Depending on the design of the method *enzyme-linked immunoabsorbant assays* (ELISA) can be used to measure either antigens or antibodies in mixtures by using enzymatic-mediated detection of the corresponding immobilized immune complex. Even though this method has proved to be most useful for the rapid estimation of antibodies or antigens in complex biological mixtures, it has also been used for the quantitation of small molecules in a manner analogous to radioimmunoassays.

*Inhibitory antibodies*  are frequently used in studies of xenobiotic metabolism, usually to estimate the contribution of particular enzymes in multienzyme mixtures. An important example is the use of antibodies to estimate the contribution of individual cytochrome P450 isoforms to the overall metabolism of a xenobiotic in microsomal preparations.

## SUGGESTED READING

Hodgson, E., and R. C. Smart, eds. *Introduction to Biochemical Toxicology*, 3rd ed. New York: Wiley, 2001. Relevant chapters include:

Chapter 2. Smart, R. C. Overview of molecular techniques in toxicology: genes/transgenes.

Chapter 3. LeBlanc, G. A. Immunochemical techniques in toxicology.

Chapter 4. Meyer, S. A. Overview of cellular techniques in toxicology.

# Toxicant Analysis and Quality Assurance Principles

ROSS B. LEIDY

## 3.1 INTRODUCTION

Today's analytical chemist is armed with an amazing number of tools for the determination and quantitation of potentially toxic compounds at concentrations thought to be impossible to detect 10 years ago. Those involved in the analyses of samples containing, for example, carcinogens in tissue samples, pesticide residues from soils, and polycyclic aromatic hydrocarbons (PAHs) from air, must have a solid understanding of all procedural aspects related to the samples that they analyze to ensure that the data generated are accurate and complete. For example, the ability to determine residues of toxicants in environmental matrices (e.g., food, soil, and water) is crucial to support efforts that are designed to protect both the environment and human health. Toxicologists who are involved in interpreting or reviewing data from studies conducted in the pharmaceutical or agrochemical industries must be familiar with the associated quality assurance and quality control aspects to ensure that the completed data package contains all of the information required by federal and state regulations. To provide the required data, certain approaches to chemical analysis must be followed. This chapter is designed to provide a brief overview of these approaches, including the importance of each step and how a failure to properly consider and follow accepted practices can result in flawed enforcement or regulatory decisions. A more detailed description of sampling, extraction and cleanup procedures, including the primary analytical instruments used to quantitate concentrations, will be discussed in Chapter 25.

## 3.2 GENERAL POLICIES RELATED TO ANALYTICAL LABORATORIES

Every analytical laboratory, governmental, private, or university, has a standard set of procedures that provide both general and specific information to laboratory members. These fall into certain categories, including the laboratory's standard operating procedures (SOPs), quality assurance/quality control manuals (QA/QC manuals), procedural manuals, analytical method files, and laboratory information management systems

*A Textbook of Modern Toxicology, Third Edition,* edited by Ernest Hodgson
ISBN 0-471-26508-X Copyright © 2004 John Wiley & Sons, Inc.

(LIMSs). These documents/computer programs are designed to standardize all of the aspects of analyses so that a logical, planned, series of events can be followed. Policies generally are reviewed yearly, and the input provided by the laboratory team enhances the efficiency and proficiency of the members to generate accurate results in a timely manner.

### 3.2.1  Standard Operating Procedures (SOPs)

Over the years members of the laboratory staff have developed routine ways of doing things such as the proper washing of glassware or how many fortifications are done per unit of samples (e.g., 1/10 samples). This set of documents, gathered into the standard operating procedures (SOPs) manual, provides details for both new and experienced chemists on procedures that are performed routinely by the laboratory members. The key word associated with SOPs is routine, because they deal with laboratory procedures that are ancillary to the actual analyses of samples. A laboratory should have SOPs for sample handling, storage, maintenance, replacement of laboratory chemicals, solvents and standards, and the use of laboratory equipment such as pH meters. However, as equipment such as pH meters are replaced with newer models and new systems (e.g., high-speed blenders) are introduced, set procedures must be developed for each. Thus, any SOP is a "living document" that is routinely updated (usually yearly) by the laboratory director with input by all laboratory members.

### 3.2.2  QA/QC Manuals

Sets of instructions that detail the procedures designed to reduce errors occurring during analytical procedures and ensure accurate quantitations are found in the quality assurance (QA) and quality control (QC) manuals. Quality assurance procedures are used by the laboratory to detect and correct problems in analytical processes. As newer methods and instrumentation are added to the laboratory, older procedures must be modified or changed completely. Quality control procedures are used to maintain a measurement system (i.e., a gas chromatograph) in a statistically satisfactory state to ensure the production of accurate and reliable data.

### 3.2.3  Procedural Manuals

Beginning chemists must be taught how to use the equipment that they will work with on a daily basis. Questions such as "How does one set the voltage on an electron capture detector?" will be answered initially by the more experienced chemists. However, later they might not be available, so where does one turn if unfamiliar equipment is being used? Many laboratories have procedural manuals that provide detailed information on using the various pieces of equipment that are found in a modern analytical laboratory. They provide step-by-step instructions on using these items and are a valuable reference guide to their proper use. In addition, placing the manufacturer's procedural manual with the laboratory-generated instruction sheet should provide detailed information on troubleshooting, a parts list, telephone numbers, and Internet sites. Like the other manuals listed above, these must be updated on a routine schedule, and as new equipment is procured to replace existing items, new instructions must be added and old information removed.

### 3.2.4 Analytical Methods Files

All laboratories maintain files containing analytical methods for sample matrices thought to contain specific toxicants that are analyzed routinely. If the US Environmental Protection Agency (USEPA) or manufacturer's methods are used and have been modified by the laboratory, the modifications can be highlighted followed by a brief discussion as to why the modifications were made. The objective is to provide insight into the logic of why the modification was made. Files can be arranged by sample matrix as a general category, and subdivided by compound type (e.g., organophosphate), then further subdivided by specific compound (e.g., diazinon). If kept up to date, these files provide a large amount of information at all levels of experience.

### 3.2.5 Laboratory Information Management System (LIMS)

As computers and software became more sophisticated, it was only a matter of time before attention was turned to laboratory management. Through large databases, applications were designed to manage and store information associated with a laboratory, and to include many unique features. These laboratory information management systems (LIMS) have found widespread use in regulatory laboratories. A typical LIMS contains the following functions:

1. A sample tracking module, using a unique computer-generated identification number, that allows sample tracking from the time the sample enters the laboratory until it is analyzed.
2. A sample scheduling module that automatically logs in the sample, prints barcoded labels and assigns analyses for routine projects.
3. A personnel and equipment module that maintains employee training records, tracks instrument calibration, repairs, costs, and so on.
4. A data entry module that allows the chemist to enter results into the LIMS, assign QC runs, and prepare reports.
5. A QA/QC module that allows generation of control charts and produces graphs that detail trends.
6. An electronic data transfer module that automatically transfers data generated from the electronics of the analytical instrument into the LIMS, thus reducing transcription errors.
7. A chemical and reagent inventory module that manages the purchase and use of laboratory supplies, keeps track of lot number purchases, shelf lives, and costs.
8. A maintenance module that allows laboratory managers to manage the database, monitor studies, methods, priorities of projects, and so on.

LIMS have become more sophisticated and recently have been combined with two newly developed systems, the chemical information management system (CIMS), which searches for, manipulates, and stores chemical structures, reactions, and associated data, and the analytical information management system (AIMS), which acts as a central repository for instrumental data (e.g., chromatograms).

## 3.3   ANALYTICAL MEASUREMENT SYSTEM

Toxicants are generally found at low concentrations (e.g., ppm or ppb) regardless of the sample matrix being evaluated. These concentrations are based on the measurement of a response from some instrument to the compound(s) of interest from an extract of the sample matrix. Thus it is necessary to have a system capable of measuring the compound of interest, and in order to ensure the reliability of the data, the analytical process (instrument and analytical method) must be monitored closely.

This measurement process involves much more than injecting some amount of the extracted sample and comparing its response to that of a standard of known concentration. Analytical standards must be prepared, weighed, and diluted carefully to ensure that the concentrations reported reflect those found in the sample analyzed. In addition the analytical instrument used must be calibrated properly to ensure accuracy. Essentially this involves two processes: calibration of the detector against the compound of interest in order to eliminate or minimize any deviation (bias) of response in one direction or another from that expected from previous experience or expected results. Second, calibration of the total analytical system using statistical approaches to minimize bias in determining the presence or absence of the analyte being sought.

### 3.3.1   Analytical Instrument Calibration

In setting up instrument parameters, consider what is involved in determining residue levels of an analyte. The data produced are only as good as the extract derived from the original sample. If the analyte is distributed uniformly over the area sampled, the concentrations found will be equal, regardless of where the sample is taken. Along these same lines, the analytical procedure will result in uniform residue values if all procedures and instrument parameters remain the same. Based on experience, we know that this distribution of residue over an area will vary as will the analytical procedures and instrument parameters. If we increase the number of samples collected and analyzed, the differences observed will tend to get smaller, resulting in a mean or average value that locates the center of the distribution. Ideally this distribution is called a normal distribution or Gaussian distribution, and looks like a bell (the classic "bell-shaped curve") when the parameters being measured are plotted on a graph (i.e., frequency vs. concentration). Second, the difference in individual measurement, called the standard deviation ($\sigma$), defines the variation found in individual measurements. Equations and tables have been developed to determine the significance of suspected deviations and are used to confirm the presence of a suspected problem. If an infinite number of samples from the area are collected and analyzed, the variation in 95% of the samples will cover the true population percentage.

### 3.3.2   Quantitation Approaches and Techniques

Quantitation is an extremely important part of the analysis to the residue chemist. Residue analysis involves the removal of the compound of interest from some sample matrix. Accurate results come from being thoroughly familiar with the procedures involved in order to establish and maintain appropriate tolerances. The most fundamental decision made is whether the analyte is present or absent, particularly when its concentration is at or close to its detection limit. Since the measurements are derivations of a known relationship between the analyte concentration and the magnitude of the signal made by the instrument, there is additional signal (noise) generated from the presence of co-extractives, instrument noise, and the like. The analyst uses this "contaminated" signal to decide whether the analyte is present or absent, and selects one of these choices. The decision process is subject to two types of errors: the analyte is present when actually it is not, and the analyte is absent when actually it is present. The terminology for these decision processes are commonly called "false positives" and "false negatives," respectively.

## 3.4 QUALITY ASSURANCE (QA) PROCEDURES

Over the last 20 years the reliability of data produced by analytical laboratories has increased dramatically. Strict requirements have ensured that the data were produced under defined standards of quality with a stated level of confidence. The routine day-to-day activities (e.g., matrix fortifications) to control, assess, and ensure the quality of generated data are the quality controls associated with analytical processes. The management of the system that ensures that these processes are in place and functional is the quality assurance portion of the laboratory program to produce reliable data.

Quality assurance (QA) is an essential part of analytical protocols. Each laboratory is required to detect and correct problems in analytical processes and to reduce errors to agreed-upon limits. To produce data that have acceptable quality, all laboratory members must follow established guidelines and protocols. Some of the essential elements that must be included in a QA program are as follows:

1. Laboratory practices (e.g., glass washing protocols) must be developed, reviewed, and updated with the staff's participation on a scheduled basis and followed strictly by all laboratory members;
2. Operating procedures (e.g., SOPs monitoring freezer temperatures daily) must be standardized, documented, and supplied to each member of the laboratory staff and updated on a set schedule;
3. Monitoring programs (e.g., surface water monitoring of supplies furnishing public drinking water) must be carefully designed.
4. Maintenance of equipment and instruments must be documented in LIMS or appropriate maintenance books kept with the equipment.
5. Expiration dates of analytical standards, chemicals, and solvents must be observed and replacements made prior to their expiration date.
6. Good laboratory practices (GLPs) must be implemented as needed.
7. Audits must be performed on a scheduled basis to verify that all aspects of the QA program are operating sufficiently.

## 3.5 QUALITY CONTROL (QC) PROCEDURES

Quality control (QC) concerns procedures that maintain a measurement system in a state of statistical control. This does not mean that statistics control the analytical procedures but that statistical evidence is used to ensure that the procedure is working under the conditions set by protocol. *The accuracy of an analytical method depends on statistical control being conducted prior to determining any other parameter.* How well the basic method will work with the sample matrix being evaluated will depend on the way the QC samples are examined. A comprehensive QC analytical procedure would include the following:

1. Replicated environmental samples to test the precision of the sampling or analytical procedures.
2. Replicated analyses conducted on the same sample multiple times in order to determine analytical precision.
3. Trip blanks to determine if contaminants are introduced the processes of collecting, shipping, or storing of samples.
4. Matrix-fortified laboratory blanks consisting of solvent and reagent blanks to determine levels of bias due to matrix effects or analytical method problems.
5. Sample blanks (a sample matrix that does not contain the toxicant, although this is sometimes difficult to obtain) to ensure no extraneous or interfering peaks; the peaks indicate where a problem might exist in the method used.
6. Fortified field blanks to determine the effects that the matrix might have on analyte recovery.

## 3.6  SUMMARY

Planning the essential elements of an analytical protocol is critical to ensuring that the data generated answer what is requested, provide information affecting regulatory action, or enable some decision affecting environmental or human welfare. Essential decision criteria must be included in the protocol that describes the analytical process in detail, including the objective(s) of the study, the QA/QC requirements, the sample plan, methods of analysis, calculations, documentation, and data reporting. Meaningful data can only be generated with the proper method of analysis.

## SUGGESTED READING

### Statistical Principles

Middlebrooks, E. J. *Statistical Calculations "How to Solve Statistical Problems."* Ann Arbor, MI: Ann Arbor Science Publishers, 1977. (No theory but a lot of practical exercises)

Snedecor, G. W., and W. G. Cochran. 1996 and Numerous editions. *Statistical Methods.* Ames: Iowa State University Press. (One of the classical textbooks on theory and use of statistics)

Steele, R. G. D., and J. H. Torre. 1992 and Numerous editions. *Principles and Procedures of Statistics.* New York: McGraw-Hill. (Another classical textbook)

Youden, W. J., and E. H. Steiner. 1975. Statistical Manual of the AOAC. Arlington, VA: AOAC, 1975. (A presentation of statistical techniques used in collaborative testing of analytical methods)

## Environmental Analyses

Currie, L. A. *Detection in Analytical Chemistry: Importance, Theory and Practice*. American Chemical Society (ACS) Symposium Series 361. Washington, DC: ACS, 1988. (Contains both theory and considerable information on detection principles)

Keith, L. H. *Environmental Sampling and Analysis*. Chelsea, MI: Lewis Publishers, 1991.

## Quality Assurance

Garfield, F. M. *Quality Assurance Principles for Analytical Laboratories*. Arlington, VA: AOAC, 1984. (An excellent volume containing all phases of QA, including forms, equipment, and audits)

Taylor, J. K. *Quality Assurance of Chemical Measurements*. Chelsea, MI: Lewis Publishers, 1987. (Excellent summary of principles and extensive bibliography)

## Analytical Methods

1. University Libraries (many scientific journals contain pesticide residue studies and can be searched through databases available on computer)
   a. Journals include: *Journal of the Association of Official Analytical Chemists* (*JAOAC*); *Journal of Agricultural and Food Chemistry* (*J Agric. Food Chem.*); *Environmental Monitoring and Assessment* (*Environ. Monitoring Assess.*)
   b. Databases include: AGRICOLA, MEDLINE, TOXLINE
2. Official methods, Association of Official Analytical Chemists (AOAC)
3. State and federal agencies and laboratories (e.g., the US EPA repository at Ft. Meade, MD, has many)
4. Web Sites
5. US EPA Methods of Analysis
6. American Society of Testing and Materials (ASTM)
7. Chemical Manufacturers (contact local technical representatives for assistance)

# PART II

# CLASSES OF TOXICANTS

# CHAPTER 4

# Exposure Classes, Toxicants in Air, Water, Soil, Domestic and Occupational Settings

W. GREGORY COPE

## 4.1 AIR POLLUTANTS

### 4.1.1 History

Air pollution probably occurred as soon as humans started to use wood fires for heat and cooking. For centuries fire was used in such a way that living areas were filled with smoke. After the invention of the chimney, combustion products and cooking odors were removed from living quarters and vented outside. Later, when soft coal was discovered and used for fuel, coal smoke became a problem in the cities. By the thirteenth century, records show that coal smoke had become a nuisance in London, and in 1273 Edward I made the first antipollution law, one that prohibited the burning of coal while Parliament was in session: "Be it known to all within the sound of my voice, whosoever shall be found guilty of burning coal shall suffer the loss of his head." Despite this and various other royal edicts, however, smoke pollution continued in London.

Increasing domestic and industrial combustion of coal caused air pollution to get steadily worse, particularly in large cities. During the twentieth century the most significant change was the rapid increase in the number of automobiles, from almost none at the turn of the century to millions within only a few decades. During this time few attempts were made to control air pollution in any of the industrialized countries until after World War II. Action was then prompted, in part, by two acute pollution episodes in which human deaths were caused directly by high levels of pollutants. One incident occurred in 1948 in Donora, a small steel mill town in western Pennsylvania. In late October, heavy smog settled in the area, and a weather inversion prevented the movement of pollutants out of the valley. Twenty-one deaths were attributed directly to the effects of the smog. The "Donora episode" helped focus attention on air pollution in the United States.

In London, in December 1952, a now infamous killer smog occurred. A dense fog at ground level coupled with smoke from coal fireplaces caused severe smog lasting more than a week. The smog was so heavy that daylight visibility was only a few

*A Textbook of Modern Toxicology, Third Edition,* edited by Ernest Hodgson
ISBN 0-471-26508-X  Copyright © 2004 John Wiley & Sons, Inc.

meters, and bus conductors had to walk in front of the buses to guide the drivers through the streets. Two days after the smog began, the death rate began to climb, and between December 5 and December 9, there were an estimated 4000 deaths above the normal daily count. The chief causes of death were bronchitis, pneumonia, and associated respiratory complaints. This disaster resulted in the passage in Britain of the Clean Air Act in 1956.

In the United States the smog problem began to occur in large cities across the country, becoming especially severe in Los Angeles. In 1955 federal air pollution legislation was enacted, providing federal support for air pollution research, training, and technical assistance. Responsibility for the administration of the federal program now lies with the US Environmental Protection Agency (EPA). Technological interest since the mid-1950s has centered on automobile air pollution, pollution by oxides of sulfur and nitrogen, and the control of these emissions. Attention is also being directed toward the problems that may be caused by a possible greenhouse effect resulting from increased concentrations of carbon dioxide ($CO_2$) in the atmosphere, possible depletion of the stratospheric ozone layer, long-range transport of pollution, and acid deposition.

### 4.1.2  Types of Air Pollutants

What is clean air? Unpolluted air is a concept of what the air would be if humans and their works were not on earth, and if the air were not polluted by natural point sources such as volcanoes and forest fires. The true composition of "unpolluted" air is unknown because humans have been polluting the air for thousands of years. In addition there are many natural pollutants such as terpenes from plants, smoke from forest fires, and fumes and smoke from volcanoes. Table 4.1 lists the components that, in the absence of such pollution, are thought to constitute clean air.

***Gaseous Pollutants.***  These substances are gases at normal temperature and pressure as well as vapors evaporated from substances that are liquid or solid. Among pollutants of greatest concern are carbon monoxide (CO), hydrocarbons, hydrogen sulfide ($H_2S$)

**Table 4.1  Gaseous Components of Normal Dry Air**

| Compound | Percent by Volume | Concentration (ppm) |
|---|---|---|
| Nitrogen | 78.09 | 780,900 |
| Oxygen | 20.94 | 209,400 |
| Argon | 0.93 | 9300 |
| Carbon dioxide | 0.0325 | 325 |
| Neon | 0.0018 | 18 |
| Helium | 0.0005 | 5.2 |
| Methane | 0.0001 | 1.1 |
| Krypton | 0.0001 | 1.0 |
| Nitrous oxide | | 0.5 |
| Hydrogen | | 0.5 |
| Xenon | | 0.008 |
| Nitrogen dioxide | | 0.02 |
| Ozone | | 0.01–0.04 |

nitrogen oxides ($N_xO_y$), ozone ($O_3$) and other oxidants, sulfur oxides ($S_xO_y$), and $CO_2$. Pollutant concentrations are usually expressed as micrograms per cubic meter ($\mu g/m^3$) or for gaseous pollutants as parts per million (ppm) by volume in which 1 ppm = 1 part pollutant per million parts ($10^6$) of air.

***Particulate Pollutants.*** Fine solids or liquid droplets can be suspended in air. Some of the different types of particulates are defined as follows:

- *Dust.* Relatively large particles about 100 $\mu$m in diameter that come directly from substances being used (e.g., coal dust, ash, sawdust, cement dust, grain dust).
- *Fumes.* Suspended solids less than 1 $\mu$m in diameter usually released from metallurgical or chemical processes, (e.g., zinc and lead oxides).
- *Mist.* Liquid droplets suspended in air with a diameter less than 2.0 $\mu$m, (e.g., sulfuric acid mist).
- *Smoke.* Solid particles (0.05–1.0 $\mu$m) resulting from incomplete combustion of fossil fuels.
- *Aerosol.* Liquid or solid particles ($<1.0$ $\mu$m) suspended in air or in another gas.

### 4.1.3 Sources of Air Pollutants

***Natural Pollutants.*** Many pollutants are formed and emitted through natural processes. An erupting volcano emits particulate matter as well as gases such as sulfur dioxide, hydrogen sulfide, and methane; such clouds may remain airborne for long periods of time. Forest and prairie fires produce large quantities of pollutants in the form of smoke, unburned hydrocarbons, CO, nitrogen oxides, and ash. Dust storms are a common source of particulate matter in many parts of the world, and oceans produce aerosols in the form of salt particles. Plants and trees are a major source of hydrocarbons on the planet, and the blue haze that is so familiar over forested mountain areas is mainly from atmospheric reactions with volatile organics produced by the trees. Plants also produce pollen and spores, which cause respiratory problems and allergic reactions.

***Anthropogenic Pollutants.*** These substances come primarily from three sources: (1) combustion sources that burn fossil fuel for heating and power, or exhaust emissions from transportation vehicles that use gasoline or diesel fuels; (2) industrial processes; and (3) mining and drilling.

The principal pollutants from combustion are fly ash, smoke, sulfur, and nitrogen oxides, as well as CO and $CO_2$. Combustion of coal and oil, both of which contain significant amounts of sulfur, yields large quantities of sulfur oxides. One effect of the production of sulfur oxides is the formation of acidic deposition, including acid rain. Nitrogen oxides are formed by thermal oxidation of atmospheric nitrogen at high temperatures; thus almost any combustion process will produce nitrogen oxides. Carbon monoxide is a product of incomplete combustion; the more efficient the combustion, the higher is the ratio of $CO_2$ to CO.

Transportation sources, particularly automobiles, are a major source of air pollution and include smoke, lead particles from tetraethyl lead additives, CO, nitrogen oxides, and hydrocarbons. Since the mid-1960s there has been significant progress in reducing exhaust emissions, particularly with the use of low-lead or no-lead gasoline as well as the use of oxygenated fuels—for example, fuels containing ethanol or MTBE (methyl *t*-butyl ether).

Industries may emit various pollutants relating to their manufacturing processes—acids (sulfuric, acetic, nitric, and phosphoric), solvents and resins, gases (chlorine and ammonia), and metals (copper, lead, and zinc).

***Indoor Pollutants.*** In general, the term "indoor air pollution" refers to home and nonfactory public buildings such as office buildings and hospitals. Pollution can come from heating and cooking, pesticides, tobacco smoking, radon, gases, and microbes from people and animals.

Although indoor air pollution has increased in developed nations because of tighter building construction and the use of building materials that may give off gaseous chemicals, indoor air pollution is a particular problem in developing countries. Wood, crop residues, animal dung, and other forms of biomass are used extensively for cooking and heating—often in poorly ventilated rooms. For women and children, in particular, this leads to high exposures of air pollutants such as CO and polycyclic aromatic hydrocarbons.

### 4.1.4 Examples of Air Pollutants

Most of the information on the effects of air pollution on humans comes from acute pollution episodes such as the ones in Donora and London. Illnesses may result from chemical irritation of the respiratory tract, with certain sensitive subpopulations being more affected: (1) very young children, whose respiratory and circulatory systems are poorly developed, (2) the elderly, whose cardiorespiratory systems function poorly, and (3) people with cardiorespiratory diseases such as asthma, emphysema, and heart disease. Heavy smokers are also affected more adversely by air pollutants. In most cases the health problems are attributed to the combined action of particulates and sulfur dioxides ($SO_2$); no one pollutant appears to be responsible. Table 4.2 summarizes some of the major air pollutants and their sources and effects.

***Carbon Monoxide.*** Carbon monoxide combines readily with hemoglobin (Hb) to form carboxyhemoglobin (COHb), thus preventing the transfer of oxygen to tissues. The affinity of hemoglobin for CO is approximately 210 times its affinity for oxygen. A blood concentration of 5% COHb, equivalent to equilibration at approximately 45 ppm CO, is associated with cardiovascular effects. Concentrations of 100 ppm can cause headaches, dizziness, nausea, and breathing difficulties. An acute concentration of 1000 ppm is invariably fatal. Carbon monoxide levels during acute traffic congestion have been known to be as high as 400 ppm; in addition, people who smoke elevate their total body burden of CO as compared with nonsmokers. The effects of low concentrations of CO over a long period are not known, but it is possible that heart and respiratory disorders are exacerbated.

***Sulfur Oxides.*** Sulfur dioxide is a common component of polluted air that results primarily from the industrial combustion of coal, with soft coal containing the highest levels of sulfur. The sulfur oxides tend to adhere to air particles and enter the inner respiratory tract, where they are not effectively removed. In the respiratory tract, $SO_2$ combines readily with water to form sulfurous acid, resulting in irritation of mucous

Table 4.2   **Principal Air Pollutants, Sources, and Effects**

| Pollutant | Sources | Significance |
|---|---|---|
| Sulfur oxides, particulates | Coal and oil power plants | Main component of acid deposition |
| | Oil refineries, smelters | Damage to vegetation, materials |
| | Kerosene heaters | Irritating to lungs, chronic bronchitis |
| Nitrogen oxides | Automobile emissions | Pulmonary edema, impairs lung defenses |
| | Fossil fuel power plants | Important component of photochemical smog and acid deposition |
| Carbon monoxide | Motor vehicle emissions | Combines with hemoglobin to form carboxyhemoglobin, poisonous |
| | Burning fossil fuels | |
| | Incomplete combustion | Asphyxia and death |
| Carbon dioxide | Product of complete combustion | May cause "greenhouse effect" |
| Ozone ($O_3$) | Automobile emissions | Damage to vegetation |
| | Photochemical smog | Lung irritant |
| Hydrocarbons, $C_xH_y$ | Smoke, gasoline fumes | Contributes to photochemical smog |
| | Cigarette smoke, industry | Polycyclic aromatic hydrocarbons, lung cancer |
| | Natural sources | |
| Radon | Natural | Lung cancer |
| Asbestos | Asbestos mines | Asbestosis |
| | Building materials | Lung cancer, mesothelioma |
| | Insulation | |
| Allergens | Pollen, house dust | Asthma, rhinitis |
| | Animal dander | |
| Arsenic | Copper smelters | Lung cancer |

membranes and bronchial constriction. This irritation in turn increases the sensitivity of the airway to other airborne toxicants.

**Nitrogen Oxides.** Nitrogen dioxide ($NO_2$), a gas found in photochemical smog, is also a pulmonary irritant and is known to lead to pulmonary edema and hemorrhage. The main issue of concern is its contribution to the formation of photochemical smog and ozone, although nitrogen oxides also contribute to acid deposition.

**Ozone.** A highly irritating and oxidizing gas is formed by photochemical action of ultraviolet (UV) light on nitrogen dioxide in smog. The resulting ozone can produce pulmonary congestion, edema, and hemorrhage.

$$NO_2 + UV \text{ light} \longrightarrow NO + O^{\bullet}$$

$$O^{\bullet} + O_2 \longrightarrow O_3$$

At this point it is worth distinguishing between "good" and "bad" ozone. *Tropospheric ozone* occurs from 0 to 10 miles above the earth's surface, and is harmful. *Stratospheric ozone*, located about 30 miles above the earth's surface, is responsible for filtering out incoming UV radiation and thus is beneficial. It is the decrease in the stratospheric ozone layer that has been of much concern recently. It is estimated that a 1% decrease in stratospheric ozone will increase the amount of UV radiation reaching the earth's surface by 2% and cause a 10% increase in skin cancer. Major contributors to damage to stratospheric ozone are thought to be the chlorofluorocarbons (CFCs). Chlorine is removed from the CFC compounds in the upper atmosphere by reaction with UV light and is then able to destroy the stratospheric ozone through self-perpetuating free radical reactions.

$$Cl + O_3 \longrightarrow ClO + O_2$$
$$ClO + O \longrightarrow Cl + O_2$$

Before being inactivated by nitrogen dioxide or methane, each chlorine atom can destroy up to 10,000 molecules of ozone. Use of CFC compounds is now being phased out by international agreements.

***Hydrocarbons (HCs) or Volatile Organic Compounds (VOCs).*** These are derived primarily from two sources: approximately 50% are derived from trees as a result of the respiration process (biogenic); the other 45% to 50% comes from the combustion of fuel and from vapor from gasoline. Many gasoline pumps now have VOC recovery devices to reduce pollution.

***Lead.*** One of the most familiar of the particulates in air pollutants is lead, with young children and fetuses being the most susceptible. Lead can impair renal function, interfere with the development of red blood cells, and impair the nervous system, leading to mental retardation and even blindness. The two most common routes of exposure to lead are inhalation and ingestion. It is estimated that approximately 20% of the total body burden of lead comes from inhalation.

***Solid Particles.*** Dust and fibers from coal, clay, glass, asbestos, and minerals can lead to scarring or fibrosis of the lung lining. Pneumoconiosis, a condition common among coal miners that breathe coal dust, silicosis caused by breathing silica-containing dusts, and asbestosis from asbestos fibers are all well-known industrial pollution diseases.

### 4.1.5  Environmental Effects

***Vegetation.*** Pollutants may visibly injure vegetation by bleaching, other color changes, and necrosis, or by more subtle changes such as alterations in growth or reproduction. Table 4.3 lists some of the more common visual effects of air pollutants on vegetation. Air pollution can also result in measurable effects on forest ecosystems, such as reduction in forest growth, change in forest species, and increased susceptibility to forest pests. High-dose exposure to pollutants, which is associated with point source emissions such as smelters, frequently results in complete destruction of trees and shrubs in the surrounding area.

***Domestic Animals.*** Although domestic animals can be affected directly by air pollutants, the main concern is chronic poisoning as a result of ingestion of forage that has been contaminated by airborne pollutants. Pollutants important in this connection are

**Table 4.3  Examples of Air Pollution Injury to Vegetation**

| Pollutant | Symptoms |
| --- | --- |
| Sulfur dioxide | Bleached spots, interveinal bleaching |
| Ozone | Flecking, stippling, bleached spotting |
| Peroxyacetylnitrate (PAN) | Glazing, silvering, or bronzing on lower leaf surfaces |
| Nitrogen dioxide | White or brown collapsed lesion near leaf margins |
| Hydrogen fluoride | Tip and margin burns, dwarfing |

arsenic, lead, and molybdenum. Fluoride emissions from industries producing phosphate fertilizers and derivatives have damaged cattle throughout the world. The raw material, phosphate rock, can contain up to 4% fluoride, some of which is released into the air and water. Farm animals, particularly cattle, sheep, and swine, are susceptible to fluoride toxicity (fluorosis), which is characterized by mottled and soft teeth, and osterofluoritic bone lesions, which lead to lameness and, eventually, death.

***Materials and Structures.*** Building materials have become soiled and blackened by smoke, and damage by chemical attack from acid gases in the air has led to the deterioration of many marble statues in western Europe. Metals are also affected by air pollution; for example, $SO_2$ causes many metals to corrode at a faster rate. Ozone is known to oxidize rubber products, and one of the effects of Los Angeles smog is cracking of rubber tires. Fabrics, leather, and paper are also affected by $SO_2$ and sulfuric acid, causing them to crack, become brittle, and tear more easily.

***Atmospheric Effects.*** The presence of fine particles (0.1–1.0 mm in diameter) or $NO_2$ in the atmosphere can result in atmospheric haze or reduced visibility due to light scattering by the particles. The major effect of atmospheric haze has been degradation in visual air quality and is of particular concern in areas of scenic beauty, including most of the major national parks such as Great Smoky Mountain, Grand Canyon, Yosemite, and Zion Parks.

There is also concern over the increase in $CO_2$ in the atmosphere because $CO_2$ absorbs heat energy strongly and retards the cooling of the earth. This is often referred to as the greenhouse effect; theoretically an increase in $CO_2$ levels would result in a global increase in air temperatures. In addition to $CO_2$, other gases contributing to the greenhouse effect include methane, CFCs, nitrous oxide, and ozone.

***Acidic Deposition.*** Acidic deposition is the combined total of wet and dry deposition, with wet acidic deposition being commonly referred to as acid rain. Normal uncontaminated rain has a pH of about 5.6, but acid rain usually has a pH of less than 4.0. In the eastern United States, the acids in acid rain are approximately 65% sulfuric, 30% nitric, and 5% other, whereas in the western states, 80% of the acidity is due to nitric acid.

Many lakes in northeastern North America and Scandinavia have become so acidic that fish are no longer able to live in them. The low pH not only directly affects fish but also contributes to the release of potentially toxic metals, such as aluminum, from the soil. The maximum effect occurs when there is little buffering of the acid by soils or rock components. Maximum fish kills occur in early spring due to the "acid shock" from the melting of winter snows. Much of the acidity in rain may be neutralized by dissolving minerals in the soil such as aluminum, calcium, magnesium, sodium, and potassium, which are leached from the soil into surface waters. The ability of the soil to neutralize or buffer the acid rain is very dependent on the alkalinity of the soil. Much of the area in eastern Canada and the northeastern United States is covered by thin soils with low acid neutralizing capacity. In such areas the lakes are more susceptible to the effects of acid deposition leading to a low pH and high levels of aluminum, a combination toxic to many species of fish.

A second area of concern is that of reduced tree growth in forests. The leaching of nutrients from the soil by acid deposition may cause a reduction in future growth rates or changes in the type of trees to those able to survive in the altered environment. In addition to the change in soil composition, there are the direct effects on the trees from sulfur and nitrogen oxides as well as ozone.

## 4.2   WATER AND SOIL POLLUTANTS

With three-quarters of the earth's surface covered by water and much of the remainder covered by soil, it is not surprising that water and soil serve as the ultimate sinks for most anthropogenic chemicals. Until recently the primary concern with water pollution was that of health effects due to pathogens, and in fact this is still the case in most developing countries. In the United States and other developed countries, however, treatment methods have largely eliminated bacterial disease organisms from the water supply, and attention has been turned to chemical contaminants.

### 4.2.1   Sources of Water and Soil Pollutants

Surface water can be contaminated by *point* or *nonpoint* sources. An effluent pipe from an industrial plant or a sewage-treatment plant is an example of a point source; a field from which pesticides and fertilizers are carried by rainwater into a river is an example of a nonpoint source. Industrial wastes probably constitute the greatest single pollution problem in soil and water. These contaminants include organic wastes such as solvents, inorganic wastes, such as chromium and many unknown chemicals. Contamination of soil and water results when by-product chemicals are not properly disposed of or conserved. In addition industrial accidents may lead to severe local contamination. For a more in-depth discussion of sources and movements of water pollutants, see Chapter 27.

Domestic and municipal wastes, both from sewage and from disposal of chemicals, are another major source of chemical pollutants. At the turn of the twentieth century, municipal wastes received no treatment and were discharged directly into rivers or oceans. Even today, many older treatment plants do not provide sufficient treatment, especially plants in which both storm water and sewage are combined. In addition to organic matter, pesticides, fertilizers, detergents, and metals are significant pollutants discharged from urban areas.

Contamination of soil and water also results from the use of pesticides and fertilizers. Persistent pesticides applied directly to the soil have the potential to move from the soil into the water and thus enter the food chain from both soil and water. In a similar way fertilizers leach out of the soil or runoff during rain events and flow into the natural water systems.

Pollution from petroleum compounds has been a major concern since the mid-1960s. In 1967 the first major accident involving an oil tanker occurred. The *Torrey Canyon* ran onto rocks in the English Channel, spilling oil that washed onto the shores of England and France. It is estimated that at least 10,000 serious oil spills occur in the United States each year. In addition, flushing of oil tankers plays a major role in marine pollution. Other sources, such as improper disposal of used oil by private car owners and small garages, further contribute to oil pollution.

### 4.2.2 Examples of Pollutants

Metals that are of environmental concern fall into three classes: (1) metals that are suspected carcinogens, (2) metals that move readily in soil, and (3) metals that move through the food chain.

*Lead*. The heavy metals of greatest concern for health with regard to drinking water exposure are lead and arsenic. The sources of lead in drinking water that are most important are from lead pipes and lead solder. Also of concern is the seepage of lead from soil contaminated with the fallout from leaded gasoline and seepage of lead from hazardous-waste sites. Lead poisoning has been common in children, particularly in older housing units and inner city dwellings, in which children may consume chips of lead contaminated paint. Lead and associated toxic effects are discussed more fully in Chapter 5.

*Arsenic*. Drinking water is at risk for contamination by arsenic from the leaching of inorganic arsenic compounds formerly used in pesticide sprays, from the combustion of arsenic-containing fossil fuels, and from the leaching of mine tailings and smelter runoff. Chronic high-level exposures can cause abnormal skin pigmentation, hyperkeratosis, nasal congestion, and abdominal pain. At lower levels of chronic exposure, cancer is the major concern. Epidemologic studies have linked chronic arsenic exposure to various cancers, including skin, lungs, and lymph glands.

*Cadmium*. One of the most significant effects of metal pollution is that aquatic organisms can accumulate metals in their tissues, leading to increased concentrations in the food chain. Concern about long-term exposure to cadmium intensified after recognition of the disease Itai-Itai (painful-painful) in certain areas of Japan. The disease is a combination of severe kidney damage and painful bone and joint disease and occurs in areas where rice is contaminated with high levels of cadmium. This contamination resulted from irrigation of the soil with water containing cadmium released from industrial sources. Cadmium toxicity in Japan has also resulted from consumption of cadmium-contaminated fish taken from rivers near smelting plants.

*Mercury*. In Japan in the 1950s and 1060s, wastes from a chemical and plastics plant containing mercury were discharged into Minamata Bay. The mercury was converted to the readily absorbed methylmercury by bacteria in the aquatic sediments. Consumption of fish and shellfish by the local population resulted in numerous cases of mercury poisoning, or Minamata disease. By 1970, at least

107 deaths had been attributed to mercury poisoning, and 800 cases of Minamata disease were confirmed. Even though the mothers appeared healthy, many infants born to these mothers who had eaten contaminated fish developed cerebral palsy-like symptoms and mental deficiency.

Pesticides are also a major source of concern as water and soil pollutants. Because of their stability and persistence, the most hazardous pesticides are the organochlorine compounds such as DDT, aldrin, dieldrin, and chlordane. Persistent pesticides can accumulate in food chains; for example, shrimp and fish can concentrate some pesticides as much as 1000- to 10,000-fold. This bioaccumulation has been well documented with the pesticide DDT, which is now banned in many parts of the world. In contrast to the persistent insecticides, the organophosphorus (OP) pesticides, such as malathion, and the carbamates, such as carbaryl, are short-lived and generally persist for only a few weeks to a few months. Thus these compounds do not usually present as serious a problem as the earlier insecticides. Herbicides, because of the large quantity used, are also of concern as potential toxic pollutants. Pesticides are discussed in more detail in Chapter 5.

Nitrates and phosphates are two important nutrients that have been increasing markedly in natural waters since the mid-1960s. Sources of nitrate contamination include fertilizers, discharge from sewage treatment plants, and leachate from septic systems and manure. Nitrates from fertilizers leach readily from soils, and it has been estimated that up to 40% of applied nitrates enter water sources as runoff and leachate. Fertilizer phosphates, however, tend to be absorbed or bound to soil particles, so that only 20% to 25% of applied nitrates are leached into water. Phosphate detergents are another source of phosphate, one that has received much media attention in recent years.

The increase in these nutrients, particularly phosphates, is of environmental concern because excess nutrients can lead to "algal blooms" or eutrophication, as it is known, in lakes, ponds, estuaries, and very slow moving rivers. The algal bloom reduces light penetration and restricts atmospheric reoxygenation of the water. When the dense algal growth dies, the subsequent biodegradation results in anaerobic conditions and the death of many aquatic organisms. High phosphate concentrations and algal blooms are generally not a problem in moving streams, because such streams are continually flushed out and algae do not accumulate.

There are two potential adverse health effects from nitrates in drinking water: (1) nitrosamine formation and (2) methemoglobinemia. Ingested nitrates can be converted to nitrites by intestinal bacteria. After entering the circulatory system, nitrite ions combine with hemoglobin to form methemoglobin, thus decreasing the oxygen-carrying capacity of the blood and resulting in anemia or blue-baby disease. It is particularly severe in young babies who consume water and milk-formula prepared with nitrate-rich water. Older children and adults are able to detoxify the methemoglobin as a result of the enzyme methemoglobin reductase, which reverses the formation of methemoglobin. In infants, however, the enzyme is not fully functional. Certain nitrosamines are known carcinogens.

Oils and petroleum are ever-present pollutants in the modern environment, whether from the used oil of private motorists or spillage from oil tankers. At sea, oil slicks are responsible for the deaths of many birds. Very few birds that are badly contaminated recover, even after de-oiling and hand feeding. Oil is deposited on rocks and sand as well, thus preventing the beaches from being used for recreation until after costly clean up. Shore animals, such as crabs, shrimp, mussels, and barnacles, are also affected by the toxic hydrocarbons they ingest. The subtle and perhaps potentially more harmful long-term effects on aquatic life are not yet fully understood.

Volatile organic compounds (VOCs) are other common groundwater contaminants. They include halogenated solvents and petroleum products, collectively referred to as VOCs. Both groups of compounds are used in large quantities by a variety of industries, such as degreasing, dry cleaning, paint, and the military. Historically petroleum products were stored in underground tanks that would erode, or were spilled onto soil surfaces. The EPA's National Priority List includes 11 VOCs: trichloroethylene, toluene, benzene, chloroform, tetrachloroethylene, 1,1,1-trichloroethane, ethylbenzene, trans-1,2-dichloroethane, xylene, dichloromethane, and vinyl chloride.

The physical and chemical properties of VOCs permit them to move rapidly into groundwater, and almost all of the previously mentioned chemicals have been detected in groundwater near contaminant sites. High levels of exposure can cause headache, impaired cognition, and kidney toxicities. At levels of exposure most frequently encountered, cancer and reproductive effects are of most concern, particularly childhood leukemia.

Low molecular weight chlorinated hydrocarbons are a by-product of the chlorination of municipal water. Chlorine reacts with organic substances commonly found in water to generate trihalomethanes (THMs), such as chloroform. The main organics that have been detected are chloroform, bromodichloromethane, dibromochloromethane, bromoform, carbon tetrachloride, and 1,2-dichloroethane. These compounds are associated with an increased risk of cancer. Studies in New Orleans in the mid-1970s showed that tap water in New Orleans contained more chlorinated hydrocarbons than did untreated Mississippi River water or well water. In addition chlorinated hydrocarbons, including carbon tetrachloride, were detected in blood plasma from volunteers who drank treated tap water. Epidemiologic studies indicated that the cancer death rate was higher among white males who drank tap water that among those who drank well water.

Radioactive contamination as some background radiation from natural sources, such as radon, occurs in some regions of the world, but there is particular concern over the contamination of surface water and groundwater by radioactive compounds generated by the production of nuclear weapons and by the processing of nuclear fuel. Many of these areas have remained unrecognized because of government secrecy.

Acids present in rain or drainage from mines, are major pollutants in many freshwater rivers and lakes. Because of their ability to lower the pH of the water to toxic levels and release toxic metals into solution, acids are considered particularly hazardous (see Chapter 5).

PCB organic compounds found as soil and water contaminants continue to grow each year. They include polychlorinated biphenyls (PCBs), phenols, cyanides, plasticizers, solvents, and numerous industrial chemicals. PCBs were historically used as coolants in electrical transformers and are also known by-products of the plastic, lubricant, rubber, and paper industries. They are stable, lipophilic, and break down only slowly in tissues. Because of these properties they accumulate to high concentrations in fish and waterfowl; in 1969 PCBs were responsible for the death of thousands of birds in the Irish Sea.

Dioxin has contaminated large areas of water and soil in the form of extremely toxic TCDD (2,3,7,8-tetrachlorodibenzo-p-dioxin) through industrial accidents and through widespread use of the herbicide 2,4,5-T. Small amounts of TCDD were contained as a contaminant in herbicide manufacturing. The US Army used this herbicide, known as Agent Orange, extensively as a defoliant in Vietnam. TCDD is one of the most toxic synthetic substances known for laboratory animals: LD50 for male rats, 0.022 mg/kg; LD50 for female rats, 0.045 mg/kg; LD50 for female guinea pigs (the most sensitive species tested), 0.0006 mg/kg. In addition it is fetotoxic to pregnant rats at a dose of only 1/400 of the LD50, and has been shown to cause birth defects at levels of 1 to 3 ng/kg. TCDD is a proven carcinogen in both mice and rats, with the liver being the primary target. Although TCDD does not appear to be particularly acutely toxic to humans, chronic low-level exposure is suspected of contributing to reproductive abnormalities and carcinogenicity.

## 4.3   OCCUPATIONAL TOXICANTS

Assessment of hazards in the workplace is a concern of occupational/industrial toxicology and has a history that dates back to ancient civilizations. The Greek historian Strabo, who lived in the first century AD, gave a graphic description of the arsenic mines in Pantus: "The air in mines is both deadly and hard to endure on account of the grievous odor of the ore, so that the workmen are doomed to a quick death." With the coming of the industrial revolution in the nineteenth century, industrial diseases increased, and new ones, such as chronic mercurialism caused by exposure to mercuric nitrate used in "felting" animal furs, were identified. Hat makers, who were especially at risk, frequently developed characteristic tremors known as "hatters' shakes," and the expression "mad as a hatter" was coined. In recent years concern has developed over the carcinogenic potential of many workplace chemicals.

### 4.3.1   Regulation of Exposure Levels

The goal of occupational toxicology is to ensure work practices that do not entail any unnecessary health risks. To do this, it is necessary to define suitable permissible levels of exposure to industrial chemicals, using the results of animal studies and epidemiological studies. These levels can be expressed by the following terms for allowable concentrations.

Threshold limit values (TLVs) refer to airborne concentrations of substances and represent conditions under which it is believed that nearly all workers may be repeatedly exposed day after day without adverse effect. Because of wide variation in individual susceptibility, a small percentage of workers may experience discomfort from some substances at or below the threshold limit; a smaller percentage may be affected more seriously by aggravation of a preexisting condition or by development of an occupational illness. Threshold limits are based on the best available information from industrial experience, from experimental human and animal studies, and when possible, from a combination of the three. The basis on which the values are established may differ from substance to substance; protection against impairment of health may be a guiding factor for some, whereas reasonable freedom from irritation, narcosis, nuisance, or other forms of stress may form the basis for others. Three categories of TLVs follow:

*Threshold limit value–time-weighted average (TLV-TWA)* is the TWA concentration for a normal 8-hour workday or 40-hour workweek to which nearly all workers may be repeatedly exposed, day after day, without adverse effect. Time-weighted averages allow certain permissible excursions above the limit provided that they are compensated by equivalent excursions below the limit during the workday. In some instances the average concentration is calculated for a workweek rather than for a workday.

*Threshold limit value–short-term exposure limit (TLV-STEL)* is the maximal concentration to which workers can be exposed for a period up to 15 minutes continuously without suffering from (1) irritation, (2) chronic or irreversible tissue change, or (3) narcosis of sufficient degree that would increase accident proneness, impair self-rescue, or materially work efficiency, provided that no more than four excursions per day are permitted, with at least 60 minutes between exposure periods, and provided that the daily TLV-TWA is not exceeded.

*Threshold limit value–ceiling (TLV-C)* is the concentration that should not be exceeded even instantaneously. For some substances—for instance, irritant gases—only one category, the TLV-ceiling, may be relevant. For other substances, two or three categories may be relevant.

Biologic limit values (BLVs) represent limits of amounts of substances (or their affects) to which the worker may be exposed without hazard to health or well-being as determined by measuring the worker's tissues, fluids, or exhaled breath. The biologic measurements on which the BLVs are based can furnish two kinds of information useful in the control of worker exposure: (1) measure of the worker's overall exposure and (2) measure of the worker's individual and characteristic response. Measurements of response furnish a superior estimate of the physiological status of the worker, and may consist of (1) changes in amount of some critical biochemical constituent, (2) changes in activity or a critical enzyme, and (3) changes in some physiological function. Measurement of exposure may be made by (1) determining in blood, urine, hair, nails, or body tissues and fluids the amount of substance to which the worker was exposed; (2) determining the amount of the metabolite(s) of the substance in tissues and fluids; and (3) determining the amount of the substance in the exhaled breath. The biologic limits may be used as an adjunct to the TLVs for air, or in place of them.

Immediately dangerous to life or health (IDLH) conditions pose a threat of severe exposure to contaminants, such as radioactive materials, that are likely to have adverse cumulative or delayed effects on health. Two factors are considered when establishing IDLH concentrations. The worker must be able to escape (1) without loss of life or without suffering permanent health damage within 30 minutes and (2) without severe eye or respiratory irritation or other reactions that could inhibit escape. If the concentration is above the IDLH, only highly reliable breathing apparatus is allowed.

## 4.3.2  Routes of Exposure

The principal routes of industrial exposure are dermal and inhalation. Occasionally toxic agents may be ingested, if food or drinking water is contaminated. Exposure to the skin often leads to localized effects known as "occupation dermatosis" caused by either irritating chemicals or allergenic chemicals. Such effects include scaling, eczema, acne, pigmentation changes, ulcers, and neoplasia. Some chemicals may also pass through the skin; these include aromatic amines such as aniline and solvents such as carbon tetrachloride and benzene.

Toxic or potentially toxic agents may be inhaled into the respiratory tract where they may cause localized effects such as irritation (e.g., ammonia, chlorine gas), inflammation, necrosis, and cancer. Chemicals may also be absorbed by the lungs into the circulatory system, thereby leading to systemic toxicity (e.g., CO, lead).

### 4.3.3  Examples of Industrial Toxicants

Carcinogen exposure is largely due to lifestyle, such as cigarette smoking, but occupation is an important source of exposure to carcinogens. Table 4.4 lists some occupational chemical hazards and the cancers associated with them.

Cadmium is a cumulative toxicant with a biologic half-life of up to 30 years in humans. More than 70% of the cadmium in the blood is bound to red blood cells; accumulation occurs mainly in the kidney and the liver, where cadmium is bound to metallothionein. In humans the critical target organ after long-term exposure to cadmium is the kidney, with the first detectable symptom of kidney toxicity being an increased excretion of specific proteins.

Chromium toxicity results from compounds of hexavalent chromium that can be readily absorbed by the lung and gastrointestinal (GI) tract and to a lesser extent by the skin. Occupational exposure to chromium ($Cr^{6+}$) causes dermatitis, ulcers on the hands and arms, perforation of the nasal septum (probably caused by chromic acid), inflammation of the larynx and liver, and bronchitis. Chromate is a carcinogen causing bronchogenic carcinoma; the risk to chromate plant workers for lung cancer is 20 times greater than that for the general population. Compounds of trivalent chromium

**Table 4.4  Some Occupational Hazards and Associated Cancers**

| Agent | Tumor Sites | Occupation |
|---|---|---|
| Asbestos | Lung, pleura, peritoneum | Miners, manufacturers, users |
| Arsenic | Skin, lung, liver | Miners and smelters, oil refinery, pesticide workers |
| Benzene | Hemopoietic tissue | Process workers, textile workers |
| Cadmium | Lung, kidney, prostate | Battery workers, smelters |
| Chloroethers | Lung | Chemical plant workers, process workers |
| Chromium | Lung, nasal cavity, sinuses | Process and production workers, pigment workers |
| Mustard gas | Bronchi, lung, larynx | Production workers |
| Naphthylamines | Bladder | Dyestuff makers and workers, Chemical workers, printers |
| Nickel | Lung, nasal sinuses | Smelters and process workers |
| Polycyclic aromatic hydrocarbons | Respiratory system, bladder | Furnace, foundry, shale, and gas workers; chimney sweeps |
| Radon, radium, uranium | Skin, lung, bone tissue, bone marrow | Medical and industrial chemists, miners |
| UV radiation | Skin | Outdoor exposure |
| X rays | Bone marrow, skin | Medical and industrial workers |

are poorly absorbed. Chromium is not a cumulative chemical, and once absorbed, it is rapidly excreted into the urine.

Lead is a ubiquitous toxicant in the environment, and consequently the normal body concentration of lead is dependent on environmental exposure conditions. Approximately 50% of lead deposited in the lung is absorbed, whereas usually less than 10% of ingested lead passes into the circulation. Lead is not a major occupational problem today, but environmental pollution is still widespread. Lead interferes in the biosynthesis of porphyrins and heme, and several screening tests for lead poisoning make use of this interaction by monitoring either inhibition of the enzyme $\delta$-aminolevulinic acid dehydratase (ALAD) or appearance in the urine of aminolevulinic acid (ALA) and coproporphorin (UCP). The metabolism of inorganic lead is closely related to that of calcium, and excess lead can be deposited in the bone where it remains for years. Inorganic lead poisoning can produce fatigue, sleep disturbances, anemia, colic, and neuritis. Severe exposure, mainly of children who have ingested lead, may cause encephalopathy, mental retardation, and occasionally, impaired vision.

Organic lead has an affinity for brain tissue; mild poisoning may cause insomnia, restlessness, and GI symptoms, whereas severe poisoning results in delirium, hallucinations, convulsions, coma, and even death.

Mercury is widely used in scientific and electrical apparatus, with the largest industrial use of mercury being in the chlorine-alkali industry for electrolytic production of chlorine and sodium hydroxide. Worldwide, this industry has been a major source of mercury contaminations. Most mercury poisoning, however, has been due to methylmercury, particularly as a result of eating contaminated fish. Inorganic and organic mercury differ in their routes of entry and absorption. Inhalation is the principal route of uptake of metallic mercury in industry, with approximately 80% of the mercury inhaled as vapor being absorbed; metallic mercury is less readily absorbed by the GI route. The principal sites of deposition are the kidney and brain after exposure to inorganic mercury salts. Organic mercury compounds are readily absorbed by all routes. Industrial mercurialism produces features such as inflammation of the mouth, muscular tremors (hatters' shakes), psychic irritation, and a nephritic syndrome characterized by proteinuria. Overall, however, occupational mercurialism is not a significant problem today.

Benzene was used extensively in the rubber industry as a solvent for rubber latex in the latter half of the nineteenth century. The volatility of benzene, which made it so attractive to the industry, also caused high atmospheric levels of the solvent. Benzene-based rubber cements were used in the canning industry and in the shoe manufacturing industry. Although cases of benzene poisoning had been reported as early as 1897 and additional reports and warnings were issued in the 1920s, the excellent solvent properties of benzene resulted in its continued extensive use. In the 1930s cases of benzene toxicity occurred in the printing industry in which benzene was used as an ink solvent. Today benzene use exceeds 11 billion gallons per year.

Benzene affects the hematopoietic tissue in the bone marrow and also appears to be an immunosuppressant. There is a gradual decrease in white blood cells, red blood cells, and platelets, and any combination of these signs may be seen. Continued exposure to benzene results in severe bone marrow damage and aplastic anemia. Benzene exposure has also been associated with leukemia.

Asbestos and other fibers of naturally occurring silicates will separate into flexible fibers. Asbestos is the general name for this group of fibers. Chrysotile is the most important commercially and represents about 90% of the total used. Use of asbestos has been extensive, especially in roofing and insulation, asbestos cements, brake linings, electrical appliances, and coating materials. Asbestosis, a respiratory disease, is characterized by fibrosis, calcification, and lung cancer. In humans, not only is there a long latency period between exposure and development of tumors but other factors also influence the development of lung cancer. Cigarette smoking, for example, enhances tumor formation. Recent studies have shown that stomach and bowel cancers occur in excess in workers (e.g., insulation workers) exposed to asbestos. Other fibers have been shown to cause a similar disease spectrum, for instance, zeolite fibers.

## SUGGESTED READING

### Air Pollutants

Costa, D. L. Air pollution. In *Casarett and Doull's Toxicology: The Basic Science of Poisons*, 6th ed., C. D. Klaassen, ed. New York: McGraw-Hill, 2001, pp. 979–1012.

Holgate, S. T., J. M. Samet, H. Koren, and R. Maynard, eds. *Air Pollution and Health*. San Diego: Academic Press, 1999.

### Water and Soil Pollutants

Abel, P. D., ed. *Water Pollution Biology*. London: Taylor and Francis, 1996.

Hoffman, D. J., B. A. Rattner, G. A. Burton, and J. Cairns, eds. *Handbook of Ecotoxicology*, 2nd ed. Boca Raton: Lewis, 2002.

Larson, S. J., P. D. Capel, and M. S. Majewski, eds. *Pesticides in Surface Waters*. Chelsea, MI: Ann Arbor Press, 1998.

### Occupational Toxicants

Thorne, P. S. Occupational toxicology. In *Casarett and Doull's Toxicology: The Basic Science of Poisons*, 6th ed., C. D. Klaassen, ed. New York: McGraw-Hill, 2001, pp. 1123–1140.

Doull, J. Recommended limits for occupational exposure to chemicals. In *Casarett and Doull's Toxicology: The Basic Science of Poisons*, 6th ed., C. D. Klaassen, ed. New York: McGraw-Hill, 2001, pp. 1155–1176.

# CHAPTER 5

# Classes of Toxicants: Use Classes

W. GREGORY COPE, ROSS B. LEIDY, and ERNEST HODGSON

## 5.1 INTRODUCTION

As discussed in Chapter 1, use classes include not only chemicals currently in use but also the toxicological aspects of the development of new chemicals for commercial use, chemicals produced as by-products of industrial processes, and chemicals resulting from the use and/or disposal of chemicals. Because any use class may include chemicals from several different chemical classes, this classification is not sufficient for mechanistic considerations. It is, however, essential for an understanding of the scope of toxicology and, in particular, is essential for many applied branches of toxicology such as exposure assessment, industrial hygiene, public health toxicology and regulatory toxicology.

## 5.2 METALS

### 5.2.1 History

Although most metals occur in nature in rocks, ores, soil, water, and air, levels are usually low and widely dispersed. In terms of human exposure and toxicological significance, it is anthropogenic activities that are most important because they increase the levels of metals at the site of human activities.

Metals have been used throughout much of human history to make utensils, machinery, and so on, and mining and smelting supplied metals for these uses. These activities increased environmental levels of metals. More recently metals have found a number of uses in industry, agriculture, and medicine. These activities have increased exposure not only to metal-related occupational workers but also to consumers of the various products.

Despite the wide range of metal toxicity and toxic properties, there are a number of toxicological features that are common to many metals. Some of the more important aspects are discussed briefly in the following sections. For a metal to exert its toxicity, it must cross the membrane and enter the cell. If the metal is in a lipid soluble form such as methylmercury, it readily penetrates the membrane; when bound to proteins

*A Textbook of Modern Toxicology, Third Edition*, edited by Ernest Hodgson
ISBN 0-471-26508-X  Copyright © 2004 John Wiley & Sons, Inc.

such as cadmium-metallothionein, the metal is taken into the cell by endocytosis; other metals (e.g., lead) may be absorbed by passive diffusion. The toxic effects of metals usually involve interaction between the free metal and the cellular target. These targets tend to be specific biochemical processes and/or cellular and subcellular membranes.

### 5.2.2 Common Toxic Mechanisms and Sites of Action

***Enzyme Inhibition/Activation.*** A major site of toxic action for metals is interaction with enzymes, resulting in either enzyme inhibition or activation. Two mechanisms are of particular importance: inhibition may occur as a result of interaction between the metal and sulfhydryl (SH) groups on the enzyme, or the metal may displace an essential metal cofactor of the enzyme. For example, lead may displace zinc in the zinc-dependent enzyme $\delta$-aminolevulinic acid dehydratase (ALAD), thereby inhibiting the synthesis of heme, an important component of hemoglobin and heme-containing enzymes, such as cytochromes.

***Subcellular Organelles.*** Toxic metals may disrupt the structure and function of a number of organelles. For example, enzymes associated with the endoplasmic reticulum may be inhibited, metals may be accumulated in the lysosomes, respiratory enzymes in the mitochondria may be inhibited, and metal inclusion bodies may be formed in the nucleus.

***Carcinogenicity.*** A number of metals have been shown to be carcinogenic in humans or animals. Arsenic, certain chromium compounds, and nickel are known human carcinogens; beryllium, cadmium, and cisplatin are probable human carcinogens. The carcinogenic action, in some cases, is thought to result from the interaction of the metallic ions with DNA (see Chapter 11 for a detailed discussion of carcinogenesis).

***Kidney.*** Because the kidney is the main excretory organ of the body, it is a common target organ for metal toxicity. Cadmium and mercury, in particular, are potent nephrotoxicants and are discussed more fully in the following sections and in Chapter 15.

***Nervous System.*** The nervous system is also a common target of toxic metals; particularly, organic metal compounds (see Chapter 16). For example, methylmercury, because it is lipid soluble, readily crosses the blood-brain barrier and enters the nervous system. By contrast, inorganic mercury compounds, which are more water soluble, are less likely to enter the nervous system and are primarily nephrotoxicants. Likewise organic lead compounds are mainly neurotoxicants, whereas the first site of inorganic lead is enzyme inhibition (e.g., enzymes involved in heme synthesis).

***Endocrine and Reproductive Effects.*** Because the male and female reproductive organs are under complex neuroendocrine and hormonal control, any toxicant that alters any of these processes can affect the reproductive system (see Chapters 17 and 20). In addition metals can act directly on the sex organs. Cadmium is known to produce testicular injury after acute exposure, and lead accumulation in the testes is associated with testicular degeneration, inhibition of spermatogenesis, and Leydig-cell atrophy.

***Respiratory System.*** Occupational exposure to metals in the form of metal dust makes the respiratory system a likely target. Acute exposure may cause irritations and inflammation of the respiratory tract, whereas chronic exposure may result in fibrosis (aluminum) or carcinogenesis (arsenic, chromium, nickel). Respiratory toxicants are discussed more fully in Chapter 18.

***Metal-Binding Proteins.*** The toxicity of many metals such as cadmium, lead, and mercury depends on their transport and intracellular bioavailability. This availability is regulated to a degree by high-affinity binding to certain cytosolic proteins. Such ligands usually possess numerous SH binding sites that can outcompete other intracellular proteins and thus mediate intracellular metal bioavailability and toxicity. These intracellular "sinks" are capable of partially sequestering toxic metals away from sensitive organelles or proteins until their binding capacity is exceeded by the dose of the metal. *Metallothionein* (MT) is a low molecular weight metal-binding protein (approximately 7000 Da) that is particularly important in regulating the intracellular bioavailability of cadmium, copper, mercury, silver, and zinc. For example, in vivo exposure to cadmium results in the transport of cadmium in the blood by various high molecular weight proteins and uptake by the liver, followed by hepatic induction of MT. Subsequently cadmium can be found in the circulatory system bound to MT as the cadmium-metallothionein complex (CdMT).

## 5.2.3 Lead

Because of the long-term and widespread use of lead, it is one of the most ubiquitous of the toxic metals. Exposure may be through air, water, or food sources. In the United States the major industrial uses, such as in fuel additives and lead pigments in paints, have been phased out, but other uses, such as in batteries, have not been reduced. Other sources of lead include lead from pipes and glazed ceramic food containers.

Inorganic lead may be absorbed through the GI tract, the respiratory system, and the skin. Ingested inorganic lead is absorbed more efficiently from the GI tract of children than that of adults, readily crosses the placenta, and in children penetrates the blood-brain barrier. Initially, lead is distributed in the blood, liver, and kidney; after prolonged exposure, as much as 95% of the body burden of lead is found in bone tissue.

The main targets of lead toxicity are the hematopoietic system and the nervous system. Several of the enzymes involved in the synthesis of heme are sensitive to inhibition by lead, the two most susceptible enzymes being ALAD and heme synthetase (HS). Although clinical anemia occurs only after moderate exposure to lead, biochemical effects can be observed at lower levels. For this reason inhibition of ALAD or appearance in the urine of ALA can be used as an indication of lead exposure.

The nervous system is another important target tissue for lead toxicity, especially in infants and young children in whom the nervous system is still developing (Chapter 16). Even at low levels of exposure, children may show hyperactivity, decreased attention span, mental deficiencies, and impaired vision. At higher levels, encephalopathy may occur in both children and adults. Lead damages the arterioles and capillaries, resulting in cerebral edema and neuronal degeneration. Clinically this damage manifests itself as ataxia, stupor, coma, and convulsions.

Another system affected by lead is the reproductive system (Chapter 20). Lead exposure can cause male and female reproductive toxicity, miscarriages, and degenerate offspring.

### 5.2.4 Mercury

Mercury exists in the environment in three main chemical forms: elemental mercury ($Hg^0$), inorganic mercurous ($Hg^+$) and mercuric ($Hg2^+$) salts, and organic methylmercury ($CH_3Hg$) and dimethylmercury ($CH_3HgCH_3$) compounds. Elemental mercury, in the form of mercury vapor, is almost completely absorbed by the respiratory system, whereas ingested elemental mercury is not readily absorbed and is relatively harmless. Once absorbed, elemental mercury can cross the blood-brain barrier into the nervous system. Most exposure to elemental mercury tends to be from occupational sources.

Of more concern from environmental contamination is exposure to organic mercury compounds. Inorganic mercury may be converted to organic mercury through the action of sulfate-reducing bacteria, to produce methylmercury, a highly toxic form readily absorbed across membranes. Several large episodes of mercury poisoning have resulted from consuming seed grain treated with mercury fungicides or from eating fish contaminated with methylmercury. In Japan in the 1950s and 1960s wastes from a chemical and plastics plant containing mercury were drained into Minamata Bay. The mercury was converted to the readily absorbed methylmercury by bacteria in the aquatic sediments. Consumption of fish and shellfish by the local population resulted in numerous cases of mercury poisoning or Minamata disease. By 1970 at least 107 deaths had been attributed to mercury poisoning, and 800 cases of Minamata disease were confirmed. Even though the mothers appeared healthy, many infants born to mothers who had eaten contaminated fish developed cerebral palsy-like symptoms and mental deficiency. Organic mercury primarily affects the nervous system, with the fetal brain being more sensitive to the toxic effects of mercury than adults.

Inorganic mercury salts, however, are primarily nephrotoxicants, with the site of action being the proximal tubular cells. Mercury binds to SH groups of membrane proteins, affecting the integrity of the membrane and resulting in aliguria, anuria, and uremia.

### 5.2.5 Cadmium

Cadmium occurs in nature primarily in association with lead and zinc ores and is released near mines and smelters processing these ores. Industrially cadmium is used as a pigment in paints and plastics, in electroplating, and in making alloys and alkali storage batteries (e.g., nickel-cadmium batteries). Environmental exposure to cadmium is mainly from contamination of groundwater from smelting and industrial uses as well as the use of sewage sludge as a food-crop fertilizer. Grains, cereal products, and leafy vegetables usually constitute the main source of cadmium in food. Reference has already been made to the disease Itai-Itai resulting from consumption of cadmium-contaminated rice in Japan (see Chapter 4, Section 4.2.2).

Acute effects of exposure to cadmium result primarily from local irritation. After ingestion, the main effects are nausea, vomiting, and abdominal pain. Inhalation exposure may result in pulmonary edema and chemical pneumonitis.

Chronic effects are of particular concern because cadmium is very slowly excreted from the body, with a half-life of about 30 years. Thus low levels of exposure can result in considerable accumulation of cadmium. The main organ of damage following long-term exposure is the kidney, with the proximal tubules being the primary site of action. Cadmium is present in the circulatory system bound primarily to the metal-binding protein, metallothionein, produced in the liver. Following glomerular filtration in the kidney, CdMT is re-absorbed efficiently by the proximal tubule cells, where it accumulates within the lysosomes. Subsequent degradation of the CdMT complex releases $Cd^{+2}$, which inhibits lysosomal function, resulting in cell injury.

### 5.2.6 Chromium

Because chromium occurs in ores, environmental levels are increased by mining, smelting, and industrial uses. Chromium is used in making stainless steel, various alloys, and pigments. The levels of this metal are generally very low in air, water, and food, and the major source of human exposure is occupational. Chromium occurs in a number of oxidation states from $Cr^{+2}$ to $Cr^{+6}$, but only the trivalent ($Cr^{+3}$) and hexavalent ($Cr^{+6}$) forms are of biological significance. Although the trivalent compound is the most common form found in nature, the hexavalent form is of greater industrial importance. In addition hexavalent chromium, which is not water soluble, is more readily absorbed across cell membranes than is trivalent chromium. In vivo the hexavalent form is reduced to the trivalent form, which can complex with intracellular macromolecules, resulting in toxicity. Chromium is a known human carcinogen and induces lung cancers among exposed workers. The mechanism of chromium ($Cr^{+6}$) carcinogenicity in the lung is believed to be its reduction to $Cr^{+3}$ and generation of reactive intermediates, leading to bronchogenic carcinoma.

### 5.2.7 Arsenic

In general, the levels of arsenic in air and water are low, and the major source of human exposure is food. In certain parts of Taiwan and South America, however, the water contains high levels of this metalloid, and the inhabitants often suffer from dermal hyperkeratosis and hyperpigmentation. Higher levels of exposure result in a more serious condition; gangrene of the lower extremities or "blackfoot disease." Cancer of the skin also occurs in these areas.

Approximately 80% of arsenic compounds are used in pesticides. Other uses include glassware, paints, and pigments. Arsine gas is used in the semiconductor industry. Arsenic compounds occur in three forms: (1) pentavalent, $As^{+5}$, organic or arsenate compounds (e.g., alkyl arsenates); (2) trivalent, $As^{+3}$, inorganic or arsenate compounds (e.g., sodium arsenate, arsenic trioxide); and (3) arsine gas, $AsH_3$, a colorless gas formed by the action of acids on arsenic. The most toxic form is arsine gas with a TLV-TWA of 0.05 ppm. Microorganisms in the environment convert arsenic to dimethylarsenate, which can accumulate in fish and shellfish, providing a source for human exposure. Arsenic compounds can also be present as contaminants in well water. Arsenite ($As^{+3}$) compounds are lipid soluble and can be absorbed following ingestion, inhalation, or skin contact. Within 24 hours of absorption, arsenic distributes over the body, where it binds to SH groups of tissue proteins. Only a small amount crosses the blood-brain barrier. Arsenic may also replace phosphorus in bone tissue and be stored for years.

After acute poisoning, severe GI gastrointestinal symptoms occur within 30 minutes to 2 hours. These include vomiting, watery and bloody diarrhea, severe abdominal pain,

**Table 5.1 Examples of Chelating Drugs Used to Treat Metal Toxicity**

British antilewisite (BAL[2,3–dimercaptopropanol]), dimercaprol
DMPS (2,3-dimercapto-1-propanesulfonic acid)
DMSA (meso-2,3-dimercaptosuccinic acid)
EDTA (ethylenediaminetetraacetic acid, calcium salt)
DTPA (diethylenetriaminepentaacetic acid, calcium salt)
DTC (dithiocarbamate)
Penicillamine ($\beta$-$\beta$-dimethylcysteine), hydrolytic product of penicillin

and burning esophageal pain. Vasodilatation, myocardial depression, cerebral edema, and distal peripheral neuropathy may also follow. Later stages of poisoning include jaundice and renal failure. Death usually results from circulatory failure within 24 hours to 4 days.

Chronic exposure results in nonspecific symptoms such as diarrhea, abdominal pain, hyperpigmentation, and hyperkeratosis. A symmetrical sensory neuropathy often follows. Late changes include gangrene of the extremities, anemia, and cancer of the skin, lung, and nasal tissue.

### 5.2.8 Treatment of Metal Poisoning

Treatment of metal exposure to prevent or reverse toxicity is done with chelating agents or antagonists. Chelation is the formation of a metal ion complex, in which the metal ion is associated with an electron donor ligand. Metals may react with O-, S-, and N-containing ligands (e.g., –OH, –COOH, –S–S–, and –NH$_2$). Chelating agents need to be able to reach sites of storage, form nontoxic complexes, not readily bind essential metals (e.g., calcium, zinc), and be easily excreted.

One of the first clinically useful chelating drugs was British antilewisite (BAL [2,3-dimercaptopropanol]), which was developed during World War II as an antagonist to arsenical war gases. BAL is a dithiol compound with two sulfur atoms on adjacent carbon atoms that compete with critical binding sites involved in arsenic toxicity. Although BAL will bind a number of toxic metals, it is also a potentially toxic drug with multiple side effects. In response to BAL's toxicity, several analogues have now been developed. Table 5.1 lists some of the more common chelating drugs in therapeutic use.

### 5.3 AGRICULTURAL CHEMICALS (PESTICIDES)

### 5.3.1 Introduction

Chemicals have been used to kill or control pests for centuries. The Chinese used arsenic to control insects, the early Romans used common salt to control weeds and sulfur to control insects. In the 1800s pyrethrin (i.e., compounds present in the flowers of the chrysanthemum, *Pyrethrum cineraefolium*) was found to have insecticidal properties. The roots of certain Derris plant species, (*D. elliptica* and *Lonchocarpus* spp.) were used by the Chinese and by South American natives as a fish poison. The active ingredient, rotenone, was isolated in 1895 and used for insect control. Another material developed for insect control in the 1800s was Paris Green, a mixture of copper and arsenic salts. Fungi were controlled with Bordeaux Mixture, a combination of lime and copper sulfate.

However, it was not until the 1900s that the compounds we identify today as having pesticidal properties came into being. Petroleum oils, distilled from crude mineral oils were introduced in the 1920s to control scale insects and red spider mites. The 1940s saw the introduction of the chlorinated hydrocarbon insecticides such as DDT and the phenoxy acid herbicides such as 2,4-*D*). Natural compounds such as Red Squill, derived from the bulbs of red squill, *Urginea (Scilla) maritima*, were effective in controlling rodents. Triazine herbicides, such as atrazine, introduced in the late 1950s, dominated the world herbicide market for years. Synthetic pyrethrins or pyrethroid insecticides (e.g., resmethrin) became and continue to be widely used insecticides due to their low toxicity, enhanced persistence compared to the pyrethrins and low application rates. New families of fungicides, herbicides, and insecticides continue to be introduced into world markets as older compounds lose their popularity due to pest resistance or adverse health effects.

Pesticides are unusual among environmental pollutants in that they are used deliberately for the purpose of killing some form of life. Ideally pesticides should be highly selective, destroying target organisms while leaving nontarget organisms unharmed. In reality, most pesticides are not so selective. In considering the use of pesticides, the benefits must be weighed against the risk to human health and environmental quality. Among the benefits of pesticides are control of vector-borne diseases, increased agricultural productivity, and control of urban pests. A major risk is environmental contamination, especially translocation within the environment where pesticides might enter both food chains and natural water systems. Factors to be considered in this regard are persistence in the environment and potential for bioaccumulation.

### 5.3.2  Definitions and Terms

The term "agricultural chemicals" has largely been replaced by the term "pesticides," defined as economic poisons, regulated by federal and state laws, that are used to control, kill, or repel pests. Depending on what a compound is designed to do, pesticides have been subclassified into a number of categories (Table 5.2). The primary classes of pesticides in use today are fumigants, fungicides, herbicides, and insecticides with total US production of 1.2 billion pounds (1997: US Environmental Protection Agency's latest figures) and production of some 665 million pounds of wood preservatives. Table 5.3 describes the relative use of different classes of pesticides in the United States.

Generally, it takes some five to seven years to bring a pesticide to market once its pesticidal properties have been verified. Many tests must be conducted to determine such things as the compound's synthesis, its chemical and physical properties, and its efficacy. In addition, in order for registration for use by the US EPA, numerous toxicity tests are undertaken including those for acute toxicity, those for chronic effects such as reproductive anomalies, carcinogenesis, and neurological effects and those for environmental effects.

The mandated pesticide label contains a number of specified items, including the concentration and/or percentage of both the active (A.I.) and inert ingredients; proper mixing of the formulation with water to obtain the application rate of A.I., what the A.I.

**Table 5.2 Classification of Pesticides, with Examples**

| Class | Principal Chemical Type | Example, Common Name |
|---|---|---|
| Algicide | Organotin | Brestar |
| Fungicide | Dicarboximide | Captan |
| | Chlorinated aromatic | Pentchlorophenol |
| | Dithiocarbamate | Maneb |
| | Mercurial | Phenylmercuric acetate |
| Herbicide | Amides, acetamides | Propanil |
| | Bipyridyl | Paraquat |
| | Carbamates, thiocarbamates | Barban |
| | Phenoxy | 2,4-D |
| | Dinitrophenol | DNOC |
| | Dinitroaniline | Trifluralin |
| | Substitute urea | Monuron |
| | Triazine | Atrazine |
| Nematocide | Halogenated alkane | Ethylene dibromide (EDB) |
| Molluscicide | Chlorinated hydrocarbon | Bayluscide |
| Insecticide | Chlorinated hydrocarbons | |
| |    DDT analogous | DDT |
| |    Chlorinated alicyclic | BHC |
| |    Cyclodiene | Aldrin |
| | Chlorinated terpenes | Toxaphene |
| | Organophosphorus | Chlorpyrifos |
| | Carbamate | Carbaryl |
| | Thiocyanate | Lethane |
| | Dinitrophenols | DNOC |
| | Fluoroacetate | Nissol |
| | Botanicals | |
| |    Nicotinoids | Nicotine |
| |    Rotenoids | Rotenone |
| |    Pyrethroids | Pyrethrin |
| | Synthetic pyrethroids | Fenvalerate |
| | Synthetic nicotinoids | Imidacloprid |
| | Fiproles | Fipronil |
| | Juvenile hormone analogs | Methroprene |
| | Growth regulators | Dimilin |
| | Inorganics | |
| |    Arsenicals | Lead arsenate |
| |    Fluorides | Sodium fluoride |
| | Microbials | Thuricide, avermectin |
| Insecticide synergists | Methylenedioxyphenyl | Piperonyl butoxide |
| | Dicarboximides | MGK-264 |
| Acaricides | Organosulfur | Ovex |
| | Formamidine | Chlordimeform |
| | Dinitrophenols | Dinex |
| | DDT analogs | Chlorbenzilate |
| Rodenticides | Anticoagulants | Warfarin |
| | Botanicals | |
| |    Alkaloids | Strychine sulfate |
| |    Glycosides | Scillaren A and B |
| | Fluorides | Fluoroacetate |
| | Inorganics | Thallium sulfate |
| | Thioureas | ANTU |

will control, and how and when to apply it. In addition the label describes environmental hazards, proper storage of the material, re-entry intervals (REIs) for application sites, and the personal protective equipment (PPE) that must be worn during application or harvesting.

Depending on the toxicity, formulation concentration, and use patterns, pesticides can be classified as "general" or "restricted" use. A general use pesticide will cause no unreasonable, adverse effects when used according to the label and can be purchased and applied by anyone. A restricted use pesticide, defined as generally causing undesirable effects on the environment, applicator, or workers can only be purchased and applied by an individual who is licensed by the state.

The US EPA has developed "category use" definitions based on toxicity. Category I pesticides are highly hazardous, are classified as restricted use and have an oral LD50 less than or equal to 1.0/kg of body weight; category II pesticides are moderately toxic and have an oral LD50 less than or equal to 500 mg/kg; category III pesticides are generally nontoxic and have an oral LD50 less than or equal to 15,000 mg/kg. In addition the US EPA has developed a "carcinogenicity categorization" to classify pesticides for carcinogenicity.

### 5.3.3 Organochlorine Insecticides

The chlorinated hydrocarbon insecticides were introduced in the 1940s and 1950s and include familiar insecticides such as DDT, methoxychlor, chlordane, heptachlor, aldrin, dieldrin, endrin, toxaphene, mirex, and lindane. The structures of two of the more familiar ones, DDT and dieldrin, are shown in Figure 5.1. The chlorinated hydrocarbons are neurotoxicants and cause acute effects by interfering with the transmission of nerve impulses. Although DDT was synthesized in 1874, its insecticidal properties were not noted until 1939, when Dr. Paul Mueller, a Swiss chemist, discovered its effectiveness as an insecticide and was awarded a Nobel Prize for his work. During World War II the United States used large quantities of DDT to control vector-borne diseases, such as typhus and malaria, to which US troops were exposed. After the war DDT use became widespread in agriculture, public health, and households. Its persistence, initially considered a desirable attribute, later became the basis for public concern. The publication of Rachel Carson's book *The Silent Spring* in 1962 stimulated this concern and eventually led to the ban of DDT and other chlorinated insecticides in the United States in 1972.

**Table 5.3   Use Patterns of Pesticides in the United States**

| Class | Percentage of Total Pesticide Use |
|---|---|
| Herbicides | 47 |
| Insecticides | 19 |
| Fungicides | 13 |
| Others[a] | 21 |

*Note*: Most recent data: for 1997, published by US EPA in 2001.
[a] Includes fumigants and wood preservates.

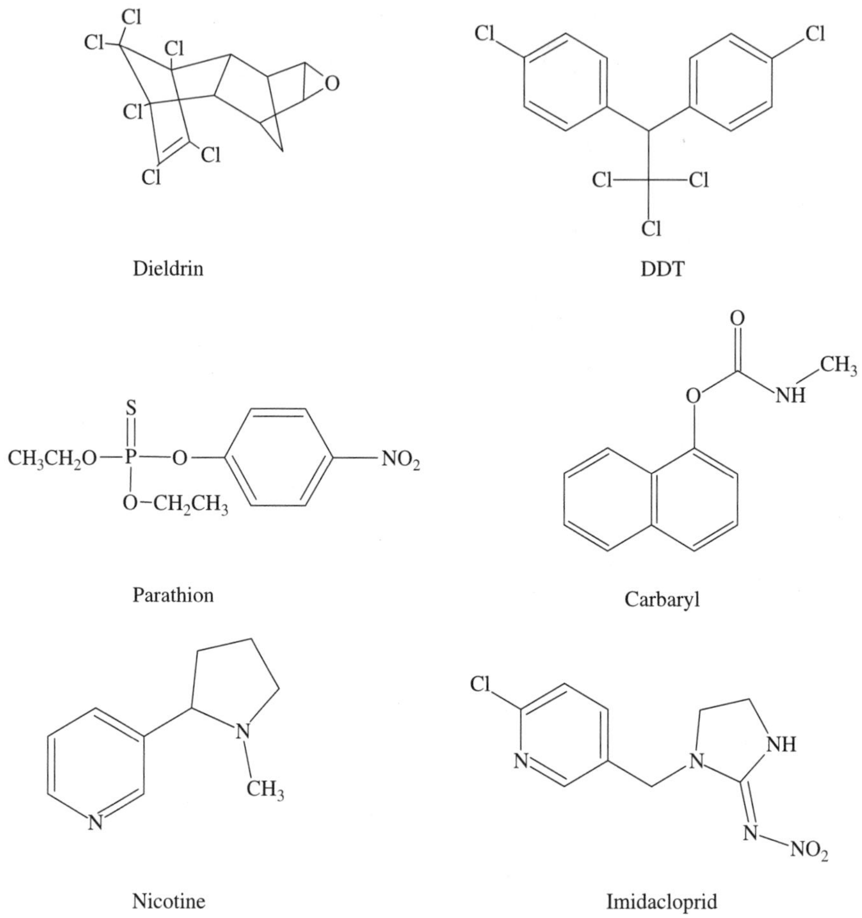

**Figure 5.1** Some examples of chemical structures of common pesticides.

DDT, as well as other organochlorines, were used extensively from the 1940s through the 1960s in agriculture and mosquito control, particularly in the World Health Organization (WHO) malaria control programs. The cyclodiene insecticides, such as chlordane were used extensively as termiticides into the 1980s but were removed from the market due to measurable residue levels penetrating into interiors and allegedly causing health problems. Residue levels of chlorinated insecticides continue to be found in the environment and, although the concentrations are now so low as to approach the limit of delectability, there continues to be concern.

### 5.3.4 Organophosphorus Insecticides

Organophosphorus pesticides (OPs) are phosphoric acid esters or thiophosphoric acid esters (Figure 5.1) and are among the most widely used pesticides for insect control. During the 1930s and 1940s Gerhard Schrader and coworkers began investigating OP compounds. They realized that the insecticidal properties of these compounds and by the end of the World War II had made many of the insecticidal OPs in use today,

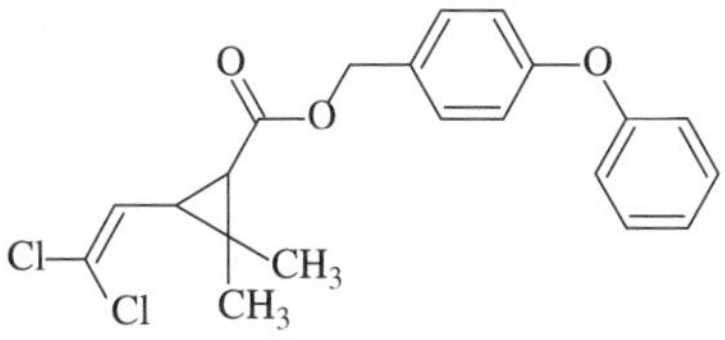

Permethrin

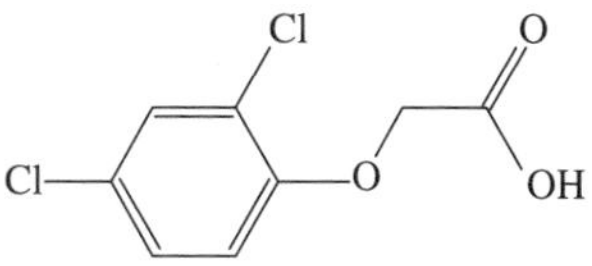

2,4-D

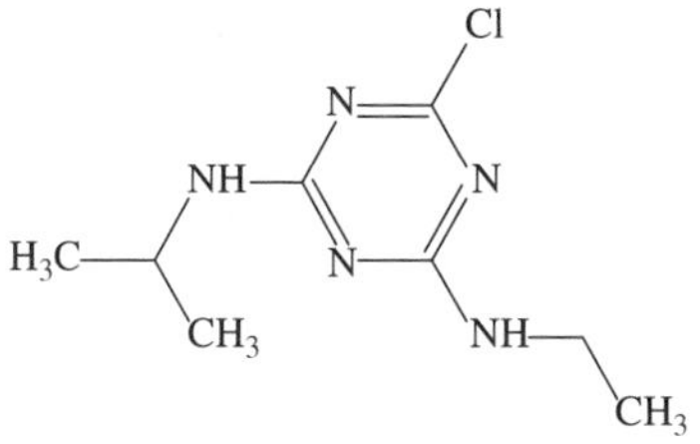

Atrazine

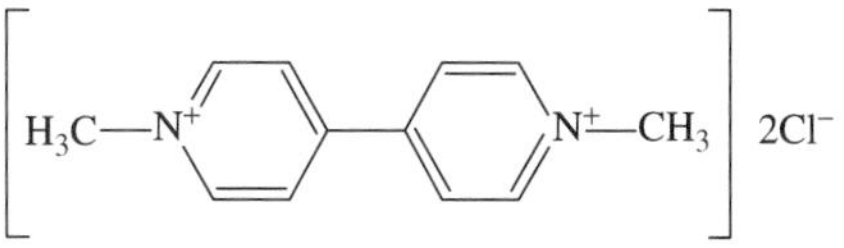

Paraquat

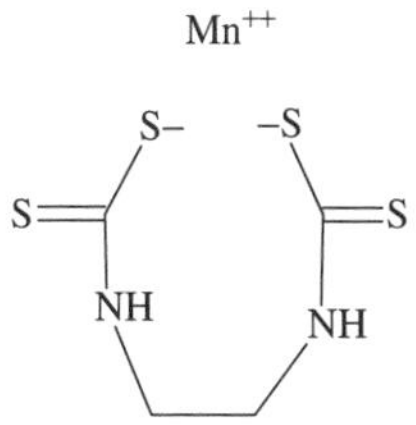

Maneb

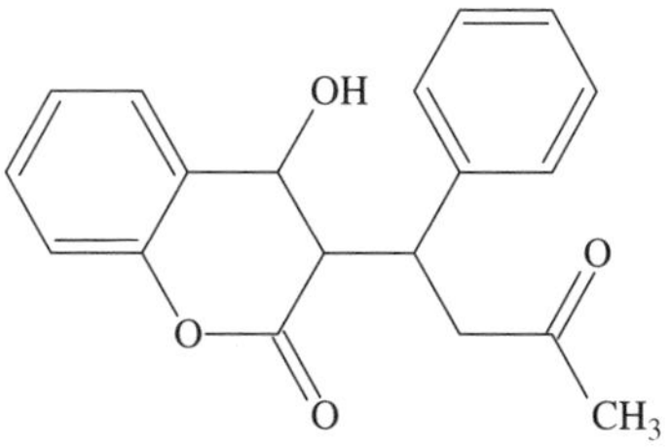

Warfarin

**Figure 5.1**    (*continued*)

such as ethyl parathion [$O,O$-diethyl $O$-(4-nitrophenyl)phosphorothioate]. The first OP insecticide to find widespread use was tetraethylpyrophosphate (TEPP), approved in Germany in 1944 and marketed as a substitute for nicotine to control aphids. Because of its high mammalian toxicity and rapid hydrolysis in water, TEPP was replaced by other OP insecticides.

Chlorpyrifos [$O,O$-diethyl $O$-(3,5,6-trichloro-2-pyridinyl) phosphorothioate] became one of the largest selling insecticides in the world and had both agricultural and urban uses. The insecticide could be purchased for indoor use by homeowners, but health-related concerns caused USEPA to cancel home indoor and lawn application uses in 2001. The only exception is its continued use as a termiticide.

Parathion was another widely used insecticide due to its stability in aqueous solutions and its broad range of insecticidal activity. However, its high mammalian toxicity through all routes of exposure led to the development of less hazardous compounds. Malathion [diethyl (dimethoxythiophosphorylthio)succinate], in particular, has low mammalian toxicity because mammals possess certain enzymes, the carboxylesterases, that readily hydrolyze the carboxyester link, detoxifying the compound. Insects, by contrast, do not readily hydrolyze this ester, and the result is its selective insecticidal action.

OPs are toxic because of their inhibition of the enzyme acetylcholinesterase. This enzyme inhibition results in the accumulation of acetylcholine in nerve tissue and effector organs, with the principal site of action being the peripheral nervous system (PNS) (see Chapter 16). In addition to acute effects, some OP compounds have been associated with delayed neurotoxicity, known as organophosphorus-induced delayed neuropathy (OPIDN). The characteristic clinical sign is bilateral paralysis of the distal muscles, predominantly of the lower extremities, occurring some 7 to 10 days following ingestion (see Chapter 16). Not all OP compounds cause delayed neuropathy. Among the pesticides associated with OPIDN are leptophos, mipafox, EPN, DEF, and trichlorfon. Testing is now required for OP substances prior to their use as insecticides.

The OP and carbamate insecticides are relatively nonpersistent in the environment. They are applied to the crop or directly to the soil as systemic insecticides, and they generally persist from only a few hours to several months. Thus these compounds, in contrast to the organochlorine insecticides, do not represent a serious problem as contaminants of soil and water and rarely enter the human food chain. Being esters, the compounds are susceptible to hydrolysis, and their breakdown products are generally nontoxic. Direct contamination of food by concentrated compounds has been the cause of poisoning episodes in several countries.

### 5.3.5   Carbamate Insecticides

The carbamate insecticides are esters of $N$-methyl (or occasionally $N,N$-dimethyl) carbamic acid ($H_2NCOOH$). The toxicity of the compound varies according to the phenol or alcohol group. One of the most widely used carbamate insecticides is carbaryl (1-napthyl methylcarbamate), a broad spectrum insecticide (Figure 5.1). It is used widely in agriculture, including home gardens where it generally is applied as a dust. Carbaryl is not considered to be a persistent compound, because it is readily hydrolyzed. Based on its formulation, it carries a toxicity classification of II or III with an oral LD50 of 250 mg/kg (rat) and a dermal LC50 of >2000 mg/kg.

An example of an extremely toxic carbamate is aldicarb [2-methyl-2-(methylthio) propionaldehyde]. Both oral and dermal routes are the primary portals of entry, and it has an oral LD50 of 1.0 mg/kg (rat)and a dermal LD50 of 20 mg/kg (rabbit). For this reason it is recommended for application to soils on crops such as cotton, citrus, and sweet potatoes. This compound moves readily through soil profiles and has contaminated groundwater supplies.

Like the OP insecticides, the mode of action of the carbamates is acetylcholinesterase inhibition with the important difference that the inhibition is more rapidly reversed than with OP compounds.

### 5.3.6   Botanical Insecticides

Extracts from plants have been used for centuries to control insects. Nicotine [($S$)-3-(1-methyl-2-pyrrolidyl)pyridine] (Figure 5.1) is an alkaloid occurring in a number of plants and was first used as an insecticide in 1763. Nicotine is quite toxic orally as well as dermally. The acute oral LD50 of nicotine sulfate for rats is 83 mg/kg and the dermal LD50 is 285 mg/kg. Symptoms of acute nicotine poisoning occur rapidly, and death may occur with a few minutes. In serious poisoning cases death results from respiratory failure due to paralysis of respiratory muscles. In therapy attention is focused primarily on support of respiration.

Pyrethrin is an extract from several types of chrysanthemum, and is one of the oldest insecticides used by humans. There are six esters and acids associated with this botanical insecticide. Pyrethrin is applied at low doses and is considered to be nonpersistent.

Mammalian toxicity to pyrethrins is quite low, apparently due to its rapid breakdown by liver microsomal enzymes and esterases. The acute LD50 to rats is about 1500 mg/kg. The most frequent reaction to pyrethrins is contact dermatitis and allergic respiratory reactions, probably as a result of other constituents in the formulation. Synthetic mimics of pyrethrins, known as the pyrethroids, were developed to overcome the lack of persistence.

### 5.3.7  Pyrethroid Insecticides

As stated, pyrethrins are not persistent, which led pesticide chemists to develop compounds of similar structure having insecticidal activity but being more persistent. This class of insecticides, known as pyrethroids, have greater insecticidal activity and are more photostable than pyrethrins. There are two broad classes of pyrethroids depending on whether the structure contains a cyclopropane ring [e.g., cypermethrin {($\pm$)-$\alpha$-cyano-3-phenoxybenzyl ($\pm$)-*cis,trans*-3-(2,2-dichlorovinyl_2,2-dimethyl cyclopropanecarboxylate)}] or whether this ring is absent in the molecule [e.g., fenvalerate{($RS$)-$\alpha$-cyano-3-phenoxybenzyl($RS$)-2-(4-chlorophenyl)-3-methylbutyrate}]. They are generally applied at low doses (e.g., 30 g/Ha) and have low mammalian toxicities [e.g., cypermethrin, oral (aqueous suspension) LD50 of 4,123 mg/kg (rat) and dermal LD50 of >2000 mg/kg (rabbit)]. Pyrethroids are used in both agricultural and urban settings (e.g., termiticide; Figure 5.1).

Pyrethrins affect nerve membranes by modifying the sodium and potassium channels, resulting in depolarization of the membranes. Formulations of these insecticides frequently contain the insecticide synergist piperonyl butoxide [5-{2-(2-butoxyethoxy) ethoxymethyl}-6-propyl-1,3-benzodioxole], which acts to increase the efficacy of the insecticide by inhibiting the cytochrome P450 enzymes responsible for the breakdown of the insecticide.

### 5.3.8  New Insecticide Classes

There are new classes of insecticides that are applied at low dosages and are extremely effective but are relatively nontoxic to humans. One such class is the fiproles, and one of these receiving major attention is fipronil [(5-amino-1-(2,6-dichloro-4-(trifluoromethyl) phenyl)-4-((1,$R$,$S$)-(trifluoromethyl)su-1-$H$-pyrasole-3-carbonitrile)]. Although it is used on corn, it is becoming a popular termiticide because of its low application rate (ca. 0.01%) and long-term effectiveness. Another class of insecticides, the chloronicotinoids, is represented by imidacloprid [1-(6-chloro-3-pyridin-3-ylmethyl)-$N$-nitroimidazolidin-2-ylidenamine] (Figure 5.1), which also is applied at low dose rates to soil and effectively controls a number of insect species, including termites.

### 5.3.9  Herbicides

Herbicides control weeds and are the most widely used class of pesticides. The latest US EPA data show that some 578 million pounds of herbicides were used in the United States in 1997 and accounts for some 47% of pesticides used. This class of pesticide can be applied to crops using many strategies to eliminate or reduce weed populations. These include preplant incorporation, pre- and postemergent applications. New families of herbicides continue to be developed, and are applied at low doses, are relatively nonphytotoxic to beneficial plants and are environmentally friendly. Some of the newer families such as the imidazolinones inhibit the action of acetohydroxyacid synthase that produces branched-chain amino acids in plants. Because this enzyme is produced only in plants, these herbicides have low toxicities to mammals, fish, insects, and birds.

The potential for environmental contamination continues to come from families of herbicides that have been used for years. The chlorophenoxy herbicides such as 2,4-*D* (2,4-dichlorophenoxy acetic acid) and 2,4,5-*T* (2,4,5-trichlorophenoxy-acetic acid) (Figure 5.1) are systemic acting compounds to control broadleaf plants and have been in use since the 1940s. The oral toxicities of these compounds are low.

A mixture of 2,4-*D* and 2,4,5-*T*, known as Agent Orange, was used by the US military as a defoliant during the Vietnam conflict, and much controversy has arisen over claims by military personnel of long-term health effects. The chemical of major toxicological concern was identified as a contaminant, TCDD (2,3,7,8-tetrachlorodibenzo-*p*-dioxin), that was formed during the manufacturing process. TCDD is one of the most toxic synthetic substances known in laboratory animals. The LD50 for male rats is 0.022 mg/kg, and for female guinea pigs (the most sensitive species tested) the LD50 is 0.0006 mg/kg. In addition it is toxic to developing embryos in pregnant rats at a dose of only 1/400 of the LD50, and has been shown to cause birth defects at levels of 1 to 3 ng/kg of body weight. TCDD is a proven carcinogen in both mice and rats, with the liver being the primary target. This chemical has also been shown to alter the immune system and enhance susceptibility in exposed animals.

Another family of herbicides, the triazines, continues to cause concern to environmentalists and toxicologists because of the contamination of surface and groundwater supplies that become public drinking water. The herbicide, atrazine [6-chloro-$N$-ethyl-$N$-(1-methylethyl)-1,2,5-triazine-2,4-diamine (Figure 5.1) is used primarily on corn and has an MCL of 3.0 μg/L. This herbicide has been found in surface and groundwaters worldwide with widely varying concentrations (e.g., 1 to >130 μg/L). It, along with two other triazines, cyanazine [2-{{4-chloro-6-(ethylamino)-1,3,5-triazin-2-yl}amino}-2-methylpropanenitrile] and simazine (6-chloro-$N,N$-diethyl-1,3,5-triazine-2,4-diamine) (MCL of 4.0 μg/L). The uses of cyanazine were canceled in 2001 and no further use was permitted after 2002. Although relatively nontoxic [e.g., atrazine, oral LD50 of 3,100 mg/kg (rat)], the major concern with these types of compounds is their carcinogenic effects, and US EPA considers these three triazines as possible human carcinogens (category C).

A member of the bipyridylium family of herbicides is the compound paraquat (1,1-dimethyl-4,4-bipyridinium ion as the chloride salt) (Figure 5.1). It is a very water-soluble contact herbicide that is active against a broad range of plants and is used as a defoliant on many crops. The compound binds tightly to soil particles following application and becomes inactivated. However, this compound is classified as a class I toxicant with an oral LD50 of 150 mg/kg (rat). Most poisoning cases, which are often fatal, are due to accidental or deliberate ingestion of paraquat. Toxicity results from lung injury resulting from both the preferential uptake of paraquat by the lungs and the redox cycling mechanism.

### 5.3.10  Fungicides

Every year fungi cause crop losses in the United States amounting to millions of dollars. In addition recent studies have shown that toxins and other airborne organic compounds released from fungi inhabiting the interior of dwellings probably are responsible for a number of adverse health effects. Compounds produced to combat these losses and adverse health effects are called fungicides, and a number of these families have been around for years.

The fungicide, chlorothalonil (tetrachloroisophthalonitrile), is a broad-spectrum fungicide which is used widely in urban environments. It is relatively cheap and controls some 140 species of organisms. As a result of the popularity of this compound, it is found routinely in surface waters entering public drinking water supplies. In the formulation that can be purchased by the general public, it is relatively nontoxic.

One family of fungicides that is of concern are the dithiocarbamates, sulfur derivatives of dithiocarbamic acid and include the metallic dimethyldithiocarbamates. The latter group includes mancozeb (a coordination product of zinc ion and manganese ethylene bisdithiocarbamate), maneb (manganese ethylenebisdithiocarbamate)(Figure 5.1), and zineb (zinc ethylenebisdithiocarbamate). All are effective fungicides and are used on a variety of crops including grapes, sugar beets, and ornamental plants. Although relatively nontoxic, they do hydrolyze producing known carcinogens such as ethylthiourea (ETU).

### 5.3.11  Rodenticides

This class of compounds is used to control rodents that cause yearly losses of 20% to 30% in grain and other food storage facilities. These pests harbor diseases in the form of fleas that carry bacteria and other organisms. A number of the rodenticides have been used for years and include warfarin [3-($\alpha$-acetonylbenzyl)-4-hydroxycoumarin] (Figure 5.1), an anticoagulant. This is a potent toxicant with an oral LD50 of 3.0 mg/kg (rat). As the rats navigate through narrow passages, they bruise themselves, developing small hemorrhages. Anticoagulants prevent the blood from clotting, and the animals bleed to death in about a week. Humans who are exposed to this class of compounds are given vitamin K, and if the poisoning is severe, blood transfusions as a treatment. Other rodenticides poison the animal and many times are applied along with an attractant such as peanut butter to overcome bait shyness. Fluoroacetamide is a fast acting poison with an oral LD50 (rat) of 15 mg/kg. This material is supplied as bait pellets or grains. ANTU ($\alpha$-naphthylthiourea), strychnine, and thallium salts are other fast acting poisons, and have been on the market for many years. Most of the rodenticides are classified as restricted use and are applied only by licensed pest

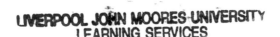

control operators. Human poisonings associated with rodenticides usually result from accidental or suicidal ingestion of the compounds.

### 5.3.12   Fumigants

Fumigants are extremely toxic gases used to protect stored products, especially grains, and to kill soil nematodes. These materials are applied to storage warehouses, freight cars, and houses infested with insects such as powder post beetles. They present a special hazard due to inhalation exposure and rapid diffusion into pulmonary blood; thus extreme care must be taken when handling and applying this class of pesticides. All fumigants are classified as restricted use compounds and require licensed applicators to handle them.

One of the most effective fumigants is methyl bromide. It essentially sterilizes soil when applied under a ground covering, because it kills insects, nematodes, and weed seed but also is used to fumigate warehouses. Overexposure to this compound causes respiratory distress, cardiac arrest, and central nervous effects. The inhalation LC50 is 0.06 mg/L (15 min) of air (rat) and 7900 ppm (1.5 h) (human). Methyl bromide has been classified as an ozone depleter under the Clean Air Act and is due to be phased out of use by 2005.

Chloropicrin (trichloronitromethane) is another soil/space fumigant that has been used for many years. It has an inhalation LC50 of 150 ppm (15 min). Thus it is highly toxic by inhalation, can injure the heart, and cause severe eye damage.

### 5.3.13   Conclusions

This section has covered only a few of the pesticides available today on the United States and world markets. An understanding of the basic chemical processes affected by pesticides has led to the discovery and production of new families of chemicals. Today's modern pesticide is generally safe to use if the directions on the label are followed. Advances in instrumentation and an understanding of how adverse health effects are produced have resulted in the production of many environmentally friendly but effective pesticides.

## 5.4   FOOD ADDITIVES AND CONTAMINANTS

Chemicals are added to food for a number of reasons: as preservatives with antibacterial, antifungal, or antioxidant properties; to change physical characteristics, particularly for processing; to change taste; to change color; and to change odor. In general, food additives have proved to be safe and without chronic toxicity. Many were introduced when toxicity testing was relatively unsophisticated, however, and some of these have been subsequently shown to be toxic. Table 5.4 gives examples of different types of organic food additives. Inorganics, the most important of which are nitrate and nitrite, are discussed later. Certainly hundreds, and possibly thousands, of food additives are in use worldwide, many with inadequate testing. The question of synergistic interactions between these compounds has not been explored adequately. Not all toxicants in food are synthetic; many examples of naturally occurring toxicants in the human diet are known, including carcinogens and mutagens.

**Table 5.4   Examples of Organic Chemicals Used as Food Additives**

| Function | Class | Example |
|---|---|---|
| Preservatives | Antioxidants | Butylatedhydroxyanisole |
| | | Ascorbic acid |
| | Fungistatic agents | Methyl $p$-benzoic acid |
| | | Propionates |
| | | Bactericides   Sodium nitrite |
| Processing aids | Anticaking agents | Calcium silicate |
| | | Sodium aluminosilicate |
| | Emulsifiers | Propylene glycol |
| | | Monoglycerides |
| | Chelating agents | EDTA |
| | | Sodium tartrate |
| | Stabilizers | Gum ghatti |
| | | Sodium alginate |
| | Humectants | Propylene glycol |
| | | Glycerol |
| Flavor and taste modification | Synthetic sweeteners | Saccharin |
| | | Mannitol |
| | | Aspartame |
| | Synthetic flavors | Piperonal |
| | | Vanillin |
| Color modification | Synthetic dyes | Tartrazine (FD&C yellow5) |
| | | Sunset Yellow |
| Nutritional supplements | Vitamins | Thiamin |
| | | Vitamin D3 |
| | Amino acids | Alanine |
| | | Aspartic acid |
| | Inorganics | Manganese sulfate |
| | | Zinc sulfate |

## 5.5   TOXINS

### 5.5.1   History

A discussion of toxins first necessitates the understanding and distinction between the toxicological terms toxicant and toxin. A *toxicant* is any chemical, of natural or synthetic origin, capable of causing a deleterious effect on a living organism. A *toxin* is a toxicant that is produced by a living organism and is not used as a synonym for toxicant—all toxins are toxicants, but not all toxicants are toxins. Toxins, whether produced by animals, plants, insects, or microbes are generally metabolic products that have evolved as defense mechanisms for the purpose of repelling or killing predators or pathogens. The action of natural toxins has long been recognized and understood throughout human history. For example, ancient civilizations used natural toxins for both medicinal (therapeutic) and criminal purposes. Even today, we continue to discover and understand the toxicity of natural products, some for beneficial pharmaceutical or therapeutic purposes whose safety and efficacy are tested, and some for other less laudable purposes like biological or chemical warfare. Toxins may be classified in various ways depending on interest and need, such as by target organ toxicity or mode of action, but are commonly classified according to source.

### 5.5.2  Microbial Toxins

The term "microbial toxin" is usually reserved by microbiologists for toxic substances produced by microorganisms that are of high molecular weight and have antigenic properties; toxic compounds produced by bacteria that do not fit these criteria are referred to simply as poisons. Many of the former are proteins or mucoproteins and may have a variety of enzymatic properties. They include some of the most toxic substances known, such as tetanus toxin, botulinus toxin, and diphtheria toxin. Bacterial toxins may be extremely toxic to mammals and may affect a variety of organ systems, including the nervous system and the cardiovascular system. A detailed account of their chemical nature and mode of action is beyond the scope of this volume.

The range of poisonous chemicals produced by bacteria is also large. Again, such compounds may also be used for beneficial purposes, for example, the insecticidal properties of *Bacillus thuringiensis*, due to a toxin, have been utilized in agriculture for some time.

### 5.5.3  Mycotoxins

The range of chemical structures and biologic activity among the broad class of fungal metabolites is large and cannot be summarized briefly. Mycotoxins do not constitute a separate chemical category, and they lack common molecular features.

Mycotoxins of most interest are those found in human food or in the feed of domestic animals. They include the ergot alkaloids produced by *Claviceps* sp., aflatoxins and related compounds produced by *Aspergillus* sp., and the tricothecenes produced by several genera of fungi imperfecti, primarily *Fusarium* sp.

The ergot alkaloids are known to affect the nervous system and to be vasoconstrictors. Historically they have been implicated in epidemics of both gangrenous and convulsive ergotism (St. Anthony's fire), although such epidemics no longer occur in humans due to increased knowledge of the cause and to more varied modern diets. Outbreaks of ergotism in livestock do still occur frequently, however. These compounds have also been used as abortifacients. The ergot alkaloids are derivatives of ergotine, the most active being, more specifically, amides of lysergic acid.

Aflatoxins are products of species of the genus *Aspergillus*, particularly *A flavus*, a common fungus found as a contaminant of grain, maize, peanuts, and so on. First implicated in poultry diseases such as Turkey-X disease, they were subsequently shown to cause cancer in experimental animals and, from epidemiological studies, in humans. Aflatoxin B1, the most toxic of the aflatoxins, must be activated enzymatically to exert its carcinogenic effect.

Tricothecenes are a large class of sesquiterpenoid fungal metabolites produced particularly by members of the genera *Fusarium* and *Tricoderma*. They are frequently acutely toxic, displaying bactericidal, fungicidal, and insecticidal activity, as well as causing various clinical symptoms in mammals, including diarrhea, anorexia, and ataxia. They have been implicated in natural intoxications in both humans and animals, such as Abakabi disease in Japan and Stachybotryotoxicosis in the former USSR, and are the center of a continuing controversy concerning their possible use as chemical warfare agents.

Mycotoxins may also be used for beneficial purposes. The mycotoxin avermectin is currently generating considerable interest both as an insecticide and for the control of nematode parasites of domestic animals.

### 5.5.4 Algal Toxins

Algal toxins are broadly defined to represent the array chemicals derived from many species of cyanobacteria (blue-green bacteria), dinoflagellates, and diatoms. The toxins produced by these freshwater and marine organisms often accumulate in fish and shellfish inhabiting the surrounding waters, causing both human and animal poisonings, as well as overt fish kills. Unlike many of the microbial toxins, algal toxins are generally heat stable and, therefore, not altered by cooking methods, which increases the likelihood of human exposures and toxicity. Many of the more common algal toxins responsible for human poisonings worldwide are summarized herein.

*Amnesic Shellfish Poisoning (ASP)* was first identified in 1987 from Prince Edward Island, Canada after four people died from eating contaminated mussels. It is caused by domoic acid produced by several species of *Pseudonitzschia* diatoms. The main contamination problems include mussels, clams, and crabs of the Pacific Northwest of the United States and Canada.

*Paralytic Shellfish Poisoning (PSP)* was first determined to be a problem in 1942 after three people and many seabirds died from eating shellfish on the west coast of the United States, near the Columbia River. It is caused by the saxitoxin family (saxitoxin + 18 related compounds) produced by several species of *Alexandrium* dinoflagellates. The main contamination problems include mussels, clams, crabs, and fish of the Pacific Northwest and Northeast Atlantic.

*Neurotoxic Shellfish Poisoning (NSP)* is caused by a red-tide producer that was first identified in 1880 from Florida, with earlier historical references. It causes sickness in humans lasting several days. NSP is not fatal to humans; however, it is known to kill fish, invertebrates, seabirds, and marine mammals (e.g., manatees). It is caused by the brevetoxin family (brevetoxin + 10 related compounds produced by the dinoflagellate *Karenia brevis* a.k.a. *Gymnodinium breve*. The main contamination problems include oysters, clams, and other filter feeders of the Gulf of Mexico and southeast Atlantic, including North Carolina.

*Diarrheic Shellfish Poisoning (DSP)*. Human poisonings were first identified in the 1960s. It causes sickness in humans lasting several days but is not fatal. It is caused by chemicals of the okadaic acid family (okadaic acid + 4 related compounds) produced by several species of *Dinophysis* dinoflagellates. The main contamination problems include mussels, clams, and other bivalves of the cold and warm temperate areas of the Atlantic and Pacific Oceans, mainly in Japan and Europe. Only two cases of DSP have been documented in North America.

*Ciguatera Fish Poisoning (CFP)* was first identified in 1511, CFP is a tropical-subtropical seafood poisoning that affects up to 50,000 people each year and is the most often reported foodborne disease of a chemical origin in the United States. Caused by consumption of reef fishes (e.g., grouper, snapper), sickness in humans lasts several days to weeks, but the human fatality rate is low. It is caused by the ciguatoxin family (ciguatoxin + 3 or more related compounds) and produced by several species of dinoflagellates including *Gambierdiscus, Prorocentrum, and Ostreopsis*. The main contamination problems include herbivorous tropical reef fish worldwide.

*Cyanobacterial (Blue-Green Bacteria) Toxins*. Cyanobacterial poisonings were first recognized in the late 1800s. Human poisonings are rare; however, kills of livestock, other mammals, birds, fish, and aquatic invertebrates are common. It is caused by a variety of biotoxins and cytotoxins, including anatoxin, microcystin, and nodularin produced by several species of cyanobacteria, including *Anabaena, Aphanizomenon, Nodularia, Oscillatoria, and Microcystis*. The main contamination problems include all eutrophic freshwater rivers, lakes, and streams.

*Ambush Predator (Pfiesteria piscicida and Toxic Pfiesteria Complex) Toxins*. Members belonging to this group of organisms were first identified in 1991 from estuaries in North Carolina. They were believed to produce a toxin that has been implicated in several large fish kills and is suspect in causing adverse human health effects. However, the toxin or toxins are not yet identified and toxicity tests are not universally conclusive. Produced by several dinoflagellate species including, *Pfiesteria piscicida, Pfiesteria shumwayae*, and perhaps several other unidentified, un-named dinoflagellates belonging to the potentially toxic *Pfiesteria* complex. Main problems include major fish kills in North Carolina and Maryland and potential human health problems. The range may extend from the Gulf of Mexico to the Atlantic estuarine waters, including Florida, North Carolina, Maryland, and Delaware, and possibly outward to Europe.

### 5.5.5 Plant Toxins

The large array of toxic chemicals produced by plants (phytotoxins), usually referred to as secondary plant compounds, are often held to have evolved as defense mechanisms against herbivorous animals, particularly insects and mammals. These compounds may be repellent but not particularly toxic, or they may be acutely toxic to a wide range of organisms. They include sulfur compounds, lipids, phenols, alkaloids, glycosides, and many other types of chemicals. Many of the common drugs of abuse such as cocaine, caffeine, nicotine, morphine, and the cannabinoids are plant toxins. Many chemicals that have been shown to be toxic are constituents of plants that form part of the human diet. For example, the carcinogen safrole and related compounds are found in black pepper. Solanine and chaconine, which are cholinesterase inhibitors and possible teratogens, are found in potatoes, and quinines and phenols are widespread in food. Livestock poisoning by plants is still an important veterinary problem in some areas.

### 5.5.6 Animal Toxins

Some species from practically all phyla of animals produce toxins for either offensive or defensive purposes. Some are passively venomous, often following inadvertent ingestion, whereas others are actively venomous, injecting poisons through specially adapted stings or mouthparts. It may be more appropriate to refer to the latter group only as venomous and to refer to the former simply as poisonous. The chemistry of animal toxins extends from enzymes and neurotoxic and cardiotoxic peptides and proteins to many small molecules such as biogenic amines, alkaloids, glycosides, terpenes, and others. In many cases the venoms are complex mixtures that include both proteins and small molecules and depend on the interaction of the components for the full expression of their toxic effect. For example, bee venom contains a biogenic amine, histamine, three peptides, and two enzymes (Table 5.5). The venoms and defensive secretions of insects may also contain many relatively simple toxicants or irritants such as formic acid, benzoquinone, and other quinines, or terpenes such as citronellal. Bites and stings from the Hymenoptera (ants, bees, wasps, and hornets) result in 5 to 60 fatal anaphylactic reactions each year in the United States. According to experts, about 0.3% to 3.0% of the US population experiences anaphylactic reactions from insect stings and bites.

Snake venoms have been studied extensively; their effects are due, in general, to toxins that are peptides with 60 to 70 amino acids. These toxins are cardiotoxic or neurotoxic, and their effects are usually accentuated by the phospholipases, peptidases, proteases, and other enzymes present in venoms. These enzymes may affect the blood-clotting mechanisms and damage blood vessels. Snake bites are responsible for less than 10 deaths per year in the United States but many thousand worldwide.

Many fish species, over 700 species worldwide, are either directly toxic or upon ingestion are poisonous to humans. A classic example is the toxin produced by the puffer fishes (*Sphaeroides* spp.) called tetrodotoxin (TTX). Tetrodotoxin is concentrated in the gonads, liver, intestine, and skin, and poisonings occurs most frequently in Japan and other Asian countries where the flesh, considered a delicacy, is eaten as "fugu." Death occurs within 5 to 30 minutes and the fatality rate is about 60%. TTX is an inhibitor of the voltage-sensitive Na channel (like saxitoxin); it may also be found in some salamanders and may be bacterial in origin.

Toxins and other natural products generally provide great benefit to society. For example, some of the most widely used drugs and therapeutics like streptomycin, the aminoglycoside antibiotic from soil bacteria, and acetylsalicylic acid (aspirin), the nonsteroidal anti-inflammatory from willow tree bark, are used by millions of people everyday to improve health and well-being. On the other hand, adverse encounters with toxins like fish and shellfish toxins, plant, and insect toxins do result in harm to humans.

**Table 5.5  Some Components of Bee Venom**

| Compound | Effect |
| --- | --- |
| Biogenic amine | |
|   Histamine | Pain, vasodilation, increased capillary permeability |
| Peptides | |
|   Apamine | CNS effects |
|   Melittin | Hemolytic, serotonin release, cardiotoxic |
|   Mast cell degranulating peptide | Histamine release from mast cells |
| Enzymes | |
|   Phospholipase A | Increased spreading and penetration of tissues |
|   Hyaluronidase | |

## 5.6 SOLVENTS

Although solvents are more a feature of the workplace they are also found in the home. In addition to cutaneous effects, such as defatting and local irritation, many have systemic toxic effects, including effects on the nervous system or, as with benzene, on the blood-forming elements. Commercial solvents are frequently complex mixtures and may include nitrogen- or sulfur-containing organics—gasoline and other oil-based products are examples of this. The common solvents fall into the following classes:

*Aliphatic Hydrocarbons*, such as hexane. These may be straight or branched-chain compounds and are often present in mixtures.

*Halogenated Aliphatic Hydrocarbons*. The best-known examples are methylene dichloride, chloroform, and carbon tetrachloride, although chlorinated ethylenes are also widely used.

*Aliphatic Alcohols*. Common examples are methanol and ethanol.

*Glycols and Glycol Ethers*. Ethylene and propylene glycols, for example, in antifreeze give rise to considerable exposure of the general public. Glycol ethers, such as methyl cellosolve, are also widely used.

*Aromatic Hydrocarbons*. Benzene is probably the one of greatest concern, but others, such as toluene, are also used.

## 5.7 THERAPEUTIC DRUGS

Although the study of the therapeutic properties of chemicals fall within the province of pharmacology, essentially all therapeutic drugs can be toxic, producing deleterious effects at some dose. The danger to the individual depends on several factors, including the nature of the toxic response, the dose necessary to produce the toxic response, and the relationship between the therapeutic dose and the toxic dose. Drug toxicity is affected by all of factors that affect the toxicity of other xenobiotics, including individual (genetic) variation, diet, age, and the presence of other exogenous chemicals.

Even when the risk of toxic side effects from a particular drug has been evaluated, it must be weighed against the expected benefits. The use of a very dangerous drug with only a narrow tolerance between the therapeutic and toxic doses may still be justified if it is the sole treatment for an otherwise fatal disease. However, a relatively safe drug may be inappropriate if safer compounds are available or if the condition being treated is trivial.

The three principal classes of cytotoxic agents used in the treatment of cancer all contain carcinogens, for example, Melphalen, a nitrogen mustard, adriamycin, an antitumor antibiotic, and methotrexate, an antimetabolite. Diethylstilbestrol (DES), a drug formerly widely used, has been associated with cancer of the cervix and vagina in the offspring of treated women.

Other toxic effects of drugs can be associated with almost every organ system. The stiffness of the joints accompanied by damage to the optic nerve (SMON—subacute myelo-optic neuropathy) that was common in Japan in the 1960s was apparently a toxic side effect of chloroquinol (Enterovioform), an antidiarrhea drug. Teratogenosis can also be caused by drugs, with thalidomide being the most alarming example. Skin effects (dermatitis) are common side effects of drugs, an example being topically applied corticosteroids.

A number of toxic effects on the blood have been documented, including agranulocytosis caused by chlorpromazine, hemolytic anemia caused by methyldopa, and megaloblastic anemia caused by methotrexate. Toxic effects on the eye have been noted and range from retinotoxicity caused by thioridazine to glaucoma caused by systemic corticosteroids.

## 5.8   DRUGS OF ABUSE

All drugs are toxic at some dose. Drugs of abuse, however, either have no medicinal function or are taken at dose levels higher than would be required for therapy. Although some drugs of abuse may affect only higher nervous functions—mood, reaction time, and coordination—many produce physical dependence and have serious physical effects, with fatal overdoses being a frequent occurrence.

The drugs of abuse include central nervous system depressants such as ethanol, methaqualone (Quaalude), and secobarbital; central nervous system stimulants, such as cocaine, methamphetamine (speed), caffeine, and nicotine; opioids, such as heroin and mependine (demerol); and hallucinogens such as lysergic acid diethylamide (LSD), phencyclidine (PCP), and tetrahydrocannabinol, the most active principal of marijuana. A further complication of toxicological significance is that many drugs of abuse are synthesized in illegal and poorly equipped laboratories with little or no quality control. The resultant products are therefore often contaminated with compounds of unknown, but conceivably dangerous, toxicity. The structures of some of these chemicals are shown in Figure 5.2.

## 5.9   COMBUSTION PRODUCTS

While many air pollutants (see Chapter 4) are the products of natural or anthropomorphic combustion, some of the most important from the point of view of human health are polycyclic aromatic hydrocarbons. Although also found in natural products such as coal and crude oil, they are generally associated with incomplete combustion of organic materials and are found in smoke from wood, coal, oil, and tobacco, for example, as well as in broiled foods. Because some of them are carcinogens, they have been studied intensively from the point of view of metabolic activation, interactions with DNA, and other aspects of chemical carcinogenesis. Some are heterocyclic, containing nitrogen in at least one of the rings. Some representative structures of the most studied polycyclic aromatic hydrocarbons are shown in Figure 5.3.

## 5.10   COSMETICS

The most common deleterious effects of modern cosmetics are occasional allergic reactions and contact dermatitis. The highly toxic and/or carcinogenic azo or aromatic

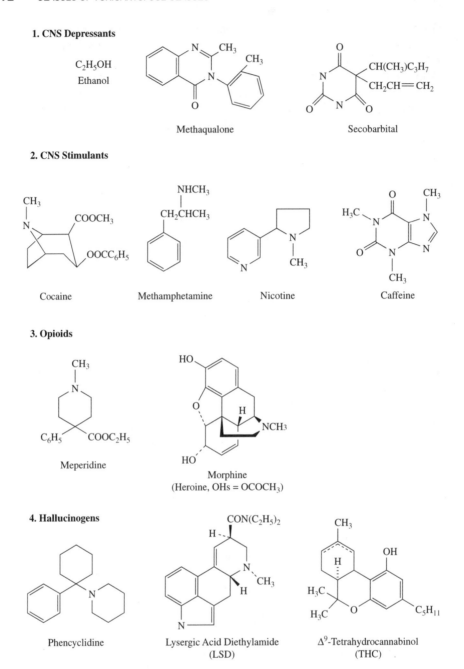

**Figure 5.2**   Some common drugs of abuse.

amine dyes are no longer in use, nor are the organometallics, used in even earlier times. Bromates, used in some cold-wave neutralizers, may be acutely toxic if ingested, as may the ethanol used as a solvent in hair dyes and perfumes. Thioglycolates and thioglycerol used in cold-wave lotion and depilatories and sodium hydroxide used in hair straighteners are also toxic on ingestion. Used as directed, cosmetics appear to

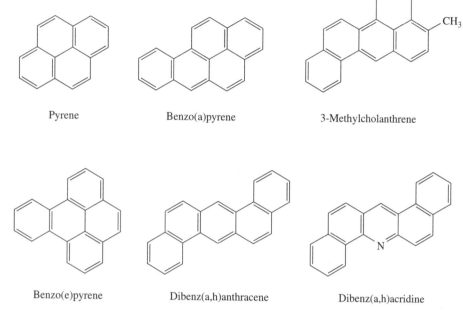

**Figure 5.3**  Some common polycyclic aromatic hydrocarbons.

present little risk of systemic poisoning, due in part to the deletion of ingredients now known to be toxic and in part to the small quantities absorbed.

## SUGGESTED READING

### General

Hodgson, E., R. B. Mailman, and J. E. Chambers. *Dictionary of Toxicology*, 2nd ed. Norwalk, CT: Appleton and Lange, 1994.

Klaassen, C. D., ed. *Casarett and Doull's Toxicology: The Basic Science of Poisons*, 6th ed. New York: McGraw-Hill, 2001.

### Metals

Goyer, R. A., and T. W. Clarkson. Toxic effects of metals. In *Casarett and Doull's Toxicology: The Basic Science of Poisons*, 6th ed., C. D. Klaassen, ed. New York: McGraw-Hill, 2001, pp. 811–867.

### Pesticides

Ecobichon, D. J. Toxic Effects of Pesticides. In *Casarett and Doull's Toxicology: The Basic Science of Poisons*, 6th ed., C. D. Klaassen, ed. New York: McGraw-Hill, 2001.

Krieger, R. *Handbook of Pesticide Toxicology*. San Diego: Academic Press, 2001.

### Toxins

Burkholder, J. M., and H. B. Glasgow *Pfiesteria piscicida* and other *Pfiesteria*-like dinoflagellates: Behavior, impacts, and environmental controls. *Limnol. Oceanogr.* **42**: 1052–1075, 1997.

Falconer, I., ed. *Algal Toxins in Seafood and Drinking Water.* New York: Academic Press, 1993.

Kotsonis, F. N., G. A. Burdock, and W. G. Flamm. Food toxicology. In *Casarett and Doull's Toxicology: The Basic Science of Poisons*, 6th ed., C. D. Klaassen, ed. New York: McGraw-Hill, 2001, pp. 1049–1088.

Norton, S. Toxic effects of plants. In *Casarett and Doull's Toxicology: The Basic Science of Poisons*, 6th ed., C. D. Klaassen, ed. New York: McGraw-Hill, 2001, pp. 965–976.

Russell, F. E. Toxic effects of terrestrial animal venoms and poisons. In *Casarett and Doull's Toxicology: The Basic Science of Poisons*, 6th ed., C. D. Klaassen, ed. New York: McGraw-Hill, 2001, pp. 945–964.

### Solvents

Bruckner, J. V., and D. A. Warren. Toxic effects of solvents and vapors. In *Casarett and Doull's Toxicology: The Basic Science of Poisons*, 6th ed., C. D. Klaassen, ed. New York: McGraw-Hill, 2001, pp. 945–964.

### Therapeutic Drugs

Hardman, J. G., L. E. Limbird, P. B. Molinoff, R. W. Ruddon, and A. G. Gilman, eds. *Goodman and Gilman's The Pharmacological Basis of Therapeutics*, 9th ed. New York: McGraw-Hill, 1996.

# PART III

# TOXICANT PROCESSING IN VIVO

# CHAPTER 6

# Absorption and Distribution of Toxicants

RONALD E. BAYNES and ERNEST HODGSON

## 6.1 INTRODUCTION

As illustrated in the previous chapter, the human body can be exposed to a variety of toxicants that may be present in various environmental media such as air, soil, water, or food. However, just simply being exposed to these hazardous chemicals does not necessarily translate into a toxicological response. The mammalian body has several inherent defense mechanisms and membrane barriers that tend to prevent the entry or absorption and distribution of these toxicants once an exposure event has occurred. However, if the toxicant is readily absorbed into the body, there are still other anatomical and physiological barriers that may prevent distribution to the target tissue to elicit a toxic response. As the toxicological response is often related to the exposed dose, interactions between the toxicant and the body's barriers and defense mechanisms will have an effect on toxicant movement in the body, and ultimately modulate the rate and extent of toxicant absorption and distribution to the target tissue.

The skin represents the largest organ in the human body, and one of its primary functions can be seen as a physical barrier to absorption of toxicants. The other major routes of toxicant entry into the body are through the respiratory and gastrointestinal tract, which can be seen to offer less resistance to toxicant absorption than the skin. In general, the respiratory tract offers the most rapid route of entry, and the dermal the least rapid. One reason for this major difference is primarily because membrane thickness, which is really the physical distance between the external environment (skin surface, air in the lung, or lumen of the gut) and the blood capillaries, varies across these portals of entry. The overall entry depends on both the amount present and the saturability of the transport processes involved.

Liver metabolism will have the most significant effect on toxicant bioavailability following gastrointestinal absorption, but microbial activity and various enzymes in the gastrointestinal tract and the skin can play a significant role in oral and dermal absorption, respectively. Physicochemical characteristics of the toxicant such as the chemical form can be a useful indicator of whether the toxicant will be absorbed and distributed in the body. In this regard toxicant molecular weight, ionization (pKa), and octanol/water partition coefficient ($\log P$) are useful indexes of predicting chemical

*A Textbook of Modern Toxicology, Third Edition,* edited by Ernest Hodgson
ISBN 0-471-26508-X Copyright © 2004 John Wiley & Sons, Inc.

transport from an environmental media across biological membranes to the blood stream. The reader should also be aware that for those toxicants that are readily ionized, the pH gradient across membranes can determine the extent of toxicant transport and accumulation in tissues.

Once the toxicant has been absorbed, the toxicant molecules can move around the body in two ways: (1) by bulk flow transfer (i.e., in the blood stream) and (2) by diffusional transfer (i.e., molecule-by-molecule over short distances). Disposition is the term often used to describe the simultaneous effects of distribution and elimination processes subsequent to absorption. The cardiovascular system provides distribution of all toxicants, regardless of their chemical nature, to various organs and tissues with various levels of affinities for toxicants. It should be remembered that organ mass and blood perfusion can vary, which can account for differential distribution of toxicants. Toxicant disposition can also be influenced by plasma protein binding in the blood stream. The nature of this toxicant-protein interaction is dependent on the chemical nature of the toxicant, the presence of other toxicants or drugs in the blood stream, as well as plasma protein levels. However, what distinguishes one toxicant pharmacokinetically from another is its diffusional characteristics. That is, its ability to cross nonaqueous diffusional barriers (e.g., cell membranes) from an aqueous compartment. This usually involves movement across several compartments separated by lipid membranes. It is therefore important to understand the *mechanisms by which drugs cross membranes* and the *physicochemical properties* of molecules and membranes that influence the movement of drugs from the environment to the body via either oral, inhalation, or dermal routes. These factors also influence movement from one compartment to another within the body during distribution as well as metabolism, and excretion.

We can quantitate this movement or transport from one compartment to another using mathematical models to describe transport rates. This in fact is what we do in pharmacokinetic analysis and modeling. *Pharmaco- or toxicokinetics* is therefore the quantitation of the time course of toxicants in the body during the various processes of absorption, distribution, and elimination or clearance (metabolism and/or excretion) of the toxicant. Stated differently, this is a study of how the body "handles" the toxicant as it is reflected in the plasma concentration at various time points. The two most important pharmacokinetic parameters that describe the disposition of a chemical are volume of distribution and systemic (body) clearance. *Pharmaco- and toxicodynamics* is the study of the biochemical and physiological effects of drugs and toxicants and determines their mechanism of action. Physiologically based pharmaco- or toxicokinetic models are used to integrate this information and to predict disposition of toxicants for a given exposure scenario. These concepts will be introduced at the end of this chapter.

## 6.2 CELL MEMBRANES

During absorption, distribution, and elimination processes the toxicant will encounter various cell membranes before interacting with the target tissue. Each step of these process involves translocation of the chemical across various membrane barriers, from the skin or mucosa through the capillary membranes, and through the cellular and organelle membranes (Figure 6.1). These membrane barriers vary from the relatively thick areas of the skin to the relatively thin lung membranes. In all cases, however, the membranes of tissue, cell, and cell organelle are relatively similar.

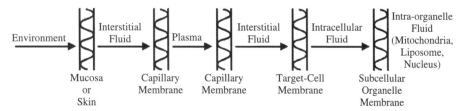

**Figure 6.1**   Schematic showing membranes that a chemical may need to cross during passage from the environment to the site of action. (Adapted from E. Hodgson and P. E. Levi, eds. *Introduction to Biochemical Toxicology*, 2nd ed., Appleton and Lange, 1994, p. 12.)

The cell membranes are predominantly a lipid matrix or can be considered a lipid barrier with an average width of a membrane being approximately 75 Å. The membrane is described as the fluid mosaic model (Figure 6.2) which consist of (1) a bilayer of phospholipids with hydrocarbons oriented inward (hydrophobic phase), (2) hydrophilic heads oriented outward (hydrophilic phase), and (3) associated intra- and extracellular proteins and transverse the membrane. The ratio of lipid to protein varies from 5:1 for the myelin membrane to 1:5 for the inner structure of the mitochondria. However, 100% of the myelin membrane surface is lipid bilayer, whereas the inner membrane of the mitochondria may have only 40% lipid bilayer surface. In this example the proportion of membrane surface that is lipid will clearly influence distribution of toxicants of varying lipophilicity.

The lipid constituents in the membrane permit considerable movement of macromolecules, and membrane constituents may move appreciably within membranes. Membrane fluidity, a function of lipid composition, can be altered by temperature and chemicals

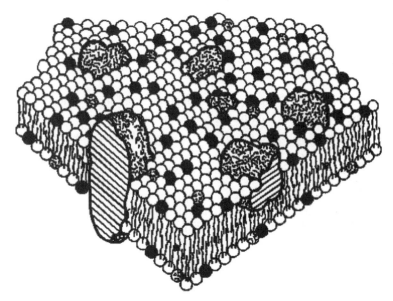

**Figure 6.2**   Schematic diagram of biological membrane. Head groups of lipids represented by spheres, tail ends by zigzag lines. Black, white, or stippled spheres indicate different kinds of lipids and illustrate asymmetry in certain cases. Large bodies are membrane-associated proteins. (Adapted from Singer and Nicolson, *Science* **175**:720, 1972.)

(e.g., anesthetics). Several types of lipids are found in membranes, with phospholipids and cholesterol predominating. Sphingolipids comprise the primary minor component. Phosphatidylcholine, phosphatidylserine, and phosphatidylethanolamine are the primary phosphatides, and their two fatty acid hydrocarbon chains (typically 16 to 18, but varying from 12 to 22) comprise the nonpolar region. Some of the fatty acids are unsaturated and contribute appreciably to the fluidity of the membrane.

Proteins, which have many physiological roles in normal cell function, are intimately associated with lipids and may be located throughout lipid bilayers. These proteins may be located on either the surface or traverse the entire structure. Hydrophobic forces are responsible for maintaining the structural integrity of proteins and lipids within membranes, but movement within the membranes may occur. External and internal membrane proteins can function as receptors. Many proteins that traverse the membrane are transport proteins, and are involved in translocation of ligands; that is, they are involved in active and facilitated transport.

Complexes of intrinsic membrane proteins and lipids can form hydrophilic or hydrophobic channels that allow transport of molecules with different physicochemical characteristics. The amphipathic nature of the membrane creates a barrier for ionized, highly polar drugs, although it does not completely exclude them. The presence of pores of approximately 4 Å are believed to allow for ready movement of small molecules such as water. Thus certain molecules that ordinarily would be excluded can rapidly traverse the highly lipid membrane barrier.

It is worth noting that differences among membranes, such as the presence of different lipids, the amount of surface lipid, differences in size and shape of proteins, or physical features of bonding, may cause differences in permeability among membranes. These biochemical and biophysical differences are thought to be responsible for permeability differences in skin from different anatomical regions of the body.

## 6.3  MECHANISMS OF TRANSPORT

In general, there are four main ways by which small molecules cross biological lipid membranes:

1. *Passive diffusion.* Diffusion occurs through the lipid membrane.
2. *Filtration.* Diffusion occurs through aqueous pores.
3. *Special transport.* Transport is aided by a carrier molecule, which act as a "ferryboat."
4. *Endocytosis.* Transport takes the form of pinocytosis for liquids and phagocytosis for solids.

The first and third routes are important in relation to pharmacokinetic mechanisms. The aqueous pores are too small in diameter for diffusion of most drugs and toxicant, although important for movement of water and small polar molecules (e.g., urea). Pinocytosis is important for some macromolecules (e.g., insulin crossing the blood-brain barrier).

### 6.3.1  Passive Diffusion

Most drugs and toxicant pass through membranes by *simple diffusion* down a concentration gradient. The driving force being the concentration gradient across the membrane.

This diffusion process can continue until equilibrium, although in reality there is always movement but the net flux is zero. Eventually the concentration of unionized or unbound (free) toxicant is the same on either side of the membrane. In other words, there is no competition of molecules and there is generally a lack of saturation. Solubility in the lipid bilayer is important, and the greater the partition coefficient, the higher is the concentration in the membrane, and the greater is the rate of diffusion across the membrane. For ionized toxicants the steady state concentration is dependent on the differences in pH across the membrane. Most membranes are relatively permeable to water either by diffusion or by flow that results from hydrostatic or osmotic differences across the membrane, and bulk flow of water can also carry with it small and water soluble molecules by this mechanism. These substances generally have a molecular weight of less than 200. Although inorganic ions are small and will readily diffuse across some membranes, their hydrated ionic radius is relatively large. In such cases active transport is required (see below). Specific ion fluxes are also controlled by specific channels that are important in nerves, muscles, and signal transduction.

We can now quantitate the *rate* at which a toxicant can be transported by passive diffusion, and this can be described by Fick's law of diffusion as follows:

$$\text{Rate of diffusion} = \frac{D \times S_a \times P_c}{d}(C_H - C_L),$$

where $D$ is the diffusion coefficient, $S_a$ is the surface area of the membrane, $P_c$ is the partition coefficient, $d$ is the membrane thickness, and $C_H$ and $C_L$ are the concentrations at both sides of the membrane (high and low, respectively). The first part of this equation $(DP_c/d)$ represents the permeability coefficient of the drug. The permeability expresses the ease of penetration of a chemical and has units of velocity, distance/time (cm/h).

The diffusion coefficient or diffusivity of the toxicant, $D$, is primarily dependent on solubility of the toxicant in the membrane and its molecular weight and molecular conformation. Depending on the membrane, there is a functional molecular size and/or weight cutoff that prevents very large molecules from being passively absorbed across any membrane. One would expect small molecular weight molecules to diffuse more rapidly than larger molecular weight toxicants. Therefore the magnitude of a toxicant's diffusion coefficient really reflects the ease with which it is able to diffuse through the membrane. The reader should also be aware that as a toxicant crosses from the donor or aqueous medium and through the membrane medium, there are really two diffusion environments and thus two diffusion coefficients to consider. Another important factor that can influence the diffusion coefficient is membrane viscosity. This physicochemical characteristic should remain constant in biological systems but can be modified in skin membranes exposed to various pharmaceutical or pesticide formulations. Formulation additives or excipients may enter the membrane barrier and reversibly or irreversibly change viscosity and thus diffusion coefficient of the drug or pesticide in the barrier membranes of the skin.

The partition coefficient, which will be described in more detail later in this chapter, is the relative solubility of the compound in lipid and water, and the compound's solubility really reflects the ability of the toxicant to move from a relatively aqueous environment across a lipid membrane. It is this factor that is often manipulated in pesticide and drug formulations to create a vehicle. Membrane permeability is therefore strongly correlated to the lipid solubility of the toxicant in the membrane as well as the aqueous environment surrounding the membrane. Please be aware that there are instances where partition coefficient or lipid solubility of the toxicant may be very large, and there may be a tendency for the drug to sequester in the membrane. Membrane surface area and membrane thickness can also vary across different organs in the body, but one does not expect these two factors in Fick's equation to vary considerably. The final component of Fick's equation is the concentration gradient ($C_H - C_L$) across the membrane, which is the **driving force for diffusion**, and as will be demonstrated below in our discussion on first-order kinetics, is the most important factor dictating the rate of transport across most biological membranes.

*First-Order Kinetics.*   When the rate of a process is dependent on a **rate constant** and a concentration gradient, a linear or first-order kinetic process will be operative. The reader should be aware that there are numerous deviations from the first-order process when chemical transport in vivo is analyzed, and this can be deemed an *approximation* since, in many barriers, penetration is slow and a long period of time is required to achieve steady state.

The rate of movement of a toxicant across a membrane may be expressed as the *change in amount of toxicant, A, (dA) or toxicant concentration, C, (dC) per unit of time (dt), which equals dA/dt*. Calculus can be used to express instantaneous rates over very small time intervals (*dt*). Thus rate processes may then be generally expressed as

$$\frac{d\mathbf{A}}{d\mathbf{t}} = \mathbf{K}\mathbf{A}^n$$

where $d\mathbf{A}/d\mathbf{t}$ is the rate of chemical ($X$) movement (e.g. absorption, distribution, elimination), $\mathbf{K}$ is the rate constant of the process, and $n$ is the kinetic order of the transport process (e.g., absorption). The $n$ either equals 1 (first order) or 0 (zero order). Thus the first-order rate equation is written as

$$\frac{d\mathbf{A}}{d\mathbf{t}} = \mathbf{K}\mathbf{A}^1 = \mathbf{K}\mathbf{A},$$

and the zero-order rate equation as

$$\frac{d\mathbf{A}}{d\mathbf{t}} = \mathbf{K}\mathbf{A}^0 = \mathbf{K}.$$

We know from Fick's law that the rate of diffusion (now expressed as $d\mathbf{A}/d\mathbf{t}$) is

$$\frac{d\mathbf{A}}{d\mathbf{t}} = \frac{\mathbf{D}\cdot\mathbf{S}_a\cdot\mathbf{P}_c(\mathbf{A}_1 - \mathbf{A}_2)}{\mathbf{d}}.$$

Once a toxicant crosses a membrane, it is rapidly removed from the "receiving side" (compartment B in Figure 6.3) either by uptake into the blood stream or elimination from the organism. Thus it is $A_1$ that is the primary driving force, and if we replace this with $A$ in all equations, then

$$\frac{d\mathbf{A}}{d\mathbf{t}} = \left(\frac{\mathbf{D}\cdot\mathbf{S}_a\cdot\mathbf{P}_c}{\mathbf{d}}\right)\mathbf{A}.$$

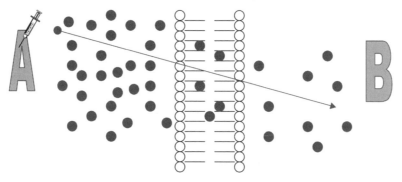

**Figure 6.3**  Illustration of concentration gradient generated by administration of a drug that can travel down this gradient from area A and across a biological membrane to area B.

If we let $\mathbf{K} = (\mathbf{D \cdot S_a \cdot P_c / d})$, then, since $A$ is present in the equation, $n$ must equal 1, so we have a first-order rate process. Fick's law of diffusion, which is important for quantitating rates of absorption, distribution, and elimination, is thus the basis for using first-order kinetics in most pharmacokinetic models.

Therefore in a first-order process, the rate of drug movement is directly proportional to the amount of drug ($A$) in the body, which is usually a function of the dose. $\mathbf{K}$ is the first-order fractional rate constant with units of liters/time (time$^{-1}$) and represents the *fraction* of drug that is transported *per unit of time*. Thus in a *first-order* process, the rate of drug movement is proportional to dose but the fraction moved per unit of time is constant and *independent of dose*.

When first-order kinetics hold, a simple relationship exists between the penetration rate constant, $K$, and $t_{0.5}$ (time necessary for one-half of the applied dose to penetrate):

$$K = \frac{0.693}{t_{0.5}},$$

where the units of $K$ are a percentage of the change/time unit. We can also derive the concentration of the toxicant if we know the volume or volume of distribution ($V_d$) of the toxicant compartment as

$$V_d(\text{volume}) = \frac{A(\text{mass})}{C(\text{mass/volume})}.$$

($V_d$ is discussed in more detail later in this chapter.)

### 6.3.2   Carrier-Mediated Membrane Transport

This mechanism is important for compounds that lack sufficient lipid solubility to move rapidly across the membrane by simple diffusion. A membrane-associated protein is usually involved, specificity, competitive inhibition, and the saturation phenomenon and their kinetics are best described by *Michaelis-Menton enzyme kinetic models*. Membrane penetration by this mechanism is more rapid than simple diffusion and, in the case of active transport, may proceed beyond the point where concentrations are equal on both sides of the membrane. Generally, there are two types of specialized carrier-mediated transport processes:

*Passive facilitated diffusion* involves movement down a concentration gradient without an input of energy. This mechanism, which may be highly selective for specific conformational structures, is necessary for transport of endogenous compounds whose rate of transport by simple diffusion would otherwise be too slow. The classical example of facilitated diffusion is transport of glucose into red blood cells.

*Active transport* requires energy, and transport is against a concentration. Maintenance against this gradient requires energy. It is often coupled to energy-producing enzymes (e.g., ATPase) or to the transport of other molecules (e.g., $Na^+$, $Cl^-$, $H^+$) that generate energy as they cross the membranes. Carrier-mediated drug transport can occur in only a few sites in the body, and the main sites are

- BBB, neuronal membranes, choroid plexus
- Renal tubular cells
- Hepatocytes, biliary tract

There are instances in which toxicants have chemical or structural similarities to endogenous chemicals that rely on these special transport mechanisms for normal physiological uptake and can thus utilize the same system for membrane transport. Useful examples of drugs known to be transported by this mechanism include levodopa, which is used in treating Parkinson's disease, and fluorouracil, a cytotoxic drug. Levodopa is taken up by the carrier that normally transports phenylalanine, and fluorouracil is transported by the system that carries the natural pyrimidines, thymine, and uracil. Iron is absorbed by a specific carrier in the mucosal cells of the jejunum, and calcium by a vitamin D-dependent carrier system. Lead may be more quickly moved by a transport system that is normally involved in the uptake of calcium.

For carrier-mediated transport, the rate of movement across a membrane will now be *constant*, since flux is dependent on the capacity of the membrane carriers and not the mass of the chemical to be transported. These processes are described by *zero-order* kinetic rate equations of the form:

$$\frac{d\mathbf{X}}{d\mathbf{t}} = \mathbf{K}\mathbf{X}^0 = \mathbf{K}_0.$$

$\mathbf{K}_0$ is now the *zero-order* rate constant and is expressed in terms of mass/time. In an active carrier-mediated transport process following zero-order kinetics, the rate of drug transport is always equal to $\mathbf{K}$ once the system is fully loaded or saturated. At subsaturation levels, the rate is *initially first order* as the carriers become loaded with the toxicant, but at concentrations normally encountered in pharmacokinetics, the rate becomes constant. Thus, as dose increases, the rate of transport does *not* increase in proportion to dose as it does with the fractional rate constant seen in first-order process. This is illustrated in the Table 6.1 where it is assumed that the first-order rate constant is 0.1 (10% per minute) and the zero-order rate is 10 mg/min.

In the case of first order, these amounts will subsequently diminish (10% of 900 is 90, etc.). In the case of zero order, the amount transported does not vary with time (constant rate of transport).

**Table 6.1 Amount of Toxicant (mg) Transported in One Minute**

| Initial Toxicant Mass (mg) | First-Order Rate | Zero-Order Rate |
|---|---|---|
| 1000 | 100 | 10 |
| 100 | 10 | 10 |
| 10 | 1 | 10 |

The plot in Figure 6.4 illustrates the differences in passive (linear) versus carrier-mediated (nonlinear) transport. At relatively low concentrations of drug, carrier-mediated processes may appear to be first order since the protein carriers are not saturated. However, at higher concentrations, zero-order behavior becomes evident. It is in plots such as this that the terms *linear* (first order) and *nonlinear* (zero order) come into existence.

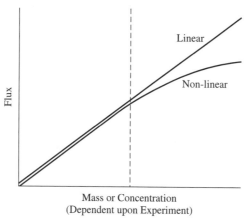

**Figure 6.4** Plot depicting a linear relationship (first order) and nonlinear relationship (zero order) between chemical flux across a membrane and the initial mass or concentration of the chemical.

## 6.4 PHYSICOCHEMICAL PROPERTIES RELEVANT TO DIFFUSION

The following physicochemical properties are important for chemical diffusion. We have discussed several of these properties in previous sections of this chapter as they relate to the passive diffusion mechanism and its impacts on rate of toxicant transport across membranes.

Molecular size and shape
Solubility at site of absorption
Degree of ionization
Relative lipid solubility of ionized and unionized forms

Although molecular weight is important, it is less important than the drug's *lipid solubility* when it comes to assessing the rate of passive diffusion across membranes.

The permeability, $P(P = P_c \times D)$, of a nonpolar substance through a cell membrane is dependent on two physicochemical factors: (1) *solubility in the membrane* ($P_c$), which can be expressed as a partition coefficient of the drug between the aqueous phase and membrane phase, and (2) *diffusivity or diffusion coefficient (D)*, which is a measure of mobility of the drug molecules within the lipid. The latter may vary only slightly among toxicants, but the former is more important. Lipid solubility is therefore one of the most important determinants of the pharmacokinetic characteristics of a chemical, and it is important to determine whether a toxicants is readily ionized or not influenced by pH of the environment. If the toxicant is readily ionized, then one needs to understand its chemicals behavior in various environmental matrices in order to adequately assess its transport mechanism across membranes.

### 6.4.1  Ionization

For the purposes of this discussion on membrane transport, chemicals can be broadly categorized into those that are ionized and those that are not ionized. Many drugs (e.g., antibiotics) and several toxicants (e.g., strychnine) are either weak acids or weak bases and can exist in solution as a mixture of nonionized and ionized forms. Generally, these drugs and toxicants must be in the uncharged or nonionized form to be transported by passive diffusion across biological membranes. This is because biological membranes are of a lipid nature and are less permeable to the ionized form of the chemical. The pH of the environment (e.g., lumen of the gastrointestinal tract and renal tubules) can influence transfer of toxicant that are ionizable by increasing or decreasing the amount of nonionized form of the toxicant. Aminoglycosides (e.g., gentamicin) are the exception to this general rule in that the uncharged species is insufficiently lipid soluble to cross the membrane appreciably. This is due to a preponderance of hydrogen-bonding groups in the sugar moiety that render the uncharged molecule hydrophilic. Note that some amphoteric drugs (e.g., tetracyclines) may be absorbed from both acidic and alkaline environments. In essence, the amount of drug or toxicant in ionized or nonionized form depends on the pKa (pH at which 50% of the drug is ionized) of the drug and the pH of the solution in which the drug is dissolved. *The pKa, which is the negative logarithm of the dissociation constant of a weak acid or weak base, is a physicochemical characteristic of the drug or toxicant.* When the pH of the solution is equal to the pKa, then 50% of the toxicant is in the ionized form and 50% is in the nonionized form. The ionized and nonionized fractions can be calculated according to the *Henderson-Hasselbach* equations listed below:

$$\text{For weak acids}: \quad \text{pKa} - \text{pH} = \log(\text{Nonionized form/Ionized form}),$$

$$\text{For weak bases}: \quad \text{pKa} - \text{pH} = \log(\text{Ionized form/Nonionized form}).$$

For an organic acid ($RCOOH \leftrightarrow RCOO^- + H^+$), acidic conditions (pH less than the pKa of the compound) will favor the formation of the nonionized RCOOH, whereas alkaline conditions (pH greater than pKa) will shift the equilibrium to the right. For an organic base ($RNH_2 + H^+ \leftrightarrow RNH_3^+$), the reverse is true, and decreasing the pH (increasing the concentration of H+) will favor formation of the ionized form, whereas increasing the pH (decreasing the concentration of $H^+$) will favor formation of the nonionized form.

**Table 6.2 Amount of Toxicant Absorbed at Various pH Values (%)**

| Compound | pKa | 3.6–4.3 | 4.7–5.0 | 7.0–7.2 | 7.8–8.0 |
|---|---|---|---|---|---|
| *Acids* | | | | | |
| Nitrosalicyclic | 2.3 | 40 | 27 | <02 | <02 |
| Salicyclic | 3.0 | 64 | 35 | 30 | 10 |
| Benzoic | 4.2 | 62 | 36 | 35 | 05 |
| *Bases* | | | | | |
| Aniline | 4.6 | 40 | 48 | 58 | 61 |
| Aminopyrene | 5.0 | 21 | 35 | 48 | 52 |
| Quinine | 8.4 | 09 | 11 | 41 | 54 |

*Memory aid*: In general, weak organic acids readily diffuse across a biological membrane in an acidic environment, and organic bases can similarly diffuse in a basic environment. This is illustrated quite well in Table 6.2 for the chemical in rat intestine. There are the usual exceptions to the generalizations concerning ionization and membrane transport, and some compounds, such as pralidoxime (2-PAM), paraquat, and diquat, are absorbed to an appreciable extent even in the ionized forms. The mechanisms allowing these exceptions are not well understood.

*Ion trapping* can occur when at equilibrium the total (ionized + nonionized) concentration of the drug will be different in each compartment, with an acidic drug or toxicant being concentrated in the compartment with the relatively high pH, and vice versa. The pH partition mechanism explains some of the qualitative effects of pH changes in different body compartment on the pharmacokinetics of weakly basic or acidic drugs or toxicant as it relates to renal excretion and penetration of the blood-brain barrier. Alkalization of urine in the lumen of renal tubules can enhance elimination of weak acids. However, this phenomenon is not the main determinant of absorption of drugs or toxicants from the gastrointestinal tract. In the gastrointestinal tract the enormous absorptive surface area of the villi and microvilli in the ileum, compared to the smaller absorptive area of the stomach, is of overriding importance.

## 6.4.2 Partition Coefficients

A second physicochemical parameter influencing chemical penetration through membranes is the relative lipid solubility of the potential toxicant that can be ascertained from its known partition coefficient. The partition coefficient is a measure of the ability of a chemical to separate between two immiscible phases. The phases consist of an organic phase (e.g., octanol or heptane) and an aqueous phase (e.g., water). The lipid solvent used for measurement is usually octanol because it best mimics the carbon chain of phospholipids, but many other systems have been reported (chloroform/water, ether/water, olive oil/water). The lipid solubility and the water solubility characteristics of the chemical will allow it to proportionately partition between the organic and water phase. The partition coefficients can be calculated using the following equation:

$$P = \frac{V_w}{V_o} \left[ \frac{(C_{wo} - C_w)}{C_w} \right],$$

where $P$ is the partition coefficient and usually expressed in terms of its logarithmic value (log $P$), $V_w$ and $V_o$ are the volumes of aqueous and oil or organic phase, respectively, and $C_{wo}$ and $C_w$ are drug or toxic concentrations in the aqueous phase before and after shaking, respectively.

The lower the partition coefficient, the more water soluble, and the least permeable the toxicant is across a membrane. Regarding dermal absorption, partition coefficients can be predictive of absorption. However, toxicants with extremely high partition coefficients tend to remain in the membrane or skin. This explains why a strong correlation between permeability and the partition coefficient can exist for a hypothetical series of analogous chemicals for a specific range of partition coefficients, but the correlation does not exists for log $P$ values greater than 6 in many instances. A log $P$ of around 1 is often taken as desirable for skin penetration. The reader should also recall that this parameter is operative as the chemical diffuses across membranes (Figure 6.1) of varying lipid content during absorption, distribution, and elimination processes.

## 6.5   ROUTES OF ABSORPTION

Primary routes of entry of toxicants to the human body are dermal, gastrointestinal, and respiratory. Methods for studying these different routes are numerous, but they are perhaps best developed for the study of dermal absorption because this route is subject to more direct methodology, whereas methods for studying respiratory or gastrointestinal absorption require more highly specialized instrumentation. Additional routes encountered in experimental studies include intraperitoneal, intramuscular, and subcutaneous routes. When direct entry into the circulatory system is desired, intravenous (IV) or intra-arterial injections can be used to bypass the absorption phase. Information from this more direct route of entry (e.g., IV) should, however, be used in addition to data from the extravascular route of interest to adequately assess the true extent of absorption of a toxicant.

### 6.5.1   Extent of Absorption

It is often useful to determine *how much of the drug* actually penetrates the membrane barrier (e.g., skin or gastrointestinal tract) and gets into the blood stream. This is usually determined experimentally for oral and dermal routes of administration. The *area under the curve (AUC)* of the concentration-time profiles for oral or dermal routes is compared with the AUC for IV routes of administration. The AUC is determined by breaking the curve up into a series of trapezoids and summing all of the areas with the aid of an appropriate computer program (Figure 6.5).

The intravenous correction is very important if absolute bioavailability is desired. The ratio of these AUC values is absolute bioavailability, $F$:

$$F = \frac{(\text{AUC})_{\text{route}}}{(\text{AUC})_{\text{IV}}}.$$

The relationship above holds if the same doses are used with both routes, but the bioavailability should be corrected if different doses are used:

$$F = \frac{\text{AUC}_{\text{route}} \times \text{Dose}_{\text{IV}}}{\text{AUC}_{\text{IV}} \times \text{Dose}_{\text{route}}}.$$

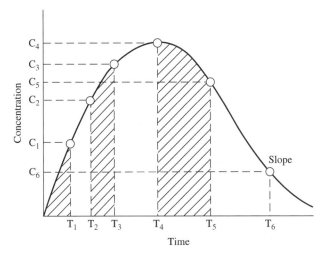

**Figure 6.5**   Plasma concentration time profile for oral exposure to a toxicant and depiction of AUCs determined by summation of trapezoids at several time periods.

Another technique is to monitor drug or toxicant excretion rather than blood concentrations, especially when blood or plasma concentrations are very low. Using the same equations, the AUC is now replaced by chemical concentrations in urine, feces, and expired air. Some chemicals are primarily excreted by the kidney and urine data alone may be necessary. The rate and extent of absorption are clearly important for therapeutic and toxicological considerations. For example, different formulations of the same pesticide can change the absorption rate in skin or gastrointestinal tract, and not bioavailability, but can result in blood concentrations near the toxic dose. Also different formulations can result in similar absorption rates but different bioavailability.

### 6.5.2   Gastrointestinal Absorption

The gastrointestinal tract (GIT) is a hollow tube (Figure 6.6a) lined by a layer of columnar cells, and usually protected by mucous, which offers minimal resistance to toxicant penetration. The distance from the outer membrane to the vasculature is about 40 μm, from which point further transport can easily occur. However, the cornified epithelium of the esophagus prevents absorption from this region of the GIT. Most of the absorption will therefore occur in the intestine (pH = 6), and to some extent in the stomach (pH = 1–3). Buccal and rectal absorption can occur in special circumstances. Note that secretions from the lachrymal duct, salivary gland, and nasal passages can enter the GIT via the buccal cavity. Therefore, following IV administration, a toxicant can enter the GIT if the drug is in these secretions.

The intestine can compensate the 2.5 log units difference between it and the stomach by the increased surface area in the small intestines. The presence of microvilli (Figure 6.6b) in the intestine is an increase of 600-fold in surface area compared to a hollow tube of comparable length. Note that there is no absorption, except for water, in the large intestine.

Most of the absorption in the GIT is by passive diffusion, except for nutrients; glucose, amino acids, and drugs that look like these substances are taken up by

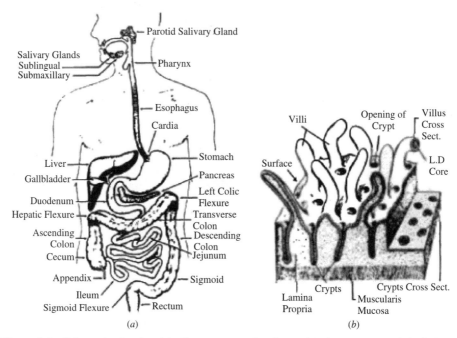

**Figure 6.6** Schematic showing (*a*) alimentary canal and associated structures and (*b*) lining of the small intestine. (*Sources:*(*a*) Scholtelius and Scholtelius in *Textbook of Physiology*, Mosby, 1973; (*b*) Ham and Cormack, in *Histology*, 8th ed., Lippincott, 1979.)

active transport. For toxicants with structural similarities to compounds normally taken up by these active transport mechanisms, entry is enhanced. For example, cobalt is absorbed by the same active transport mechanism that normally transports iron, and 5-bromouracil is absorbed by the pyrimidine transport system.

Very lipid soluble toxicants and drugs, which are not miscible in the aqueous intestinal fluid, are presented as emulsions, and brought into solution through the action of detergent-like bile acids. The product of this mixing is large surface area micelles (hydrophobic interior) that deliver the lipids to the brush border of the intestine for diffusion across the membrane. As stated previously, the rate of passive transfer will be dependent on ionization and lipid solubility. Very strong bases (e.g., tubocurarine, succinylcholine) and strong acids are not readily absorbed in the GIT. These muscle relaxants therefore are given IV. The smaller the particle size of the toxicant, the greater is the absorption, and a chemical must be in aqueous solution for it to be absorbed in the GIT. A feature of the GIT that seems to contradict basic assumptions of absorption is the penetration of certain very large molecules. Compounds such as bacterial endotoxins, large particles of azo dyes, and carcinogens are apparently absorbed by endocytotic mechanisms.

GIT motility has a significant effect on GIT absorption of a toxicant. For example, excessively rapid movement of gut contents can reduce absorption by reducing residence time in the GIT, while the presence of food in the stomach can delay the progress of drugs from the stomach to the small intestine where most of the absorption will occur. Increased splanchnic blood flow after a meal can result in absorption of several drugs (e.g., propranolol), but in hypovolemic states, absorption can be reduced.

Biotransformation in the GIT prior to absorption can have a significant impact on bioavailability of a toxicant. The resident bacterial population can metabolize drugs in the GIT. Because of microbial fermentation in the rumen of ruminants and large intestine and cecum of horses and rabbits, its is often difficult to compare drug absorption profiles with carnivores (e.g., dogs) and omnivores (e.g., humans, pigs). Acid hydrolysis of some compounds can also occur, and enzymes in the intestinal mucosa can also have an effect on oral bioavailability. If the toxicant survive these microbial and chemical reactions in the stomach and small intestine, it is absorbed in the GIT and carried by the hepatic portal vein to the liver, which is the major site of metabolism. Chapters 7, 8, and 9 will discuss liver metabolism of toxicants in more detail. In brief, this activity in the liver can result in detoxification and/or bioactivation. Some drugs and toxicant that are conjugated (e.g., glucuronidation) in the liver are excreted via the biliary system back into the GIT. Once secreted in bile by active transport and excreted from the bile duct into the small intestine, this conjugated toxicant can be subjected to microbial beta-glucuronidase activity that can result in regeneration of the parent toxicant that is more lipophilic than the conjugate. The toxicant can now be reabsorbed by the GIT, prolonging the presence of the drug or toxicant in the systemic circulation. This is called *enterohepatic circulation*, which will be covered in greater detail in subsequent chapters.

### 6.5.3 Dermal Absorption

The skin is a complex multilayered tissue with a large surface area exposed to the environment. Skin anatomy, physiology, and biochemistry vary among species, within species, and even between anatomic sites within an individual animal or human. Logically these biological factors alone can influence dermal absorption. What is consistent is that the outer layer, the *stratum corneum* (SC), can provide as much as 80% of the resistance to absorption to most ions as well as aqueous solutions. However, the skin is permeable to many toxicants, and dermal exposure to agricultural pesticides and industrial solvent can result in severe systemic toxicity.

The anatomy of the skin is depicted in the schematic diagram of Figure 6.7. In mammalian skin there are really three distinct layers, which are the epidermis, dermis, and hypodermis or subcutaneous fat layer. Human skin is 3 mm thick, but it is the epidermis, which is only 0.1 to 0.8 mm, that provides the greatest resistance to toxicant penetration. The five layers of the epidermis, starting from the outside, are the stratum corneum, stratum lucidum, stratum granulosum, stratum spinosum, and stratum basale. The basal cells of the epidermis proliferate and differentiate as they migrate outward toward the surface of the skin. It requires about 2 to 28 days for cells to migrate from the basal layer to the stratum corneum, where they are eventually sloughed off. These dead, keratinized cells are, however, very water absorbant (hydrophilic), a property that keeps the skin soft and supple. Sebum, a natural oil covering the skin, functions in maintaining the water-holding ability of the epidermis. The stratum corneum is the primary barrier to penetration, and it consists primarily of these dead keratin-filled keratinocytes embedded in an extracellular lipid matrix. The lipids are primarily sterols, other neutral lipids, and ceramides. This association between lipids and dead keratinized cells, which is often referred to as the "brick and mortar" model as depicted in Figure 6.7*b*, is used to simplify the composition of the stratum corneum that is integral to chemical transport through skin.

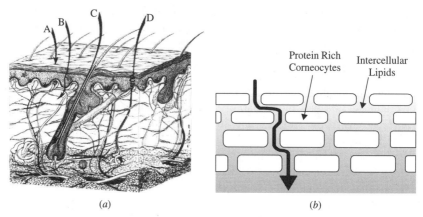

$(a)$                                    $(b)$

**Figure 6.7**  ($a$) Schematic diagram of the microstructure of mammalian skin and potential pathways for absorption by (A) intercellular, (B) transcellular, (C) transfollicular, or (D) sweat pore routes. ($b$) Brick-and-mortar" model of the stratum corneum depicting intercellular pathway (i.e., route A) between keratinocytes through the lipid domain of the stratum corneum.

A number of appendages are associated with the skin, including hair follicles, sebaceous glands, eccrine and apocrine sweat glands, and nails. Recently it was found that removal of the stratum corneum does not allow complete absorption; thus it is apparent that some role, although of lesser importance, is played by other parts of the skin. The dermis and subcutaneous areas of the skin are less important in influencing penetration, and once a toxicant has penetrated the epidermis, the other layers are traversed rather easily. The dermis is highly vascular, a characteristic that provides maximal opportunity for further transport once molecules have gained entry through the epidermis or through skin appendages. Most of the systemic absorption occurs at the capillary loops located at the epidermis-dermis junction. The blood supply of the dermis is under neural and humoral influences whose temperature-regulating functions could thus affect penetration and distribution of toxicants. Vasoactive drugs or environmental temperature can also influence absorption by altering blood flow to these capillaries. The subcutaneous layer of the skin is highly lipid in nature and serves as a shock absorber, an insulator, and a reserve depot of energy. The pH of the skin varies between 4 and 7 and is markedly affected by hydration.

Cutaneous biotransformation is mostly associated with the stratum basale layer where there can be phase I and phase II metabolism. However, the skin is not very efficient, compared to the liver. The epidermal layer accounts for the major portion of biochemical transformations in skin, although the total skin activity is low (2–6% that of the liver). Where activity is based on epidermis alone, that layer is as active as the liver or, in the case of certain toxicants, several times more active. For some chemicals, metabolism can influence absorption, and transdermal delivery systems of drugs utilize this activity. For example prodrug such as lipid esters are applied topically, and cutaneous esterases liberate the free drug. These basal cells and extracellular esterases have been shown to be involved in detoxification of several pesticides and bioactivation of carcinogens such as benzo(a)pyrene. For rapidly penetrating substances, metabolism by the skin is not presently considered to be of major significance, but skin may have an important first-pass metabolic function, especially for compounds that are absorbed slowly.

The *intercellular pathway* is now accepted as the major pathway for absorption. Recall that the rate of penetration is often correlated with the partition coefficient. In fact this is a very tortuous pathway, and the *h* (skin thickness) in Fick's first law of diffusion is really 10× the measured distance. By placing a solvent (e.g., ether, acetone) on the surface or tape stripping the surface, the stratum corneum (SC) is removed, and absorption can be significantly increased by removing this outer barrier. This may not be the case for very lipophilic chemical. This is because the viable epidermis and dermis are regarded as aqueous layers compared to the SC. Note that the more lipophilic the drug, the more likely it will form a depot in the SC and be slowly absorbed over time and thus have a prolonged half-life.

The *transcellular pathway* has been discredited as a major pathway, although some polar substances can penetrate the outer surface of the protein filaments of hydrated stratum corneum. The *transfollicular pathway* is really an invagination of the epidermis into the dermis, and the chemical still has to penetrate the epidermis to be absorbed into the blood stream. This is also a regarded as *minor* route. *Sweat pores* are not lined with the stratum corneum layer, but the holes are small, and this route is still considered a minor route for chemical absorption. In general, the epidermal surface is 100 to 1000 times the surface area of skin appendages, and it is likely that only very small and/or polar molecules penetrate the skin via these appendages.

Variations in areas of the body cause appreciable differences in penetration of toxicants. The rate of penetration is in the following order:

Scrotal > Forehead > Axilla >= Scalp > Back = Abdomen > Palm and plantar.

The palmar and plantar regions are highly cornified and are 100 to 400 times thicker than other regions of the body. Note that there are differences in blood flow and to a lesser extent, hair density, that may influence absorption of more polar toxicants.

Formulation additives used in topical drug or pesticide formulations can alter the stratum corneum barrier. Surfactants are least likely to be absorbed, but they can alter the lipid pathway by fluidization and delipidization of lipids, and proteins within the keratinocytes can become denatured. This is mostly likely associated with formulations containing anionic surfactants than non-ionic surfactants. Similar effects can be observed with solvents. Solvents can partition into the intercellular lipids, thereby changing membrane lipophilicity and barrier properties in the following order: ether/acetone > DMSO > ethanol > water. Higher alcohols and oils do not damage the skin, but they can act as a depot for lipophilic drugs on the skin surface. The presence of water in several of these formulations can hydrate the skin. Skin occlusion with fabric or transdermal patches, creams, and ointments can increase epidermal hydration, which can increase permeability.

The reader should be aware of the animal model being used to estimate dermal absorption of toxicants in humans. For many toxicants, direct extrapolation from a rodent species to human is not feasible. This is because of differences in skin thickness, hair density, lipid composition, and blood flow. Human skin is the least permeable compared to skin from rats, mice, and rabbits. Pig skin is, however, more analogous to human skin anatomically and physiologically, and pig skin is usually predictive of dermal absorption of most drugs and pesticides in human skin. Human skin is the best model, followed by skin from pigs, primates, and hairless guinea pigs, and then rats, mice, and rabbits. In preliminary testing of a transdermal drug, if the drug does not cross rabbit or mice skin, it is very unlikely that it will cross human skin. There are several in vitro experimental techniques such as static diffusion (Franz) cells or flow-through diffusion (Bronough) cells. There are several ex vivo methods including the isolated perfused porcine skin flap (IPPSF), which with its intact microvasculature makes this model unique. In vivo methods are the golden standard, but they are very expensive, and there are human ethical and animal rights issues to be considered.

There are other factors that can influence dermal absorption, and these can include environmental factors such as air flow, temperature, and humidity. Preexisting skin disease and inflammation should also be considered. The topical dose this is usually expressed in per unit surface area can vary, and relative absorption usually decreases with increase in dose.

### 6.5.4 Respiratory Penetration

As observed with the GIT and skin, the respiratory tract can be regarded as an external surface. However, the lungs, where gas/vapor absorption occurs, are preceded by protective structures (e.g., nose, mouth, pharynx, trachea, and bronchus), which can reduce the toxicity of airborne substances, especially particles. There is little or no absorption in these structures, and residual volume can occur in these sites. However, cells lining the respiratory tract may absorb agents that can cause a toxicological response. The absorption site, which is the alveoli-capillary membrane, is very thin (0.4–1.5 $\mu$m). The membranes to cross from the alveolar air space to the blood will include: *type I cells to basement membrane to capillary endothelial cells* (Figure 6.8). This short distance allows for rapid exchange of gases/vapors. The analogous absorption distance in skin is 100 to 200 $\mu$m, and in GIT it is about 30 $\mu$m. There is also a large surface area (50 times the area of skin) available for absorption as well as significant blood flow, which makes it possible to achieve rapid adjustments in plasma concentration.

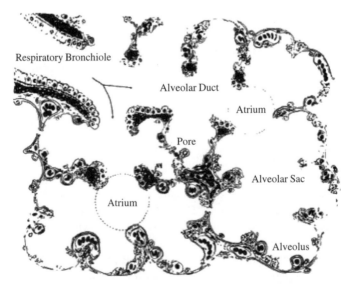

**Figure 6.8** Schematic representation of the respiratory unit of the lung. (From Bloom and Fawcett, in *A Textbook of Histology*, Philadelphia: Saunders, 1975.)

Gases/vapors must get into solution in the thin fluid film in the alveoli for systemic absorption to occur. For this reason doses are often a measurement of partial pressures, which is important for gases/vapors.

The process of respiration involves the movement and exchange of air through several interrelated passages, including the nose, mouth, pharynx, trachea, bronchi, and successively smaller airways terminating in the alveoli, where gaseous exchange occurs. These alveoli consist mainly of type I pneumocytes, which represent 40% of all cells but cover > 90% of surface area, and type II pneumocytes, which represent 60% of all cells but cover 5% of surface area. Macrophages make up 90% of cells in alveolar space. The amount of air retained in the lung despite maximum expiratory effort is known as the residual volume. Thus toxicants in the respiratory air may not be cleared immediately because of slow release from the residual volume. The rate of entry of vapor-phase toxicants is controlled by the alveolar ventilation rate, with the toxicant being presented to the alveoli in an interrupted fashion approximately 20 times/min.

Airborne toxicants can be simplified to two general types of compounds, namely gases and aerosols. Compounds such as gases, solvents, and vapors are subject to gas laws and are carried easily to alveolar air. Much of our understanding of xenobiotic behavior is with anesthetics. Compounds such as aerosols, particulates, and fumes are not subject to gas laws because they are in particulate form.

The transfer of gas from alveoli to blood is the actual absorption process. Among the most important factors that determine rate and extent of absorption of a gas in lungs is the solubility of that gas. Therefore it is not the membrane partition coefficient that necessarily affects absorption as has been described for skin and GIT membranes, but rather the blood: gas partition coefficient or blood/gas solubility of the gas. A high blood: gas partition coefficient indicates that the blood can hold a large amount of gas. Keep in mind that it is the *partial pressure* at equilibrium that is important, so the more soluble the gas is in blood, the greater the amount of gas that is needed to dissolve in the blood to raise the partial pressure or tension in blood. For example, anesthetics such as diethyl ether and methoxyflurane, which are soluble (Table 6.3), require a longer period for this partial pressure to be realized. Again, the aim is to generate the same tension in blood as in inspired air. Because these gases are very soluble, detoxification is a prolonged process. In practice, anesthetic induction is slower, and so is recovery from anesthesia. For less soluble gases (e.g., NO, isoflurane, halothane), the partial pressure or tension in blood can be raised a lot easier to that of inspired gases, and detoxification takes less time than those gases that are more soluble.

There are several other important factors that can determine whether the gas will be absorbed in blood and then transported from the blood to the perfused tissue. The concentration of the gas in inspired air influences gas tension, and partial pressure can be increased by overventilation. In gas anesthesiology we know that the effects of

**Table 6.3    Blood: Gas Partition Coefficient in Humans**

| Agent | Coefficient |
| --- | --- |
| Methoxyflurane | 13.0–15.0 |
| Halothane | 2.3–2.5 |
| Isoflurane | 1.4 |
| NO | 0.5 |

respiratory rate on speed of induction are transient for gases that have low solubility in blood and tissues, but there is a significant effect for agents that are more soluble and take a longer time for gas tensions to equilibrate. In determining how much of the gas is absorbed, its important to consider what fraction of the lung is ventilated and what fraction is perfused. However, one should be aware that due to diseased lungs, there can be differences between these fractions. For example, decreased perfusion will decrease absorption, although there is agent in the alveoli, and vice versa. The rate at which a gas passes into tissues is also dependent on gas solubility in the tissues, rate of delivery of the gas to tissues, and partial pressures of gas in arterial blood and tissues. After uptake of the gas, the blood takes the gas to other tissues. The mixed venous blood returned to the lungs progressively begins to have more of the gas, and differences between arterial (or alveolar) and mixed venous gas tensions decreases continuously.

While gases are more likely to travel freely through the entire respiratory tract to the alveoli, passage of aerosols and particles will be affected by the upper respiratory tract, which can act as an effective filter to prevent particulate matter from reaching the alveoli. Mucous traps particles to prevent entry to alveoli, and the mucociliary apparatus in the trachea traps and pushes particles up the trachea to the esophagus where they are swallowed and possibly absorbed in the GI tract.

In addition to upper pathway clearance, lung phagocytosis is very active in both upper and lower pathways of the respiratory tract and may be coupled to the mucus cilia. Phagocytes may also direct engulfed toxicants into the lymph, where the toxicants may be stored for long periods. If not phagocytized, particles $\leq 1$ $\mu$m may penetrate to the alveolar portion of the lung. Some particles do not desequamate but instead form a dust node in association with a developing network of reticular fibers. Overall, removal

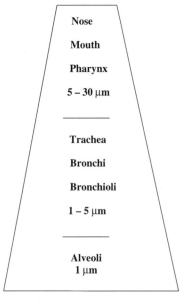

**Figure 6.9**  Schematic illustration of the regions where absorption may occur in the respiratory tract.

of alveolar particles is markedly slower than that achieved by the directed upper pulmonary mechanisms. This defense mechanism is not important for vapors/gases. The efficiency of the system is illustrated by the fact that on average, only 100 g of coal dust is found postmortem in the lungs of coal miners, although they inhale approximately 6000 g during their lifetime.

The deposition site of particles in the respiratory tract is primarily dependent on the *aerodynamic behavior* of the particles. The particle size, density, shape, hygroscopicity, breathing pattern, and lung airway structure are also important factors influencing the deposition site and efficiency. The *aerodynamic-equivalent diameter* (for particle > 0.5 μm) and *diffusion-equivalent diameter* (< 0.5 μm) are defined as the diameter of a *unit density sphere* having the same *settling velocity* (aerodynamic-equivalent) or the same *diffusion rate* (diffusion-equivalent) as the *irregularly shaped particle of interest*. Deposition occurs by five possible mechanisms: electrostatic precipitation, interception, impaction, sedimentation, impaction, and diffusion. The latter three are most important. Only particle sizes less than 10 to 20 μm that get pass the nasopharyngeal regions and reach the alveoli are of medical concern. As particle size decreases below 0.5 μm, the aerosol begins to behave like a gas (Figure 6.9). For these particles, diffusion becomes the primary mechanism of deposition in the respiratory tract before it finally reaches the alveoli.

## 6.6 TOXICANT DISTRIBUTION

### 6.6.1 Physicochemical Properties and Protein Binding

Absorption of toxicants into the blood needs to be high enough so that it will have a significant effect at the site of action in other areas of the body. The distribution process that takes the absorbed drug to other tissues is dependent on various physiological factors and physicochemical properties of the drug. This process is therefore a reversible movement of the toxicant between blood and tissues or between extracellular and intracellular compartments. There are, however, several complicating factors that can influence the distribution of a toxicant. For example, *perfusion* of tissues is an important physiological process, as some organs are better perfused (e.g., heart, brain) than others (e.g., fat). There can also be significant *protein binding* that affects delivery of drug to tissues. To further complicate the issue, elimination processes such as excretion and biotransformation (discussed at a later time) is occurring simultaneously to remove the toxicant from the blood as well as the target site.

There are several physiochemical properties of the toxicant that can influence its distribution. These include lipid solubility, pKa, and molecular weight, all of which were described earlier in this chapter (Section 6.4) and will not be described here. For many toxicants, distribution from the blood to tissues is by simple diffusion down a concentration gradient, and the absorption principles described earlier also apply here. The concentration gradient will be influenced by the partition coefficient or rather the ratio of toxicant concentrations in blood and tissue. Tissue mass and blood flow will also have a significant effect on distribution. For example, a large muscle mass can result in increased distribution to muscle, while limited blood flow to fat or bone tissue can limit distribution. The ratio of blood flow to tissue mass is also a useful indicator of how well the tissue is perfused. The well perfused tissues include liver, kidney, and brain, and the low perfused tissues include fat and bone where there is slow elimination from these tissues. Initial distribution to well-perfused tissues (e.g., heart, brain) occurs within the first few minutes, while delivery of drug to other tissues (e.g., fat, skin) is slower.

If the affinity for the target tissue is high, then the chemical will accumulate or form a depot. The advantage here is that if this is a drug, there is no need to load up the central compartment to get to the active site. However, if the reservoir for the drug has a large capacity and fills rapidly, it so alters the distribution of the drug that larger quantities of the drug are required initially to provide a therapeutic effective concentration at the target organ. If this is a toxicant, this may be an advantageous feature as toxicant levels at the target site will be reduced. In general, lipid-insoluble toxicants stay mainly in the plasma and interstitial fluids, while lipid-soluble toxicants reach all compartments, and may accumulate in fat. There are numerous examples of cellular reservoirs for toxicants and drugs to distribute. Tetracycline antibiotics have a high affinity for calcium-rich tissues in the body. The bone can become a reservoir for the slow release of chemicals such as lead, and effects may be chronic or there may be acute toxicity if the toxicant is suddenly released or mobilized from these depots. The antimalaria drug quinacrine accumulates due to reversible intracellular binding, and the concentration in the liver can be several thousand times that of plasma. Another antimalaria drug, chloroquine, has a high affinity for melanin, and this drug can be taken up by tissues such as the retina, which is rich in melanin granules, and can cause retinitis with a drug overdose. Lipophilic pesticides and toxicants (e.g., PCBs) and lipid soluble gases can be expected to accumulate in high concentration in fat tissue.

There are unique anatomical barriers that can limit distribution of toxicants. A classical example of such a unique barrier is the blood-brain barrier (BBB), which can limit the distribution of toxicants into the CNS and cerebrospinal fluid. There are three main processes or structures that keep drug or toxicant concentrations low in this region: (1) The BBB, which consist of capillary endothelial tight junctions and glial cells, surrounds the precapillaries, reduces filtration, and requires that the toxicant cross several membranes in order to get to the CSF. (Note that endothelial cells in other organs can have intercellular pores and pinocytotic vesicles.) (2) Active transport systems in the choroid plexus allow for transport of organic acids and bases from the CSF into blood. (3) The continuous process of CSF production in the ventricles and venous drainage continuously dilutes toxicant or drug concentrations. Disease processes such as meningitis can disrupt this barrier and can allow for penetration of antibiotics (e.g., aminoglycosides) that would not otherwise readily cross this barrier in a healthy individual. Other tissue/blood barriers include prostate/blood, testicles/blood, and globe of eye/blood, but inflammation or infection can increase permeability of these barriers. Toxicants can cross the placenta primarily by simple diffusion, and this is most easily accomplished if the toxicants are lipid-soluble (i.e., nonionized weak acids or bases). The view that the placenta is a barrier to drugs and toxicants is inaccurate. The fetus is, at least to some extent, exposed to essentially all drugs even if those with low lipid solubility are taken by the mother.

As was indicated earlier, the circulatory system and components in the blood stream are primarily responsible for the transport of toxicants to target tissues or reservoirs. Erythrocytes and lymph can play important roles in the transport of toxicants, but compared to plasma proteins, their role in toxicant distribution is relatively minor for most toxicants. Plasma protein binding can affect distribution because only the unbound toxicant is free or available to diffuse across the cell membranes. The toxicant-protein binding reaction is reversible and obeys the laws of mass action:

$$\text{Toxicant} + \text{Protein} \underset{k_2}{\overset{k_1}{\longleftrightarrow}} \text{Toxicant-Protein}$$
$$\text{(free)} \qquad\qquad\qquad\qquad \text{(bound)}$$

Usually the ratio of unbound plasma concentration $(C_u)$ of the toxicant to total toxicant concentration in plasma $(C)$ is the fraction of drug unbound, $f_u$, that is,

$$f_u = \frac{C_u}{C}.$$

The constants $k_1$ and $k_2$ are the specific rate constants for association and dissociation, respectively. The association constant $K_a$ will be the ratio $k_1/k_2$, and conversely, the dissociation constant, $K_d$ will be $k_2/k_1$. The constants and parameters are often used to describe and, more important, to compare the relative affinity of xenobiotics for plasma proteins.

The are many circulating proteins, but those involved in binding xenobiotics include albumin, $\alpha_1$-acid glycoprotein, lipoproteins, and globulins. Because many toxicants are lipophilic, they are likely to bind to plasma $\alpha$- and $\beta$-lipoproteins. There are mainly three classes of lipoproteins, namely high-density lipoprotein (HDL), low-density lipoprotein (LDL), and very low density lipoprotein (VLDL). Iron and copper are known to interact strongly with the metal-binding globulins transferin and ceruloplasmin, respectively. Acidic drugs bind primarily to albumin, and basic drugs are bound primarily to $\alpha_1$-acid glycoprotein and $\beta$-globulin. Albumin makes up 50% of total plasma proteins, and it reacts with a wide variety of drugs and toxicants. The $\alpha_1$-acid glycoprotein does not have as many binding sites as albumin, but it has one high-affinity binding site. The amount of toxicant drug that is bound depends on free drug concentration, and its affinity for the binding sites, and protein concentration. Plasma protein binding is nonselective, and therefore toxicants and drugs with similar physicochemical characteristics can compete with each other and endogenous substances for binding sites. Binding to these proteins does not necessarily prevent the toxicant from reaching the site of action, but it slows the rate at which the toxicant reaches a concentration sufficient to produce a toxicological effect. Again, this is related to what fraction of the toxicant is free or unbound $(f_u)$.

Toxicants complex with proteins by various mechanisms. Covalent binding may have a pronounced effect on an organism due to the modification of an essential molecule, but such binding is usually a very minor portion of the total dose. Because covalently bound molecules dissociate very slowly, if at all, they are not considered further in this discussion. However, we should recognize that these interactions are often associated with carcinogenic metabolites. Noncovalent binding is of primary importance to distribution because the toxicant or ligand can dissociate more readily than it can in covalent binding. In rare cases the noncovalent bond may be so stable that the toxicant remains bound for weeks or months, and for all practical purposes, the bond is equivalent to a covalent one. Types of interactions that lead to noncovalent binding under the proper physiological conditions include ionic binding, hydrogen bonding, van der Waals forces, and hydrophobic interactions. There are, however, some transition metals that have high association constants and dissociation is slow.

We know more about ligand-protein interactions today because of the numerous protein binding studies performed with drugs. The major difference between drugs and most toxicants is the frequent ionizability and high water solubility of drugs as compared with the non-ionizability and high lipid solubility of many toxicants. Thus experience with drugs forms an important background, but one that may not always be relevant to other potentially toxic compounds.

Variation in chemical and physical features can affect binding to plasma constituents. Table 6.4 shows the results of binding studies with a group of insecticides with greatly differing water and lipid solubilities. The affinity for albumin and lipoproteins is inversely related to water solubility, although the relation may be imperfect. Chlorinated hydrocarbons bind strongly to albumin but even more strongly to lipoproteins. Strongly lipophilic organophosphates bind to both protein groups, whereas more water-soluble compounds bind primarily to albumin. The most water-soluble compounds appear to be transported primarily in the aqueous phase. Chlordecone (Kepone) has partitioning characteristics that cause it to bind in the liver, whereas DDE, the metabolite of DDT, partitions into fatty depots. Thus the toxicological implications for these two compounds may be quite different.

Although highly specific (high-affinity, low-capacity) binding is more common with drugs, examples of specific binding for toxicants seem less common. It seems probable that low-affinity, high-capacity binding describes most cases of toxicant binding. The number of binding sites can only be estimated, often with considerable error, because of the nonspecific nature of the interaction. The number of ligand or toxicant molecules bound per protein molecule, and the maximum number of binding sites, $n$, define the definitive capacity of the protein. Another consideration is the binding affinity $K_{binding}$ (or $1/K_{diss}$). If the protein has only one binding site for the toxicant, a single value, $K_{binding}$, describes the strength of the interaction. Usually more than one binding site is present, each site having its intrinsic binding constant, $k_1, k_2, \ldots, k_n$. Rarely does one find a case where $k_1 = k_2 = \ldots = k_n$, where a single value would describe the affinity

**Table 6.4   Relative Distribution of Insecticides into Albumin and Lipoproteins**

| Insecticide | Percent Bound | Percent Distribution of Bound Insecticide | | |
|---|---|---|---|---|
| | | Albumin | LOL | HDL |
| DDT | 99.9 | 35 | 35 | 30 |
| Deildrin | 99.9 | 12 | 50 | 38 |
| Lindane | 98.0 | 37 | 38 | 25 |
| Parathion | 98.7 | 67 | 21 | 12 |
| Diazinon | 96.6 | 55 | 31 | 14 |
| Carbaryl | 97.4 | 99 | <1 | <1 |
| Carbofuran | 73.6 | 97 | 1 | 2 |
| Aldicarb | 30.0 | 94 | 2 | 4 |
| Nicotine | 25.0 | 94 | 2 | 4 |

*Source*: Adapted from B. P. Maliwal and F. E. Guthrie, *Chem Biol Interact* **35**:177–188, 1981.
*Note*: LOL, low-density lipoprotein; HOL, high-density lipoprotein.

constant at all sites. This is especially true when hydrophobic binding and van der Waals forces contribute to nonspecific, low-affinity binding. Obviously the chemical nature of the binding site is of critical importance in determining binding. The three-dimensional molecular structure of the binding site, the environment of the protein, the general location in the overall protein molecule, and allosteric effects are all factors that influence binding. Studies with toxicants, and even more extensive studies with drugs, have provided an adequate elucidation of these factors. Binding appears to be too complex a phenomenon to be accurately described by any one set of equations.

There are many methods for analyzing binding, but equilibrium dialysis is the most extensively used. Again, the focus of these studies is to determine the percentage of toxicant bound, the number of binding sites ($n$), and the affinity constant ($K_a$). The examples presented here are greatly simplified to avoid the undue confusion engendered by a very complex subject.

Toxicant-protein complexes that utilize relatively weak bonds (energies of the order of hydrogen bonds or less) readily associate and dissociate at physiological temperatures, and the law of mass action applies to the thermodynamic equilibrium:

$$K_{\text{binding}} = \frac{[TP]}{[T][P]} = \frac{1}{K_{\text{diss}}},$$

where $K_{\text{binding}}$ is the equilibrium constant for association, $[TP]$ is the molar concentration of toxicant-protein complex, $[T]$ is the molar concentration of free toxicant, and $[P]$ is the molar concentration of free protein. This equation does not describe the binding site(s) or the binding affinity. To incorporate these parameters and estimate the extent of binding, double-reciprocal plots of $1/[TP]$ versus $1/[T]$ may be used to test the specificity of binding. The $1/[TP]$ term can also be interpreted as moles of albumin per moles of toxicant. The slope of the straight line equals $1/nK_a$ and the intercept of this line with the $x$-axis equals $-K_a$. Regression lines passing through the origin imply infinite binding, and the validity of calculating an affinity constant under these circumstances is questionable. Figure 6.10 illustrates one such case with four pesticides, and the insert illustrates the low-affinity, "unsaturable" nature of binding in this example.

The two classes of toxicant-protein interactions encountered may be defined as (1) specific, high affinity, low capacity, and (2) nonspecific, low affinity, high capacity. The term high affinity implies an affinity constant ($K_{\text{binding}}$) of the order of $10^8$ $\text{M}^{-1}$, whereas low affinity implies concentrations of $10^4$ $\text{M}^{-1}$. Nonspecific, low-affinity binding is probably most characteristic of nonpolar compounds, although most cases are not as extreme as that shown in Figure 6.10.

An alternative and well-accepted treatment for binding studies is the Scatchard equation especially in situations of high-affinity binding:

$$v = \frac{nk[T]}{1 + k[T]},$$

which is simplified for graphic estimates to

$$\frac{v}{[T]} = k(n - v),$$

where $v$ is the moles of ligand (toxicant) bound per mole of protein, $[T]$ is the concentration of free toxicant, $k$ is the intrinsic affinity constant, and $n$ is the number of sites exhibiting such affinity. When $v[T]$ is plotted against $v$, a straight line is obtained

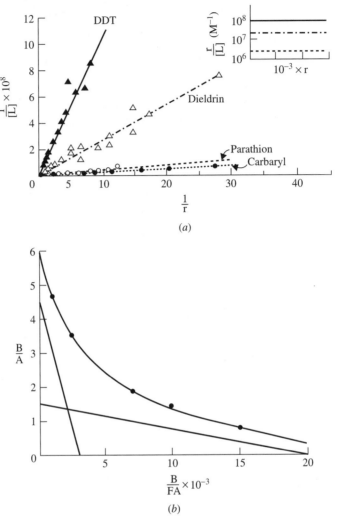

**Figure 6.10**  Binding of toxicants to blood proteins: (*a*) Double-reciprocal plot of binding of rat serum lipoprotein fraction with four insecticides. Insert illustrates magnitude of differences in slope with Scatchard plot. (*b*) Scatchard plot of binding of salicylate to human serum proteins. (Sources: (*a*) Skalsky and Guthrie, *Pest. Biochem. Physiol.* **7**: 289, 1977; (*b*) Moran and Walker, *Biochem. Pharmacol.* **17**: 153, 1968.)

if only one class of binding sites is evident. The slope is $-k$, and the intercept on the $v$-axis becomes $n$ (number of binding sites). If more than one class of sites occurs (probably the most common situation for toxicants), a curve is obtained from which the constants may be obtained. This is illustrated in Figure 6.10*b*, for which the data show not one but two species of binding sites: one with low capacity but high affinity, and another with about three times the capacity but less affinity. Commonly used computer programs usually solve such data by determining one line for the specific binding and one line for nonspecific binding, the latter being an average of many possible solutions.

When hydrophobic binding of lipid toxicants occurs, as is the case for many environmental contaminants, binding is probably not limited to a single type of plasma protein. For example, the binding of the chlorinated hydrocarbon DDT is strongest for lipoproteins and albumin, but other proteins account for a significant part of overall transport. Similar results have been observed for several compounds with a range of physiochemical properties.

The presence of another toxicant and/or drug that can bind at the same site can also increase the amount of free or unbound drug. This is an example of drug interaction that can have serious toxicological or pharmacological consequences. In general, when bound concentrations are less than 90% of the total plasma concentrations, plasma protein binding has little clinical importance. Plasma protein binding *becomes important when it is more than 90%*. For example, if a toxicant is 99% bound to plasma proteins, then 1% is free, but if there is toxicant interaction (e.g., competitive binding) that results in 94% bound, 6% is now free. Note that because of this interaction, the amount of available toxicant to cause a toxicological response *has increased sixfold*. Such a scenario may result in severe acute toxicity. Extensive plasma protein binding can influence renal clearance if glomerular filtration is the major elimination process in the kidney, but not if it is by active secretion in the kidney. Binding can also affect drug clearance if the extraction ratio (*ER*) in the liver is low, but not if the *ER* is high for that toxicant. Plasma protein binding can vary between and within chemical classes, and it is also species specific. For example, humans tend to bind acidic drugs more extensively than do other species.

There are several other variables that can alter plasma protein concentrations. These include malnutrition, pregnancy, cancer, liver abscess, renal disease, and age can reduce serum albumin. Furthermore $\alpha_1$-glycoprotein concentrations can increase with age, inflammation, infections, obesity, renal failure, and stress. Small changes in body temperature or changes in acid-base balance may alter chemical protein-binding characteristics. Although termination of drug or toxicant effect is usually by biotransformation and excretion, it may also be associated with redistribution from its site of action into other tissues. The classical example of this is when highly lipid-soluble drugs or toxicants that act on the brain or cardiovascular system are administered by IV or by inhalation.

### 6.6.2  Volume of Distribution ($V_d$)

Usually after a toxicant or drug is absorbed it can be distributed into various physiologic fluid compartments. The total body water represents 57% of total body mass (0.57 L/kg) (Table 6.5). The plasma, interstitial fluid, extracellular fluid, and intracellular fluid represent about 5, 17, 22, and 35% body weight, respectively. The extracellular fluid comprises the blood plasma, interstitial fluid, and lymph. Intracellular fluid includes

**Table 6.5   Volume of Distribution into Physiological Fluid Compartments**

| Compartment | Volume of Distribution in L/kg Body Weight (Ls/70 kg Body weight) |
|---|---|
| Plasma | 0.05(3.5 L) |
| Interstitial fluid | 0.18(12.6 L) |
| Extracellular fluid | 0.23(16.1 L) |
| Intracellular fluid | 0.35(24.5 L) |
| Total body water | 0.55(39 L) |

the sum of fluid contents of all cells in the body. There is also transcellular fluid that represents 2% body weight, and this includes cerebrospinal, intraocular, peritoneal, pleural, and synovial fluids, and digestive secretions. Fat is about 20% body weight, while the GIT contents in monogastrics make up 1% body weight, and in ruminants it can constitute 15% body weight.

Its sometimes useful to quantitate how well a drug or toxicant is distributed into these various fluid compartments, and in this context the apparent volume of distribution can be a useful parameter. The apparent volume of distribution, $V_d$, is defined as the volume of fluid required to contain the total amount, $A$, of drug in the body at the same concentration as that present in plasma, $C_p$,

$$V_d = \frac{A}{C_p}.$$

In general, the $V_d$ for a drug is to some extent descriptive of its distribution pattern in the body. For example, drugs or toxicants with relatively small $V_d$ values may be confined to the plasma as diffusion across the capillary wall is limited. There are other toxicants that have a slightly larger $V_d$ (e.g., 0.23 L/kg), and these toxicants may be distributed in the extracellular compartment. This includes many polar compounds (e.g., tubocurarine, gentamicin, $V_d = 0.2$–$0.4$ L/kg). These toxicants cannot readily enter cells because of their low lipid solubility. If the $V_d$ for some of these toxicants is in excess of the theoretical value, this may be due to limited degree of penetration into cells or from the extravascular compartment. Finally there are many toxicants distributed throughout the body water ($V_d \geq 0.55$ L/kg) that may have $V_d$ values much greater than that for total body water. This distribution is achieved by relatively lipid-soluble toxicants and drugs that readily cross cell membranes (e.g., ethanol, diazepam; $V_d = 1$ to $2$ L/kg). Binding of the toxicant anywhere outside of the plasma compartment, as well as partitioning into body fat, can increase $V_d$ beyond the absolute value for total body water. In general, toxicants with a large $V_d$ can even reach the brain, fetus, and other transcellular compartments. In general, toxicants with large $V_d$ are a consequence of extensive tissue binding. The reader should be aware that we are talking about tissue binding, and not plasma protein binding where distribution is limited to plasma for obvious reasons.

The fraction of toxicant located in plasma is dependent on whether a toxicant binds to both plasma and tissue components. Plasma binding can be measured directly, but not tissue binding. It can, however, be inferred from the following relationship:

$$\text{Amount in body} \quad = \quad \text{Amount in plasma} \quad + \quad \text{Amount outside plasma}$$
$$V_d \times C \quad = \quad V_p \times C \quad + \quad V_{TW} \times C_{TW}$$

where $V_d$ is the apparent volume of distribution, $V_p$ the volume of plasma, $V_{TW}$ the apparent volume of tissue, and $C_{TW}$ the tissue concentration. If the preceding equation is divided by $C$, it now becomes

$$V_d = V_p + V_{TW} \times \frac{C_{TW}}{C}$$

Recall that $f_u = C_u/C$ occurs with plasma, and also that the fraction unbound in tissues is $f_{uT} = C_{uT}/C_{TW}$.

Assuming at equilibrium that unbound concentration in tissue and plasma are equal, then we let the ratio of $f_u/f_{uT}$ replace $C_{TW}/C$ and determine the volume of distribution as follows:

$$V_d = V_p + V_{TW} \times \left( \frac{f_u}{f_{uT}} \right).$$

It is possible to predict what happens to $V_d$ when $f_u$ or $f_{uT}$ changes as a result of physiological or disease processes in the body that change plasma and/or tissue protein concentrations. For example, $V_d$ can increase with increased unbound toxicant in plasma or with a decrease in unbound toxicant tissue concentrations. The preceding equation explains why: because of both plasma and tissue binding, some $V_d$ values rarely correspond to a real volume such as plasma volume, extracellular space, or total body water. Finally interspecies differences in $V_d$ values can be due to differences in body composition of body fat and protein, organ size, and blood flow as alluded to earlier in this section. The reader should also be aware that in addition to $V_d$, there are volumes of distribution that can be obtained from pharmacokinetic analysis of a given data set. These include the volume of distribution at steady state ($V_{d,ss}$), volume of the central compartment ($V_c$), and the volume of distribution that is operative over the elimination phase ($V_{d,area}$). The reader is advised to consult other relevant texts for a more detailed description of these parameters and when it is appropriate to use these parameters.

## 6.7  TOXICOKINETICS

The explanation of the pharmacokinetics or toxicokinetics involved in absorption, distribution, and elimination processes is a highly specialized branch of toxicology, and is beyond the scope of this chapter. However, here we introduce a few basic concepts that are related to the several transport rate processes that we described earlier in this chapter. Toxicokinetics is an extension of pharmacokinetics in that these studies are conducted at higher doses than pharmacokinetic studies and the principles of pharmacokinetics are applied to xenobiotics. In addition these studies are essential to provide information on the fate of the xenobiotic following exposure by a define route. This information is essential if one is to adequately interpret the dose-response relationship in the risk assessment process. In recent years these toxicokinetic data from laboratory animals have started to be utilized in physiologically based pharmacokinetic (PBPK) models to help extrapolations to low-dose exposures in humans. The ultimate aim in all of these analyses is to provide an estimate of tissue concentrations at the target site associated with the toxicity.

Immediately on entering the body, a chemical begins changing location, concentration, or chemical identity. It may be transported independently by several components of the circulatory system, absorbed by various tissues, or stored; the chemical may effect an action, be detoxified, or be activated; the parent compound or its metabolite(s) may react with body constituents, be stored, or be eliminated—to name some of the more important actions. Each of these processes may be described by rate constants similar to those described earlier in our discussion of first-order rate processes that are associated with toxicant absorption, distribution, and elimination and occur

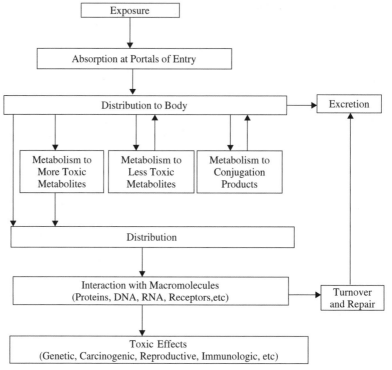

**Figure 6.11** Sequence of events following exposure of an animal to exogenous chemicals.

simultaneously. Thus at no time is the situation stable but is constantly changing as indicated in Figure 6.11.

It should be noted, however, that as the toxicant is being absorbed and distributed throughout the body, it is being simultaneously eliminated by various metabolism and/or excretion mechanisms, as will be discussed in more detail in the following chapters. However, one should mention here that an important pharmacokinetic parameter known as *clearance* ($C\ell$) can be used to quantitatively assess elimination of a toxicant. Clearance is defined as the rate of toxicant excreted relative to its plasma concentration, $C_p$:

$$C\ell = \frac{\text{Rate of toxicant excretion}}{C_p}.$$

The rate of excretion is really the administered dose times the fractional elimination rate constant $K_{el}$ described earlier. Therefore we can express the preceding equation in terms of $K_{el}$ and administered dose as volume of distribution, $V_d$:

$$C\ell = K_{el} \cdot \frac{\text{Dose}}{C_p} = K_{el} \cdot (V_d \cdot C_p)/C_p = K_{el} \cdot V_d.$$

In physiological terms we can also define clearance as the volume of blood cleared of the toxicant by an organ or body per unit time. Therefore, as the equations above indicate, the body clearance of a toxicant is expressed in units of volume per unit time (e.g., L/h), and can be derived if we know the volume of distribution of the toxicant and fractional rate constant. In many instances this can only be derived by appropriate pharmacokinetic analysis of a given data set following blood or urine sample collection and appropriate chemical analyses to determine toxicant concentrations in either of these biological matrices.

Each of the processes discussed thus far—absorption, distribution, and elimination—can be described as a rate process. In general, the process is assumed to be first order in that the rate of transfer at any time is proportional to the amount of drug in the body at that time. Recall that the rate of transport ($dC/dt$) is proportional to toxicant concentration ($C$) or stated mathematically:

$$\frac{dC}{dt} = KC,$$

where $K$ is the rate constant (fraction per unit time). Many pharmacokinetic analyses of a chemical are based primarily on toxicant concentrations in blood or urine samples. It is often assumed in these analyses that the rate of change of toxicant concentration in blood reflects quantitatively the change in toxicant concentration throughout the body (first-order principles). Because of the elimination/clearance process, which also assumed to be a first-order rate process, the preceding rate equation now needs a negative sign. This is really a decaying process that is observed as a decline of toxicant concentration in blood or urine after intravenous (IV) administration. The IV route is preferred in these initial analyses because there is no absorption phase, but only chemical depletion phase. However, one cannot measure infinitesimal change of $C$ or time, $t$; therefore there needs to be integration after rearrangement of the equation above:

$$\frac{-dC}{C} = kdt \quad \text{becomes} \quad \int \frac{-dC}{C} = k \int dt,$$

which can be expressed as

$$C = C^0 e^{-kt},$$

where $e$ is the base of the natural logarithm. We can remove $e$ by taking the ln of both sides:

$$\ln C^t = \ln C^0 - kt.$$

Note that $K$ is the slope of the straight line for a semilog plot of toxicant concentration versus time (Figure 6.12). In the preceding equation it is the elimination rate constant that is related to the half-life of the toxicant described earlier in this chapter. The derived $C^0$ can be used to calculate the volume of distribution ($V_d$) of the toxicant as follows:

$$V_d = \frac{\text{Dose}}{C^0}.$$

However, toxicokinetic data for many toxicants do not always provide a straight line when plotted as described above. More complicated equations with more than one exponential term with rate constants may be necessary to mathematically describe the concentration-time profile. These numerous rate constants are indicative of chemical transport between various compartments in the body and not only to a single central compartment as suggested in the simple equation and semilog plot described in

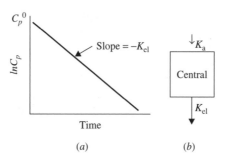

(a)                    (b)

**Figure 6.12** (a) Semilog plot of plasma concentration ($C_p$) versus time. $C_p^0$ is the intercept on the y-axis, and $K_{el}$ is the elimination rate constant. (b) Single compartment model with rate constants for absorption, $K_a$ and for elimination, $K_{el}$.

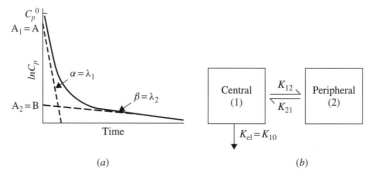

(a)                    (b)

**Figure 6.13** (a) Semilog plot of plasma concentration for ($C_p$) versus time representative of a two-compartment model. The curve can be broken down into an $\alpha$ or $\lambda_1$ distribution phase and $\beta$ or $\lambda_2$ elimination phase. (b) Two-compartment model with transfer rate constants, $K_{12}$ and $K_{21}$, and elimination rate constant, $K_{el}$.

Figure 6.12. In some instances the data may fit to a bi-exponential concentration-time profile (Figure 6.13). The equation to describe this model is

$$C = Ae^{-\alpha t} + Be^{-\beta t}.$$

In other instances, complex profiles may require a three- or multi-exponential concentration-time profile (Figure 6.14). The equation to describe the three-profile case is

$$C = Ae^{-\alpha t} + Be^{-\beta t} + Ce^{-\gamma t}.$$

In the physiological sense, one can divide the body into "compartments" that represent discrete parts of the whole-blood, liver, urine, and so on, or use a mathematical model describing the process as a composite that pools together parts of tissues involved in distribution and bioactivation. Usually pharmacokinetic compartments have no anatomical or physiological identity; they represent all locations within the body that have similar characteristics relative to the transport rates of the particular toxicant. Simple first-order kinetics is usually accepted to describe individual

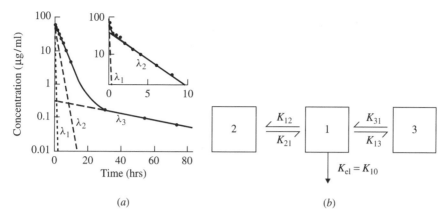

**Figure 6.14**   (*a*) Semilog plot of plasma concentration for ($C_p$) versus time representative of a three- or multi-compartment model. The curve can be broken down into three phases, $\lambda_1$, $\lambda_2$, and $\lambda_3$. (*b*) Three-compartment model with transfer rate constants, $K_{12}$, $K_{21}$, $K_{13}$, $K_{31}$, and elimination rate constant, $K_{el}$. As these models can get more complicated, the $\alpha$, $\beta$, and $\gamma$ nomenclature may get replaced with $\lambda_n$ as indicated in the profile.

rate processes for the toxicant after entry. The resolution of the model necessitates mathematical estimates (as a function of time) concerning the absorption, distribution, biotransformation, and excretion of the toxicant.

Drugs and toxicants with multi-exponential behavior depicted in Figure 6.14 require calculation of the various micro constants. An alternative method involves using model-independent pharmacokinetics to arrive at relevant parameters. Very briefly, it involves determination of the area under the curve (AUC) of the concentration-time profiles. The emergence of microcomputers in recent years has greatly facilitated this approach.

In conclusion, pharmacokinetics is a study of the time course of absorption, distribution, and elimination of a chemical. We use pharmacokinetics as a tool to analyze plasma concentration time profiles after chemical exposure, and it is the derived rates and other parameters that reflect the underlying physiological processes that determine the fate of the chemical. There are numerous software packages available today to accomplish these analyses. The user should, however, be aware of the experimental conditions, the time frame over which the data were collected, and many of the assumptions embedded in the analyses. For example, many of the transport processes described in this chapter may not obey first-order kinetics, and thus may be nonlinear especially at toxicological doses. The reader is advised to consult other texts for more detailed descriptions of these nonlinear interactions and data analyses.

## SUGGESTED READING

R. Bronaugh and H. Maibach, eds. *Percutaneous Absorption*. New York: Dekker, 1989.

A. Goodman Gilman, T. W. Rall, A. S. Nies, and P. Taylor, eds. *Goodman and Gilman's The Pharmacological Basis of Therapeutics*, 8th edn. Elmsford, NY: Pergamon Press, 1990.

P. Grandjean, ed. *Skin Penetration: Hazardous Chemicals at Work*. London: Taylor and Francis, 1990.

R. Krieger, ed. *Handbook of Pesticide Toxicology*, 2nd edn. San Diego: Academic Press, 2001.

M. Rowland and T. N. Tozer, eds. *Clinical Pharmacokinetics. Concepts and Applications*, 3rd edn. Philadelphia: Lea and Febiger, 1995.

L. Shargel and A. B. C. Yu, eds. *Applied Biopharmaceutics and Pharmacokinetics*, 4th edn. Norwalk, CT: Appleton and Lange, 1999.

# CHAPTER 7

# Metabolism of Toxicants

RANDY L. ROSE and ERNEST HODGSON

## 7.1  INTRODUCTION

One of the most important determinants of xenobiotic persistence in the body and subsequent toxicity to the organism is the extent to which they can be metabolized and excreted. Several families of metabolic enzymes, often with wide arrays of substrate specificity, are involved in xenobiotic metabolism. Some of the more important families of enzymes involved in xenobiotic metabolism include the cytochrome P450 monooxygenases (CYPs), flavin-containing monooxygenases (FMOs), alcohol and aldehyde dehydrogenases, amine oxidases, cyclooxygenases, reductases, hydrolases, and a variety of conjugating enzymes such as glucuronidases, sulfotransferases, methyltransferases, glutathione transferases, and acetyl transferases.

Most xenobiotic metabolism occurs in the liver, an organ devoted to the synthesis of many important biologically functional proteins and thus with the capacity to mediate chemical transformations of xenobiotics. Most xenobiotics that enter the body are lipophilic, a property that enables them to bind to lipid membranes and be transported by lipoproteins in the blood. After entrance into the liver, as well as in other organs, xenobiotics may undergo one or two phases of metabolism. In phase I a polar reactive group is introduced into the molecule rendering it a suitable substrate for phase II enzymes. Enzymes typically involved in phase I metabolism include the CYPs, FMOs, and hydrolases, as will be discussed later. Following the addition of a polar group, conjugating enzymes typically add much more bulky substituents, such as sugars, sulfates, or amino acids that result in a substantially increased water solubility of the xenobiotic, making it easily excreted. Although this process is generally a detoxication sequence, reactive intermediates may be formed that are much more toxic than the parent compound. It is, however, usually a sequence that increases water solubility and hence decreases the biological half life ($t_{0.5}$) of the xenobiotic in vivo.

Phase I monooxygenations are more likely to form reactive intermediates than phase II metabolism because the products are usually potent electrophiles capable of reacting with nucleophilic substituents on macromolecules, unless detoxified by some subsequent reaction. In the following discussion, examples of both detoxication and intoxication reactions are given, although greater emphasis on activation products is provided in Chapter 8.

*A Textbook of Modern Toxicology, Third Edition,* edited by Ernest Hodgson
ISBN 0-471-26508-X  Copyright © 2004 John Wiley & Sons, Inc.

## 7.2   PHASE I REACTIONS

Phase I reactions include microsomal monooxygenations, cytosolic and mitochondrial oxidations, co-oxidations in the prostaglandin synthetase reaction, reductions, hydrolyses, and epoxide hydration. All of these reactions, with the exception of reductions, introduce polar groups to the molecule that, in most cases, can be conjugated during phase II metabolism. The major phase I reactions are summarized in Table 7.1.

### 7.2.1   The Endoplasmic Reticulum, Microsomal Preparation, and Monooxygenations

Monooxygenation of xenobiotics are catalyzed either by the cytochrome P450 (CYP)-dependent monooxygenase system or by flavin-containing monooxygenases (FMO).

**Table 7.1   Summary of Some Important Oxidative and Reductive Reactions of Xenobiotics**

| Enzymes and Reactions | Examples |
|---|---|
| Cytochrome P450 | |
| Epoxidation/hydroxylation | Aldrin, benzo(a)pyrene, aflatoxin, bromobenzene |
| $N$-, $O$-, $S$-Dealkylation | Ethylmorphine, atrazine, $p$-nitroanisole, methylmercaptan |
| $N$-, $S$-, $P$-Oxidation | Thiobenzamide, chlorpromazine, 2-acetylaminofluorene |
| Desulfuration | Parathion, carbon disulfide |
| Dehalogenation | Carbon tetrachloride, chloroform |
| Nitro reduction | Nitrobenzene |
| Azo reduction | $O$-Aminoazotoluene |
| Flavin-containing monooxygenase | |
| $N$-, $S$-, $P$-Oxidation | Nicotine, imiprimine, thiourea, methimazole |
| Desulfuration | Fonofos |
| Prostaglandin synthetase cooxidation | |
| Dehydrogenation | Acetaminophen, benzidine, epinephrine |
| $N$-Dealkylation | Benzphetamine, dimethylaniline |
| Epoxidation/hydroxylation | Benzo(a)pyrene, 2-aminofluorene, phenylbutazone |
| Oxidation | FANFT, ANFT, bilirubin |
| Molybdenum hydroxylases | |
| Oxidation | Purines, pteridine, methotrexate, 6-deoxycyclovir |
| Reductions | Aromatic nitrocompounds, azo dyes, nitrosoamines |
| Alcohol dehydrogenase | |
| Oxidation | Methanol, ethanol, glycols, glycol ethers |
| Reduction | Aldehydes and ketones |
| Aldehyde dehydrogenase | |
| Oxidation | Aldehydes resulting from alcohol and glycol oxidations |
| Esterases and amidases | |
| Hydrolysis | Parathion, paraoxon, dimethoate |
| Epoxide hydrolase | |
| Hydrolysis | Benzo($a$)pyrene epoxide, styrene oxide |

Both are located in the endoplasmic reticulum of the cell and have been studied in many tissues and organisms. This is particularly true of CYPs, probably the most studied of all enzymes.

Microsomes are derived from the endoplasmic reticulum as a result of tissue homogenization and are isolated by centrifugation of the postmitochondrial supernatant fraction, described below. The endoplasmic reticulum is an anastomosing network of lipoprotein membranes extending from the plasma membrane to the nucleus and mitochrondria, whereas the microsomal fraction derived from it consists of membranous vesicles contaminated with free ribosomes, glycogen granules, and fragments of other subcellular structures such as mitochondria and Golgi apparatus. The endoplasmic reticulum, and consequently the microsomes derived from it, consists of two types, rough and smooth, the former having an outer membrane studded with ribosomes, which the latter characteristically lack. Although both rough and smooth microsomes have all of the components of the CYP-dependent monooxygenase system, the specific activity of the smooth type is usually higher.

The preparation of microsomal fractions, S9, and cytosolic fractions from tissue homogenates involves the use of two to three centrifugation steps. Following tissue extraction, careful mincing, and rinses of tissue for blood removal, the tissues are typically homogenized in buffer and centrifuged at $10,000 \times g$ for 20 minutes. The resulting supernatant, often referred to as the S9 fraction, can be used in studies where both microsomal and cytosolic enzymes are desired. More often, however, the S9 fraction is centrifuged at $100,000 \times g$ for 60 minutes to yield a microsomal pellet and a cytosolic supernatant. The pellet is typically resuspended in a volume of buffer, which will give 20 to 50 mg protein/ml and stored at $-20$ to $-70°C$. Often, the microsomal pellet is resuspended a second time and resedimented at $100,000 \times g$ for 60 minutes to further remove contaminating hemoglobin and other proteins. As described above, enzymes within the microsomal fraction (or microsomes) include CYPs, FMOs, cyclooxygenases, and other membrane-bound enzymes, including necessary coenzymes such as NADPH cytochrome P450 reductase for CYP. Enzymes found in the cytosolic fraction (derived from the supernatant of the first $100,000 \times g$ spin) include hydrolases and most of the conjugating enzymes such as glutathione transferases, glucuronidases, sulfotransferases, methyl transferases, and acetylases. It is important to note that some cytosolic enzymes can also be found in microsomal fractions, although the opposite is not generally the case.

Monooxygenations, previously known as mixed-function oxidations, are those oxidations in which one atom of a molecule of oxygen is incorporated into the substrate while the other is reduced to water. Because the electrons involved in the reduction of CYPs or FMOs are derived from NADPH, the overall reaction can be written as follows (where RH is the substrate):

$$RH + O_2 + NADPH + H^+ \longrightarrow NADP^+ + ROH + H_2O.$$

## 7.2.2   The Cytochrome P450-Dependent Monooxygenase System

The CYPs, the carbon monoxide-binding pigments of microsomes, are heme proteins of the b cytochrome type. Originally described as a single protein, there are now known to be more than 2000 CYPs widely distributed throughout animals, plants, and microorganisms. A system of nomenclature utilizing the prefix CYP has been devised for the genes and cDNAs corresponding to the different forms (as discussed later in this section), although P450 is still appropriate as a prefix for the protein products. Unlike most cytochromes, the name CYP is derived not from the absorption maximum of the reduced form in the visible region but from the unique wavelength of the absorption maximum of the carbon monoxide derivative of the reduced form, namely 450 nm.

The role of CYP as the terminal oxidase in monooxygenase reactions is supported by considerable evidence. The initial proof was derived from the demonstration of the concomitant light reversibility of the CO complex of CYP and the inhibition, by CO, of the C-21 hydroxylation of 17 $\alpha$-hydroxy-progesterone by adrenal gland microsomes. This was followed by a number of indirect, but nevertheless convincing, proofs involving the effects on both CYP and monooxygenase activity of CO, inducing agents, and spectra resulting from ligand binding and the loss of activity on degradation of CYP to cytochrome P420. Direct proof was subsequently provided by the demonstration that monooxygenase systems, reconstituted from apparently homogenous purified CYP, NADPH-CYP reductase, and phosphatidylchloline, can catalyze many monooxygenase reactions.

CYPs, like other hemoproteins, have characteristic absorptions in the visible region. The addition of many organic, and some inorganic, ligands results in perturbations of this spectrum. Although the detection and measurement of these spectra requires a high-resolution spectrophotometer, these perturbations, measured as optical difference spectra, have been of tremendous use in the characterization of CYPs, particularly in the decades preceding the molecular cloning and expression of specific CYP isoforms.

The most important difference spectra of oxidized CYP are type I, with an absorption maximum at 385 to 390 nm. Type I ligands are found in many different chemical classes and include drugs, environmental contaminants, pesticides, and so on. They appear to be generally unsuitable, on chemical grounds, as ligands for the heme iron and are believed to bind to a hydrophobic site in the protein that is close enough to the heme to allow both spectral perturbation and interaction with the activated oxygen. Although most type I ligands are substrates, it has not been possible to demonstrate a quantitative relationship between $K_S$ (concentration required for half-maximal spectral development) and $K_M$ (Michaelis constant). Type II ligands, however, interact directly with the heme iron of CYP, and are associated with organic compounds having nitrogen atoms with $sp^2$ or $sp^3$ nonbonded electrons that are sterically accessible. Such ligands are frequently inhibitors of CYP activity.

The two most important difference spectra of reduced CYP are the well-known CO spectrum, with its maximum at or about 450 nm, and the type III spectrum, with two pH-dependent peaks at approximately 430 and 455 nm. The CO spectrum forms the basis for the quantitative estimation of CYP. The best-known type III ligands for CYP are ethyl isocyanide and compounds such as the methylenedioxyphenyl synergists and SKF 525A, the last two forming stable type III complexes that appear to be related to the mechanism by which they inhibit monooxygenations.

In the catalytic cycle of CYP, reducing equivalents are transferred from NADPH to CYP by a flavoprotein enzyme known as NADPH-cytochrome P450 reductase. The evidence that this enzyme is involved in CYP monooxygenations was originally derived from the observation that cytochrome c, which can function as an artificial electron acceptor for the enzyme, is an inhibitor of such oxidations. This reductase is an essential component in CYP-catalyzed enzyme systems reconstituted from purified components. Moreover antibodies prepared from purified reductase are inhibitors of microsomal monooxygenase reactions. The reductase is a flavoprotein of approximately 80,000 daltons that contain 2 mole each of flavin mononucleotide (FMN) and flavinadenine dinucleotide (FAD) per mole of enzyme. The only other component necessary for activity in the reconstituted system is a phospholipid, phosphatidylchloline. This is not involved directly in electron transfer but appears to be involved in the coupling of the reductase to the cytochrome and in the binding of the substrate to the cytochrome.

The mechanism of CYP function has not been established unequivocally; however, the generally recognized steps are shown in Figure 7.1. The initial step consists of the binding of substrate to oxidize CYP followed by a one electron reduction catalyzed by NADPH-cytochrome P450 reductase to form a reduced cytochrome-substrate complex. This complex can interact with CO to form the CO-complex, which gives rise to the well-known difference spectrum with a peak at 450 nm and also inhibits monooxyge-nase activity. The next several steps are less well understood. They involve an initial interaction with molecular oxygen to form a ternary oxygenated complex. This ternary complex accepts a second electron, resulting in the further formation of one or more less understood complexes. One of these, however, is probably the equivalent of the peroxide anion derivative of the substrate-bound hemoprotein. Under some conditions this complex may break down to yield hydrogen peroxide and the oxidized cytochrome substrate complex. Normally, however, one atom of molecular oxygen is transferred to the substrate and the other is reduced to water, followed by dismutation reactions lead-ing to the formation of the oxygenated product, water, and the oxidized cytochrome.

The possibility that the second electron is derived from NADH through cytochrome $b_5$ has been the subject of argument for some time and has yet to be completely resolved. Cytochrome $b_5$ is a widely distributed microsomal heme protein that is involved in metabolic reactions such as fatty acid desaturation that involve endogenous substrates. It is clear, however, that cytochrome $b_5$ is not essential for all

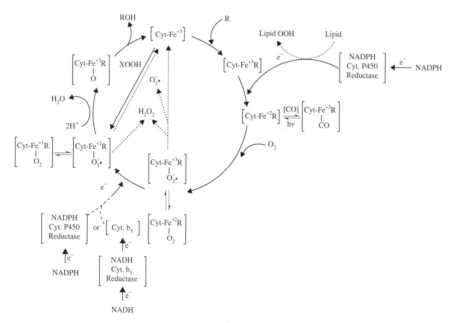

**Figure 7.1**  Generalized scheme showing the sequence of events for P450 monooxygenations.

CYP-dependent monooxygenations because many occur in systems reconstituted from NADPH, $O_2$, phosphatidylchloline, and highly purified CYP and NADPH-cytochrome P450 reductase. Nevertheless, there is good evidence that many catalytic activities by isoforms including CYP3A4, CYP3A5, and CYP2E1 are stimulated by cytochrome $b_5$. In some cases apocytochrome $b_5$ (devoid of heme) has also been found to be stimulatory, suggesting that an alternate role of cytochrome $b_5$ may be the result of conformational changes in the CYP/NADPH cytochrome P450 reductase systems. Thus cytochrome $b_5$ may facilitate oxidative activity in the intact endoplasmic reticulum. The isolation of forms of CYP that bind avidly to cytochrome $b_5$ also tends to support this idea.

***Distribution of Cytochrome P450.*** In vertebrates the liver is the richest source of CYP and is most active in the monooxygenation of xenobiotics. CYP and other components of the CYP-dependent monooxygenase system are also in the skin, nasal mucosa, lung, and gastrointestinal tract, presumably reflecting the evolution of defense mechanisms at portals of entry. In addition to these organs, CYP has been demonstrated in the kidney, adrenal cortex and medulla, placenta, testes, ovaries, fetal and embryonic liver, corpus luteum, aorta, blood platelets, and the nervous system. In humans, CYP has been demonstrated in the fetal and adult liver, the placenta, kidney, testes, fetal and adult adrenal gland, skin, blood platelets, and lymphocytes.

Although CYPs are found in many tissues, the function of the particular subset of isoforms in organ, tissue, or cell type does not appear to be the same in all cases. In the liver, CYPs oxidize a large number of xenobiotics as well as some endogenous steroids and bile pigments. The CYPs of the lung also appear to be concerned primarily with xenobiotic oxidations, although the range of substrates is more limited than that of the liver. The skin and small intestine also carry out xenobiotic oxidations, but their activities have been less well characterized. In normal pregnant females, the placental microsomes display little or no ability to oxidize foreign compounds, appearing to function as a steroid hormone metabolizing system. On induction of the CYP enzymes, such as occurs in pregnant women who smoke, CYP-catalyzed aryl hydrocarbon hydroxylase activity is readily apparent. The CYPs of the kidney are active in the $\omega$-oxidation of fatty acids, such as lauric acid, but are relatively inactive in xenobiotic oxidation. Mitochondrial CYPs, such as those of the placenta and adrenal cortex, are active in the oxidation of steroid hormones rather than xenobiotics.

Distribution of CYPs within the cell has been studied primarily in the mammalian liver, where it is present in greatest quantity in the smooth endoplasmic reticulum and in smaller but appreciable amounts in the rough endoplasmic reticulum. The nuclear membrane has also been reported to contain CYP and to have detectable aryl hydrocarbon hydroxylase activity, an observation that may be of considerable importance in studies of the metabolic activation of carcinogens.

***Multiplicity of Cytochrome P450, Purification, and Reconstitution of Cytochrome P450 Activity.*** Even before appreciable purification of CYP had been accomplished, it was apparent from indirect evidence that mammalian liver cells contained more than one CYP enzyme. Subsequent direct evidence on the multiplicity of CYPs included the separation and purification of CYP isozymes, distinguished from each other by chromatographic behavior, immunologic specificity, and/or substrate specificity after reconstitution and separation of distinct polypeptides by sodium dodecyl sulfate polyacrylamide gel electrophoresis (SDS-PAGE), which could then be related to distinct CYPs present in the original microsomes.

Purification of CYP and its usual constituent isoforms was, for many years, an elusive goal; one, however, that has been largely resolved. The problem of instability on solubilization was resolved by the use of glycerol and dithiothreitol as protectants, and the problem of reaggregation by maintaining a low concentration of a suitable detergent, such as Emulgen 911 (Kao-Atlas, Tokyo), throughout the procedure. Multiple CYP isoforms, as discussed previously, may be separated from each other and purified as separate entities, although individual isoforms are now routinely cloned and expressed as single entities. The lengthy processes of column purification of CYPs have now been largely superceded by the cloning and expression of transgenic isoforms in a variety of expression systems.

Systems reconstituted from purified CYP, NADPH-cytochrome P450 reductase and phosphatidylchloline will, in the presence of NADPH and $O_2$, oxidize xenobiotics such as benzphetamine, often at rates comparable to microsomes. Although systems reconstituted from this minimal number of components are enzymatically active, other microsomal components, such as cytochrome $b_5$, may facilitate activity either in vivo or in vitro or may even be essential for the oxidation of certain substrates.

One important finding from purification studies as well as cloning and expressing of individual isoforms is that the lack of substrate specificity of microsomes for monooxygenase activity is not an artifact caused by the presence of several specific cytochromes. Rather, it appears that many of the cytochromes isolated are still relatively nonspecific. The relative activity toward different substrates does nevertheless vary greatly from one CYP isoform to another even when both are relatively nonspecific. This lack of specificity is illustrated in Table 7.2, using human isoforms as examples.

***Classification and Evolution of Cytochrome P450.*** The techniques of molecular biology have been applied extensively to the study of CYP. More than 1925 genes have been characterized as of 2002, and the nucleotide and derived amino acid sequences compared. In some cases the location of the gene on a particular chromosome has been determined and the mechanism of gene expression investigated.

A system of nomenclature proposed in 1987 has since been updated several times, most recently in 1996. The accepted guidelines from nomenclature designate cytochrome P450 genes as CYP (or cyp in the case of mouse genes). The CYP designation is followed by an Arabic numeral to denote the gene family, followed by a letter designating the subfamily. The individual isoform is then identified using a second Arabic numeral following the subfamily designation. Polymorphic isoforms of genes are indicated by an asterisk followed by an arabic numeral. If there are no subfamilies or if there is only a single gene within the family or subfamily, the letter and/or the second numeral may be omitted (e.g., CYP17). The name of the gene is italicized, whereas the protein (enzyme) is not.

In general, enzymes within a gene family share more than 40% amino acid sequence identity. Protein sequences within subfamilies have greater than 55% similarity in the case of mammalian genes, or 46% in the case of nonmammalian genes. So far, genes in the same subfamily have been found to lie on the same chromosome within the same gene cluster and are nonsegregating, suggesting a common origin through gene duplication events. Sequences showing less than 3% divergence are arbitrarily designated allelic variants unless other evidence exists to the contrary. Known sequences fit

**Table 7.2   Some Important Human Cytochrome P450 Isozymes and Selected Substrates**

| P450 | Drugs | Carcinogens/Toxicants/ Endogenous Substrates | Diagnostic Substrates In vivo [In vitro] |
|---|---|---|---|
| 1A1 | Verlukast (very few drugs) | Benzo(a)pyrene, dimethylbenz(a)anthracene | [Ethoxyresorufin, benzo(a)pyrene] |
| 1A2 | Phenacetin, theophylline, acetaminophen, warfarin, caffeine, cimetidine | Aromatic amines, arylhydrocarbons, NNK,[3] aflatoxin, estradiol | Caffeine, [acetanilide, methoxyresorufin, ethoxyresorufin] |
| 2A6 | Coumarin, nicotine | Aflatoxin, diethylnitrosamine, NNK[3] | Coumarin |
| 2B6 | Cyclophosphamide, ifosphamide, nicotine | 6 Aminochrysene, aflatoxin, NNK[3] | [7-ethoxy-4-trifluoro-methyl coumarin] |
| 2C8 | Taxol, tolbutamide, carbamazepine | — | [Chloromethyl fluorescein diethyl ether] |
| 2C9 | Tienilic acid, tolbutamide, warfarin, phenytoin, THC, hexobarbital, diclofenac | — | [Diclofenac (4′-OH)] |
| 2C19 | S-Mephenytoin, diazepam, phenytoin, omeprazole, indomethacin, impramine, propanolol, proguanil | — | [S-Mephentoin (4′-OH)] |
| 2D6 | Debrisoquine, sparteine, bufuralol, propanolol, thioridazine, quinidine, phenytoin, fluoxetine | NNK[3] | Dextromethorphan, [bufuralol (4′-OH) |
| 2E1 | Chlorzoxazone, isoniazid, acetaminophen, halothane, enflurane, methoxyflurane | Dimethylnitrosamine, benzene, halogenated alkanes (eg, $CCl_4$) acylonitrile, alcohols, aniline, styrene, vinyl chloride | Chlorzoxazone (6-OH), [p-nitrophenol] |
| 3A4 | Nifedipine, ethylmorphine, warfarin, quinidine, taxol, ketoconazole, verapamil, erythromycin, diazepam | Aflatoxin, 1-nitropyrene, benzo(a)pyrene 7,8-diol, 6 aminochrysene, estradiol, progesterone, testosterone, other steroids, bile acids | Erythromycin, nifedipine [testosterone (6-$\beta$)] |
| 4A9/11 | (Very few drugs) | Fatty acids, prostaglandins, thromboxane, prostacyclin | [Lauric acid] |

*Note*: NNK[3] = 4(methylnitrosamino)-1-(3-pyridl)-1-butanone, a nitrosamine specific to tobacco smoke.

the classification scheme surprisingly well, with few exceptions found at the family, subfamily, or allelic variant levels, and in each case additional information is available to justify the departure from the rules set out.

In some cases a homologue of a particular CYP enzyme is found across species (e.g., CYP1A1). In other cases the genes diverged subsequent to the divergence of the species and no exact analogue is found is various species (e.g., the CYP2C subfamily). In this case the genes are numbered in the order of discovery, and the gene products from a particular subfamily may even have differing substrate specificity in different species (e.g., rodent vs. human). Relationships between different CYP families and subfamilies are related to the rate and extent of CYP evolution.

Figure 7.2 demonstrates some of the evolutionary relationships between CYP genes between some of the earliest vertebrates and humans. This dendogram compares CYP genes from the puffer fish (fugu) and 8 other fish species with human CYPs (including 3 pseudogenes). The unweighted pair group method arithmetic averaging (UPGMA) phylogenetic tree demonstrates the presence of five CYP clans (clusters of CYPs that are consistently grouped together) and delineates the 18 known human CYPs. This data set demonstrates that the defining characteristics of vertebrate CYPs have not changed much in 420 million years. Of these 18 human CYPs, only 1 family was missing in fugu (CYP39), indicating that the mammalian diversity of CYPs likely predates the tetrapod-ray finned fish divergence. The fish genome also has new CYP1C, 3B, and 7C subfamilies that are not seen in mammals.

The gene products, the CYP isoforms, may still be designated P450 followed by the same numbering system used for the genes, or the CYP designation may be used, for example, P4501A1 or CYP1A1.

As of May 16, 2002, a total of 1925 CYP sequences have been "named" with several others still awaiting classification. Of these, 977 are animal sequences, 607 from plants, 190 from lower eukaryotes and 151 are bacterial sequences. These sequences fall into more than 265 CYP families, 18 of which belong to mammals. Humans have 40 sequenced CYP genes. As the list of CYPs is continually expanding, progress in this area can be readily accessed via the internet at the Web site of the P450 Gene Super-family Nomenclature Committee (*http://drnelson. utmem.edu/nelsonhomepage.html*) or at another excellent Web site (*http://www.icgeb.trieste.it/p450*).

### Cytochrome P450 Families with Xenobiotic Metabolizing Potential. Although mammals are known to have 18 CYP families, only three families are primarily responsible for most xenobiotic metabolism. These families (families 1–3) are considered to be more recently derived from the "ancestral" CYP families. The remaining families are less promiscuous in their metabolizing abilities and are often responsible for specific metabolic steps. For example, members of the CYP4 family are responsible for the end-chain hydroxylation of long-chain fatty acids. The remaining mammalian CYP families are involved in biosynthesis of steroid hormones. In fact some of the nomenclature for some of these families is actually derived from the various positions in the steroid nucleus where the metabolism takes place. For example, CYP7 mediates hydroxylation of cholesterol at the $7\alpha$-position, while CYP17 and 21 catalyze the $17\alpha$ and 21-hydroxylations of progesterone, respectively. CYP19 is responsible for the aromatization of androgens to estrogen by the initial step of hydroxylation at the 19-position. Many of the CYPs responsible for steroidogenesis are found in the adrenal cortex, while those involved in xenobiotic metabolism are found predominantly in tissues that are more likely to be involved in exposure such as liver, kidneys, lungs, and olfactory tissues.

To simplify discussion of important CYP family members, the following discussion concentrates upon human CYP family members. However, since there is a great deal of homology among family members, many of the points of discussion are generally applicable to CYP families belonging to several species.

The CYP1 family contains three known human members, CYP1A1, CYP1A2, and CYP1B1. CYP1A1 and CYP1A2 are found in all classes of the animal kingdom.

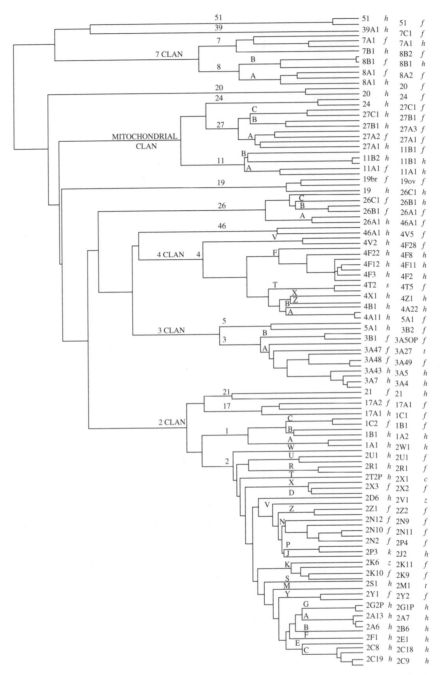

**Figure 7.2** UPGMA tree of 54 puffer fish (fugu), 60 human, and 8 other fish P450s. Species are indicated by *f*, *h*, *z*, *c*, *k*, *s*, and *t* for fugu, human, zebrafish, catfish, killifish, seabass, and trout, respectively. (Reprinted from D. R. Nelson, *Archives of Biochemistry and Biophysics* 409, pp. 18–24. 2003, with permission from Academic Press.)

Because these two highly homologous forms are so highly conserved among species, it is thought that both may possess important endogenous functions that have yet to be elucidated. CYP2E1 is the only other CYP that retains the same gene designation in many different species.

CYP1A1 and CYP1A2 possess distinct but overlapping substrate specificities: CYP1A1 preferring neutral polycyclic aromatic hydrocarbons (PAHs), and the latter preferring polyaromatic and heterocyclic amines and amides. Because of the preference of this family for molecules with highly planar molecular structures, CYP1 family members are closely associated with metabolic activation of many procarcinogens and mutagens including benzo(a)pyrene, aflatoxin B1, dimethylbenzanthracene, $\beta$-naphthylamine, 4-aminobiphenyl, 2-acetylaminoflourene, and benzidine. Figure 7.3 illustrates a typical reaction sequence leading to the formation of epoxide and the epoxide diols that are often implicated in the formation of carcinogenic metabolites formed by these enzymes.

Many of the planar PAH compounds induce their own metabolism by inducing transcription of the aryl hydrocarbon receptor (Ah receptor). Although expression of CYP1A1 and 1A2 is often coordinately induced, there are clear differences in regulation, not only with respect to substrate specificity but also in their biological expression. For example, CYP1A1 does not appear to be expressed in human liver unless induced,

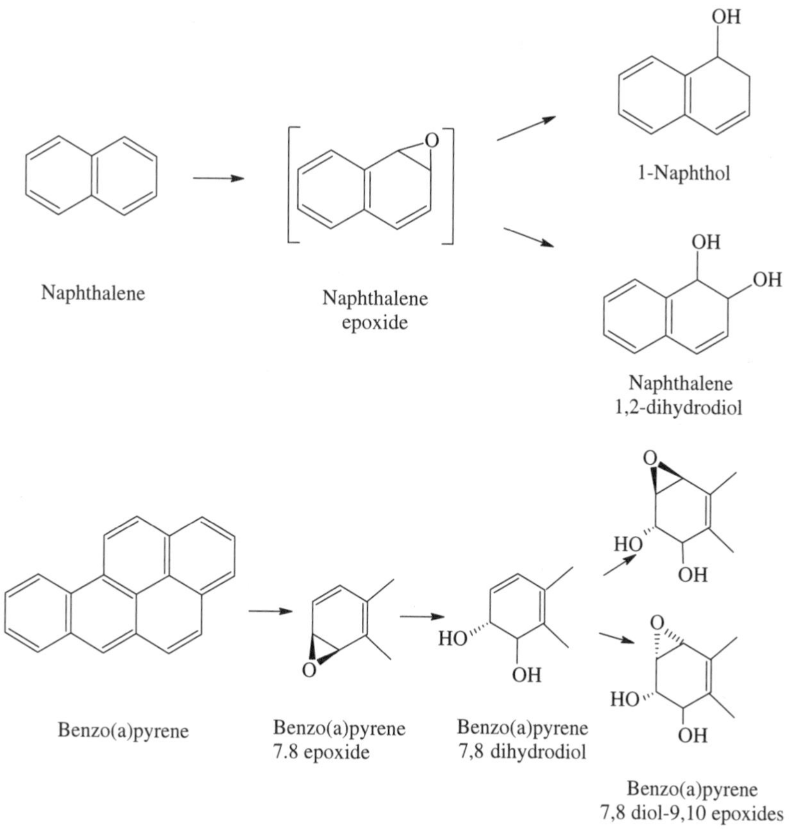

**Figure 7.3**   Examples of epoxidation reactions.

whereas CYP1A2 is endogenously expressed in the liver. CYP1A1, however, is present in many extrahepatic tissues including the lung, where there is a possible association between CYP-mediated activation of benzo(a)pyrene and other related chemicals present in cigarette smoke and lung cancer in humans.

The CYP2 family consists of 10 subfamilies, five of which are present in mammalian liver. Some of the more important isoforms found in humans within this family are CYP2A6, -2B6, -2C8, -2C9, -2C19, -2D6, and -2E1. The enzyme CYP2A6 is expressed primarily in liver tissue, where it represents 1–10% of total CYP content. CYP2A6 is responsible for the 7-hydroxylation of the naturally occurring plant compound coumarin and its activity is often phenotyped by monitoring this particular metabolic pathway. Other drugs metabolized by CYP2A6 include nicotine, 2-acetylaminofluorene, methoxyflurane, halothane, valproic acid, and disulfiram. Precarcinogens likely activated by 2A6 include aflatoxin B1, 1,3 butadiene, 2,6-dichlorobenzonitrile, and a number of nitrosamines. Because CYP2A6 is responsible for up to 80% of the human metabolism of nicotine, a number of studies have been conducted to determine whether individuals with 2A6 polymorphisms have reduced risk of lung cancers. Although theoretically individuals lacking 2A6 would be expected to smoke less and be less likely to activate carcinogens found in tobacco smoke, studies have not conclusively demonstrated any clear associations between 2A6 polymorphisms and risk of lung cancer.

Like CYP2A6, the human isoform CYP2B6 has recently gained greater recognition for its role in metabolism of many clinical drugs. Some common pharmaceutical substrates for CYP2B6 include cyclophosphamide, nevirapine, S-mephobarbitol, artemisinin, bupropion, propofol, ifosfamide, ketamine, selegiline, and methadone. CYP2B6 has also been demonstrated to have a role in the activation of the organophosphate, chlorpyrifos, and in the degradation of the commonly used insecticide repellant, diethyl toluamide (DEET). Historically it was thought that CYP2B6 is found in a small proportion of livers (<25%), but more recent data using antibodies prepared from human proteins have demonstrated that most liver samples have detectable levels of 2B6, though greater than 20-fold differences in levels of protein have been observed.

In contrast with CYP2A6 and CYP2B6, members of the CYP2C family constitute a fairly large percentage of CYP in human liver (ca. 20%) and are responsible for the metabolism of several drugs. All four members of the subfamily in humans exhibit genetic polymorphisms, many of which have important clinical consequences in affected individuals. Genetic polymorphisms in CYP2C19 were shown to be responsible for one of the first described polymorphic effects, that involving mephenytoin metabolism. This polymorphism significantly reduces the metabolism of mephenytoin, resulting in the classification of those individuals possessing this trait as poor metabolizers (PM). Among Caucasians, PMs represent only 3–5% of the populations, while in Asian and Polynesian populations 12–23% and 38–79% of the populations are represented, respectively. At least seven different mutations in this allele have been described, some of which negatively affect catalytic activity while others prevent expression of the protein. Other important drugs affected by these CYP2C19 polymorphisms include the anti-ulcer drug omeprazole, other important proton pump inhibitors, barbiturates, certain tricyclic antidepressants such as imipramine, and the antimalarial drug proguanil. Other important members of the CYP2C family in humans include CYP2C8, -2C9, and -2C18. Substrates metabolized exclusively by CYP2C8 include retinol, retinoic acid, taxol, and arachidonic acid. CYP2C9, the principal CYP2C in human liver, metabolizes several important drugs including the diabetic agent tolbutamide, the anticonvulsant phenytoin, the anticoagulant warfarin and a number of anti-inflammatory drugs including ibuprofen, diclofenac, and others. Both CYP2C9 and -2C8, which are responsible for metabolism of the anticancer drug paclitaxel, have been demonstrated to be polymorphic.

CYP2E1 is the only member of the CYP2E family in most mammals with the exception of rabbits. Substrates for this family tend to be of small molecular weight and include ethanol, carbon tetrachloride, benzene, and acetaminophen. In contrast to many other inducible CYP families, CYP2E1 is regulated by a combination of increased transcription levels and increased message and protein stabilization.

Undoubtedly the largest amount of CYP in human liver is that of the CYP3 family. CYP3A4 is the most abundant CYP in the human liver, accounting for nearly 30% of the total amount, and is known to metabolize many important drugs including cyclosporine A, nifedipine, rapamycin, ethinyl estradiol, quinidine, digitoxin, lidocaine, erythromycin, midazolam, triazolam, lovastatin, and tamoxifen. Other important oxidations ascribed to the CYP3 family include many steroid hormones, macrolide antibiotics, alkaloids, benzodiazepines, dihydropyridines, warfarin, polycyclic hydrocarbon-derived dihydrodiols, and aflatoxin $B_1$. Many chemicals are also capable of inducing this family including phenobarbital, rifampicin, and dexamethasone. Because of potential difficulties arising from CYP induction, drugs metabolized by this family must be closely examined for the possibility of harmful drug-drug interactions.

***Cytochrome P450 Reactions.*** Although microsomal monooxygenase reactions are basically similar in the role played by molecular oxygen and in the supply of electrons, the many CYP isoforms can attack a large variety of xenobiotic substrates, with both substrates and products falling into many different chemical classes. In the following sections enzyme activities are therefore classified on the basis of the overall chemical reaction catalyzed; one should bear in mind, however, that not only do these classes often overlap, but often a substrate may also undergo more than one reaction. See Table 7.1 for a listing of important oxidation and reduction reactions of CYPs.

*Epoxidation and Aromatic Hydroxylation.* Epoxidation is an extremely important microsomal reaction because not only can stable and environmentally persistent epoxides be formed (see aliphatic epoxidations, below), but highly reactive intermediates of aromatic hydroxylations, such as arene oxides, can also be produced. These highly reactive intermediates are known to be involved in chemical carcinogenesis as well as chemically induced cellular and tissue necrosis.

The oxidation of naphthalene was one of the earliest examples of an epoxide as an intermediate in aromatic hydroxylation. As shown in Figure 7.3, the epoxide can rearrange nonenzymatically to yield predominantly 1-naphthol, or interact with the enzyme epoxide hydrolase to yield the dihydrodiol, or interact with glutathione S-transferase to yield the glutathione conjugate, which is ultimately metabolized to a mercapturic acid. These reactions are also of importance in the metabolism of other xenobiotics that contain an aromatic nucleus, such as the insecticide carbaryl and the carcinogen benzo(a)pyrene.

The ultimate carcinogens arising from the metabolic activation of benzo(a)pyrene are stereoisomers of benzo(a)pyrene 7,8-diol-9,10-epoxide (Figure 7.3). These metabolites arise by prior formation of the 7,8 epoxide, which gives rise to the 7,8-dihydrodiol through the action of epoxide hydrolase. This is further metabolized by the CYP to the 7,8-diol-9,10-epoxides, which are both potent mutagens and unsuitable substrates for the further action of epoxide hydrolase. Stereochemistry is important in the final product. Of the four possible isomers of the diol epoxide, the (+)-benzo(a)pyrene diol epoxide-2 is the most active carcinogen.

*Aliphatic Hydroxylation.* Simple aliphatic molecules such as *n*-butane, *n*-pentane, and *n*-hexane, as well as alicylcic compounds such as cyclohexane, are known to be oxidized to alcohols. Likewise alkyl side chains of aromatic compounds such as cyclohexane, are known to be oxidized to alcohols, but alkyl side chains of aromatic compounds are more readily oxidized, often at more than one position, and so provide good examples of this type of oxidation. The *n*-propyl side chain of *n*-propyl benzene can be oxidized at any one of three carbons to yield 3-phenylpropan-1-ol ($C_6H_5CH_2CH_2CH_2OH$) by $\omega$-oxidation, benzylmethyl carbinol ($C_6H_5CH_2CHOHCH_3$) by $\omega$-1 oxidation, and ethyl-phenylcarbinol ($C_6H_5CHOHCH_2CH_3$) by $\alpha$-oxidation. Further oxidation of these alcohols is also possible.

*Aliphatic Epoxidation.* Many aliphatic and alicylcic compounds containing unsaturated carbon atoms are thought to be metabolized to epoxide intermediates (Figure 7.4). In the case of aldrin the product, dieldrin, is an extremely stable epoxide and represents the principle residue found in animals exposed to aldrin. Epoxide formation in the case of aflatoxin is believed to be the final step in formation of the ultimate carcinogenic species and is, therefore, an activation reaction.

*Dealkylation: O-, N-, and S-Dealkylation.* Probably the best known example of *O*-dealkylation is the demethylation of *p*-nitroanisole. Due to the ease with which the product, *p*-nitrophenol, can be measured, it is a frequently used substrate for the demonstration of CYP activity. The reaction likely proceeds through formation of an unstable methylol intermediate (Figure 7.5).

The *O*-dealkylation of organophosphorus triesters differs from that of *p*-nitroanisole in that it involves the dealkylation of an ester rather than an ether. The reaction was

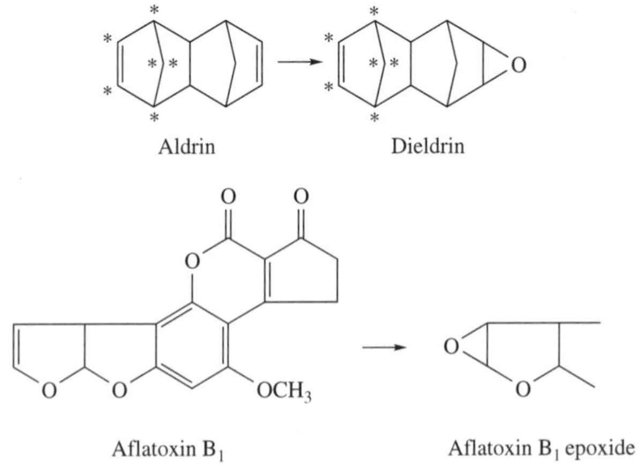

Aflatoxin B$_1$       Aflatoxin B$_1$ epoxide

**Figure 7.4** Examples of aliphatic epoxidation. * denote Cl atoms.

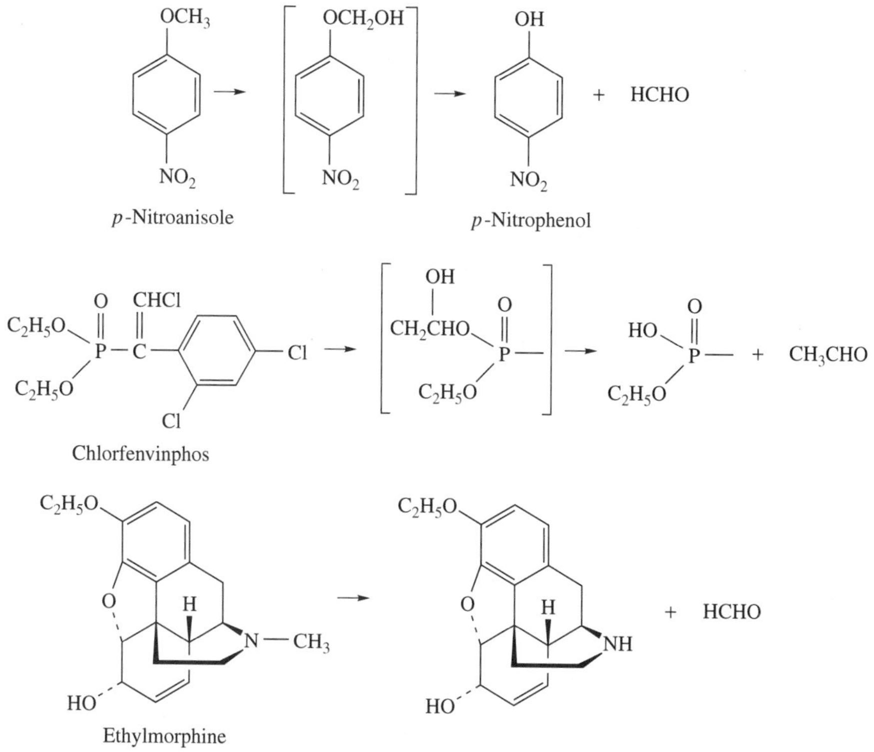

**Figure 7.5** Examples of dealkylation.

first described for the insecticide chlorfenvinphos and is known to occur with a wide variety of vinyl, phenyl, phenylvinyl, and naphthyl phosphate and thionophosphate triesters (Figure 7.5).

*N*-dealkylation is a common reaction in the metabolism of drugs, insecticides, and other xenobiotics. The drug ethylmorphine is a useful model compound for this reaction. In this case the methyl group is oxidized to formaldehyde, which can be readily detected by the Nash reaction.

*S*-dealkylation is believed to occur with a number of thioethers, including methylmercaptan and 6-methylthiopurine, although with newer knowledge of the specificity of the flavin-containing monooxygenase (see the discussion below) it is possible that the initial attack is through sulfoxidation mediated by FMO rather than CYP.

*N-Oxidation.* *N*-oxidation can occur in a number of ways, including hydroxylamine formation, oxime formation, and *N*-oxide formation, although the latter is primarily dependent on the FMO enzyme. Hydroxylamine formation occurs with a number of amines such as aniline and many of its substituted derivatives. In the case of 2-acetylaminofluorene the product is a potent carcinogen, and thus the reaction is an activation reaction (Figure 7.6).

Oximes can be formed by the *N*-hydroxylation of imines and primary amines. Imines have been suggested as intermediates in the formation of oximes from primary amines (Figure 7.6).

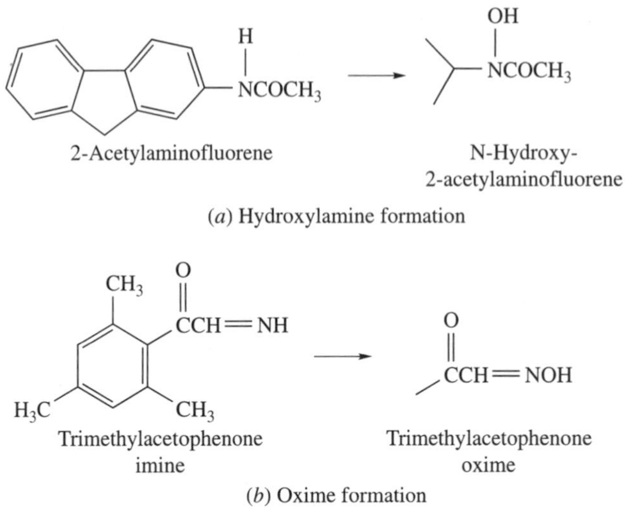

(a) Hydroxylamine formation

(b) Oxime formation

**Figure 7.6** Examples of *N*-oxidation.

*Oxidative Deamination.* Oxidative deamination of amphetamine occurs in the rabbit liver but not to any extent in the liver of either the dog or the rat, which tend to hydroxylate the aromatic ring. A close examination of the reaction indicates that it is probably not an attack on the nitrogen but rather on the adjacent carbon atom, giving rise to a carbinol amine, which eliminates ammonia, producing a ketone:

$$R_2CHNH_2 \xrightarrow{+O} R_2C(OH)NH_2 \xrightarrow{-NH_3} R_2C=O$$

The carbinol, by another reaction sequence, can also give rise to an oxime, which can be hydrolyzed to yield the ketone. The carbinol is thus formed by two different routes:

$$R_2C(OH)NH2 \xrightarrow{-H_2O} R_2C=NH \xrightarrow{+O} R_2CNOH \xrightarrow{+H_2O} R_2C=O$$

*S-Oxidation.* Thioethers in general are oxidized by microsomal monooxygenases to sulfoxides, some of which are further oxidized to sulfones. This reaction is very common among insecticides of several different chemical classes, including carbamates, organophosphates, and chlorinated hydrocarbons. Recent work suggests that members of the CYP2C family are highly involved in sulfoxidation of several organophosphate compounds including phorate, coumaphos, demeton, and others. The carbamate methiocarb is oxidized to a series of sulfoxides and sulfones, and among the chlorinated hydrocarbons endosulfan is oxidized to endosulfan sulfate and methiochlor to a series of sulfoxides and sulfones, eventually yielding the bis-sulfone. Drugs, including chlorpromazine and solvents such as dimethyl sulfoxide, are also subject to S-oxidation. The fact that FMOs are versatile sulfur oxidation enzymes capable of carrying out many of the previously mentioned reactions raises important questions as to the relative role of this enzyme versus that of CYP. Thus, a reexamination of earlier work in which many of these reactions were ascribed to CYP is required.

*P-Oxidation.* *P*-oxidation, a little known reaction, involves the conversion of trisubstituted phosphines to phosphine oxides, for example, diphenylmethylphosphine to diphenylmethylphosphine oxide. Although this reaction is described as a typical CYP-dependent monooxygenation, it too is now known to be catalyzed by the FMO also.

*Desulfuration and Ester Cleavage.* The phosphorothionates $[(R^1O)_2P(S)OR^2)]$ and phosphorodithioate $[(R^1O)_2P(S)SR^2]$ owe their insecticidal activity and their mammalian toxicity to an oxidative reaction in which the P=S group is converted to P=O, thereby converting the compounds from chemicals relatively inactive toward cholinesterase into potent inhibitors (see Chapter 11 for a discussion of the mechanism of cholinesterase inhibition). This reaction has been described for many organophosphorus compounds but has been studied most intensively in the case of parathion. Much of the splitting of the phosphorus ester bonds in organophosphorus insecticides, formerly believed to be due to hydrolysis, is now known to be due to oxidative dearylation. This is a typical CYP-dependent monooxygenation, requiring NADPH and $O_2$ and being inhibited by CO. Current evidence supports the hypothesis that this reaction and oxidative desulfuration involve a common intermediate of the "phosphooxithirane" type (Figure 7.7). Some organophosphorus insecticides, all phosphonates, are activated by the FMO as well s the CYP.

*Methylenedioxy (Benzodioxole) Ring Cleavage.* Methylenedioxy-phenyl compounds, such as safrole or the insecticide synergist, piperonyl butoxide, many of which are effective inhibitors of CYP monooxygenations, are themselves metabolized to catechols. The most probable mechanism appears to be an attack on the methylene carbon, followed by elimination of water to yield a carbene. The highly reactive carbene either reacts with the heme iron to form a CYP-inhibitory complex or breaks down to yield the catechol (Figure 7.8).

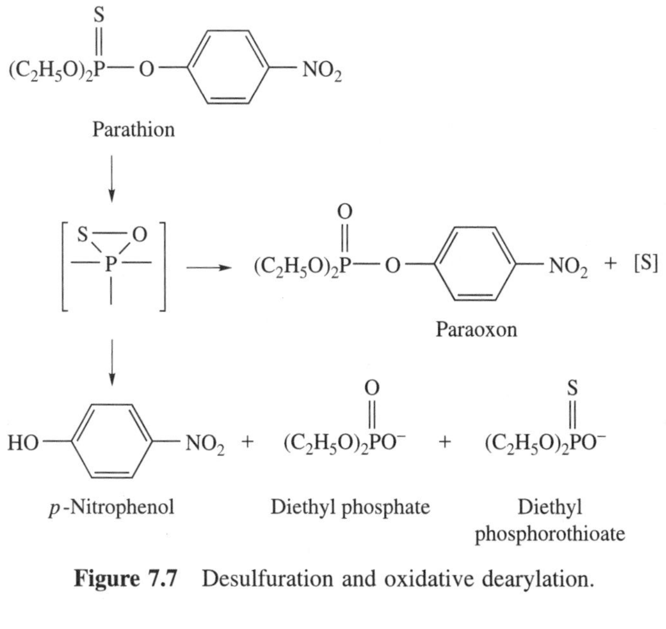

**Figure 7.7** Desulfuration and oxidative dearylation.

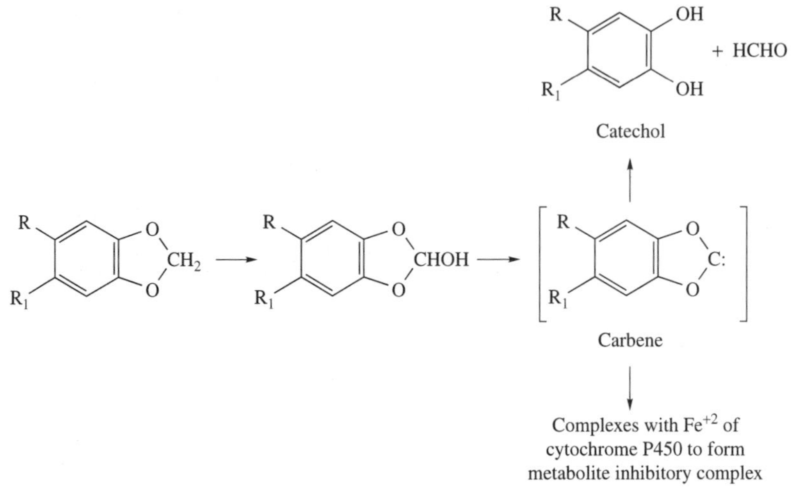

**Figure 7.8** Monooxygenation of methylenedioxyphenyl compounds.

### 7.2.3 The Flavin-Containing Monooxygenase (FMO)

Tertiary amines such as trimethylamine and dimethylamine had long been known to be metabolized to *N*-oxides by a microsomal amine oxidase that was not dependent on CYP. This enzyme, now known as the microsomal flavin-containing monooxygenase (FMO), is also dependent on NADPH and $O_2$, and has been purified to homogeneity from a number of species. Isolation and characterization of the enzyme from liver and lung samples provided evidence of clearly distinct physicochemical properties and substrate specificities suggesting the presence of at least two different isoforms. Subsequent studies have verified the presence of multiple forms of the enzyme.

At least six different isoforms have been described by amino acid or cDNA sequencing, and are classified as FMO1 to FMO6. These isoforms share approximately 50–60% amino acid identity across species lines. The identity of orthologues is greater than 82%. Although each isoform has been characterized in humans, several are essentially nonfunctional in adults. For example, FMO1, expressed in the embryo, disappears relatively quickly after birth. FMO2 in most Caucasians and Asians contains a premature stop codon, preventing the expression of functional protein. Functional FMO2 is found in 26% of the African-American population and perhaps also in the Hispanic population. FMO3, the predominant human FMO, is poorly expressed in neonatal humans but is expressed in most individuals by one year of age. Gender independent expression of FMO3 (contrasting with what is observed in other mammals) continues to increase through childhood, reaching maximal levels of expression at adulthood. Several polymorphic forms of FMO3 are responsible for the disease, trimethylamineuria, also known as "fish odor syndrome," characterized by the inability of some individuals to convert the malodorous trimethylamine, either from the diet or from metabolism, to its odorless N-oxide. Although the FMO4 transcript is found in several species, the protein has yet to be successfully expressed in any species. Although FMO5 is expressed in humans at low levels, the poor catalytic activity of FMO5 for most classical FMO substrates suggests that it has minimal participation in xenobiotic oxidation. No data are yet available on the role and abundance of the most recently discovered FMO, FMO6.

Substrates containing soft nucleopohiles (e.g., nitrogen, sulfur, phosphorus, and selenium) are good candidates for FMO oxidation (Figure 7.9). A short list of known substrates include drugs such as dimethylaniline, imipramine, thiobenzamide, chlorpromazine, promethazine, cimetidine, and tamoxifen; pesticides such as phorate, fonofos, and methiocarb; environmental agents including the carcinogen 2-aminofluorine, and the neurotoxicants nicotine and 1-methyl-4phenyl-1,2,3,6-tetrahydropyridine (MPTP). Although there is no known physiologically relevant substrate for FMO a few dietary and/or endogenous substrates have been identified, including trimethylamine, cysteamine, methionine and several cysteine-s-conjugates. In most cases metabolism by FMO results in detoxication products, although there are several examples of substrates that are bioactivated by FMO oxidation; particularly in the case of substrates involving sulfur oxidation.

Most FMO substrates are also substrates for CYP. Since both enzymes are microsomal and require NADPH and oxygen, it is difficult to distinguish which enzyme is responsible for oxidation without the use of techniques involving specific inactivation or inhibition or one or the other of these enzymes while simultaneously examining the

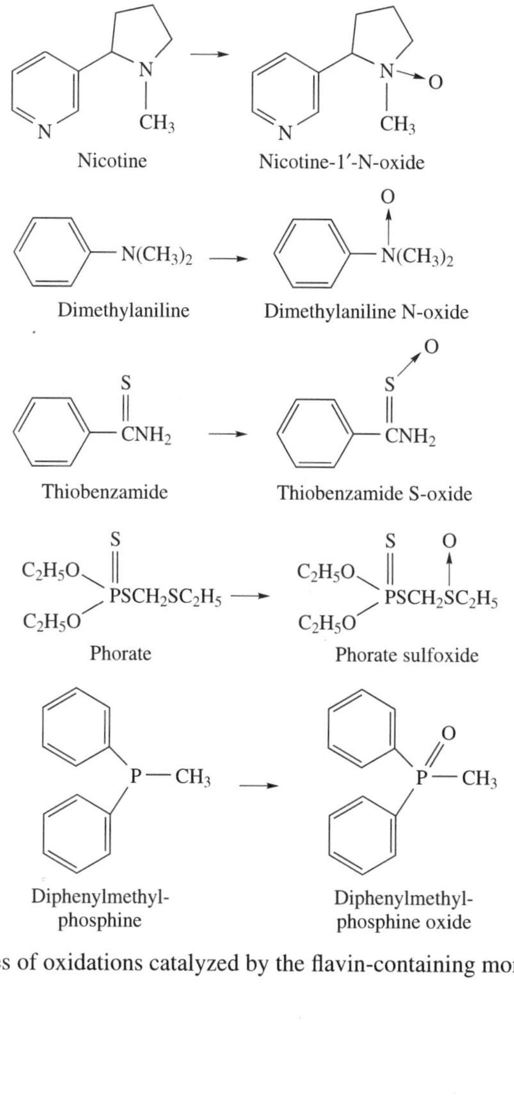

**Figure 7.9**  Examples of oxidations catalyzed by the flavin-containing monooxygenase (FMO).

metabolic contribution of the other. Since FMOs are generally heat labile, heating the microsomal preparation to $50°C$ for one minute inactivates the FMOs while having minimal effects of CYPs. Alternatively, the contribution of FMO can be assessed by use of a general CYP inhibitor such as $N$-benzylimidazole or by an inhibitory antibody to NADPH cytochrome P450 reductase, a necessary CYP coenzyme. Typically results of these tests are sought in combination so that the best estimates of CYP and FMO contribution can be obtained.

Toxicologically it is of interest that the FMO enzyme is responsible for the oxidation of nicotine to nicotine $1'$-N-oxide, whereas the oxidation of nicotine to cotinine is catalyzed by two enzymes acting in sequence: CYP followed by a soluble aldehyde dehydrogenase. Thus nicotine is metabolized by two different routes, the relative contributions of which may vary with both the extrinsic and intrinsic factors outlined in Chapter 9.

### 7.2.4 Nonmicrosomal Oxidations

In addition to the microsomal monooxygenases, other enzymes are involved in the oxidation of xenobiotics. These enzymes are located in the mitochondria or in the soluble cytoplasm of the cell.

*Alcohol Dehydrogenase.* Alcohol dehydrogenases catalyze the conversion of alcohols to aldehydes or ketones:

$$RCH_2OH + NAD^+ \longrightarrow RCHO + NADH + H^+$$

This reaction should not be confused with the monooxygenation of ethanol by CYP that occurs in the microsomes. The alcohol dehydrogenase reaction is reversible, with the carbonyl compounds being reduced to alcohols.

This enzyme is found in the soluble fraction of the liver, kidney, and lung and is probably the most important enzyme involved in the metabolism of foreign alcohols. Alcohol dehydrogenase is a dimer whose subunits can occur in several forms under genetic control, thus giving rise to a large number of variants of the enzyme. In mammals, six classes of enzymes have been described. Alcohol dehydrogenase can use either NAD or NADP as a coenzyme, but the reaction proceeds at a much slower rate with NADP. In the intact organism the reaction proceeds in the direction of alcohol consumption, because aldehydes are further oxidized to acids. Because aldehydes are toxic and are not readily excreted because of their lipophilicity, alcohol oxidation may be considered an activation reaction, the further oxidation to an acid being a detoxication step.

Primary alcohols are oxidized to aldehydes, $n$-butanol being the substrate oxidized at the highest rate. Although secondary alcohols are oxidized to ketones, the rate is less than for primary alcohols, and tertiary alcohols are not readily oxidized. Alcohol dehydrogenase is inhibited by a number of heterocyclic compounds such as pyrazole, imidazole, and their derivatives.

*Aldehyde Dehydrogense.* Aldehydes are generated from a variety of endogenous and exogenous substrates. Endogenous aldehydes may be formed during metabolism of amino acids, carbohydrates, lipids, biogenic amines, vitamins, and steroids. Metabolism of many drugs and environmental agents produces aldehydes. Aldehydes are highly reactive electrophilic compounds; they may react with thiol and amino groups to produce a variety of effects. Some aldehydes produce therapeutic effects, but more often the effects are cytotoxic, genotoxic, mutagenic, and carcinogenic. Aldehyde dehydrogenases are important in helping to alleviate some of the toxic effects of aldehyde generation. This enzyme catalyzes the formation of acids from aliphatic and aromatic aldehydes; the acids are then available as substrates for conjugating enzymes:

$$RCHO + NAD^+ \longrightarrow RCOOH + NADH + H^+$$

The aldehyde gene superfamily is large with more than 330 aldehyde dehydrogenase genes in prokaryote and eukaryotic species. The eukaryotic aldehyde dehydrogenase gene superfamily consists of 20 gene families, 9 of which contain 16 human genes and 3 pseudogenes. The importance of some of these genes in detoxication pathways is underscored by the fact that identified polymorphisms are associated with several metabolic diseases.

One especially interesting polymorphism is that which occurs at the aldehyde dehydrogenase 2 locus. When inherited as the homozygous trait, this aldehyde dehydrogenase polymorphism results in a 20-fold greater generation of acetaldehyde from ethanol, resulting in the flushing syndrome characteristic of many Asian individuals after ethanol consumption. Alcoholics are not likely to be found among individuals expressing this particular polymorphism.

Other enzymes in the soluble fraction of liver that oxidize aldehydes are aldehyde oxidase and xanthine oxidase, both flavoproteins that contain molybdenum; however, their primary role seems to be the oxidation of endogenous aldehydes formed as a result of deamination reactions.

**Amine Oxidases.** The most important function of amine oxidases appears to be the oxidation of amines formed during normal processes. Two types of amine oxidases are concerned with oxidative deamination of both endogenous and exogenous amines. Typical substrates are shown in Figure 7.10.

*Monoamine Oxidases.* The monomine oxidases are a family of flavoproteins found in the mitochondria of a wide variety of tissues: liver, kidney, brain, intestine, and

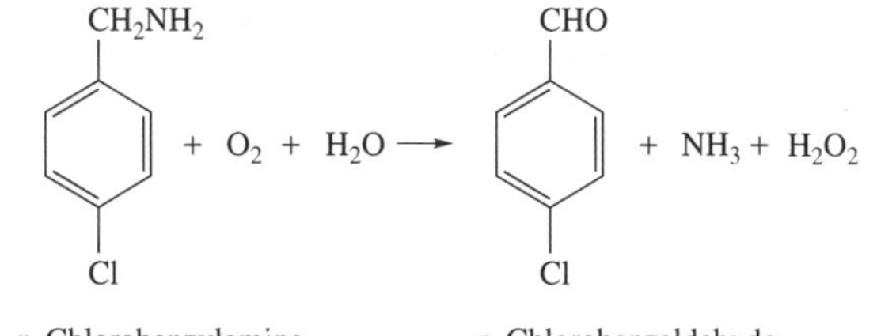

*p*-Chlorobenzylamine          *p*-Chlorobenzaldehyde

(*a*) Monoamine oxidase

$$H_2N(CH_2)_5NH_2 + O_2 + H_2O \longrightarrow H_2N(CH_2)_4CHO + NH_3 + H_2O_2$$
Cadaverine

(*b*) Diamine oxidase

**Figure 7.10**  Examples of oxidations catalyzed by amine oxidases.

blood platelets. They are a group of similar enzymes with overlapping specificities and inhibition. Although the enzyme in the central nervous system is concerned primarily with neurotransmitter turnover, that in the liver will deaminate primary, secondary, and tertiary aliphatic amines, reaction rates with the primary amines being faster. Electron-withdrawing substitutions on an aromatic ring increase the reaction rate, whereas compounds with a methyl group on the $\alpha$-carbon such as amphetamine and ephedrine are not metabolized.

*Diamine Oxidases.* Diamine oxidases are enzymes that also oxidize amines to aldehydes. The preferred substates are aliphatic diamines in which the chain length is four (putrescine) or five (cadaverine) carbon atoms. Diamines with carbon chains longer than nine will not serve as substrates but can be oxidized by monoamine oxidases. Secondary and tertiary amines are not metabolized. Diamine oxidases are typically soluble pyridoxal phosphate-containing proteins that also contain copper. They have been found in a number of tissues, including liver, intestine, kidney, and placenta.

### 7.2.5 Cooxidation by Cyclooxygenases

During the biosynthesis of prostaglandins, a polyunsaturated fatty acid, such as arachidonic acid, is first oxygenated to yield a hydroperoxy endoperoxide, prostaglandin G2. This is then further metabolized to prostaglandin H2, both reactions being catalyzed by the same enzyme, cyclooxygenase (COX), also known as prostaglandin synthase (Figure 7.11). This enzyme is located in the microsomal membrane and is found in greatest levels in respiratory tissues such as the lung. It is also common in the kidney and seminal vesicle. It is a glycoprotein with a subunit molecular mass of about 70,000 daltons, containing one heme per subunit. During the second step of the previous sequence (peroxidase), many xenobiotics can be cooxidized, and investigations of the mechanism have shown that the reactions are hydroperoxide-dependent reactions catalyzed by a peroxidase that uses prostaglandin G as a substrate. In at least some of these cases, the identity of this peroxidase has been established as a prostaglandin synthase. Many of the reactions are similar or identical to those catalyzed by other peroxidases and also by microsomal monooxygenases; they include both detoxication and activation reactions. This mechanism is important in xenobiotic metabolism, particularly in tissues that are low in CYP and/or the FMO but high in prostaglandin synthase.

The cyclooxygenase (COX) enzyme is known to exist as two distinct isoforms. COX-1 is a constitutively expressed housekeeping enzyme found in nearly all tissues and mediates physiological responses. COX-2 is an inducible form expressed primarily by cells involved in the inflammatory response. Several tissues low in CYP expression are rich in COX, which is believed to have significance in the carcinogenic effects of aromatic amines in these organs.

During cooxidation, some substrates are activated to become more toxic than they were originally. In some cases substrate oxidation results in the production of free radicals, which may initiate lipid peroxidation or bind to cellular proteins or DNA. Another activation pathway involves the formation of a peroxyl radical from subsequent metabolism of prostaglandin G2. This reactive molecule can epoxidize many substates including polycyclic aromatic hydrocarbons, generally resulting in increasing toxicity of the respective substrates.

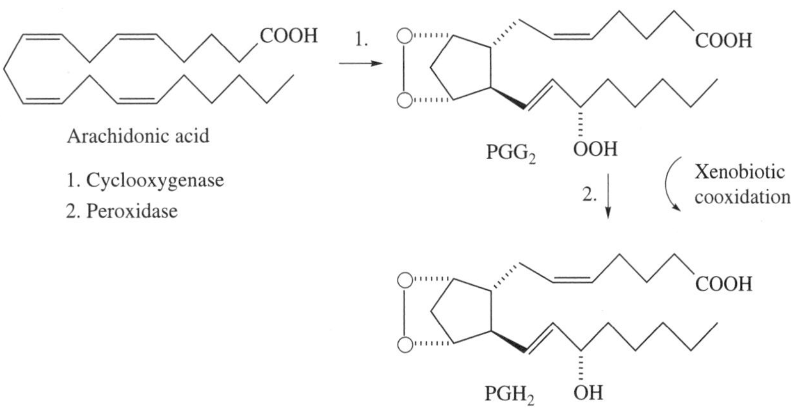

**Figure 7.11**   Cooxidation during prostaglandin biosynthesis.

To differentiate between xenobiotic oxidations by COX and CYP, in vitro microsomal incubations of the xenobiotic may be performed either in the presence of arachidonic acid (COX catalyzed) or in the presence of NADPH (CYP catalyzed). In the presence of arachidonic acid while in the absence of NADPH, substrates co-oxidized by COX will be formed while those requiring CYP will not. Specific inhibitors of PG synthase (indomethacin) and CYP (Metyrapone or SKF 525A) have also been used.

## 7.2.6   Reduction Reactions

A number of functional groups, such as nitro, diazo, carbonyl, disulfide sulfoxide, alkene, and pentavalent arsenic, are susceptible to reduction, although in many cases it is difficult to tell whether the reaction proceeds enzymatically or nonenzymatically by the action of such biologic reducing agents as reduced flavins or reduced pyridine nucleotides. In some cases, such as the reduction of the double bound in cinnamic acid ($C_6H_5CH=CHCOOH$), the reaction has been attributed to the intestinal microflora. Examples of reduction reactions are shown in Figure 7.12.

*Nitro Reduction.* Aromatic amines are susceptible to reduction by both bacterial and mammalian nitroreductase systems. Convincing evidence has been presented that this reaction sequence is catalyzed by CYP. It is inhibited by oxygen, although NADPH is still consumed. Earlier workers had suggested a flavoprotein reductase was involved, and it is not clear if this is incorrect or if both mechanisms occur. It is true, however, that high concentration of FAD or FMN will catalyze the nonenzymatic reduction of nitro groups.

*Azo Reduction.* Requirements for azo reduction are similar to those for nitroreduction, namely anaerobic conditions and NADPH. They are also inhibited by CO, and presumably they involve CYP. The ability of mammalian cells to reduce azo bonds is rather poor, and intestinal microflora may play a role.

*Disulfide Reduction.* Some disulfides, such as the drug disulfiram (Antabuse), are reduced to their sulfhydryl constituents. Many of these reactions are three-step

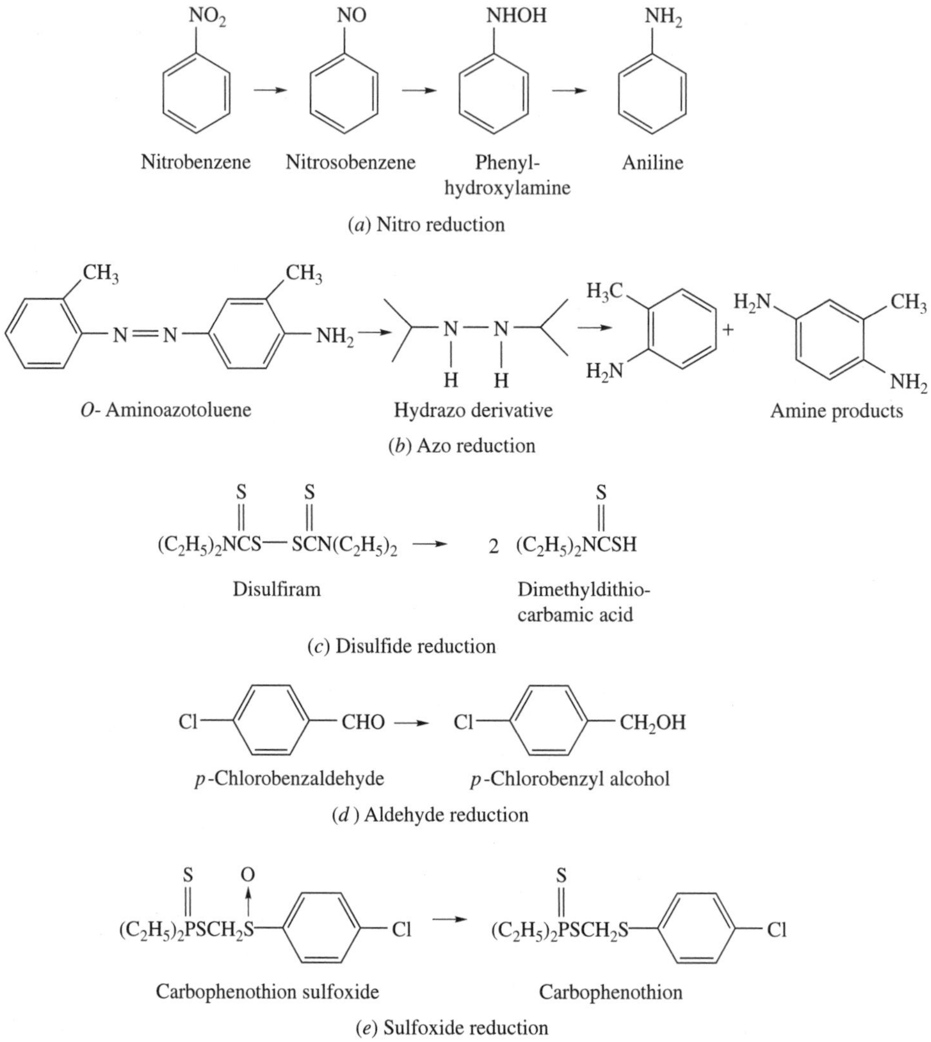

(a) Nitro reduction

(b) Azo reduction

(c) Disulfide reduction

(d) Aldehyde reduction

(e) Sulfoxide reduction

**Figure 7.12** Examples of metabolic reduction reactions.

sequences, the last reaction of which is catalyzed by glutathione reductase, using glutathione (GSH) as a cofactor:

$$RSSR + GSH \longrightarrow RSSG + RSH$$

$$RSSG + GSH \longrightarrow GSSG + RSH$$

$$GSSG + NADPH + H^+ \longrightarrow 2GSH + NADP^+$$

*Ketone and Aldehyde Reduction.* In addition to the reduction of aldehyde and ketones through the reverse reaction of alcohol dehydrogenase, a family of aldehyde reductases also reduces these compounds. These reductases are NADPH-dependent, cytoplasmic enzymes of low molecular weight and have been found in liver, brain, kidney, and other tissues.

*Sulfoxide Reduction.* The reduction of sulfoxides has been reported to occur in mammalian tissues. Soluble thioredoxin-dependent enzymes in the liver are responsible in some cases. It has been suggested that oxidation in the endoplasmic reticulum followed by reduction in the cytoplasm may be a form of recycling that could extend the in vivo half-life of certain toxicants.

### 7.2.7 Hydrolysis

Enzymes with carboxylesterase and amidases activity are widely distributed in the body, occurring in many tissues and in both microsomal and soluble fractions. They catalyze the following general reactions:

$$RC(O)OR' + H_2O \longrightarrow RCOOH + HOR' \qquad \text{Carboxylester hydrolysis}$$

$$RC(O)NR'R'' + H_2O \longrightarrow RCOOH + HNR'R'' \qquad \text{Carboxyamide hydrolysis}$$

$$RC(O)SR' + H_2O \longrightarrow RCOOH + HSR' \qquad \text{Carboxythioester hydrolysis}$$

Although carboxylesterases and amidases were thought to be different, no purified carboxylesterase has been found that does not have amidase activity toward the corresponding amide. Similarly enzymes purified on the basis of their amidase activity have been found to have esterase activity. Thus these two activities are now regarded as different manifestations of the same activity, specificity depending on the nature of R, R', and R" groups and, to a lesser extent, on the atom (O, S, or N) adjacent to the carboxyl group.

In view of the large number of esterases in many tissues and subcellular fractions, as well as the large number of substrates hydrolyzed by them, it is difficult to derive a meaningful classification scheme. The division into A-, B-, and C- esterases on the basis of their behavior toward such phosphate triesters as paraoxon, first devized by Aldridge, is still of some value, although not entirely satisfactory.

A-esterases, also referred to as arylesterases, are distinguished by their ability to hydrolyze esters derived from aromatic compounds. Organophosphates, such as the insecticide paraoxon are often used to characterize this group. B-esterases, the largest and most important group, are inhibited by organophosphates. All the B-esterases have a serine residue in their active site that is phosphorylated by this inhibitor. This group includes a number of different enzymes and their isozymes, many of which have quite different substrate specificities. For example, the group contains carboxylesterase, amidases, cholinesterases, monoacylglycerol lipases, and arylamidases. Many of these enzymes hydrolyze physiological (endogenous) substrates as well as xenobiotics. Several examples of their activity toward xenobiotic substrates are shown in Figure 7.13. C-esterases, or acetylesterases, are defined as those esterases that prefer acetyl esters as substrates, and for which paraoxon serves as neither substrate nor inhibitor.

### 7.2.8 Epoxide Hydration

Epoxide rings of alkene and arene compounds are hydrated by enzymes known as epoxide hydrolases, the animal enzyme forming the corresponding *trans*-diols, although bacterial hydrolases are known that form *cis*-diols. Although, in general, the hydration

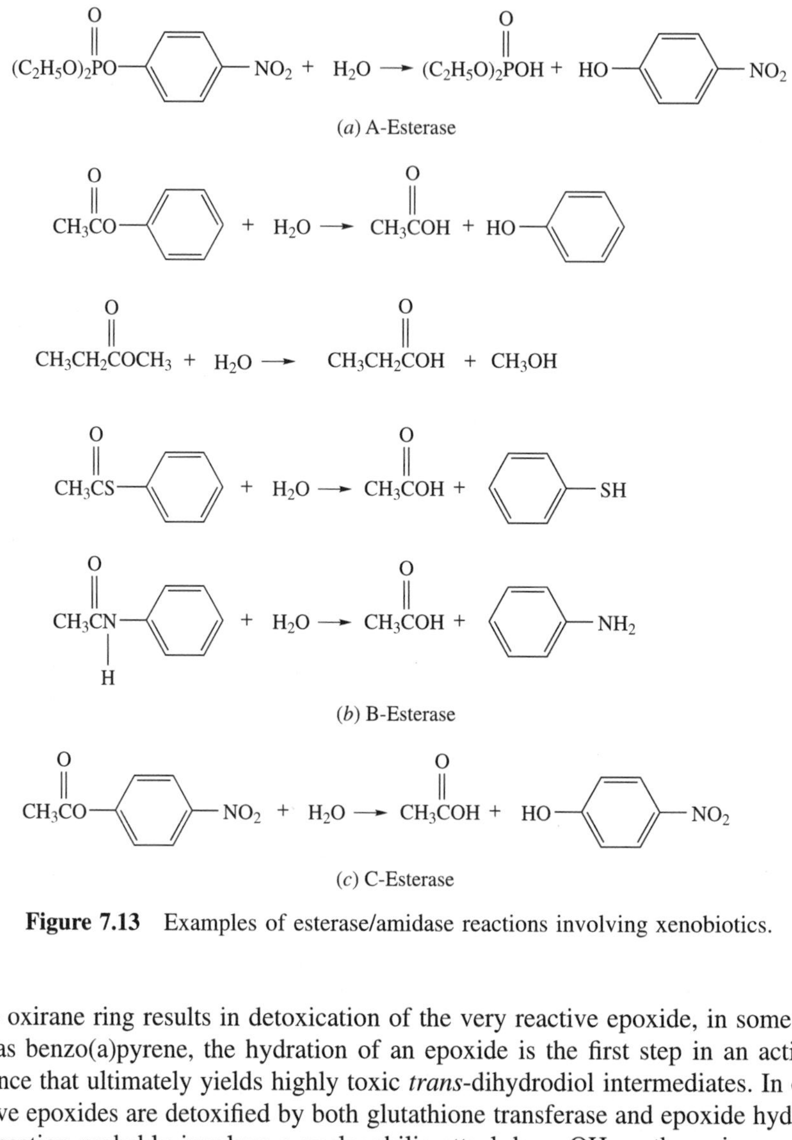

$$(C_2H_5O)_2\overset{\overset{O}{\|}}{P}O \text{—} \langle \rangle \text{—}NO_2 \;+\; H_2O \;\longrightarrow\; (C_2H_5O)_2\overset{\overset{O}{\|}}{P}OH \;+\; HO\text{—}\langle \rangle \text{—}NO_2$$

(a) A-Esterase

$$CH_3\overset{\overset{O}{\|}}{C}O\text{—}\langle \rangle \;+\; H_2O \;\longrightarrow\; CH_3\overset{\overset{O}{\|}}{C}OH \;+\; HO\text{—}\langle \rangle$$

$$CH_3CH_2\overset{\overset{O}{\|}}{C}OCH_3 \;+\; H_2O \;\longrightarrow\; CH_3CH_2\overset{\overset{O}{\|}}{C}OH \;+\; CH_3OH$$

$$CH_3\overset{\overset{O}{\|}}{C}S\text{—}\langle \rangle \;+\; H_2O \;\longrightarrow\; CH_3\overset{\overset{O}{\|}}{C}OH \;+\; \langle \rangle\text{—}SH$$

$$CH_3\overset{\overset{O}{\|}}{C}\underset{\underset{H}{|}}{N}\text{—}\langle \rangle \;+\; H_2O \;\longrightarrow\; CH_3\overset{\overset{O}{\|}}{C}OH \;+\; \langle \rangle\text{—}NH_2$$

(b) B-Esterase

$$CH_3\overset{\overset{O}{\|}}{C}O\text{—}\langle \rangle\text{—}NO_2 \;+\; H_2O \;\longrightarrow\; CH_3\overset{\overset{O}{\|}}{C}OH \;+\; HO\text{—}\langle \rangle\text{—}NO_2$$

(c) C-Esterase

**Figure 7.13**   Examples of esterase/amidase reactions involving xenobiotics.

of the oxirane ring results in detoxication of the very reactive epoxide, in some cases, such as benzo(a)pyrene, the hydration of an epoxide is the first step in an activation sequence that ultimately yields highly toxic *trans*-dihydrodiol intermediates. In others, reactive epoxides are detoxified by both glutathione transferase and epoxide hydrolase. The reaction probably involves a nucleophilic attack by −OH on the oxirane carbon. The most studied epoxide hydrolase is microsomal, and the enzyme has been purified from hepatic microsomes of several species. Although less well known, soluble epoxide hydrolases with different substrate specificities have also been described. Examples of epoxide hydrolase reactions are shown in Figure 7.14.

### 7.2.9   DDT Dehydrochlorinase

DDT-dehydrochlorinase is an enzyme that occurs in both mammals and insects and has been studied most intensively in DDT-resistant houseflies. It catalyzes the dehydrochlorination of DDT to DDE and occurs in the soluble fraction of tissue homogenates. Although the reaction requires glutathione, it apparently serves in a catalytic role

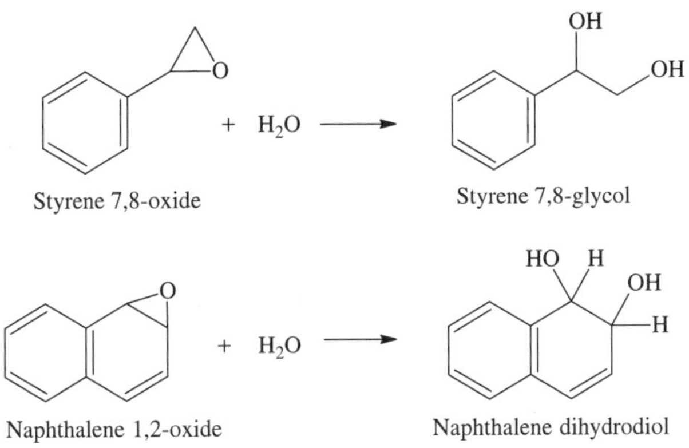

**Figure 7.14**   Examples of epoxide hydrolase reactions.

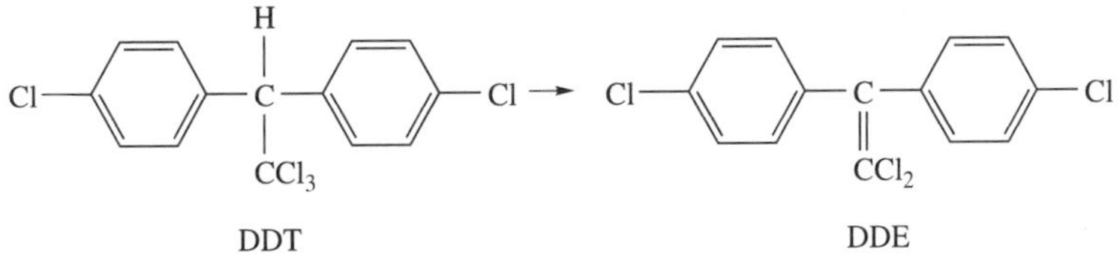

**Figure 7.15**   DDT-dehydrochlorinase.

because it does not appear to be consumed during the reaction. The Km for DDT is $5 \times 10^{-7}$ mol/L with optimum activity at pH 7.4. The monomeric form of the enzyme has a molecular mass of about 36,000 daltons, but the enzyme normally exists as a tetramer. In addition to catalyzing the dehydrochlorination of DDT to DDE and DDD (2,2-*bis*(*p*-chlorophenyl)-1,1-dichloroethane) to TDE (2,2-*bis*(*p*-chlorophenyl)-1-chlorothylen), DDT dehydrochlorinase also catalyzes the dehydrohalogenation of a number of other DDT analogues. In all cases the *p,p* configuration is required, *o,p*, and other analogues are not utilized as substrates. The reaction is illustrated in Figure 7.15.

## 7.3   PHASE II REACTIONS

Products of phase I metabolism and other xenobiotics containing functional groups such as hydroxyl, amino, carboxyl, epoxide, or halogen can undergo conjugation reactions with endogenous metabolites, these conjugations being collectively termed phase II reactions. The endogenous metabolites in question include sugars, amino acids, glutathione, sulfate, and so on. Conjugation products, with only rare exceptions, are more polar, less toxic, and more readily excreted than are their parent compounds.

Conjugation reactions usually involve metabolite activation by some high-energy intermediate and have been classified into two general types: type I, in which an activated conjugating agent combines with the substrate to yield the conjugated product, and type II, in which the substrate is activated and then combines with an amino acid to yield a conjugated product. The formation of sulfates and glycosides are examples of type I, whereas type II consists primarily of amino acid conjugation.

### 7.3.1 Glucuronide Conjugation

The glucuronidation reaction is one of the major pathways for elimination of many lipophilic xenobiotics and endobiotics from the body. The mechanism for this conjugation involves the reaction of one of many possible functional groups (R–OH, Ar–OH, R–NH2, Ar–NH$_2$, R–COOH, Ar–COOH) with the sugar derivative, uridine 5′-diphosphoglucuronic acid (UDPGA). Homogeneous glucuronosyl transferase has been isolated as a single polypeptide chain of about 59,000 daltons, apparently containing carbohydrate, whose activity appears to be dependent on reconstitution with microsomal lipid. There appears to be an absolute requirement for UDPGA: related UDP-sugars will not suffice. This enzyme, as it exists in the microsomal membrane, does not exhibit its maximal capacity for conjugation; activation by some means (e.g., detergents), is required. The reaction involves a nucleophilic displacement (SN$^2$ reaction) of the functional group of the substrate with Walden inversion. UDPGA is in the $\alpha$-configuration whereas, due to the inversion, the glucuronide formed is in the $\beta$-configuration. The enzyme involved, the UDP glucuronosyl transferase (UGT), is found in the microsomal fraction of liver, kidney, intestine, and other tissues. Examples of various types of glucuronides are shown in Figure 7.16.

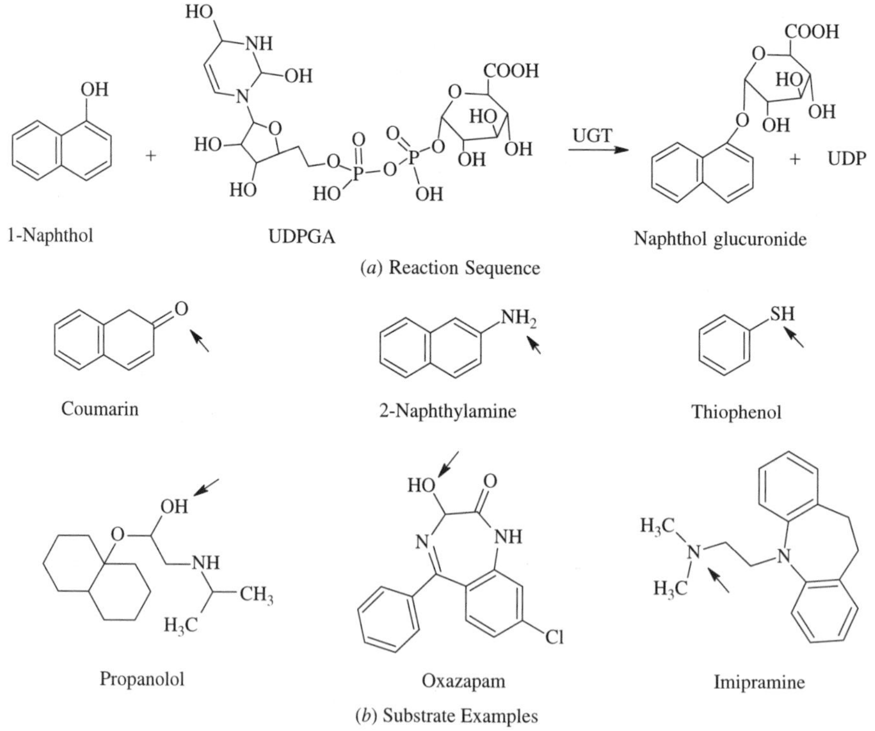

*(a)* Reaction Sequence

1-Naphthol          UDPGA          Naphthol glucuronide

Coumarin          2-Naphthylamine          Thiophenol

Propanolol          Oxazapam          Imipramine

*(b)* Substrate Examples

**Figure 7.16**   Reaction sequences of uridine diphospho glucuronosyl transferase and chemical structures of compounds that form glucuronides. Arrows indicate the position on each molecule where glucuronidation occurs.

Glucuronide conjugation generally results in the formation of products that are less biologically and chemically reactive. This, combined with their greater polarity and greater susceptibility to excretion, contributes greatly to the detoxication of most xenobiotics. However, there are now many examples where glucuronide conjugation results in greater toxicity. Perhaps the best-known example involves the bioactivation of N-hydroxy-2-acetylaminofluorine. This substrate, unlike 2-acetylaminofluorine, is unable to bind to DNA in the absence of metabolism. However, following glucuronide conjugation by linkage of the oxygen through the N-hydroxy group, this substrate becomes equipotent as a hepatocarcinogen with 2-acetylaminofluorine based on its ability to bind to DNA. Another relatively large class of xenobiotics that are often activated by glucuronide conjugation are the acyl glucuronides of carboxylic acids. Useful therapeutic drugs within this class include nonsteroidal anti-inflammatory drugs (NSAIDS), hypolipidemic drugs (clofibrate), and anticonvulsants (valproic acid). The various syndromes associated with the clinical use of some of these drugs (including cytotoxic, carcinogenic, and various immunologic effects) are thought to be the result of the ability of the glucuronide conjugates to react with nucleophilic macromolecules (protein and DNA).

A wide variety of reactions are mediated by glucuronosyltransferases, O-glucuronides, N-glucuronides, and S-glucuronides have all been identified. At this time over 35 different UGT gene products have been described from several different species. These are responsible for the biotransformation of greater than 350 different substrates. Evidence from molecular cloning suggests that the UGTs belong to one of two large superfamilies, sharing less than 50% amino acid identity. Nomenclature of these genes is similar to that of the CYP superfamily. The UGT1 gene family consists of a number of UGTs that arise from alternate splicing of multiple first exons and share common exons 2–5. Members of the UGT2 family catalyze the glucuronidation of a wide variety of substrates including steroids, bile acids, and opioids.

There are nine known human isozymes within the UGT1 family and six within the UGT2 family. Polymorphic forms of some of these enzymes are associated with diseases and significant adverse effects to some drugs.

Jaundice, a condition resulting from the failure of either transport or conjugation of bilirubin, becomes clinically evident when serum bilirubin levels exceed 35 $\mu$M. Although the human UGT1A locus encompasses nine functional transferase genes, only one isoform, UGT1A1, is involved in inherited diseases of bilirubin metabolism. All three inheritable hyperbilirubineamias are the result of either mutant UGT1A1 alleles or UGT1A1 promoter polymorphisms. To date, 33 mutant UGT1A1 alleles have been identified. For the disease to be clinically manifest, one must either be homozygous for the mutant allele or have multiple heterozygous mutant alleles.

## 7.3.2 Glucoside Conjugation

Although rare in vertebrates, glucosides formed from xenobiotics are common in insects and plants. Formed from UDP-glucose, they appear to fall into the same classes as the glucuronides.

## 7.3.3 Sulfate Conjugation

Sulfation and sulfate conjugate hydrolysis, catalyzed by various members of the sulfotransferases (SULT) and sulfatase enzyme superfamilies, play important roles in the metabolism and disposition of many xenobiotics and endogenous substrates. Reactions of the sulfotransferase enzyme with various xenobiotics, including alcohols, arylamines, and phenols, result in the production of water soluble sulfate esters that often are readily eliminated from the organism. Although generally these reactions are important in detoxication, they have also been shown to be involved in carcinogen activation, prodrug processing, cellular signaling pathways, and the regulation of several potent endogenous chemicals including thyroid hormones, steroids, and catechols. The overall sulfation pathway shown in Figure 7.17, consists of two enzyme systems: the SULTs, which catalyze the sulfation reaction, and the sulfatases, which catalyze the hydrolysis of sulfate esters formed by the action of the SULTs.

Sulfation is expensive in energy terms for the cell, since two molecules of ATP are necessary for the synthesis of one molecule of 3′-phosphoadenosine 5′-phosphosulfate (PAPS). Both enzymes involved in the synthesis of PAPS, ATP sulfurylase, and APS kinase, reside within a single bifunctional cytosolic protein of approximately 56 kDa, where substrate channeling of APS from ATP sulfurylase to APS kinase occurs. Several group VI anions other than sulfate can also serve as substrates, although the resultant anhydrides are unstable. Because this instability would lead to the overall consumption of ATP, these other anions can exert a toxic effect by depleting the cell of ATP.

In humans, there are five well-characterized SULT genes, each possessing widely different amino acid sequences and with widely different substrate specificities. Based

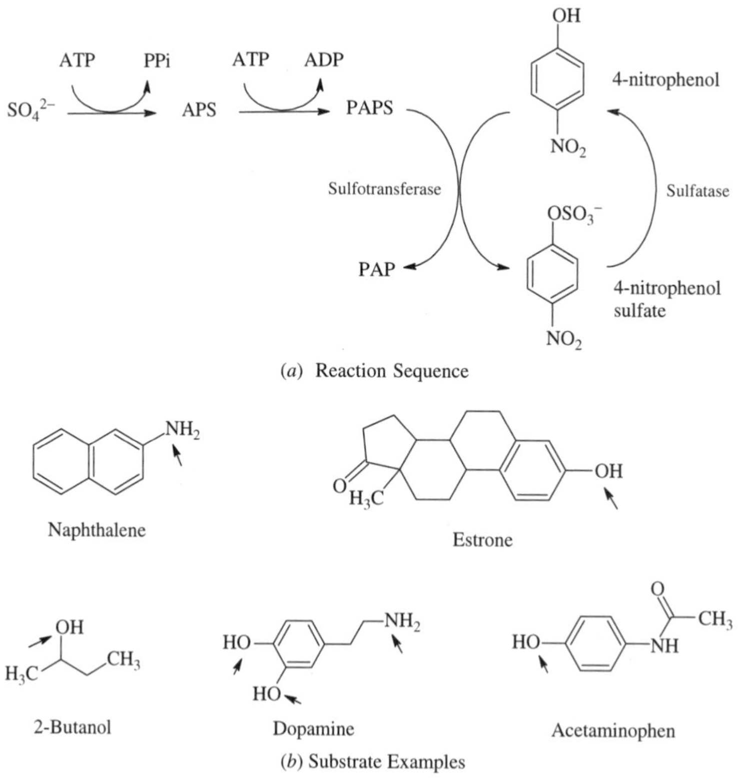

(a) Reaction Sequence

(b) Substrate Examples

**Figure 7.17** Reaction sequence of sulfotransferases and chemical structures of compounds that form sulfates. Arrows indicate positions on each molecule where sulfotransferases may attack.

on amino acid sequence identity as well as substrate preference, these can be separated into two families, phenol SULTs (P-PST, SULT1A2, M-PST and EST) and hydroxysteriod SULT (HST). Phenol SULTs from rat liver have been separated into four distinct forms, each of which catalyzes the sulfation of various phenols and catecholamines. They differ, however, in pH optimum, relative substrate specificity, and immunologic properties. The molecules of all of them are in the range of 61,000 to 64,000 daltons.

Hydroxysteroid sulfotansferase also appears to exist in several forms. This reaction is now known to be important, not only as a detoxication mechanism but also in the synthesis and possibly the transport of steroids. Hydroxysteroid sulfotransferase will react with hydroxysterols and primary and secondary alcohols but not with hydroxyl groups in the aromatic rings of steroids.

### 7.3.4  Methyltransferases

A large number of both endogenous and exogenous compounds can be methylated by several $N$-, $O$-, and $S$-methyl transferases. The most common methyl donor is $S$-adenosyl methionine (SAM), which is formed from methionine and ATP. Even though these reactions may involve a decrease in water solubility, they are generally detoxication reactions. Examples of biologic methylation reactions are seen in Figure 7.18.

*N-Methylation.* Several enzymes are known that catalyze $N$-methylation reactions. They include histamine $N$-methyltransferase, a highly specific enzyme that occurs in

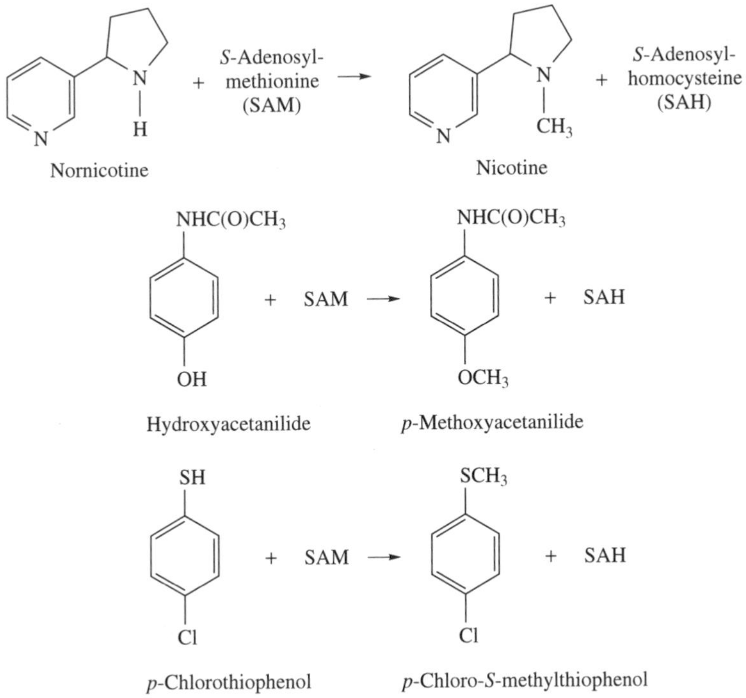

**Figure 7.18**  Examples of methyl transferase reactions.

the soluble fraction of the cell, phenylethanolamine $N$-methyltransferase, which catalyzes the methylation of noradrenaline to adrenaline as well as the methylation of other phenylethanolamine derivatives. A third $N$-methyltransferase is the indoethylamine $N$-methyltansferase, or nonspecific $N$-methyltransferase. This enzyme occurs in various tissues. It methylates endogenous compounds such as serotonin and tryptamine and exogenous compounds such as nornicotine and norcodeine. The relationship between this enzyme and phenylethanolamine $N$-methyltransferase is not yet clear.

*O-Methylation.* Catechol $O$-methyltransferase occurs in the soluble fraction of several tissues and has been purified from rat liver. The purified form has a molecular weight 23,000 daltons, requires $S$-adenosylmethionine and $Mg^+$, and catalyzes the methylation of epinephrine, norepinephrine, and other catechol derivatives. There is evidence that this enzyme exists in multiple forms.

A microsomal $O$-methyltransferase that methylates a number of alkyl-, methoxy-, and halophenols has been described from rabbit liver and lungs. These methylations are inhibited by SKF-525, $N$-ethyl-maleimide and $p$-chloromercuribenzoate. A hydroxyindole $O$-methyltransferase, which methylates $N$-acetyl-serotonin to melatonin and, to a lesser extent, other 5-hydroxyindoles and 5,6-dihydroxyindoles, has been described from the pineal gland of mammals, birds, reptiles, amphibians, and fish.

*S-Methylation.* Thiol groups of some foreign compounds are also methylated, the reaction being catalyzed by the enzyme, thiol $S$-methyltransferase. This enzyme is microsomal and, as with most methyl transferases, utilizes $S$-adenosylmethionine. It has been purified from rat liver and is a monomer of about 28,000 daltons. A wide variety of substrates are methylated, including thiacetanilide, mercaptoethanol, and diphenylsulfide. This enzyme may also be important in the detoxication of hydrogen sulfide, which is methylated in two steps, first to the highly toxic methanethiol and then to dimethylsulfide.

Methylthiolation, or the transfer of a methylthio ($CH3S-$) group to a foreign compound may occur through the action of another recently discovered enzyme, cysteine conjugate $\beta$-lyase. This enzyme acts on cysteine conjugates of foreign compounds as follows:

$$RSCH_2CH(NH_2)COOH \longrightarrow RSH + NH_3 + CH_3C(O)COOH$$

The thiol group can then be methylated to yield the methylthio derivative of the original xenobiotic.

*Biomethylation of Elements.* The biomethylation of elements is carried out principally by microorganisms and is important in environmental toxicology, particularly in the case of heavy metals, because the methylated compounds are absorbed through the membranes of the gut, the blood-brain barrier, and the placenta more readily than are the inorganic forms. For example, inorganic mercury can be methylated first to monomethylmercury and subsequently, to dimethylmercury:

$$Hg^{2+} \longrightarrow CH_3HG^+ \longrightarrow (CH_3)_2 Hg$$

The enzymes involved are reported to use either $S$-adenosylmethionine or vitamin $B_{12}$ derivatives as methyl donors, and in addition to mercury, the metals, lead, tin, and thallium as well as the metalloids, arsenic, selenium, tellurium, and sulfur are methylated. Even the unreactive metals, gold and platinum, are reported as substrates for these reactions.

### 7.3.5  Glutathione S-Transferases (GSTs) and Mercapturic Acid Formation

Although mercapturic acids, the $N$-acetylcysteine conjugates of xenobiotics, have been known since the early part of the twentieth century, only since the early 1960s has the source of the cysteine moiety (glutathione) and the enzymes required for the formation of these acids been identified and characterized. The overall pathway is shown in Figure 7.19.

The initial reaction is the conjugation of xenobiotics having electrophilic substituents with glutathione, a reaction catalyzed by one of the various forms of GST. This is followed by transfer of the glutamate by $\gamma$-glutamyltranspeptidase, by loss of glycine through cysteinyl glycinase, and finally by acetylation of the cysteine amino group. The overall sequence, particularly the initial reaction is extremely important in toxicology because, by removing reactive electrophiles, vital nucleophilic groups in macromolecules such as proteins and nucleic acids are protected. The mercapturic acids formed can be excreted either in the bile or in the urine.

The GSTs, the family of enzymes that catalyzes the initial step, are widely distributed, being found in essentially all groups of living organisms. Although the best-known examples have been described from the soluble fraction of mammalian liver, these enzymes have also been described in microsomes. All forms appear to be highly specific with respect to glutathione but nonspecific with respect to xenobiotic substrates, although the relative rates for different substrates can vary widely from one form to another. The types of reactions catalyzed include the following: alkyltransferase,

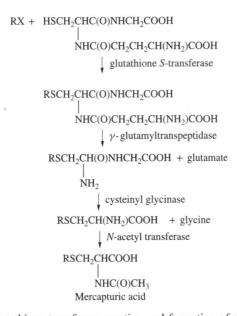

**Figure 7.19**  Glutathione transferase reaction and formation of mercapturic acids.

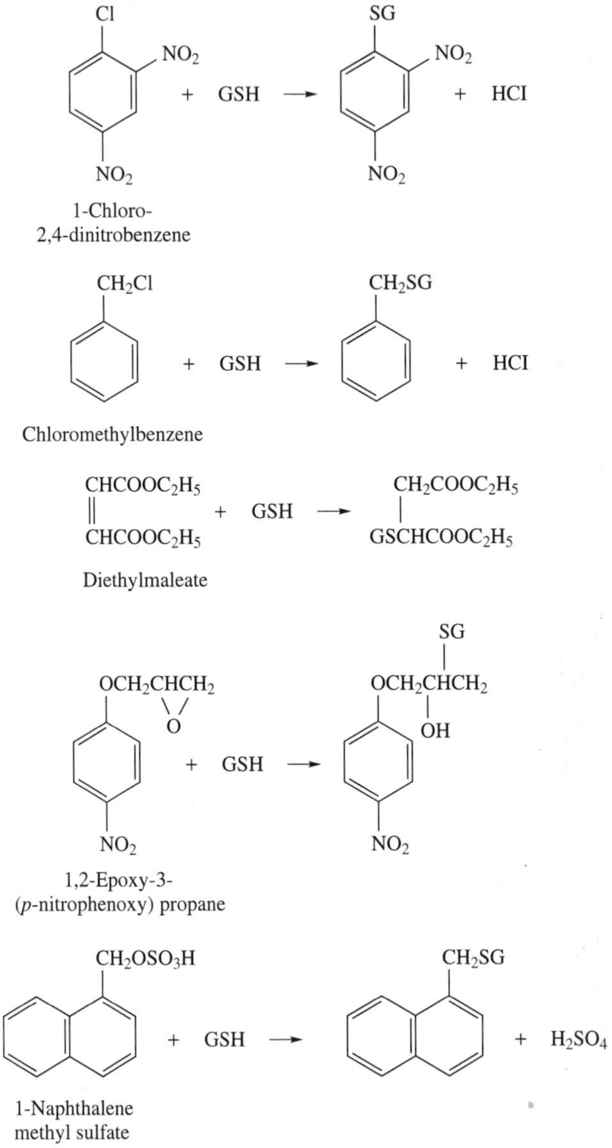

**Figure 7.20**  Examples of glutathione transferases reactions.

aryltransferase, aralkyltransferase, alkenetransferase, and epoxidetransferase. Examples are shown in Figure 7.20.

Multiple forms of GST have been demonstrated in the liver of many mammalian species; multiple forms also occur in insects. Most GSTs are soluble dimeric proteins with molecular weights ranging between 45,000 and 50,000 daltons. All forms appear to be nonspecific with respect to the reaction types described, although the kinetic constants for particular substrates vary from one form to another. They are usually named from their chromatographic behavior. At least two are membrane-bound glutathione transferases, one of which is involved in metabolism of xenobiotics and is designated the microsomal GST. The cytosolic GSTs are divided into six families (historically called classes): the $\alpha$ (alpha), $\kappa$ (kappa), $\mu$ (mu), $\pi$ (pi), $\sigma$ (sigma), and $\theta$ (theta) families. A new system of nomenclature proposes the term GST for the enzyme, preceded by the use of a small roman letter for the species (m for mouse, h for humans, etc.) followed by a capital roman letter for the family (A for $\alpha$, K for $\kappa$, etc.). Subunits are to be designated by arabic numbers, with the two subunits represented with a hyphen between them. For example, hGSTM1-2 designates a heterodimer of the human family mu, which possesses subunits one and two.

Glutathione conjugation dramatically increases the water solubility of the metabolites compared to the parent compounds. The metabolites are released from the cell by an active transport system belonging to the multi-drug resistance (mdr) protein. Prior to excretion, the metabolites are usually processed by multiple enzymes to release the substrate conjugated to a mercapturic acid (Figure 7.19). The enzymes involved in this process are $\gamma$-glutamyltranspeptidase, cysteinyl glycinase, and $N$-acetyl transferase.

$\gamma$-Glutamyltranspeptidase is a membrane-bound glycoprotein that has been purified from both the liver and kidney of several species. Molecular weights for the kidney enzyme are in the range of 68,000 to 90,000 daltons, and the enzyme appears to consist of two unequal subunits; the different forms appear to differ in the degree of sialalylation. This enzyme, which exhibits wide specificity toward $\gamma$-glutamyl peptides and has a number of acceptor amino acids, catalyzes two types of reactions:

| | | |
|---|---|---|
| Hydrolysis | $\gamma$-Glu-R + $H_2O$ | Glu + HR |
| Transpeptidation | $\gamma$-Glu-R + Acceptor | $\gamma$-Glu-Acceptor + HR |
| | $\gamma$-Glu-R + $\gamma$-Glu-R | $\gamma$-Glu$\gamma$-Glu-R + HR |

Aminopeptidases that catalyze the hydrolysis of cysteinyl peptides are known. The membrane-bound aminopeptidases are glycoproteins, usually with molecular weights of about 100,000 daltons. They appear to be metalloproteins, one of the better known being a zinc-containing enzyme. Other enzymes, such as the leucine aminopeptidase, are cytosolic but, at least in this case, are also zinc-containing. The substrate specificity of these enzymes varies but most are relatively nonspecific.

Little is known of the $N$-acetyltransferase(s) responsible for the acetylation of the S-substituted cysteine. It is found in the microsomes of the kidney and the liver, however, and is specific for acetyl CoA as the actyl donor. It is distinguished from other N-acetyltransferases by its substrate specificity and subcellular location.

### 7.3.6  Cysteine Conjugate $\beta$-Lyase

This enzyme uses cysteine conjugates as substrates, releasing the thiol of the xenobiotic, pyruvic acid, and ammonia, with subsequent methylation giving rise to the methylthio derivative. The enzyme from the cytosolic fraction of rat liver is pyridoxal phosphate requiring protein of about 175,000 daltons. Cysteine conjugates of aromatic compounds are the best substrates, and it is necessary for the cysteine amino and carboxyl groups to be unsubstituted for enzyme activity.

### 7.3.7  Acylation

Acylation reactions are of two general types, the first involving an activated conjugation agent, coenzyme A (CoA), and the second involving activation of the foreign

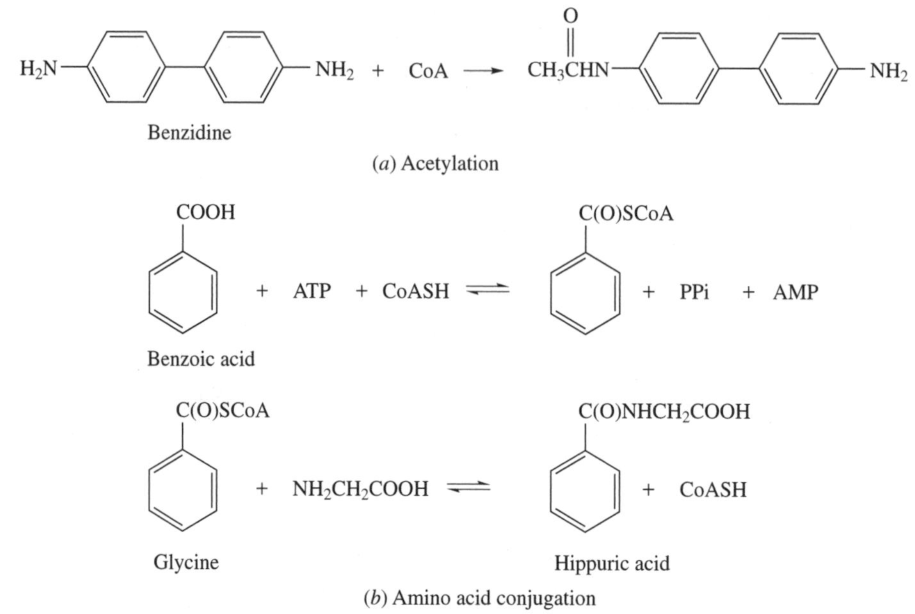

**Figure 7.21** Examples of acylation reactions.

compounds and subsequent acylation of an amino acid. This type of conjugation is commonly undergone by exogenous carboxylic acids and amides, and although the products are often less water soluble than the parent compound, they are usually less toxic. Examples of acylation reactions are shown in Figure 7.21.

*Acetylation.* Acetylated derivatives of foreign exogenous amines are acetylated by $N$-acetyl transferase, the acetyl donor being CoA. This enzyme is cytosolic, has been purified from rat liver, and is known to occur in several other organs. Evidence exists for the existence of multiple forms of this enzyme. Although endogenous amino, hydroxy, and thiol compounds are acetylated in vivo, the acetylation of exogenous hydroxy and thiol groups is presently unknown.

Acetylation of foreign compounds is influenced by both development and genetics. Newborn mammals generally have a low level of the transferase, whereas due to the different genes involved, fast and slow acetylators have been identified in both rabbit and human populations. Slow acetylators are more susceptible to the effects of compounds detoxified by acetylation.

*N,O-Acyltransferase.* The $N$-acyltransferase enzyme is believed to be involved in the carcinogenicity of arylamines. These compounds are first $N$-oxidized, and then, in species capable of their $N$-acetylation, acetylated to arylhydroxamic acids. The effect of $N,O$-transacetylation is shown in Figure 7.22. The $N$-acyl group of the hydroxamic acid is first removed and is then transferred, either to an amine to yield a stable amide or to the oxygen of the hydroxylamine to yield a reactive $N$-acyloxyarylamine. These compounds are highly reactive in the formation of adducts with both proteins and nucleic acids, and $N,O$-acyltransferase, added to the medium in the Ames test, increases the mutagenicity of compounds such as $N$-hydroxy-2-acetylaminofluorene.

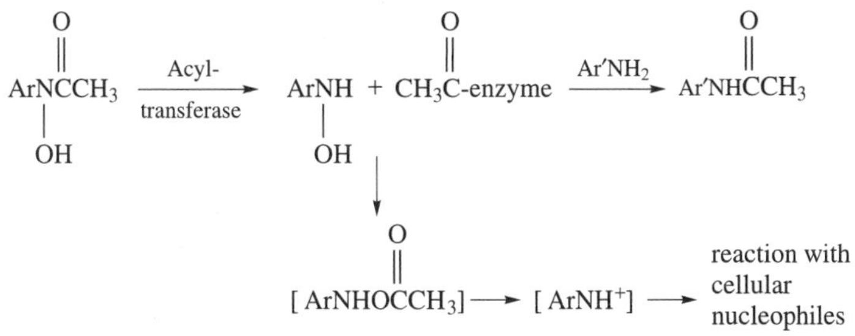

$$\underset{\underset{OH}{|}}{\overset{\overset{O}{\|}}{ArNCCH_3}} \xrightarrow[\text{transferase}]{\text{Acyl-}} \underset{\underset{OH}{|}}{ArNH} + CH_3\overset{\overset{O}{\|}}{C}\text{-enzyme} \xrightarrow{Ar'NH_2} Ar'NH\overset{\overset{O}{\|}}{C}CH_3$$

$$[\,Ar NH O\overset{\overset{O}{\|}}{C}CH_3] \longrightarrow [\,Ar NH^+\,] \longrightarrow \begin{array}{l}\text{reaction with}\\ \text{cellular}\\ \text{nucleophiles}\end{array}$$

**Figure 7.22**   $N$-, $O$-Acyltransferase reactions of arylhydroxamic acid. Ar = aryl group.

Despite its great instability this enzyme has been purified from the cytosolic fraction of the rat liver.

*Amino Acid Conjugation.* In the second type of acylation reaction, exogenous carboxylic acids are activated to form S-CoA derivative in a reaction involving ATP and CoA. These CoA derivatives then acylate the amino group of a variety of amino acids. Glycine and glutamate appear to be the most common acceptor of amino acids in mammals; in other organisms, other amino acids are involved. These include ornithine in reptiles and birds and taurine in fish.

The activating enzyme occurs in the mitochondria and belongs to a class of enzymes known as the ATP-dependent acid: CoA ligases (AMP) but has also been known as acyl CoA synthetase and acid-activating enzyme. It appears to be identical to the intermediate chain length fatty acyl-CoA-synthetase.

Two acyl-CoA: amino acid $N$-acyltransferases have been purified from liver mitochondria of cattle, Rhesus monkeys, and humans. One is a benzoyltransferase CoA that utilizes benzyl-CoA, isovaleryl-CoA, and tiglyl-CoA, but not phenylacetyl CoA, malonyl-CoA, or indolacetyl-CoA. The other is a phenylacetyl transferase that utilizes phenylacetyl-CoA and indolacetyl-CoA but is inactive toward benzoyl-CoA. Neither is specific for glycine, as had been supposed from studies using less defined systems; both also utilize asparagine and glutamine, although at lesser rates than glycine.

Bile acids are also conjugated by a similar sequence of reactions involving a microsomal bile acid: CoA ligase and a soluble bile acid $N$-acyl-transferase. The latter has been extensively purified, and differences in acceptor amino acids, of which taurine is the most common, have been related to the evolutionary history of the species.

*Deacetylation.* Deacetylation occurs in a number of species, but there is a large difference between species, strains, and individuals in the extent to which the reaction occurs. Because acetylation and deacetylation are catalyzed by different enzymes, the levels of which vary independently in different species, the importance of deacetylation as a xenobiotic metabolizing mechanism also varies between species. This can be seen in a comparison of the rabbit and the dog. The rabbit, which has high acetyltransferase activity and low deacetylase, excretes significant amounts of acetylated amines. The dog, in which the opposite situation obtains, does not.

A typical substrate for the aromatic deacetylases of the liver and kidney is acetanilide, which is deacylated to yield aniline.

## 7.3.8 Phosphate Conjugation

Phosphorylation of xenobiotics is not a widely distributed conjugation reaction, insects being the only major group of animals in which it is found. The enzyme from the gut of cockroaches utilizes ATP, requires $Mg^+$, and is active in the phosphorylation of 1-naphthol and $p$-nitrophenol.

## SUGGESTED READING

Benedetti, M. S. Biotransformation of xenobiotics by amine oxidases. *Fundam. Clinic. Pharmacol.* **15**: 75–84, 2001.

Coughtrie, M. W. H., S. Sharp, K. Maxwell, and N. P. Innes. Biology and function of the reversible sulfation pathway catalyzed by human sulfotransferases and sulfatases. *Chemico-Biol. Inter.* **109**: 3–27, 1998.

Duffel, M. W., A. D. Marshall, P. McPhie, V. Sharma, and W. B. Jakoby. Enzymatic aspects of the phenol (aryl) sulfotransferases. *Drug Metab. Rev.* **33**: 369–395, 2001.

Hayes, J. D., and D. J. Pulford. The glutathione *S*-transferase supergene family: Regulation of GST and the contribution of the isoenzymes to cancer chemoprotection and drug resistance. *Crit. Rev. Biochem. Mol. Biol.* **30**: 445–600, 1995.

Hodgson, E., and J. A. Goldstein. Metabolism of toxicants: Phase I reactions and pharmacogenetics. In *Introduction to Biochemical Toxicology*, 3rd ed., E. Hodgson and R. C. Smart, eds. New York: Wiley, 2001, pp. 67–113.

Koukouritaki, S. B., P. Simpson, C. K. Yeung, A. E. Rettie, and R. N. Hines. Human hepatic flavin-containing monooxygenases 1 (FMO1) and 3 (FMO3) developmental expression. *Pediatric Res.* **51**: 236–243, 2002.

Lawton, M. P., J. R. Cashman, T. Cresteil, C. T. Dolphin, A. A. Elfarra, R. N. Hines, E. Hodgson, T. Kimura, J. Ozols, I. R. Phillips, R. M. Philpot, L. L. Poulsen, A. E. Rettie, E. A. Shephard, D. E. Williams and D. M. Ziegler. A nomenclature for the mammalian flavin-containing monooxygenase gene family based on amino acid sequence identities. *Arch. Biochem. Biophys.* **308**: 254–257, 1994.

LeBlanc, G. A., and W. A. Dauterman. Conjugation and elimination of toxicants: In *Introduction to Biochemical Toxicology*, 3rd edn., E. Hodgson and R. C. Smart, eds. New York: Wiley, 2001, pp. 115–135.

Nelson, D. R., L. Koymans, T. Kamataki, J. J. Stegeman, R. Feyereisen, D. J. Waxman, M. R. Waterman, O. Gotoh, M. J. Coon, R. W. Estabrook, I. C. Gunsalus, and D. W. Nebert. P450 superfamily: Update on new sequences, gene mapping, accession numbers and nomenclature. *Pharmacogenetics* **6**: 1–42, 1996.

Oesch, F, and M. Arand. Xenobiotic metabolism. In *Toxicology*, H. Marquardt, S. G. Shafer, R. McClellan, and F. Welsch, eds. New York: Academic Press, 1999, pp. 83–109.

Ritter, J. K. Roles of glucuronidation and UDP-glucuronosyltransferases in xenobiotic bioactivation reactions. *Chemico-Biol. Inter.* **129**: 171–193, 2001.

Tukey, R. H., and C. P. Strassburg. Human UDP-glucuronosyltransferases: Metabolism, expression, and disease. *Ann. Rev. Pharmacol. Toxicol.* **40**: 581–616, 2000.

Vasiliou, V., A. Pappa, and D. R. Petersen. Role of aldehyde dehydrogenases in endogenous and xenobiotic metabolism. *Chemico-Biol. Inter.* **129**: 1–19, 2000.

# CHAPTER 8

# Reactive Metabolites

RANDY L. ROSE and PATRICIA E. LEVI

## 8.1  INTRODUCTION

Between uptake from the environment and excretion from the body, many exogenous compounds (xenobiotics) undergo metabolism to highly reactive intermediates. These metabolites may interact with cellular constituents in numerous ways, such as binding covalently to macromolecules and/or stimulating lipid peroxidation. This biotransformation of relatively inert chemicals to highly reactive intermediary metabolites is commonly referred to as metabolic activation or bioactivation, and it is known to be the initial event in many chemically induced toxicities. Some toxicants are direct acting and require no activation, whereas other chemicals may be activated nonenzymatically. The focus of this chapter, however, is on toxicants requiring metabolic activation and to those processes involved in activation.

In the 1940s and 1950s the pioneering studies of James and Elizabeth Miller provided early evidence for in vivo conversion of chemical carcinogens to reactive metabolites. They found that reactive metabolites of the aminoazo dye $N,N$-dimethyl-4-aminoazobenzene (DAB), a hepatocarcinogen in rats, would bind covalently to proteins and nucleic acids. The term, metabolic activation, was coined by the Millers to describe this process. Moreover they demonstrated that covalent binding of these chemicals was an essential part of the carcinogenic process.

The overall scheme of metabolism for potentially toxic xenobiotics is outlined in Figure 8.1. As illustrated by this diagram, xenobiotic metabolism can produce not only nontoxic metabolites, which are more polar and readily excreted (detoxication), but also highly reactive metabolites, which can interact with vital intracellular macromolecules, resulting in toxicity. In addition reactive metabolites can be detoxified—for example, by interaction with glutathione. In general, reactive metabolites are electrophiles (molecules containing positive centers). These electrophiles in turn can react with cellular nucleophiles (molecules containing negative centers), such as glutathione, proteins, and nucleic acids. Other reactive metabolites may be free radicals or act as radical generators that interact with oxygen to produce reactive oxygen species that are capable of causing damage to membranes, DNA, and other macromolecules.

*A Textbook of Modern Toxicology, Third Edition,* edited by Ernest Hodgson
ISBN 0-471-26508-X  Copyright © 2004 John Wiley & Sons, Inc.

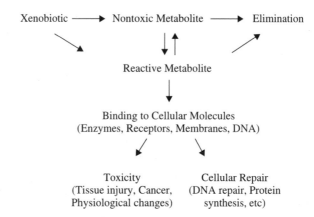

**Figure 8.1**   The relationship between metabolism, activation, detoxication, and toxicity of a chemical.

Although a chemical can be metabolized by several routes, the activation pathway is often a minor route with the remainder of the pathways resulting in detoxication. Activation, however, may become a more dominant pathway in certain situations, thus leading to toxicity. Several examples illustrating these situations are discussed later in this chapter. Some important terms that are often used when discussing activation include parent compound, sometimes referred to as procarcinogen in the case of a carcinogen or prodrug for pharmaceutical compounds; proximate toxic metabolite or proximate carcinogen for one or more of the intermediates; and ultimate toxic metabolite or ultimate carcinogen for the reactive species that binds to macromolecules and DNA.

## 8.2   ACTIVATION ENZYMES

Whereas most, if not all, of the enzymes involved in xenobiotic metabolism can form reactive metabolites (Table 8.1), the enzyme systems most frequently involved in the activation of xenobiotics are those which catalyze oxidation reactions. The cytochrome P450 monooxygenases (CYP) are by far the most important enzymes involved in the oxidation of xenobiotics. This is because of the abundance of CYP (especially in the liver), the numerous isozymes of CYP, and the ability of CYP to be induced by xenobiotic compounds.

Although the CYP enzymes are the most abundant in the liver, they are also present in other tissues including the skin, kidney, intestine, lung, placenta, and nasal mucosa. Because CYP exists as multiple isozymes with different substrate specificities, the presence or absence of a particular CYP isozyme may contribute to tissue-specific toxicities. Many drugs and other foreign compounds are known to induce one or more of the CYP isozymes, resulting in an increase, decrease, or an alteration in the metabolic pathway of chemicals metabolized by the CYP isozymes involved. Specific examples of these types of interactions are given later in this section.

In addition to activations catalyzed by CYPs and FMOs, phase two conjugations, cooxidation by COX during prostaglandin biosynthesis, and metabolism by intestinal

**Table 8.1 Enzymes Important in Catalyzing Metabolic Activation Reactions**

| Type of Reaction | Enzyme |
| --- | --- |
| Oxidation | Cytochrome P450s |
| | Prostaglandin synthetase |
| | Flavin-containing monooxygenases |
| | Alcohol and aldehyde dehydrogenases |
| Reduction | Reductases |
| | Cytochromes P450 |
| | Gut microflora |
| Conjugation | Glutathione transferases |
| | Sulfotransferases |
| | Glucuronidases |
| Deconjugation | Cysteine $S$-conjugate $\beta$-lyase |
| Hydrolysis | Gut microflora, hydrolyses |

microflora may also lead to the formation of reactive toxic products. With some chemicals only one enzymatic reaction is involved, whereas with other compounds, several reactions, often involving multiple pathways, are necessary for the production of the ultimate reactive metabolite.

## 8.3 NATURE AND STABILITY OF REACTIVE METABOLITES

Reactive metabolites include such diverse groups as epoxides, quinones, free radicals, reactive oxygen species, and unstable conjugates. Figure 8.2 gives some examples of activation reactions, the reactive metabolites formed, and the enzymes catalyzing their bioactivation.

As a result of their high reactivity, reactive metabolites are often considered to be short-lived. This is not always true, however, because reactive intermediates can be transported from one tissue to another, where they may exert their deleterious effects. Thus reactive intermediates can be divided into several categories depending on their half-life under physiological conditions and how far they may be transported from the site of activation.

### 8.3.1 Ultra-short-lived Metabolites

These are metabolites that bind primarily to the parent enzyme. This category includes substrates that form enzyme-bound intermediates that react with the active site of the enzyme. Such chemicals are known as "suicide substrates." A number of compounds are known to react in this manner with CYP, and such compounds are often used experimentally as CYP inhibitors (see the discussion of piperonyl butoxide, Section 7.2.2). Other compounds, although not true suicide substrates, produce reactive metabolites that bind primarily to the activating enzyme or adjacent proteins altering the function of the protein.

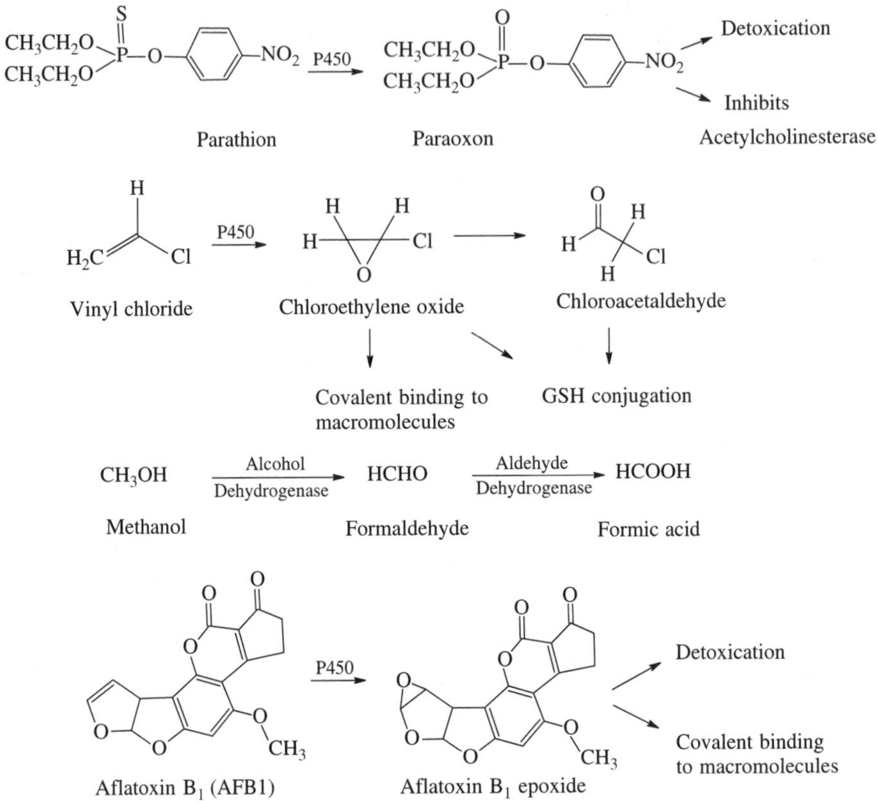

**Figure 8.2**  Examples of some activation reactions.

## 8.3.2  Short-lived Metabolites

These metabolites remain in the cell or travel only to nearby cells. In this case covalent binding is restricted to the cell of origin and to adjacent cells. Many xenobiotics fall into this group and give rise to localized tissue damage occurring at the sites of activation. For example, in the lung, the Clara cells contain high concentrations of CYP and several lung toxicants that require activation often result in damage primarily to Clara cells.

## 8.3.3  Longer-lived Metabolites

These metabolites may be transported to other cells and tissues so that although the site of activation may be the liver, the target site may be in a distant organ. Reactive intermediates may also be transported to other tissues, not in their original form but as conjugates, which then release the reactive intermediate under the specific conditions in the target tissue. For example, carcinogenic aromatic amines are metabolized in the liver to the N-hydroxylated derivatives that, following glucuronide conjugation, are transported to the bladder, where the N-hydroxy derivative is released under the acidic conditions of urine.

## 8.4    FATE OF REACTIVE METABOLITES

If production of reactive metabolites is the initial process in the role of reactive metabolites in toxicity, then the fate of these reactive metabolites is the next step to understand in the process. Within the tissue a variety of reactions may occur depending on the nature of the reactive species and the physiology of the organism.

### 8.4.1    Binding to Cellular Macromolecules

As mentioned previously, most reactive metabolites are electrophiles that can bind covalently to nucleophilic sites on cellular macromolecules such as proteins, polypeptides, RNA, and DNA. This covalent binding is considered to be the initiating event for many toxic processes such as mutagenesis, carcinogenesis, and cellular necrosis, and is discussed in greater detail in the chapters in Parts IV and V.

### 8.4.2    Lipid Peroxidation

Radicals such as $CCl_3{}^\bullet$, produced during the oxidation of carbon tetrachloride, may induce lipid peroxidation and subsequent destruction of lipid membranes (Figure 8.3). Because of the critical nature of various cellular membranes (nuclear, mitochondrial, lysosomal, etc.), lipid peroxidation can be a pivotal event in cellular necrosis.

### 8.4.3    Trapping and Removal: Role of Glutathione

Once reactive metabolites are formed, mechanisms within the cell may bring about their rapid removal or inactivation. Toxicity then depends primarily on the balance

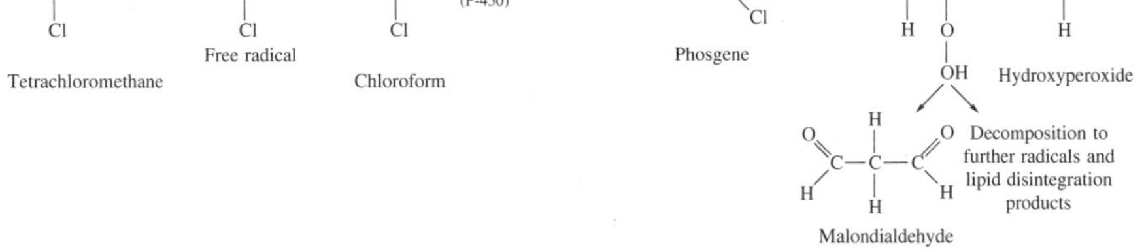

**Figure 8.3**    Metabolism of tetrachloromethane. Upon metabolic activation a $CCl_3$ radical is formed. This radical extracts protons from unsaturated fatty acids to form a free fatty-acid radical. This leads to diene conjugates. At the same time, $O_2$ forms a hydroperoxide with the C radical. Upon its decomposition, malondialdehyde and other disintegration products are formed. In contrast, the $CCl_3$ radical is converted to chloroform, which undergoes further oxidative metabolism. (Reprinted from H. M. Bolt and J. T. Borlak, in *Toxicology*, pp. 645–657, copyright 1999, with permission from Elsevier.)

between the rate of metabolite formation and the rate of removal. With some compounds, reduced glutathione plays an important protective role by trapping electrophilic metabolites and preventing their binding to hepatic proteins and enzymes. Although conjugation reactions occasionally result in bioactivation of a compound, the acetyl-, glutathione-, glucuronyl-, or sulfotransferases usually result in the formation of a nontoxic, water-soluble metabolite that is easily excreted. Thus availability of the conjugating chemical is an important factor in determining the fate of the reactive intermediates.

## 8.5   FACTORS AFFECTING TOXICITY OF REACTIVE METABOLITES

A number of factors can influence the balance between the rate of formation of reactive metabolites and the rate of removal, thereby affecting toxicity. The major factors discussed in this chapter are summarized in the following subsections. A more in-depth discussion of other factors affecting metabolism and toxicity are presented in Chapter 9.

### 8.5.1   Levels of Activating Enzymes

Specific isozymes of CYPs are often important in determining metabolic activation of a foreign compound. As mentioned previously, many xenobiotics induce specific forms of cytochrome P450. Frequently the CYP forms induced are those involved in the metabolism of the inducing agent. Thus a carcinogen or other toxicant has the potential for inducing its own activation. In addition there are species and gender differences in enzyme levels as well as specific differences in the expression of particular isozymes.

### 8.5.2   Levels of Conjugating Enzymes

Levels of conjugating enzymes, such as glutathione transferases, are also known to be influenced by gender and species differences as well as by drugs and other environmental factors. All of these factors will in turn affect the detoxication process.

### 8.5.3   Levels of Cofactors or Conjugating Chemicals

Treatment of animals with $N$-acetylcysteine, a precursor of glutathione, protects animals against acetaminophen-induced hepatic necrosis, possibly by reducing covalent binding to tissue macromolecules. However, depletion of glutathione potentiates covalent binding and hepatotoxicity.

## 8.6   EXAMPLES OF ACTIVATING REACTIONS

The following examples have been selected to illustrate the various concepts of activation and detoxication discussed in the previous sections.

### 8.6.1 Parathion

Parathion is one of several organophosphorus insecticides that has had great economic importance worldwide for several decades. Organophosphate toxicity is the result of excessive stimulation of cholinergic nerves, which is dependent on their ability to inhibit acetylcholinesterases. Interestingly the parent organophosphates are relatively poor inhibitors of acetylcholinesterases, requiring metabolic conversion of a P=S bond to a P=O bond for acetylcholinesterase inhibition (Figure 8.2; see Chapters 11 and 16 for a discussion of the mechanism of acetylcholinesterase inhibition). In vitro studies of rat and human liver have demonstrated that CYP is inactivated by the electrophilic sulfur atom released during oxidation of parathion to paraoxon. Some have shown that the specific isoforms responsible for the metabolic activation of parathion are destroyed in the process. For example, preincubations of NADPH-supplemented human liver microsomes with parathion resulted in the inhibition of some isoform-specific metabolites including testosterone (CYP3A4), tolbutamide (CYP2C9), and 7-ethylresorufin (CYP1A2) but not aniline (CYP2E1). These losses of metabolic activity were also associated with the loss of CYP content as measured by the CO-difference spectra. These results suggest that parathion acts as a suicide substrate, in that its metabolism results in the destruction of the particular isoforms involved in its metabolism. This becomes particularly important because the principal CYP involved in parathion metabolism is CYP3A4, which is the dominant CYP in humans; accounting for between 30–50% of the total liver CYP. Because of this enzyme's importance in drug metabolism, the strong potential for inhibition by organophosphate compounds may have serious consequences in individuals undergoing drug therapy.

### 8.6.2 Vinyl Chloride

A second example of a suicide inhibitor is vinyl chloride. The first step in the biotransformation of vinyl chloride involves the CYP-mediated oxidation of the double bond leading to the formation of an epoxide, or oxirane, which is highly reactive and can easily bind to proteins and nucleic acids. Following activation by CYP, reactive metabolites such as those formed by vinyl chloride bind covalently to the pyrrole nitrogens present in the heme moiety, resulting in destruction of the heme and loss of CYP activity. The interaction of the oxirane structure with nucleic acids results in mutations and cancer. The first indications that vinyl chloride was a human carcinogen involved individuals who cleaned reactor vessels in polymerization plants who were exposed to high concentrations of vinyl chloride and developed angiosarcomas of the liver as a result of their exposure (Figure 8.2).

### 8.6.3 Methanol

Ingestion of methanol, particularly during the prohibition era, resulted in significant illness and mortality. Where epidemics of methanol poisoning have been reported, one-third of the exposed population recovered with no ill effects, one-third have severe visual loss or blindness, and one-third have died. Methanol itself is not responsible for the toxic effects but is rapidly metabolized in humans by alcohol dehydrogenase to formaldehyde, which is subsequently metabolized by aldehyde dehydrogenase to form the highly toxic formic acid (Figure 8.2). The aldehyde dehydrogenase is so efficient in its metabolism of formaldehyde that it is actually difficult to detect formaldehyde in post mortem tissues. Accumulation of formic acid in the tissues results first in blindness through edema of the retina, and eventually to death as a result of acidosis. Successful treatment of acidosis by treatment with base was often still unsuccessful in preventing mortality due to subsequent effects on the central nervous system. Treatment generally consists of hemodialysis to remove the methanol, but where this option is not available, administration of ethanol effectively competes with the production of formic acid by competing with methanol for the alcohol dehydrogenase pathway.

### 8.6.4 Aflatoxin $B_1$

Aflatoxin $B_1$ (AFB1) is one of the mycotoxins produced by Aspergillus flavus and A. parasiticus and is a well-known hepatotoxicant and hepatocarcinogen. It is generally accepted that the activated form of AFB1 that binds covalently to DNA is the 2,3-epoxide (Figure 8.2). AFB1-induced hepatotoxicity and carcinogenicity is known to vary among species of livestock and laboratory animals. The selective toxicity of AFB1 appears to be dependent on quantitative differences in formation of the 2,3-epoxide, which is related to the particular enzyme complement of the organism. Table 8.2 shows the relative rates of AFB1 metabolism by liver microsomes from different species. Because the epoxides of foreign compounds are frequently further metabolized by epoxide hydrolases or are nonenzymatically converted to the corresponding dihydro-diols, existence of the dihydrodiol is considered as evidence for prior formation of the epoxide. Because epoxide formation is catalyzed by CYP enzymes, the amount of AFB1-dihydrodiol produced by microsomes is reflective of the CYP isozyme comple-ment involved in AFB1 metabolism. In Table 8.2, for example, it can be seen that in rat microsomes in which specific CYP isozymes have been induced by phenobarbital (PB), dihydrodiol formation is considerably higher than that in control microsomes.

### 8.6.5 Carbon Tetrachloride

Carbon tetrachloride has long been known to cause fatty acid accumulation and hepatic necrosis. Extraction of a chlorine atom by CYP from carbon tetrachloride results in

**Table 8.2 Formation of Aflatoxin $B_1$ Dihydrodiol by Liver Microsomes**

| Source of Microsomes | Dihydrodiol Formation[a] |
| --- | --- |
| Rat | 0.7 |
| C57 mouse | 1.3 |
| Guinea pig | 2.0 |
| Phenobarbital-induced rat | 3.3 |
| Chicken | 4.8 |

*Source*: Adapted from G. E. Neal et al., *Toxicol. Appl. Pharma-col.* **58**: 431–437, 1981.
[a] μg formed/mg microsomal protein/30 min.

the formation of a trichloromethyl radical that extracts protons from esterified desaturated fatty acids resulting in the production of chloroform (Figure 8.3). Chloroform also undergoes subsequent metabolism by CYP leading to the production of phosgene, which covalently binds to sulfhydryl containing enzymes and proteins leading to toxicity. Differences between hepatic and renal effects of carbon tetrachloride and chloroform toxicity suggest that each tissue produces its own toxic metabolites from these chemicals.

In the case of hepatic toxicity due to carbon tetrachloride, the extraction of protons from fatty acids by the trichloromethyl radical results in the formation of highly unstable lipid radicals that undergo a series of transformations, including rearrangement of double bonds to produce conjugated dienes (Figure 8.3). Lipid radicals also readily react with oxygen, with the subsequent process, termed lipid peroxidation, producing damage to the membranes and enzymes. The resulting lipid peroxyl radicals decompose to aldehydes, the most abundant being malondialdehyde and 4-hydroxy-2,3-nonenal (Figure 8.3).

Since desaturated fatty acids are highly susceptible to free radical attack, neighboring fatty acids are readily affected, and the initial metabolic transformation results in a cascade of detrimental effects on the tissue. The initial production of the trichloromethyl radical from carbon tetrachloride also results in irreversible covalent binding to CYP, resulting in its inactivation. In cases of carbon tetrachloride poisoning, preliminary sublethal doses actually become protective to an organism in the event of further poisoning, since the metabolic activating enzymes are effectively inhibited by the first dose.

### 8.6.6   Acetylaminofluorene

In the case of the hepatocarcinogen, 2-acetylaminofluorene (2-AAF), two activation steps are necessary to form the reactive metabolites (Figure 8.4). The initial reaction, $N$-hydroxylation, is a CYP-dependent phase I reaction, whereas the second reaction, resulting in the formation of the unstable sulfate ester, is a phase II conjugation reaction that results in the formation of the reactive intermediate. Another phase II reaction, glucuronide conjugation, is a detoxication step, resulting in a readily excreted conjugation product.

In some animal species, 2-AAF is known to be carcinogenic, whereas in other species it is noncarcinogenic. The species- and sex-specific carcinogenic potential of

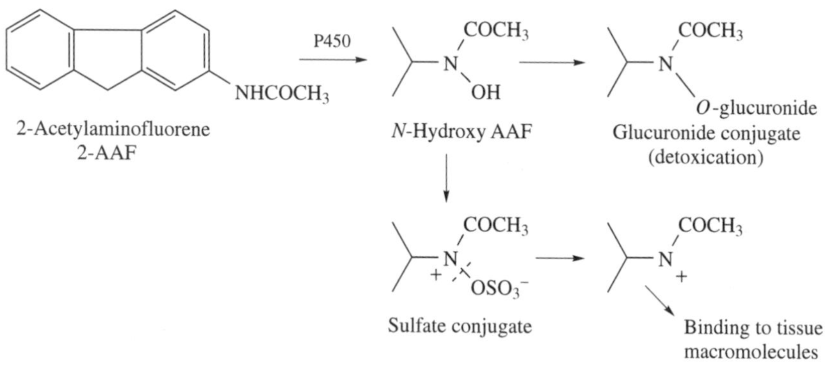

**Figure 8.4**   Bioactivation of 2-acetylaminofluorene.

2-AAF is correlated with the ability of the organism to sequentially produce the *N*-hydroxylated metabolite followed by the sulfate ester. Therefore in an animal such as the guinea pig, which does not produce the *N*-hydroxylated metabolite, 2-AAF is not carcinogenic. In contrast, both male and female rats produce the *N*-hydroxylated metabolite, but only male rats have high rates of tumor formation. This is because male rats have up to 10-fold greater expression of sulfotransferase 1C1 than female rats, which has been implicated in the sulfate conjugation of 2-AAF resulting in higher production of the carcinogenic metabolite.

### 8.6.7  Benzo(*a*)pyrene

The polycyclic aromatic hydrocarbons are a group of chemicals consisting of two or more condensed aromatic rings that are formed primarily from incomplete combustion of organic materials including wood, coal, mineral oil, motor vehicle exhaust, and cigarette smoke. Early studies of cancer in the 1920s involving the fractionation of coal tar identified the carcinogenic potency of pure polycyclic aromatic hydrocarbons, including dibenz(a,h)anthracene and benzo(*a*)pyrene. Although several hundred different polycyclic aromatic hydrocarbons are known, environmental monitoring usually only detects a few compounds, one of the most important of which is benzo(*a*)pyrene. Benzo(*a*)pyrene is also one of the most prevalent polycyclic aromatic hydrocarbons found in cigarette smoke.

Extensive studies of metabolism of benzo(*a*)pyrene have identified at least 15 phase I metabolites. The majority of these are the result of CYP1A1 and epoxide hydrolase reactions. Many of these metabolites are further metabolized by phase II enzymes to produce numerous different metabolites. Studies examining the carcinogenicity of this compound have identified the 7,8-oxide and 7,8-dihydrodiol as proximate carcinogens and the 7,8-diol-9,10 epoxide as a strong mutagen and ultimate carcinogen. Because of the stereoselective metabolizing abilities of CYP isoforms, the reactive 7,8-diol-9,10-epoxide can appear as four different isomers. (Figure 8.5). Interestingly only one of these isomers(+)-benzo(a)pyrene 7,8-diol-9,10 epoxide-2 has significant carcinogenic potential. Comparative studies with several other polycyclic aromatic hydrocarbons have demonstrated that only those substances that are epoxidized in the bay region of the ring system possess carcinogenic properties.

### 8.6.8  Acetaminophen

A good example of the importance of tissue availability of the conjugating chemical is found with acetaminophen. At normal therapeutic doses, acetaminophen is safe, but can be hepatotoxic at high doses. The major portion of acetaminophen is conjugated with either sulfate or glucuronic acid to form water-soluble, readily excreted metabolites and only small amounts of the reactive intermediate, believed to be quinoneimine, are formed by the CYP enzymes (Figure 8.6).

When therapeutic doses of acetaminophen are ingested, the small amount of reactive intermediate forms is efficiently deactivated by conjugation with glutathione. When large doses are ingested, however, the sulfate and glucuronide cofactors (PAPS and UDPGA) become depleted, resulting in more of the acetaminophen being metabolized to the reactive intermediate.

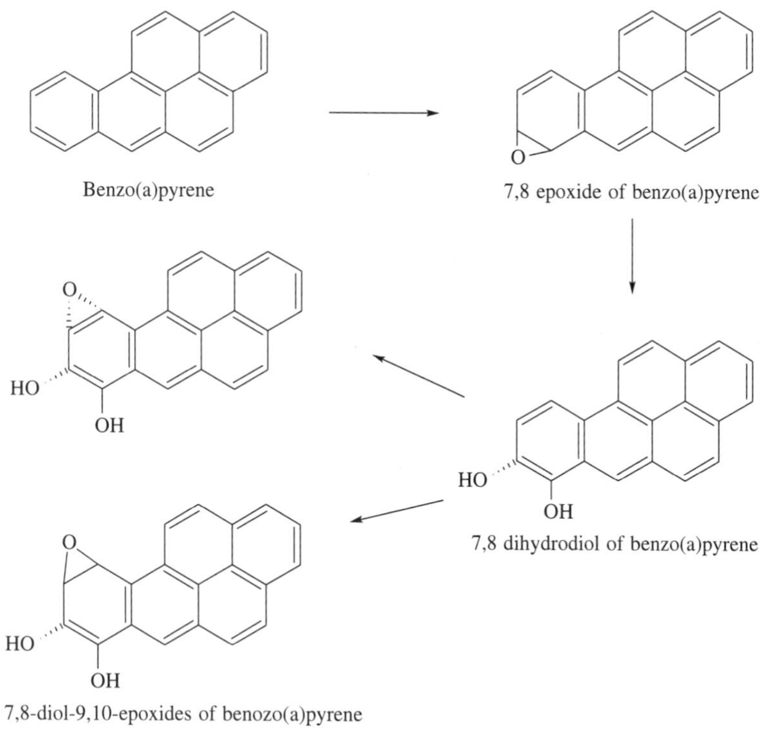

**Figure 8.5**    Selected stages of biotransformation of benzo(a)pyrene. The diol epoxide can exist in four diastereoisomeric forms of which the key carcinogenic metabolite is (+)-benzo(a)pyrene 7,8-diol-9,10-epoxide.

As long as glutathione (GSH) is available, most of the reactive intermediate can be detoxified. When the concentration of GSH in the liver also becomes depleted, however, covalent binding to sulfhydryl (-SH) groups of various cellular proteins increases, resulting in hepatic necrosis. If sufficiently large amounts of acetaminophen are ingested, as in drug overdoses and suicide attempts, extensive liver damage and death may result.

### 8.6.9  Cycasin

When flour from the cycad nut, which is used extensively among residents of South Pacific Islands, is fed to rats, it leads to cancers of the liver, kidney, and digestive tract. The active compound in cycasin is the $\beta$-glucoside of methylazoxymethanol (Figure 8.7). If this compound is injected intraperitoneally rather than given orally, or if the compound is fed to germ-free rats, no tumors occur. Intestinal microflora possess the necessary enzyme, $\beta$-glucosidase, to form the active compound methylazoxymethanol, which is then absorbed into the body. The parent compound, cycasin, is carcinogenic only if administered orally because $\beta$-glucosidases are not present in mammalian tissues but are present in the gut. However, it can be demonstrated that the metabolite, methylazoxymethanol, will lead to tumors in both normal and germ-free animals regardless of the route of administration.

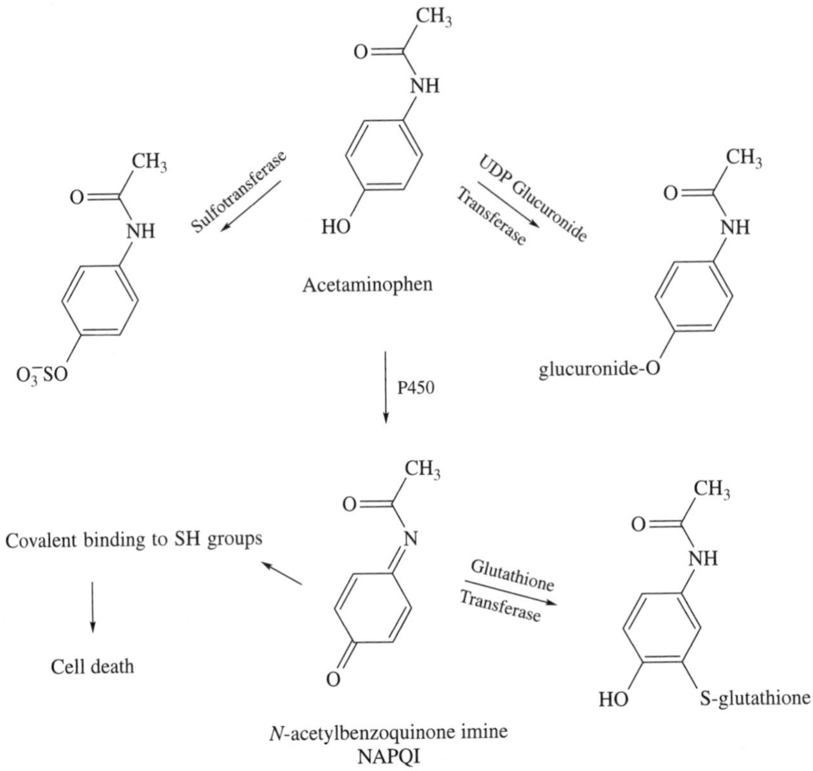

**Figure 8.6**   Metabolism of acetaminophen and formation of reactive metabolites.

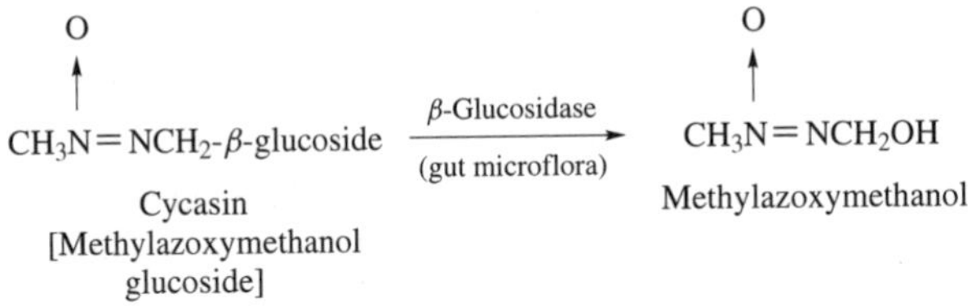

**Figure 8.7**   Bioactivation of cycasin by intestinal microflora to the carcinogen methylazoxy-methanol.

## 8.7   FUTURE DEVELOPMENTS

The current procedures for assessing safety and carcinogenic potential of chemicals using whole animal studies are expensive as well as becoming less socially acceptable. Moreover the scientific validity of such tests for human risk assessment is also being questioned. Currently a battery of short-term mutagenicity tests are used extensively as early predictors of mutagenicity and possible carcinogenicity.

Most of these systems use test organisms—for example, bacteria—that lack suitable enzyme systems to bioactivate chemicals, and therefore an exogenous activating system is used. Usually the postmitochondrial fraction from rat liver, containing both phase I and phase II enzymes, is used as the activating system. The critical question is, To what extent does this rat system represent the true in vivo situation, especially in humans? If not this system, then what is the better alternative? As some of the examples in this chapter illustrate, a chemical that is toxic or carcinogenic in one species or gender may be inactive in another, and this phenomenon is often related to the complement of enzymes, either activation or detoxication, expressed in the exposed organism.

Another factor to consider is the ability of many foreign compounds to selectively induce the CYP enzymes involved in their metabolism, especially if this induction results in the activation of the compound. With molecular techniques now available, considerable progress is being made in defining the enzyme and isozyme complements of human and laboratory species and understanding their mechanisms of control. Another area of active research is the use of in vitro expression systems to study the oxidation of foreign chemicals (e.g., bacteria containing genes for specific human CYP isozymes).

In summary, in studies of chemical toxicity, pathways and rates of metabolism as well as effects resulting from toxicokinetic factors and receptor affinities are critical in the choice of the animal species and experimental design. Therefore it is important that the animal species chosen as a model for humans in safety evaluations metabolize the test chemical by the same routes as humans and, furthermore, that quantitative differences are considered in the interpretation of animal toxicity data. Risk assessment methods involving the extrapolation of toxic or carcinogenic potential of a chemical from one species to another must consider the metabolic and toxicokinetic characteristics of both species.

## SUGGESTED READING

Anders, M. W., W. Dekant, and S. Vamvakas, Glutathione-dependent toxicity. *Xenobiotics* **22**: 1135–1145, 1992.

Coughtrie, M. W. H., S. Sharp, K. Maxwell, and N. P. Innes. Biology and function of the reversible sulfation pathway catalysed by human sulfotransferases and sulfatases. *Chemico-Biol. Interact.* **109**: 3–27, 1998.

Gonzalez, F. J., and H. V. Gelboin. Role of human cytochromes P450 in the metabolic activation of chemical carcinogens and toxins. *Drug Metabol. Rev.* **26**: 165–183, 1994.

Guengerich, F. P. Bioactivation and detoxication of toxic and carcinogenic chemicals. *Drug Metabol. Disp.* **21**: 1–6, 1993.

Guengerich, F. P. Metabolic activation of carcinogens. *Pharmac. Ther.* **54**: 17–61, 1992.

Levi, P. E., and E. Hodgson. Reactive metabolites and toxicity. In *Introduction to Biochemical Toxicology*, 3rd ed., E. Hodgson and R. C. Smart, eds. New York: Wiley, 2001, pp. 199–220.

Omiecinski, C. J., R. P. Remmel, and V. P. Hosagrahara. Concise review of the cytochrome P450s and their roles in toxicology. *Toxicol. Sci.* **48**: 151–156, 1999.

Rinaldi, R., E. Eliasson, S. Swedmark, and R. Morganstern. Reactive intermediates and the dynamics of glutathione transferases. *Drug Metabol. Disp.* **30**: 1053–1058, 2002.

Ritter, J. K. Roles of glucuronidation and UDP-glucuronosyltransferases in xenobiotic bioactivation reactions. *Chemico-Biol. Interact.* **129**: 171–193, 2000.

Vasiliou, V., A. Pappa, and D. R. Petersen. Role of aldehyde dehydrogenases in endogenous and xenobiotic metabolism. *Chemico-Biol. Interact.* **129**: 1–19, 2000.

■■■■■■■ CHAPTER 9

# Chemical and Physiological Influences on Xenobiotic Metabolism

RANDY L. ROSE and ERNEST HODGSON

## 9.1  INTRODUCTION

The metabolism of toxicants and their overall toxicity can be modified by many factors both extrinsic and intrinsic to the normal functioning of the organism. It is entirely possible that many changes in toxicity are due to changes in metabolism, because most sequences of events that lead to overt toxicity involve activation and/or detoxication of the parent compound. In many cases the chain of cause and effect is not clear, due to the difficulty of relating single events measured in vitro to the complex and interrelated effects that occur in vivo. This relationship between in vitro and in vivo studies is important and is discussed in connection with enzymatic inhibition and induction (see Section 9.5). It is important to note that the chemical, nutritional, physiological, and other effects noted herein have been described primarily from experiments carried out on experimental animals. These studies indicate that similar effects may occur in humans or other animals, but not that they must occur or that they occur at the same magnitude in all species if they occur at all.

## 9.2  NUTRITIONAL EFFECTS

Many nutritional effects on xenobiotic metabolism have been noted, but the information is scattered and often appears contradictory. This is one of the most important of several neglected areas of toxicology. This section is concerned only with the effects of nutritional constituents of the diet; the effects of other xenobiotics in the diet are discussed under chemical effects (see Section 9.5).

### 9.2.1  Protein

Low-protein diets generally decrease monooxygenase activity in rat liver microsomes, and gender and substrate differences may be seen in the effect. For example, aminopyrine $N$-demethylation, hexobarbital hydroxylation, and aniline hydroxylation are all

*A Textbook of Modern Toxicology, Third Edition,* edited by Ernest Hodgson
ISBN 0-471-26508-X  Copyright © 2004 John Wiley & Sons, Inc.

decreased, but the effect on the first two is greater in males than in females. In the third case, aniline hydroxylation, the reduction in males is equal to that in females. Tissue differences may also be seen. These changes are presumably related to the reductions in the levels of cytochrome P450 and NADPH-cytochrome P450 reductase that are also noted. One might speculate that the gender and other variations are due to differential effects on P450 isozymes. Even though enzyme levels are reduced by low-protein diets, they can still be induced to some extent by compounds such as phenobarbital. Such changes may also be reflected in changes in toxicity. Changes in the level of azoreductase activity in rat liver brought about by a low-protein diet are reflected in an increased severity in the carcinogenic effect of dimethylaminoazobenzene. The liver carcinogen dimethylnitrosamine, which must be activated metabolically, is almost without effect in protein-deficient rats.

Strychnine, which is detoxified by microsomal monooxygenase action, is more toxic to animals on low-protein diets, whereas octamethylpyrophosphoramide, carbon tetrachloride, and heptachlor, which are activated by monooxygenases, are less toxic. Phase II reactions may also be affected by dietary protein levels. Chloramphenicol glucuronidation is reduced in protein-deficient guinea pigs, although no effect is seen on sulfotransferase activity in protein-deficient rats.

### 9.2.2   Carbohydrates

High dietary carbohydrate levels in the rat tend to have much the same effect as low dietary protein, decreasing such activities as aminopyrine $N$-demethylase, pentobarbital hydroxylation, and $p$-nitrobenzoic acid reduction along with a concomitant decrease in the enzymes of the cytochrome P450 monooxygenase system. Because rats tend to regulate total caloric intake, this may actually reflect low-protein intake.

In humans it has been demonstrated that increasing the ratio of protein to carbohydrate in the diet stimulates oxidation of antipyrine and theophylline, while changing the ratio of fat to carbohydrate had no effect. In related studies, humans fed charcoal-broiled beef (food high in polycyclic hydrocarbon content) for several days had significantly enhanced activities of CYPs 1A1 and 1A2, resulting in enhanced metabolism of phenacetin, theophylline, and antipyrine. Studies of this nature indicate that there is significant interindividual variability in these observed responses.

### 9.2.3   Lipids

Dietary deficiencies in linoleic or in other unsaturated fats generally bring about a reduction in P450 and related monooxygenase activities in the rat. The increase in effectiveness of breast and colon carcinogens brought about in animals on high fat diets, however, appears to be related to events during the promotion phase rather than the activation of the causative chemical.

Lipids also appear to be necessary for the effect of inducers, such as phenobarbital, to be fully expressed.

### 9.2.4   Micronutrients

Vitamin deficiencies, in general, bring about a reduction in monooxygenase activity, although exceptions can be noted. Riboflavin deficiency causes an increase in P450 and aniline hydroxylation, although at the same time it causes a decrease in P450 reductase and benzo(a)pyrene hydroxylation. Ascorbic acid deficiency in the guinea pig not only causes a decrease in P450 and monooxygenase activity but also causes a reduction in microsomal hydrolysis of procaine. Deficiencies in vitamins A and E cause a decrease in monooxygenase activity, whereas thiamine deficiency causes an increase. The effect of these vitamins on different P450 isozymes has not been investigated. Changes in mineral nutrition have also been observed to affect monooxygenase activity. In the immature rat, calcium or magnesium deficiency causes a decrease, whereas, quite unexpectedly, iron deficiency causes an increase. This increase is not accompanied by a concomitant increase in P450, however. An excess of dietary cobalt, cadmium, manganese, and lead all cause an increase in hepatic glutathione levels and a decrease in P450 content.

## 9.2.5  Starvation and Dehydration

Although in some animals starvation appears to have effects similar to those of protein deficiency, this is not necessarily the case. For example, in the mouse, monooxygenation is decreased but reduction of $p$-nitrobenzoic acid is unaffected. In male rats, hexobarbital and pentobarbital hydroxylation as well as aminopyrine $N$-demethylation are decreased, but aniline hydroxylation is increased. All of these activities are stimulated in the female. Water deprivation in gerbils causes an increase in P450 and a concomitant increase in hexobarbital metabolism, which is reflected in a shorter sleeping time.

## 9.2.6  Nutritional Requirements in Xenobiotic Metabolism

Because xenobiotic metabolism involves many enzymes with different cofactor requirements, prosthetic groups, or endogenous cosubstrates, it is apparent that many different nutrients are involved in their function and maintenance. Determination of the effects of deficiencies, however, is more complex because reductions in activity of any particular enzyme will be effective only if it affects a change in a rate-limiting step in a process. In the case of multiple deficiencies, the nature of the rate-limiting step may change with time

*Phase I Reactions.* Nutrients involved in the maintenance of the cytochrome P450 monooxygenase system are shown in Figure 9.1. The B complex vitamins niacin and riboflavin are both involved, the former in the formation of NADPH and the latter in the formation of FAD and FMN. Essential amino acids are, of course, required for the synthesis of all of the proteins involved. The heme of the cytochrome requires iron, an essential inorganic nutrient. Other nutrients required in heme synthesis include pantothenic acid, needed for the synthesis of the coenzyme A used in the formation of acetyl Co-A, pyridoxine, a cofactor in heme synthesis and copper, required in the ferroxidase system that converts ferrous to ferric iron prior to its incorporation into heme. Although it is clear that dietary deficiencies could reduce the ability of the P450 system to metabolize xenobiotics, it is not clear how this effect will be manifested in vivo unless there is an understanding of the rate-limiting factors involved, which is a considerable task in such a complex of interrelated reactions. Similar considerations

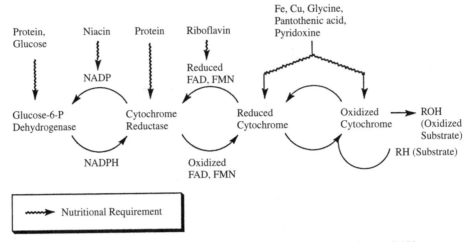

**Figure 9.1** Nutritional requirements with potential effects on the cytochrome P450 monooxygenase system (From W. E. Donaldeson Nutritional factors, in *Introduction to Biochemical Toxicology*, 3rd ed., E. Hodgson and R. C. Smart, Wiley, 2001.)

could be made for other phase I reaction systems such as arachidonic acid cooxidations, the glutathione peroxidase system, and so on.

***Phase II Reactions.*** As with phase I reactions, phase II reactions usually depend on several enzymes with different cofactors and different prosthetic groups and, frequently, different endogenous cosubstrates. All of these many components can depend on nutritional requirements, including vitamins, minerals, amino acids, and others. Mercapturic acid formation can be cited to illustrate the principles involved. The formation of mercapturic acids starts with the formation of glutathione conjugates, reactions catalyzed by the glutathione *S*-transferases.

This is followed by removal of the glutamic acid and the glycine residues, which is followed by acetylation of the remaining cysteine. Essential amino acids are required for the synthesis of the proteins involved, pantothenic acid for coenzyme A synthesis, and phosphorus for synthesis of the ATP needed for glutathione synthesis. Similar scenarios can be developed for glucuronide and sulfate formation, acetylation, and other phase II reaction systems.

## 9.3  PHYSIOLOGICAL EFFECTS

### 9.3.1  Development

Birth, in mammals, initiates an increase in the activity of many hepatic enzymes, including those involved in xenobiotic metabolism. The ability of the liver to carry out monooxygenation reactions appears to be very low during gestation and to increase after birth, with no obvious differences being seen between immature males and females. This general trend has been observed in many species, although the developmental pattern may vary according to gender and genetic strain. The component enzymes of the P450 monooxygenase system both follow the same general trend, although there may be differences in the rate of increase. In the rabbit, the postnatal increase in P450 and its reductase is parallel; in the rat, the increase in the reductase is slower than that of the cytochrome.

Phase II reactions may also be age dependent. Glucuronidation of many substrates is low or undetectable in fetal tissues but increases with age. The inability of new-born mammals of many species to form glucuronides is associated with deficiencies in both glucuronosyltransferase and its cofactor, uridine diphosphate glucuronic acid (UDPGA). A combination of this deficiency, as well as slow excretion of the biliru-bin conjugate formed, and the presence in the blood of pregnanediol, an inhibitor of glucuronidation, may lead to neonatal jaundice. Glycine conjugations are also low in the newborn, resulting from a lack of available glycine, an amino acid that reaches normal levels at about 30 days of age in the rat and 8 weeks in the human. Glutathione conjugation may also be impaired, as in fetal and neonatal guinea pigs, because of a deficiency of available glutathione. In the serum and liver of perinatal rats, glutathione transferase is barely detectable, increasing rapidly until all adult levels are reached at about 140 days (Figure 9.2). This pattern is not followed in all cases, because sulfate conjugation and acetylation appear to be fully functional and at adult levels in the guinea pig fetus. Thus some compounds that are glucuronidated in the adult can be acetylated or conjugated as sulfates in the young.

An understanding of how these effects may be related to the expression of individual isoforms is now beginning to emerge. It is known that in immature rats of either gender, P450s 2A1, 2D6, and 3A2 predominate, whereas in mature rats, the males show a predominance of P450s 2C11, 2C6, and 3A2 and the females P450s 2A1, 2C6, and 2C12.

The effect of senescence on the metabolism of xenobiotics has yielded variable results. In rats monooxygenase activity, which reaches a maximum at about 30 days

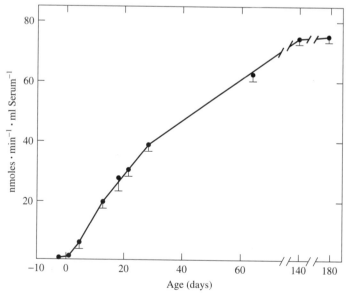

**Figure 9.2**  Developmental pattern of serum glutathione $S$-transferase activity in female rats. (Adapted from H. Mukhtar and J. R. Bend, *Life Sci.* **21**: 1277, 1977.)

of age, begins to decline some 250 days later, a decrease that may be associated with reduced levels of sex hormones. Glucuronidation also decreases in old animals, whereas monoamine oxidase activity increases. These changes in the monooxygenase activities are often reflected by changes in drug efficacy or overall toxicity.

In humans, age-related impairment of enzyme activity is highly controversial. Age-related declines in activity were not detected with respect to the activity of CYP2C and CYP3A isoforms among 54 liver samples from donors ranging in age from 9 to 89 years. Studies involving an erythromycin breath test in humans also suggested that there were no age-related declines associated with CYP3A4 activity. However, a study of CYP content and antipyrine clearance in liver biopsies obtained from 226 closely matched subjects indicated that subjects older than 70 had significantly less activity and clearance than younger subjects. Likewise, in older subjects, clearance of the drug omeprazole, a CYP2C19 substrate, was nearly half the rates observed in younger subjects.

### 9.3.2  Gender Differences

Metabolism of xenobiotics may vary with the gender of the organism. Gender differences become apparent at puberty and are usually maintained throughout adult life. Adult male rats metabolize many compounds at rates higher than females, for example, hexobarbital hydroxylation, aminopyrine $N$-demethylation, glucuronidation of $o$-aminophenol, and glutathione conjugation of aryl substrates; however, with other substrates, such as aniline and zoxazolamine, no gender differences are seen. In other species, including humans, the gender difference in xenobiotic metabolism is less pronounced. The differences in microsomal monooxygenase activity between males and females have been shown to be under the control of sex hormones, at least in some species. Some enzyme activities are decreased by castration in the male and administration of androgens to castrated males increases the activity of these sex-dependent enzyme activities without affecting the independent ones. Procaine hydrolysis is faster in male than female rats, and this compound is less toxic to the male. Gender differences in enzyme activity may also vary from tissue to tissue. Hepatic microsomes from adult male guinea pigs are less active in the conjugation of $p$-nitrophenol than are those from females, but no such gender difference is seen in the microsomes from lung, kidney, and small intestines.

Many differences in overall toxicity between males and females of various species are known (Table 9.1). Although it is not always known whether metabolism is the only or even the most important factor, such differences may be due to gender-related differences in metabolism. Hexobarbital is metabolized faster by male rats; thus female rats have longer sleeping times. Parathion is activated to the cholinesterase inhibitor paraoxon more rapidly in female than in male rats, and thus is more toxic to females. Presumably many of the gender-related differences, as with the developmental differences, are related to quantitative or qualitative differences in the isozymes of the xenobiotic-metabolizing enzymes that exist in multiple forms, but this aspect has not been investigated extensively.

In the rat, sexually dimorphic P450s appear to arise by programming, or imprinting, that occurs in neonatal development. This imprinting is brought about by a surge of testosterone that occurs in the male, but not the female, neonate and appears to imprint the developing hypothalamus so that in later development the growth hormone

**Table 9.1  Gender-Related Differences in Toxicity**

| Species | Toxicant | Susceptibility |
|---|---|---|
| Rat | EPN, warfarin, strychnine, hexobarbital, parathion | F > M |
|  | Aldrin, lead, epinephrine, ergot alkaloids | M > F |
| Cat | Dinitrophenol | F > M |
| Rabbit | Benzene | F > M |
| Mouse | Folic acid | F > M |
|  | Nicotine | M > F |
| Dog | Digitoxin | M > F |

is secreted in a gender-specific manner. Growth hormone production is pulsatile in adult males with peaks of production at approximately 3-hour intervals and more continuous in females, with smaller peaks. This pattern of growth hormone production and the higher level of circulating testosterone in the male maintain the expression of male-specific isoforms such as P450 2C11. The more continuous pattern of growth hormone secretion and the lack of circulating testosterone appears to be responsible for the expression of female specific isoforms such as P450 2C12. The high level of sulfotransferases in the female appears to be under similar control, raising the possibility that this is a general mechanism for the expression of gender-specific xenobiotic-metabolizing enzymes or their isoforms. A schematic version of this proposed mechanism is seen in Figure 9.3.

Gender-specific expression is also seen in the flavin-containing monooxygenases. In mouse liver FMO1 is higher in the female than in the male, and FMO3, present at high levels in female liver, is not expressed in male liver (Figure 9.4). No gender-specific differences are observed for FMO5. The important role of testosterone in the regulation of FMO1 and FMO3 was demonstrated in gonadectomized animals with and without testosterone implants. In males, castration increased FMO1 and FMO3 expression to levels similar to those observed in females, and testosterone replacement to castrated males resulted in ablation of FMO3 expression. Similarly, administration of testosterone to females caused ablation of FMO3 expression. Although these results clearly indicate a role for testosterone in the regulation of these isoforms, the physiological reasons for their gender-dependent expression remain unknown.

### 9.3.3  Hormones

Hormones other than sex hormones are also known to affect the levels of xenobiotic metabolizing enzymes, but these effects are much less studied or understood.

***Thyroid Hormone.*** Treatment of rats with thyroxin increases hepatic microsomal NADPH oxidation in both male and female rats, with the increase being greater in females. Cytochrome P450 content decreases in the male but not in the female. Hyperthyroidism causes a decrease in gender-dependent monooxygenase reactions and appears to interfere with the ability of androgens to increase the activity of the enzymes responsible. Gender differences are not seen in the response of mice and rabbits to

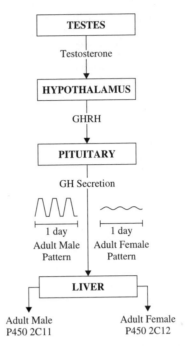

**Figure 9.3** Hypothetical scheme for neonatal imprinting of the hypothalamus–pituitary–liver axis resulting in sexually dimorphic expression of hepatic enzymes in the adult rat. Neonatal surges of testosterone appear to play a role in imprinting. (From M. J. J. Ronis and H. C. Cunny, in *Introduction to Biochemical Toxicology*, 2nd ed. E. Hodgson and P. E. Levi, eds., Appleton and Lange, 1994, p. 136.)

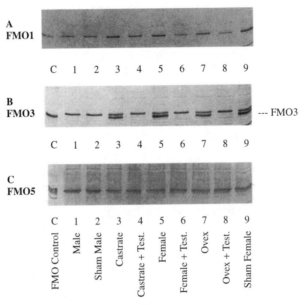

**Figure 9.4** Immunoreactivity of liver microsomes from sexually intact control, sham control, gonadectomized mice, or mice undergoing gonadectomy and/or receiving testosterone implants (5 mg). (From J. G. Falls et al., *Arch. Biochem. Biophys.* **342**: 212–223, 1997.)

thyroxin. In mice, aminopyrine $N$-demethylase, aniline hydroxylase, and hexobarbital hydroxylase are decreased, whereas $p$-nitrobenzoic acid reduction is unchanged. In rabbits, hexobarbital hydroxylation is unchanged, whereas aniline hydroxylation and $p$-nitrobenzoic acid reduction increase. Thyroid hormone can also affect enzymes other than microsomal monooxygenases. For example, liver monoamine oxidase activity is decreased, whereas the activity of the same enzymes in the kidney is increased.

*Adrenal Hormones.* Removal of adrenal glands from male rats results in a decrease in the activity of hepatic microsomal enzymes, impairing the metabolism of aminopyrine and hexobarbital, but the same operation in females has no effect on their metabolism. Cortisone or prednisolone restores activity to normal levels.

*Insulin.* The effect of diabetes on xenobiotic metabolism is quite varied and, in this regard, alloxan-induced diabetes may not be a good model for the natural disease. The in vitro metabolism of hexobarbital and aminopyrine is decreased in alloxan-diabetic male rats but is increased in similarly treated females. Aniline hydroxylase is increased in both males and females with alloxan diabetes. The induction of P450 2D1 in diabetes (and in fasting) is believed to be due to the high circulating levels of endogenously generated ketones. Studies of activity of the enzymes mentioned show no gender differences in the mouse; both sexes show an increase. Some phase II reactions, such as glucuronidation, are decreased in diabetic animals. This appears to be due to a lack of UDPGA caused by a decrease in UDPG dehydrogenase, rather than a decrease in transferase activity, and the effect can be reversed by insulin.

*Other Hormones.* Pituitary hormones regulate the function of many other endocrine glands, and hypophysectomy in male rats' results in a decrease in the activity of xenobiotic metabolizing enzymes. Administration of adrenocorticotropic hormone (ACTH) also results in a decrease of those oxidative enzyme activities that are gender dependent. In contrast, ACTH treatment of female rats causes an increase in aminopyrine $N$-demethylase but no change in other activities.

### 9.3.4  Pregnancy

Many xenobiotic metabolizing enzyme activities decrease during pregnancy. Catechol $O$-methyltransferase and monoamine oxidase decrease, as does glucuronide conjugation. The latter may be related to the increasing levels of progesterone and pregnanediol, both known to be inhibitors of glucuronosyltransferase in vitro. A similar effect on sulfate conjugation has been seen in pregnant rats and guinea pigs. In some species, liver microsomal monooxygenase activity may also decrease during pregnancy, this decrease being accompanied by a concomitant decrease in P450 levels. An increased level of FMO2 is seen in the lung of pregnant rabbits.

### 9.3.5  Disease

Quantitatively, the most important site for xenobiotic metabolism is the liver; thus effects on the liver are likely to be pronounced in the organism's overall capacity in this regard. At the same time, effects on other organs can have consequences no less serious for the organism. Patients with acute hepatitis frequently have an impaired ability to oxidize drugs, with a concomitant increase in plasma half-life. Impaired oxidative metabolism has also been shown in patients with chronic hepatitis or cirrhosis. The decrease in drug metabolism that occurs in obstructive jaundice may be a consequence of the accumulation of bile salts, which are known inhibitors of some of the enzymes involved. Phase II reactions may also be affected, decreases in acetylation, glucuronidation, and a variety of esterase activities having been seen in various liver diseases. Hepatic tumors, in general, have a lower ability to metabolize foreign compounds than does normal liver tissue, although in some cases the overall activity of tumor bearing livers may be no lower than that of controls. Kidney diseases may also affect the overall ability to handle xenobiotics, because this organ is one of the main routes for elimination of xenobiotics and their metabolites. The half-lives of tolbutamide, thiopental, hexobarbital, and chloramphenicol are all prolonged in patients with renal impairment.

### 9.3.6 Diurnal Rhythms

Diurnal rhythms, both in P450 levels and in the susceptibility to toxicants, have been described, especially in rodents. Although such changes appear to be related to the light cycle, they may in fact be activity dependent because feeding and other activities in rodents are themselves markedly diurnal.

## 9.4 COMPARATIVE AND GENETIC EFFECTS

Comparative toxicology is the study of the variation in toxicity of exogenous chemicals toward different organisms, either of different genetic strains or of different taxonomic groups. Thus the comparative approach can be used in the study of any aspect of toxicology, such as absorption, metabolism, mode of action, and acute or chronic effects. Most comparative data for toxic compounds exist in two areas—acute toxicity and metabolism. The value of the comparative approach can be summarized under four headings:

1. *Selective toxicity.* If toxic compounds are to be used for controlling diseases, pests, and parasites, it is important to develop selective biocides, toxic to the target organism but less toxic to other organisms, particularly humans.
2. *Experimental models.* Comparative studies of toxic phenomena are necessary to select the most appropriate model for extrapolation to humans and for testing and development of drugs and biocides. Taxonomic proximity does not necessarily indicate which will be the best experimental animal because in some cases primates are less valuable for study than are other mammals.
3. *Environmental xenobiotic cycles.* Much concern over toxic compounds springs from their occurrence in the environment. Different organisms in the complex ecological food webs metabolize compounds at different rates and to different products; the metabolic end products are released back to the environment, either to be further metabolized by other organisms or to exert toxic effects of their own. Clearly, it is desirable to know the range of metabolic processes possible.

Laboratory micro ecosystems have been developed, and with the aid of $^{14}$C-labeled compounds, chemicals and their metabolites can be followed through the plants and terrestrial and aquatic animals involved.

4. *Comparative biochemistry.* Some researchers believe that the proper role of comparative biochemistry is to put evolution on a molecular basis, and that detoxication enzymes, like other enzymes, are suitable subjects for study. Xenobiotic-metabolizing enzymes were probably essential in the early stages of animal evolution because secondary plant products, even those of low toxicity, are frequently lipophilic and as a consequence would, in the absence of such enzymes, accumulate in lipid membranes and lipid depots. The evolution of cytochrome P450 isoforms, with more than 2000 isoform cDNA sequences known, is proving a useful tool for the study of biochemical evolution.

### 9.4.1  Variations Among Taxonomic Groups

There are few differences in xenobiotic metabolism that are specific for large taxonomic groups. The formation of glucosides by insects and plants rather than the glucuronides of other animal groups is one of the most distinct. Although differences among species are common and of toxicologic significance, they are usually quantitative rather than qualitative in nature and tend to occur within as well as between taxonomic groups. Although the ultimate explanation of such differences must be at the level of biochemical genetics, they are manifested at many other levels, the most important of which are summarized in the following sections.

*In vivo Toxicity.* Toxicity is a term used to describe the adverse effects of chemicals on living organisms. Depending on the degree of toxicity, an animal may die, suffer injury to certain organs, or have a specific functional derangement in a subcellular organelle. Sublethal effects of toxicants may be reversible. Available data on the toxicity of selected pesticides to rats suggest that herbicide use, in general, provides the greatest human safety factor by selectively killing plants. As the evolutionary position of the target species approaches that of humans, however, the human safety factor is narrowed considerably. Thus the direct toxicity to humans and other mammals of biocide toxicity seems to be in the following progression: herbicides = fungicides < molluscicides < acaricides < nematocides < insecticides < rodenticides. This formula is obviously oversimplified because marked differences in lethality are observed when different members of each group of biocides is tested against laboratory test animals and target species. One should also bear in mind that any chemical can be environmentally dangerous if misused because many possible targets are interrelated in complex ecological systems.

Interspecific differences are also known for some naturally occurring poisons. Nicotine, for instance, is used as an insecticide and kills many insect pests at low doses, yet tobacco leaves constitute a normal diet for several species. As indicated earlier, most strains of rabbit eat Belladonna leaves without ill effects, whereas other mammals are easily poisoned. Natural tolerance to cyanide poisoning in millipedes and the high resistance to the powerful axonal blocking tetrodotoxin in puffer fish are examples of the tolerance of animals to the toxins they produce.

The specific organ toxicity of chemicals also exhibits wide species differences. Carbon tetrachloride, a highly potent hepatotoxicant, induces liver damage in many species, but chickens are almost unaffected by it. Dinitrophenol causes cataracts in humans, ducks, and chickens but not in other experimental animals. The eggshell thinning associated with DDT poisoning in birds is observed in falcons and mallard ducks, whereas this reproductive toxicity is not observed in gallinaceous species. Delayed neurotoxicity caused by organophosphates such as leptophos and tri-*o*-cresyl phosphate occurs in humans and can be easily demonstrated in chickens, but can be produced only with difficulty in most common laboratory mammals.

***In vivo Metabolism.*** Many ecological and physiological factors affect the rates of penetration, distribution, biotransformation, and excretion of chemicals, and thus govern their biological fate in the body. In general, the absorption of xenobiotics, their tissue distribution, and penetration across the blood-brain barrier and other barriers are dictated by their physicochemical nature and, therefore, tend to be similar in various animal species. The biologic effect of a chemical depends on the concentration of its binding to tissue macromolecules. Thus substantial differences in these variables should confer species specificity in the biologic response to any metabolically active xenobiotic. The biologic half-life is governed by the rates of metabolism and excretion and thus reflects the most important variables explaining interspecies differences in toxic response. Striking differences among species can be seen in the biologic half-lives of various drugs. Humans, in general, metabolize xenobiotics more slowly than do various experimental animals. For example, phenylbutazone is metabolized slowly in humans, with a half-life averaging 3 days. In the monkey, rat, guinea pig, rabbit, dog, and horse, however, this drug is metabolized readily, with half-lives ranging between 3 and 6 hours. The interdependence of metabolic rate, half-life, and pharmacologic action is well illustrated in the case of hexobarbital. The duration of sleeping time is directly related to the biologic half-life and is inversely proportional to the in vitro degradation of liver enzymes from the respective species. Thus mice inactivate hexobarbital readily, as reflected in a brief biologic half-life in vivo and short sleeping time, whereas the reverse is true in dogs.

Xenobiotics, once inside the body, undergo a series of biotransformations. Those reactions that introduce a new functional group into the molecule by oxidation, reduction, or hydrolysis are designated phase I reactions, whereas the conjugation reactions by which phase I metabolites are combined with endogenous substrates in the body are referred to as phase II reactions. Chemicals may undergo any one of these reactions or any combination of them, either simultaneously or consecutively. Because biotransformations are catalyzed by a large number of enzymes, it is to be expected that they will vary among species. Qualitative differences imply the occurrence of different enzymes, whereas quantitative differences imply variations in the rate of biotransformation along a common metabolic pathway, the variations resulting from differences in enzyme levels, in the extent of competing reactions or in the efficiency of enzymes capable of reversing the reaction.

Even in the case of a xenobiotic undergoing oxidation primarily by a single reaction, there may be remarkable species differences in relative rates. Thus in humans, rats, and guinea pigs, the major route of papaverine metabolism is *O*-demethylation to yield phenolic products, but very little of these products is formed in dogs. Aromatic hydroxylation of aniline is another example. In this case, both *ortho* and *para* positions are susceptible to oxidative attack yielding the respective aminophenols. The biological fate of aniline has been studied in many species and striking selectivity in hydroxylation position has been noted (Table 9.2). These data show a trend,

**Table 9.2  In vivo Hydroxylation of Aniline in Females of Various Species**

| Species | Percent Dose Excreted as Aminophenol | | |
|---|---|---|---|
| | *Ortho* | *Para* | *P/O* Ratio |
| Dog | 18.0 | 9.0 | 0.5 |
| Cat | 32.0 | 14.0 | 0.4 |
| Ferret | 26.0 | 28.0 | 1.0 |
| Rat | 19.0 | 48.0 | 2.5 |
| Mouse | 4.0 | 12.0 | 3.0 |
| Hamster | 5.5 | 53.0 | 10.0 |
| Guinea pig | 4.2 | 46.0 | 11.0 |
| Rabbit | 8.8 | 50.0 | 6.0 |
| Hen | 10.5 | 44.0 | 4.0 |

*Source*: Adapted from D. V. Parke, *Biochem. J.* **77**: 493, 1960.

in that carnivores generally display a high aniline *ortho*-hydroxylase ability with a *para/ortho* ratio of $\leq 1$ whereas rodents exhibit a striking preference for the *para* position, with a *para/ortho* ratio of from 2.5 to 11. Along with extensive *p*-aminophenol, substantial quantities of *o*-aminophenol are also produced from aniline administered to rabbits and hens. The major pathway is not always the same in any two animal species. 2-Acetylaminofluorene may be metabolized in mammals by two alternative routes: *N*-hydroxylation, yielding the carcinogenic *N*-hydroxy derivative, and aromatic hydroxylation, yielding the noncarcinogenic 7-hydroxy metabolite. The former is the metabolic route in the rat, rabbit, hamster, dog, and human in which the parent compound is known to be carcinogenic. In contrast, the monkey carries out aromatic hydroxylation and the guinea pig appears to deacetylate the *N*-hydroxy derivative; thus both escape the carcinogenic effects of this compound.

The hydrolysis of esters by esterases and of amides by amidases constitutes one of the most common enzymatic reactions of xenobiotics in humans and other animal species. Because both the number of enzymes involved in hydrolytic attack and the number of substrates for them is large, it is not surprising to observe interspecific differences in the disposition of xenobiotics due to variations in these enzymes. In mammals the presence of carboxylesterase that hydrolyzes malathion but is generally absent in insects explains the remarkable selectivity of this insecticide. As with esters, wide differences exist between species in the rates of hydrolysis of various amides in vivo. Fluoracetamide is less toxic to mice than to the American cockroach. This is explained by the faster release of the toxic fluoroacetate in insects as compared with mice. The insecticide dimethoate is susceptible to the attack of both esterases and amidases, yielding nontoxic products. In the rat and mouse, both reactions occur, whereas sheep liver contains only the amidases and that of guinea pig only the esterase. The relative rates of these degradative enzymes in insects are very low as compared with those of mammals, however, and this correlates well with the high selectivity of dimethoate.

The various phase II reactions are concerned with the conjugation of primary metabolites of xenobiotics produced by phase I reactions. Factors that alter or govern the rates of phase II reactions may play a role in interspecific differences in xenobiotic metabolism. Xenobiotics, frequently in the form of conjugates, can be eliminated through urine, feces, lungs, sweat, saliva, milk, hair, nails, or placenta, although comparative data are generally available only for the first two routes. Interspecific variation in the pattern of biliary excretion may determine species differences in the relative extent to which compounds are eliminated in the urine or feces. Fecal excretion of a chemical or its metabolites tends to be higher in species that are good biliary excretors, such as the rat and dog, than in species that are poor biliary excretors, such as the rabbit, guinea pig, and monkey. For example, the fecal excretion of stilbestrol in the rat accounts for 75% of the dose, whereas in the rabbit about 70% can be found in the urine. Dogs, like humans, metabolize indomethacin to a glucuronide but, unlike humans that excrete it in the urine, dogs excrete it primarily in the feces—apparently due to inefficient renal and hepatic blood clearance of the glucuronide. These differences may involve species variation in enterohepatic circulation, plasma level, and biologic half-life.

Interspecific differences in the magnitude of biliary excretion of a xenobiotic excretion product largely depend on molecular weight, the presence of polar groups in the molecule, and the extent of conjugation. Conjugates with molecular weights of less than 300 are poorly excreted in bile and tend to be excreted with urine, whereas the reverse is true for those with molecular weights higher than 300. The critical molecular weight appears to vary between species, and marked species differences are noted for biliary excretion of chemicals with molecular weights of about 300. Thus the biliary excretion of succinylsulfathioazole is 20- to 30-fold greater in the rat and the dog than in the rabbit and the guinea pig, and more than 100-fold greater than in the pig and the rhesus monkey. The cat and sheep are intermediate and excrete about 7% of the dose in the bile.

The evidence reported in a few studies suggests some relationship between the evolutionary position of a species and its conjugation mechanisms (Table 9.3). In humans and most mammals, the principal mechanisms involve conjugations with glucuronic acid, glycine, glutamine, glutathione and sulfate. Minor conjugation mechanisms in

**Table 9.3 Occurrence of Common and Unusual Conjugation Reactions**

| Conjugating Group | Common | Unusual |
|---|---|---|
| Carbohydrate | Glucuronic acid (animals) | N-Acetylglucosamine (rabbits) |
| | Glucose (insects, plants) | Ribose (rats, mice) |
| Amino acids | Glycine | Glutamine (insects, humans) |
| | Glutathione | Ornithine (birds) |
| | Methionine | Arginine (ticks, spiders) |
| | | Glycyltaurine (cats) |
| | | Glycylglycine (cats) |
| | | Serine (rabbits) |
| Acetyl | Acetyl group from acetyl-0CoA | |
| Formyl | | Formylation (dogs, rats) |
| Sulfate | Sulfate group from PAPS | |
| Phosphate | | Phosphate monoester formation (dogs, insects) |

*Source*: Modified from A. P. Kulkarni and E. Hodgson, Comparative toxicology, in *Introduction to Biochemical Toxicology*. E. Hodgson and F. E. Guthrie, eds., New York: Elsevier, 1980, p. 115.

mammals include acetylation and methylation pathways. In some species of birds and reptiles, ornithine conjugation replaces glycine conjugation; in plants, bacteria, and insects, conjugation with glucose instead of glucuronic acid results in the formation of glucosides. In addition to these predominant reactions, certain other conjugative processes are found involving specific compounds in only a few species. These reactions include conjugation with phosphate, taurine, *N*-acetyl-glucosamine, ribose, glycyltaurine, serine, arginine, and formic acid.

From the standpoint of evolution, similarity might be expected between humans and other primate species as opposed to the nonprimates. This phylogenic relationship is obvious from the relative importance of glycine and glutamine in the conjugation of arylacetic acids. The conjugating agent in humans is exclusively glutamine, and the same is essentially true with Old World monkeys. New World monkeys, however, use both the glycine and glutamine pathways. Most nonprimates and lower primates carry out glycine conjugation selectively. A similar evolutionary trend is also observed in the *N*-glucuronidation of sulfadimethoxine and in the aromatization of quinic acid; both reactions occur extensively in human, and their importance decreases with increasing evolutionary divergence from humans. When the relative importance of metabolic pathways is considered, one of the simplest cases of an enzyme-related species difference in the disposition of a substrate undergoing only one conjugative reaction is the acetylation of 4-aminohippuric acid. In the rat, guinea pig, and rabbit, the major biliary metabolite is 4-aminohippuric acid; the cat excretes nearly equal amounts of free acid and its acetyl derivative; and the hen excretes mainly the unchanged compound. In the dog, 4-aminohippuric acid is also passed into the bile unchanged because this species is unable to acetylate aromatic amino groups.

Defective operation of phase II reactions usually causes a striking species difference in the disposition pattern of a xenobiotic. The origin of such species variations is usually either the absence or a low level of the enzyme(s) in question and/or its cofactors. Glucuronide synthesis is one of the most common detoxication mechanisms in most mammalian species. The cat and closely related species have a defective glucuronide-forming system, however. Although cats form little or no glucuronide from *o*-aminophenol, phenol, *p*-nitrophenol, 2-amino-4-nitrophenol, 1-or 2-naphthol, and morphine, they readily form glucuronides from phenolphthalein, bilirubin, thyroxine, and certain steroids. Recently polymorphisms of UDP glucuronyl-transferase have been demonstrated in rat and guinea pig liver preparations; thus defective glucuronidation in the cat is probably related to the absence of the appropriate transferase rather than that of the active intermediate, UDPGA or UDP glucose dehydrogenase, which converts UDP glucose into UDPGA.

Studies on the metabolic fate of phenol in several species have indicated that four urinary products are excreted (Figure 9.5). Although extensive phenol metabolism takes place in most species, the relative proportions of each metabolite produced varies from species to species. In contrast to the cat, which selectively forms sulfate conjugates, the pig excretes phenol exclusively as the glucuronide. This defect in sulfate conjugation in the pig is restricted to only a few substrates, however, and may be due to the lack of a specific phenyl sulfotransferase because the formation of substantial amounts of the sulfate conjugate of 1-naphthol clearly indicates the occurrence of other forms of sulfotransferases.

Certain unusual conjugation mechanisms have been uncovered during comparative investigations, but this may be a reflection of inadequate data on other species. Future

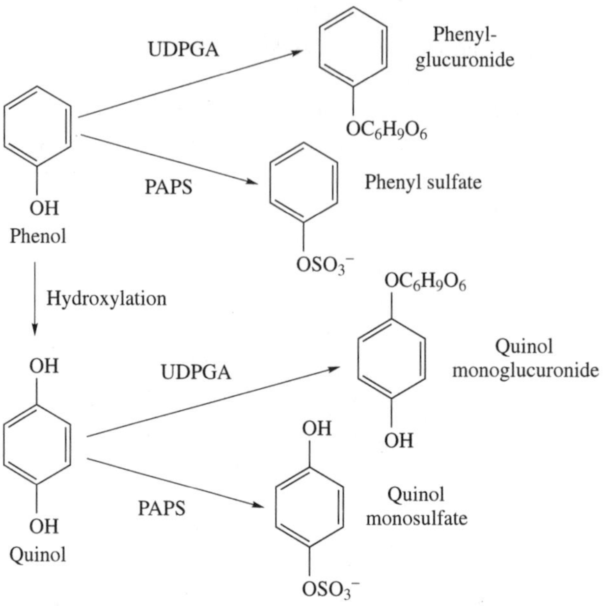

| Species | Percent of 24-hr Excretion as Glucuronide | | Percent of 24-hr Excretion as Sulfate | |
|---|---|---|---|---|
| | *Phenol* | *Quinol* | *Phenol* | *Quinol* |
| Pig | 100 | 0 | 0 | 0 |
| Indian fruit bat | 90 | 0 | 10 | 0 |
| Rhesus monkey | 35 | 0 | 65 | 0 |
| Cat | 0 | 0 | 87 | 13 |
| Human | 23 | 7 | 71 | 0 |
| Squirrel monkey | 70 | 19 | 10 | 0 |
| Rat-tail monkey | 65 | 21 | 14 | 0 |
| Guinea pig | 78 | 5 | 17 | 0 |
| Hamster | 50 | 25 | 25 | 0 |
| Rat | 25 | 7 | 68 | 0 |
| Ferret | 41 | 0 | 32 | 28 |
| Rabbit | 46 | 0 | 45 | 9 |
| Gerbil | 15 | 0 | 69 | 15 |

**Figure 9.5** Species variation in the metabolic conversion of phenol in vivo.

investigations may demonstrate a wider distribution. A few species of birds and reptiles use ornithine for the conjugation of aromatic acids rather than glycine, as do mammals. For example, the turkey, goose, duck, and hen excrete ornithuric acid as the major metabolite of benzoic acid, whereas pigeons and doves excrete it exclusively as hippuric acid.

Taurine conjugation with bile acids, phenylacetic acid, and indolylacetic acid seems to be a minor process in most species, but in the pigeon and ferret, it occurs extensively. Other infrequently reported conjugations include serine conjugation of xanthurenic acid in rats; excretion of quinaldic acid as quinaldylglycyltaurine and quinaldylglycylglycine in the urine of the cat, but not of the rat or rabbit; and conversion of furfural to furylacrylic acid in the dog and rabbit, but not in the rat, hen, or human. The dog and human but not the guinea pig, hamster, rabbit, or rat excrete the carcinogen 2-naphthyl hydroxylamine as a metabolite of 2-naphthylamine, which, as a result, has carcinogenic activity in the bladder of humans and dogs.

***In vitro Metabolism.*** Numerous variables simultaneously modulate the in vivo metabolism of xenobiotics; therefore their relative importance cannot be studied easily. This problem is alleviated to some extent by in vitro studies of the underlying enzymatic mechanisms responsible for qualitative and quantitative species differences. Quantitative differences may be related directly to the absolute amount of active enzyme present and the affinity and specificity of the enzyme toward the substrate in question. Because many other factors alter enzymatic rates in vitro, caution must be exercised in interpreting data in terms of species variation. In particular, enzymes are often sensitive to the experimental conditions used in their preparation. Because this sensitivity varies from one enzyme to another, their relative effectiveness for a particular reaction can be sometimes miscalculated.

Species variation in the oxidation of xenobiotics, in general, is quantitative (Table 9.4), whereas qualitative differences, such as the apparent total lack of parathion oxidation by lobster hepatopancreas microsomes, are seldom observed. Although the amount of P450 or the activity of NADPH-cytochrome P450 reductase seems to be related to the oxidation of certain substrates, this explanation is not always satisfactory

**Table 9.4    Species Variation in Hepatic Microsomal Oxidation of Xenobiotics In vitro**

| Substrate Oxidation | Rabbit | Rat | Mouse | Guinea Pig | Hamster | Chicken | Trout | Frog |
|---|---|---|---|---|---|---|---|---|
| Coumarin 7-hydroxylase[a] | 0.86 | 0.00 | 0.00 | 0.45 | — | — | — | — |
| Biphenyl 4-hydroxylase[b] | 3.00 | 1.50 | 5.70 | 1.40 | 3.80 | 1.70 | 0.22 | 1.15 |
| Biphenyl-2-hydroxylase[b] | 0.00 | 0.00 | 2.20 | 0.00 | 1.80 | 0.00 | 0.00 | 1.15 |
| 2-Methoxybiphenyl demethylase[a] | 5.20 | 1.80 | 3.40 | 2.20 | 2.30 | 2.00 | 0.60 | 0.40 |
| 4-Methoxybiphenyl demethylase[a] | 8.00 | 3.0 | 3.20 | 2.30 | 2.30 | 1.70 | 0.40 | 0.90 |
| p-Nitroanisole O-demethylase[b] | 2.13 | 0.32 | 1.35 | — | — | 0.76 | — | — |
| 2-Ethoxybiphenyl demethylase[a] | 5.30 | 1.60 | 1.40 | 2.10 | 2.50 | 1.70 | 0.60 | 0.40 |
| 4-Ethoxybiphenyl demethylase[a] | 7.80 | 2.80 | 1.80 | 2.30 | 1.80 | 1.50 | 0.40 | 0.90 |
| Ethylmorphine N-Demethylase[b] | 4.0 | 11.60 | 13.20 | 5.40 | — | — | — | — |
| Aldrin epoxidase[b] | 0.34 | 0.45 | 3.35 | — | — | 0.46 | 0.006 | — |
| Parathion desulfurase[b] | 2.11 | 4.19 | 5.23 | 8.92 | 7.75 | — | — | — |

*Source*: Modified from A. P. Kulkarni and E. Hodgson, Comparative toxicology, in *Introduction to Biochemical Toxicology*, E. Hodgson and F. E. Guthrie eds., New York: Elsevier, 1980, p. 120.
[a] nmol/mg/h.
[b] nmol/mg/min.

because the absolute amount of cytochrome P450 is not necessarily the rate-limiting characteristic. It is clear that there are multiple forms of P450 isozymes in each species, and that these forms differ from one species to another. Presumably both quantitative and qualitative variation in xenobiotic metabolism depends on the particular isoforms expressed and the extent of this expression.

Reductive reactions, like oxidation, are carried out at different rates by enzyme preparations from different species. Microsomes from mammalian liver are 18 times or more higher in azoreductase activity and more than 20 times higher in nitroreductase activity than those from fish liver. Although relatively inactive in nitroreductase, fish can reduce the nitro group of parathion, suggesting multiple forms of reductase enzymes.

Hydration of epoxides catalyzed by epoxide hydrolase is involved in both detoxication and intoxication reactions. With high concentrations of styrene oxide as a substrate, the relative activity of hepatic microsomal epoxide hydrolase in several animal species is rhesus monkey > human = guinea pig > rabbit > rat > mouse. With some substrates, such as epoxidized lipids, the cytosolic hydrolase may be much more important than the microsomal enzyme.

Blood and various organs of humans and other animals contain esterases capable of acetylsalicylic acid hydrolysis. A comparative study has shown that the liver is the most active tissue in all animal species studied except for the guinea pig, in which the kidney is more than twice as active as the liver. Human liver is least active; the enzyme in guinea pig liver is the most active. The relatively low toxicity of some of the new synthetic pyrethroid insecticides appears to be related to the ability of mammals to hydrolyze their carboxyester linkages. Thus mouse liver microsomes catalyzing (+)-*trans*-resmethrin hydrolysis are more than 30-fold more active than insect microsomal preparations. The relative rates of hydrolysis of this substrate in enzyme preparations from various species are mouse >> milkweed bug >> cockroach >> cabbage looper > housefly.

The toxicity of the organophosphorus insecticide dimethoate depends on the rate at which it is hydrolyzed in vivo. This toxicant undergoes two main metabolic detoxication reactions, one catalyzed by an esterase and the other by an amidases. Although rat and mouse liver carry out both reactions, only the amidase occurs in sheep liver and the esterase in guinea pig liver. The ability of liver preparations from different animal species to degrade dimethoate is as follows: rabbit > sheep > dog > rat > cattle > hen > guinea pig > mouse > pig, these rates being roughly inversely proportioned to the toxicity of dimethoate to the same species. Insects degrade this compound much more slowly than do mammals and hence are highly susceptible to dimethoate.

Hepatic microsomes of several animal species possess UDP glucuronyltransferase activity and with *p*-nitrophenol as a substrate, a 12-fold difference in activity due to species variation is evident. Phospholipase-A activates the enzyme and results of activation experiments indicate that the amount of constraint on the activity of this enzyme is variable in different animal species.

Glutathione *S*-transferase in liver cytosol from different animal species also shows a wide variation in activity. Activity is low in humans, whereas the mouse and guinea pig appear to be more efficient than other species. The ability of the guinea pig to form the initial glutathione conjugate contrasts with its inability to readily *N*-acetylate cysteine conjugates; consequently mercapturic acid excretion is low in guinea pigs.

## 9.4.2  Selectivity

Selective toxic agents have been developed to protect crops, animals of economic importance, and humans from the vagaries of pests, parasites, and pathogens. Such selectivity is conferred primarily through distribution and comparative biochemistry.

Selectivity through differences in uptake permits the use of an agent toxic to both target and nontarget cells provided that lethal concentrations accumulate only in target cells, leaving nontarget cells unharmed. An example is the accumulation of tetracycline by bacteria but not by mammalian cells, the result being drastic inhibition of protein synthesis in the bacteria, leading to death.

Certain schistosome worms are parasitic in humans and their selective destruction by antimony is accounted for by the differential sensitivity of phosphofructokinase in the two species, the enzyme from schistosomes being more susceptible to inhibition by antimony than is the mammalian enzyme.

Sometimes both target and nontarget species metabolize a xenobiotic by the same pathways but differences in rate determine selectivity. Malathion, a selective insecticide, is metabolically activated by P450 enzymes to the cholinesterase inhibitor malaoxon. In addition to this activation reaction, several detoxication reactions also occur. Carboxylesterase hydrolyzes malathion to form the monoacid, phosphatases hydrolyze the P–O–C linkages to yield nontoxic products, and glutathione $S$-alkyltransferase converts malathion to desmethylmalathion. Although all of these reactions occur in both insects and mammals, activation is rapid in both insects and mammals, whereas hydrolysis to the monoacid is rapid in mammals but slow in insects. As a result malaoxon accumulates in insects but not in mammals, resulting in selective toxicity.

A few examples are also available in which the lack of a specific enzyme in some cells in the human body has enabled the development of a therapeutic agent. For example, guanine deaminase is absent from the cells of certain cancers but is abundant in healthy tissue; as a result 8-azaguanine can be used therapeutically.

Distinct differences in cells with regard to the presence or absence of target structures or metabolic processes also offer opportunities for selectivity. Herbicides such as phenylureas, simazine, and so on, block the Hill reaction in chloroplasts, thereby killing plants without harm to animals. This is not always the case because paraquat, which blocks photosynthetic reactions in plants, is a pulmonary toxicant in mammals, due apparently to analogous free-radical reactions (see Figure 18.4) involving enzymes different from those involved in photosynthesis.

## 9.4.3  Genetic Differences

Just as the xenobiotic-metabolizing ability in different animal species seems to be related to evolutionary development and therefore to different genetic constitution, different strains within a species may differ from one another in their ability to metabolize xenobiotics. One reason for differences among strains is that many genes are polymorphic, or exist in multiple forms. A polymorphism is defined as an inherited monogenetic trait that exists in the population in at least two genotypes (two or more stable alleles) and is stably inherited. They arise as the result of a mutational event, and generally result in an altered gene product. The frequency of genetic polymorphisms is arbitrarily defined as having a population frequency of greater than 1%. Many polymorphisms are somewhat race specific, arising with greater frequency in one race than in another.

Observed differences between strains of rats and mice, as described below, may be the result of gene polymorphisms. In cases involving insecticide selection pressure, resistant populations may arise as a result of direct mutations of insecticide-metabolizing enzymes and/or insecticide target sites that are passed on to succeeding generations.

*In vivo Toxicity.* The toxicity of organic compounds has been found to vary between different strains of laboratory animals. For example, mouse strain C3H is resistant to histamine, the LD50 being 1523 mg/kg in C3H/Jax mice as compared with 230 in Swiss/ICR mice; that is, the animals of the former strain are 6.6 times less susceptible to the effects of histamine. Striking differences in the toxicity of thiourea, a compound used in the treatment of hyperthyroidism, are seen in different strains of the Norway rat. Harvard rats were 11 times more resistant, and wild Norway rats were 335 times more resistant than were rats of the Hopkins strain.

The development of strains resistant to insecticides is an extremely widespread phenomenon that is known to have occurred in more than 200 species of insects and mites, and resistance of up to several 100-fold has been noted. The different biochemical and genetic factors involved have been studied extensively and well characterized. Relatively few vertebrate species are known to have developed pesticide resistance and the level of resistance in vertebrates is low compared to that often found in insects. Susceptible and resistant strains of pine voles exhibit a 7.4-fold difference in endrin toxicity. Similarly pine mice of a strain resistant to endrin were reported to be 12-fold more tolerant than a susceptible strain. Other examples include the occurrence of organochlorine insecticide-resistant and susceptible strains of mosquito fish, and resistance to Belladonna in certain rabbit strains.

Several genetic polymorphisms have been recently described and characterized with respect to CYP enzymes. The first and best known example involves CYP2D6. In the course of a clinical trial for debrisoquine, a potential drug for use in lowering blood pressure, Dr. Robert Smith, one of the investigators who used himself as a volunteer, developed severe orthostatic hypotension with blood pressure dropping to 70/50. The effects of the drug persisted for two days, while in other volunteers no adverse effects were noted. Urine analysis demonstrated that in Dr. Smith, debrisoquine was excreted unchanged, while in the other volunteers the primary metabolite was 4-hydroxy debrisoquine. Subsequent studies demonstrated that CYP2D6 was responsible for the formation of 4-hydroxy debrisoquine and that the polymorphic form of 2D6 is prevalent in Caucasians and African-Americans, in which approximately 7% are poor metabolizers. In Asian populations the frequency of poor metabolizers is only 1%.

Another well-known genetic polymorphism has been described in the metabolism of drugs such as isoniazid. "Slow acetylators" are homozygous for a recessive gene; this is believed to lead to the lack of the hepatic enzyme acetyltransferase, which in normal homozygotes or heterozygotes (rapid acetylators) acetylates isoniazid as a step in the metabolism of this drug. This effect is seen also in humans, the gene for slow acetylation showing marked differences in distribution between different human populations. It is very low in Eskimos and Japanese, with 80% to 90% of these populations being rapid acetylators, whereas 40% to 60% of African and some European populations are rapid acetylators. Rapid acetylators often develop symptoms of hepatotoxicity and polyneuritis at the dosage necessary to maintain therapeutic blood levels of isoniazid.

Many other significant polymorphisms in xenobiotic metabolizing enzymes have been described, including those for several CYP genes, alcohol and aldehyde dehydrogenases, epoxide hydrolase, and paraoxonase. One interesting polymorphism affecting metabolism of dietary trimethylamines involves FMO3. Individuals with FMO3 polymorphisms have a condition known as fish odor syndrome, or trimethylaminurea. Individuals with this syndrome exhibit an objectionable body odor resembling rotting fish due to their inability to $N$-oxidize trimethylamines, which are found in many foods including meat, eggs, and soybeans. This syndrome often leads to social isolation, clinical depression, and even suicide. Other toxicological implications of this polymorphism are still not known.

***Metabolite Production.*** Strain variations to hexobarbital are often dependant on its degradation rate. For example, male mice of the AL/N strain are long sleepers, and this trait is correlated with slow inactivation of the drug. The reverse is true in CFW/N mice, which have short sleeping time due to rapid hexobarbital oxidation. This close relationship is further evidenced by the fact that the level of brain hexobarbital at awakening is essentially the same in all stains. Similar strain differences have been reported for zoxazolamine paralysis in mice.

Studies on the induction of arylhydrocarbon hydroxylase by 3-methylcholanthrene have revealed several responsive and nonresponsive mouse strains, and it is now well established that the induction of this enzyme is controlled by a single gene. In the accepted nomenclature, Ah$^b$ represents the allele for responsiveness, whereas Ah$^d$ denotes the allele for nonresponsiveness.

In rats, both age and gender seem to influence strain variation in xenobiotic metabolism. Male rats exhibit about twofold variation between strains in hexobarbital metabolism, whereas female rats may display up to sixfold variation. In either gender the extent of variations depend on age. The ability to metabolize hexobarbital is related to the metabolism of other substrates and the interstrain differences are maintained.

A well-known interstrain difference in phase II reactions is that of glucuronidation in Gunn rats. This is a mutant strain of Wistar rats that is characterized by a severe, genetically determined defect of bilirubin glucuronidation. Their ability to glucuronidate $o$-aminophenol, o-aminobenzoic acid, and a number of other substrates is also partially defective. This deficiency does not seem to be related to an inability to form UDPGA but rather to the lack of a specific UDP glucuronosyl-transferase. It has been demonstrated that Gunn rats can conjugate aniline by $N$-glucuronidation and can form the $O$-glucuronide of $p$-nitrophenol.

Rabbit strains may exhibit up to 20-fold variation, particularly in the case of hexobarbital, amphetamine, and aminopyrine metabolism. Relatively smaller differences between strains occur with chlorpromazine metabolism. Wild rabbits and California rabbits display the greatest differences from other rabbit strains in hepatic drug metabolism.

***Enzyme Differences.*** Variation in the nature and amount of constitutively expressed microsomal P450s have not been studied extensively in different strains of the same vertebrate. The only thorough investigations, those of the Ah Locus, which controls aryl hydrocarbon hydroxylase induction, have shown that in addition to quantitative differences in the amount of P450 after induction in different strains of mice, there may also be a qualitative difference in the P450 isozymes induced (see Section 9.5.2).

## 9.5 CHEMICAL EFFECTS

With regard to both logistics and scientific philosophy, the study of the metabolism and toxicity of xenobiotics must be initiated by considering single compounds. Unfortunately, humans and other living organisms are not exposed in this way; rather, they are exposed to many xenobiotics simultaneously, involving different portals of entry, modes of action, and metabolic pathways. Some estimation of the number of chemicals in use in the United States are given in Table 9.5. Because it bears directly on the problem of toxicity-related interaction among different xenobiotics, the effect of chemicals on the metabolism of other exogenous compounds is one of the more important areas of biochemical toxicology.

Xenobiotics, in addition to serving as substrates for a number of enzymes, may also serve as inhibitors or inducers of these or other enzymes. Many examples are known of compounds that first inhibit and subsequently induce enzymes such as the microsomal monooxygenases. The situation is even further complicated by the fact that although some substances have an inherent toxicity and are detoxified in the body, others without inherent toxicity can be metabolically activated to potent toxicants. The following examples are illustrative of the situations that might occur involving two compounds:

- Compound A, without inherent toxicity, is metabolized to a potent toxicant. In the presence of an inhibitor of its metabolism, there would be a reduction in toxic effect.
- Compound A, given after exposure to an inducer of the activating enzymes, would appear more toxic.
- Compound B, a toxicant, is metabolically detoxified. In the presence of an inhibitor of the detoxifying enzymes, there would be an increase in the toxic effect.
- Compound B, given after exposure to an inducer of the detoxifying enzymes, would appear less toxic.

In addition to the previously mentioned cases, the toxicity of the inhibitor or inducer, as well as the time dependence of the effect, must also be considered because, as

**Table 9.5   Some Estimates of the Number of Chemicals in Use in the United States**

| Number | Type | Source of Estimate[a] |
|---|---|---|
| 1500 | Active ingredients of pesticides | EPA |
| 4000 | Active ingredients of drugs | FDA |
| 2000 | Drug additives (preservatives, stabilizers, etc.) | FDA |
| 2500 | Food additives (nutritional value) | FDA |
| 3000 | Food additives (preservatives, stabilizers, etc.) | FDA |
| 50,000 | Additional chemicals in common use | EPA |

[a]EPA, Environmental Protection Agency; FDA, Food and Drug Administration.

mentioned, many xenbiotics that are initially enzyme inhibitors ultimately become inducers.

### 9.5.1 Inhibition

As previously indicated, inhibition of xenobiotic-metabolizing enzymes can cause either an increase or a decrease in toxicity. Several well-known inhibitors of such enzymes are shown in Figure 9.6 and are discussed in this section. Inhibitory effects can be demonstrated in a number of ways at different organizational levels.

#### *Types of Inhibition: Experimental Demonstration*

*In vivo Symptoms.* The measurement of the effect of an inhibitor on the duration of action of a drug in vivo is the most common method of demonstrating its action. These methods are open to criticism, however, because effects on duration of action can be mediated by systems other than those involved in the metabolism of the drug. Furthermore they cannot be used for inhibitors that have pharmacological activity similar or opposite to the compound being used.

Previously the most used and most reliable of these tests involved the measurement of effects on the hexobarbital sleeping time and the zoxazolamine paralysis time. Both of these drugs are fairly rapidly deactivated by the hepatic microsomal monooxygenase

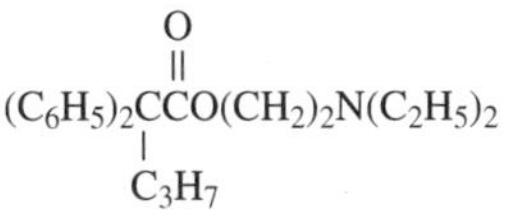

SKF-525A [P450]
2-(Diethylamino)ethyl-
2,2-diphenylpentanoate

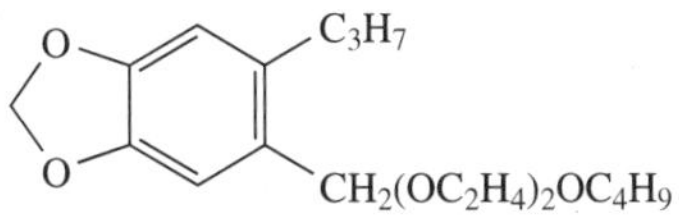

Piperonyl Butoxide [P450]
3,4-Methylenedioxy-6-propylbenzyl
n-butyl diethyleneglycol ether

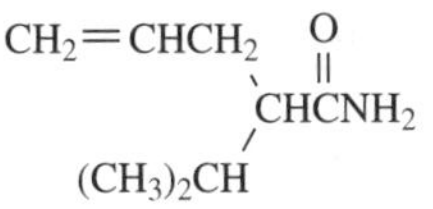

Allylisopropylacetamide
[P450]

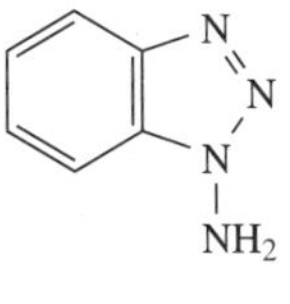

1-Aminobenzotriazole (1-ABT)
[P450]

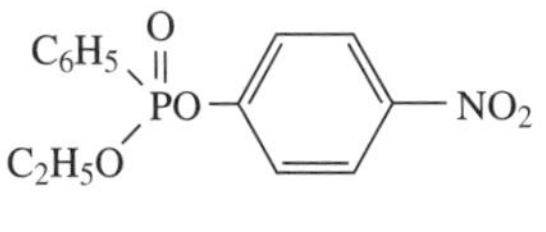

EPN [esterases]
O-Ethyl-O-*p*-nitrophenyl
phenylphosphonothioate

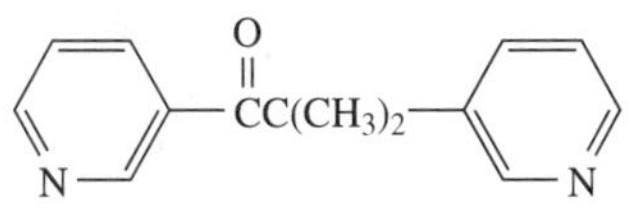

Metyrapone
[P450]

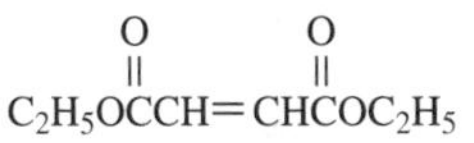

Diethyl maleate
[glutathione S-transferase]

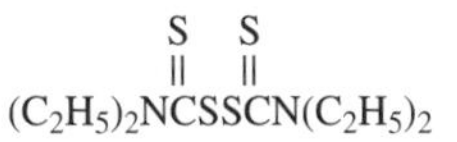

Disulfiram (Antabuse)
[aldehyde dehydrogenase]

**Figure 9.6** Some common inhibitors of xenobiotic-metabolizing enzymes.

system; thus, inhibitors of this system prolong their action. For example, treatment of mice with chloramphenicol 0.5 to 1.0 hour before pentobarbital treatment prolongs the duration of the pentobarbital sleeping time in a dose-related manner; it is effective at low doses ($< 5$ mg/kg) and has a greater than 10-fold effect at high doses (100–200 mg/kg). The well-known inhibitor of drug metabolism, SKF-525A (Figure 9.6), causes an increase in both hexobarbital sleeping time and zoxazolamine paralysis time in rats and mice, as do the insecticide synergists piperonyl butoxide and tropital, the optimum pretreatment time being about 0.5 hour before the narcotic is given. As a consequence of the availability of single expressed isoforms for direct studies of inhibitory mechanisms, these methods are now used much less often.

In the case of activation reactions, such as the activation of the insecticide azinphosmethyl to its potent anticholinesterase oxon derivative, a decrease in toxicity is apparent when rats are pretreated with the P450 inhibitor SKF-525A.

Cocarcinogenicity may also be an expression of inhibition of a detoxication reaction, as in the case of the cocarcinogenicity of piperonyl butoxide, a P450 inhibitor, and the carcinogens, freons 112 and 113.

*Distribution and Blood Levels.* Treatment of an animal with an inhibitor of foreign compound metabolism may cause changes in the blood levels of an unmetabolized toxicant and/or its metabolites. This procedure may be used in the investigation of the inhibition of detoxication pathways; it has the advantage over in vitro methods of yielding results of direct physiological or toxicological interest because it is carried out in the intact animal. For example, if animals are first treated with either SKF-525A, glutethimide, or chlorcyclizine, followed in 1 hour or less by pentobarbital, it can be shown that the serum level of pentobarbital is considerably higher in treated animals than in controls within 1 hour of its injection. Moreover the time sequence of the effects can be followed in individual animals, a factor of importance when inhibition is followed by induction—a not uncommon event.

*Effects on Metabolism In vivo.* A further refinement of the previous technique is to determine the effect of an inhibitor on the overall metabolism of a xenobiotic in vivo, usually by following the appearance of metabolites in the urine and/or feces. In some cases the appearance of metabolites in the blood or tissue may also be followed. Again, the use of the intact animal has practical advantages over in vitro methods, although little is revealed about the mechanisms involved.

Studies of antipyrine metabolism may be used to illustrate the effect of inhibition on metabolism in vivo; in addition, these studies have demonstrated variation among species in the inhibition of the metabolism of xenobiotics. In the rat, a dose of piperonyl butoxide of at least 100 mg/kg was necessary to inhibit antipyrine metabolism, whereas in the mouse a single intraperitoneal (IP) or oral dose of 1 mg/kg produced a significant inhibition. In humans an oral dose of 0.71 mg/kg had no discernible effect on the metabolism of antipyrine.

Disulfiram (Antabuse) inhibits aldehyde dehydrogenase irreversibly, causing an increase in the level of acetaldehyde, formed from ethanol by the enzyme alcohol dehydrogenase. This results in nausea, vomiting, and other symptoms in the human—hence its use as a deterrent in alcoholism. Inhibition by disulfiram appears to be irreversible, the level returning to normal only as a result of protein synthesis.

Use of specific metabolic enzyme inhibitors may often provide valuable information with respect to the metabolism of a particular drug. For example, quinidine is a potent and selective inhibitor of CYP2D6. This drug has been used in clinical studies as a pharmacological tool to mimic the lack of CYP2D6 in humans. By demonstrating that quinidine substantially slows the metabolism of trimipramine (a tricyclic antidepressant), investigators have implicated CYP2D6 in its metabolism.

*Effects on In vitro Metabolism Following In vivo Treatment.* This method of demonstrating inhibition is of variable utility. The preparation of enzymes from animal tissues usually involves considerable dilution with the preparative medium during homogenization, centrifugation, and re-suspension. As a result inhibitors not tightly bound to the enzyme in question are lost, either in whole or in part, during the preparative processes. Therefore negative results can have little utility because failure to inhibit and loss of the inhibitor give identical results. Positive results, however, not only indicate that the compound administered is an inhibitor but also provide a clear indication of excellent binding to the enzyme, most probably due to the formation of a covalent or slowly reversible inhibitory complex. The inhibition of esterases following treatment of the animal with organophosphorus compounds, such as paraoxon, is a good example, because the phosphorylated enzyme is stable and is still inhibited after the preparative procedures. Inhibition by carbamates, however, is greatly reduced by the same procedures because the carbamylated enzyme is unstable and, in addition, the residual carbamate is highly diluted.

Microsomal monooxygenase inhibitors that form stable inhibitory complexes with P450, such as SKF-525A, piperonyl butoxide, and other methylenedioxphenyl compounds, and amphetamine and its derivatives, can be readily investigated in this way. This is because the microsomes isolated from pretreated animals have a reduced capacity to oxidize many xenobiotics.

Another form of chemical interaction, resulting from inhibition in vivo, that can then be demonstrated in vitro involves those xenobiotics that function by causing destruction of the enzyme in question, so-called suicide substrates. Exposure of rats to vinyl chloride results in a loss of cytochrome P450 and a corresponding reduction in the capacity of microsomes subsequently isolated to metabolize foreign compounds. Allyl isopropylacetamide and other allyl compounds have long been known to have a similar effect.

*In vitro Effects.* In vitro measurement of the effect of one xenobiotic on the metabolism of another is by far the most common type of investigation of interactions involving inhibition. Although it is the most useful method for the study of inhibitory mechanisms, particularly when purified enzymes are used, it is of limited utility in assessing the toxicological implications for the intact animal. The principal reason for this is that in vitro measurement does not assess the effects of factors that affect absorption, distribution, and prior metabolism, all of which occur before the inhibitory event under consideration.

Although the kinetics of inhibition of xenobiotic-metabolizing enzymes can be investigated in the same ways as any other enzyme mechanism, a number of problems arise that may decrease the value of this type of investigation. They include the following:

- The P450 system, a particulate enzyme system, has been investigated many times, but using methods developed for single soluble enzymes. As a result Lineweaver-Burke or other reciprocal plots are frequently curvilinear, and the same reaction may appear to have quite a different characteristics from laboratory to laboratory, species to species, and organ to organ.

- The nonspecific binding of substrate and/or inhibitor to membrane components is a further complicating factor affecting inhibition kinetics.
- Both substrates and inhibitors are frequently lipophilic, with low solubility in aqueous media.
- Xenobiotic-metabolizing enzymes commonly exist in multiple forms (e.g., glutathione $S$-transferases and P450s). These isozymes are all relatively nonspecific but differ from one another in the relative affinities of the different substrates.

The primary considerations in studies of inhibition mechanisms are reversibility and selectivity. The inhibition kinetics of reversible inhibition give considerable insight into the reaction mechanisms of enzymes and, for that reason, have been well studied. In general, reversible inhibition involves no covalent binding, occurs rapidly, and can be reversed by dialysis or, more rapidly, by dilution. Reversible inhibition is usually divided into competitive inhibition, uncompetitive inhibition, and noncompetitive inhibition. Because these types are not rigidly separated, many intermediate classes have been described.

*Competitive inhibition* is usually caused by two substrates competing for the same active site. Following classic enzyme kinetics, there should be a change in the apparent $K_m$ but not in $V_{max}$. In microsomal monooxygenase reaction, type I ligands, which often appear to bind as substrates but do not bind to the heme iron, might be expected to be competitive inhibitors, and this frequently appears to be the case. Examples are the inhibition of the $O$-demethylation of $p$-nitroanisole by aminopyrine, aldrin epoxidation by dihydroaldrin, and $N$-demethylation of aminopyrine by nicotinamide. More recently some of the polychlorinated biphenyls (PCBs), notably dichlorbiphenyl, have been shown to have a high affinity as type I ligands for rabbit liver P450 and to be competitive inhibitors of the $O$-demethylation of $p$-nitroanisole.

*Uncompetitive inhibition* has seldom been reported in studies of xenobiotic metabolism. It occurs when an inhibitor interacts with an enzyme-substrate complex but cannot interact with free enzyme. Both $K_m$ and $V_{max}$ change by the same ratio, giving rise to a family of parallel lines in a Lineweaver-Burke plot.

*Noncompetitive inhibitors* can bind to both the enzyme and enzyme-substrate complex to form either an enzyme-inhibitor complex or an enzyme-inhibitor-substrate complex. The net result is a decrease in $V_{max}$ but no change in $K_m$. Metyrapone (Figure 9.6), a well-known inhibitor of monooxygenase reactions, can also, under some circumstances, stimulate metabolism in vitro. In either case the effect is noncompetitive, in that the $K_m$ does not change, whereas $V_{max}$ does, decreasing in the case of inhibition and increasing in the case of stimulation.

*Irreversible inhibition*, which is much more important toxicologically, can arise from various causes. In most cases the formation of covalent or other stable bonds or the disruption of the enzyme structure is involved. In these cases the effect cannot be readily reversed in vitro by either dialysis or dilution. The formation of stable inhibitory complexes may involve the prior formation of a reactive intermediate that then interacts with the enzyme. An excellent example of this type of inhibition is the effect of the insecticide synergist piperonyl butoxide (Figure 9.6) on hepatic microsomal monooxygenase activity. This methylenedioxyphenyl compound can form a stable inhibitory complex that blocks CO binding to P450 and also prevents substrate oxidation. This complex results from the formation of a reactive intermediate, which is shown by the fact that the type of inhibition changes from competitive to irreversible as metabolism, in the presence of NADPH and oxygen, proceeds. It appears probable that the metabolite in question is a carbine formed spontaneously by elimination of water following hydroxylation of the methylene carbon by the cytochrome (see Figure 7.8 for metabolism of methylenedioxyphenyl compounds). Piperonyl butoxide inhibits the in vitro metabolism of many substrates of the monooxygenase system, including aldrin, ethylmorphine, aniline, and aminopyrine, as well as carbaryl, biphenyl, hexobarbital, and $p$-nitroanisole among many others. Although most of the studies carried out on piperonyl butoxide have involved rat or mouse liver microsomes, they have also been carried out on pig, rabbit, and carp liver microsomes, and in various preparations from houseflies, cockroaches, and other insects. Certain classes of monooxygenase inhibitors, in addition to methylenedioxyphenyl compounds, are now known to form "metabolite inhibitory complexes," including amphetamine and its derivatives and SKF-525A and its derivatives.

The inhibition of the carboxylesterase that hydrolyzes malathion by organophosphorus compounds, such as EPN is a further example of xenobiotic interaction resulting from irreversible inhibition. In this case the enzyme is phosphorylated by the inhibitor.

Another class of irreversible inhibitors of toxicological significance consists of those compounds that bring about the destruction of the xenobiotic-metabolizing enzymes, hence the designation "suicide substrates." The drug allylisopropylacetamide (Figure 9.6), as well as other allyl compounds, has long been known to cause the breakdown of P450 and the resultant release of heme. More recently the hepatocarcinogen vinyl chloride has also been shown to have a similar effect, probably also mediated through the generation of a highly reactive intermediate (see Figure 8.2). Much information has accumulated since the mid-1970s on the mode of action of the hepatotoxicant carbon tetrachloride, which effects a number of irreversible changes in both liver proteins and lipids, such changes being generated by reactive intermediates formed during its metabolism (Figure 8.3).

The less specific disruptors of protein structure, such as urea, detergents, strong acids, and so on, are probably of significance only in vitro experiments.

***Synergism and Potentiation.*** The terms synergism and potentiation have been used and defined in various ways, but in any case, they involve a toxicity that is greater when two compounds are given simultaneously or sequentially than would be expected from a consideration of the toxicities of the compounds given alone. Some toxicologists have used the term synergism for cases that fit this definition, but only when one compound is toxic alone whereas the other has little or no intrinsic toxicity. For example, the nontoxic synergist, piperonyl butoxide is often included in pesticide formulations because of its ability to significantly increase the toxicity of the active pesticide ingredient by inhibiting its detoxication in the target species.

The term potentiation is then reserved for those cases where both compounds have appreciable intrinsic toxicity, such as in the case of malathion and EPN. Malathion has a low mammalian toxicity due primarily to its rapid hydrolysis by a carboxylesterase. EPN (Figure 9.6) another organophosphate insecticide, causes a dramatic increase in malathion toxicity to mammals at dose levels, which, given alone, cause essentially no inhibition of acetylcholinesterase. The increase in toxicity as a result of coadminstration of these two toxicants is the result of the ability of EPN, at low concentrations, to inhibit the carboxylesterase responsible for malathion degradation.

Unfortunately, the terms synergist and potentiation have often been used by some toxicologists in precisely the opposite manner. Historically, the term synergist has been used by pharmacologists to refer to simple additive toxicity and potentiation either as a synonym or for examples of greater than additive toxicity or efficacy. In an attempt to make uniform the use of these terms, it is suggested that insofar as toxic effects are concerned, the terms be used according to the following: *Both synergism and potentiation involve toxicity greater than would be expected from the toxicities of the compounds administered separately, but in the case of synergism one compound has little or no intrinsic toxicity when administered alone, whereas in the case of potentiation both compounds have appreciable toxicity when administered alone. It is further suggested that no special term is needed for simple additive toxicity of two or more compounds.*

**Antagonism.** In toxicology, antagonism may be defined as that situation where the toxicity of two or more compounds administered together or sequentially is less than would be expected from a consideration of their toxicities when administered individually. Strictly speaking, this definition includes those cases in which the lowered toxicity results from induction of detoxifying enzymes (this situation is considered separately in Section 9.5.2). Apart from the convenience of treating such antagonistic phenomena together with the other aspects of induction, they are frequently considered separately because of the significant time that must elapse between treatment with the inducer and subsequent treatment with the toxicant. The reduction of hexobarbital sleeping time and the reduction of zoxazolamine paralysis time by prior treatment with phenobarbital to induce drug—metabolizing enzymes are obvious examples of such induction effects at the acute level of drug action, whereas protection from the carcinogenic action of benzo(a)pyrene, aflatoxin B1, and diethylnitrosamine by phenobarbital treatment are examples of inductive effects at the level of chronic toxicity. In the latter case the P450 isozymes induced by phenobarbital metabolize the chemical to less toxic metabolites.

Antagonism not involving induction is a phenomenon often seen at a marginal level of detection and is consequently both difficult to explain and of marginal significance. In addition several different types of antagonism of importance to toxicology that do not involve xenobiotic metabolism are known but are not appropriate for discussion in this chapter. They include competition for receptor sites, such as the competition between CO and $O_2$ in CO poisoning or situations where one toxicant combines nonenzymatically with another to reduce its toxic effects, such as in the chelation of metal ions. Physiological antagonism, in which two agonists act on the same physiological system but produce opposite effects, is also of importance.

### 9.5.2 Induction

In the early 1960s, during investigations on the *N*-demethylation of aminoazo dyes, it was observed that pretreatment of mammals with the substrate or, more remarkably, with other xenobiotics, caused an increase in the ability of the animal to metabolize these dyes. It was subsequently shown that this effect was due to an increase in the microsomal enzymes involved. A symposium in 1965 and a landmark review by Conney in 1967 established the importance of induction in xenobiotic interactions. Since then, it has become clear that this phenomenon is widespread and nonspecific. Several hundred compounds of diverse chemical structure have been shown to induce monooxygenases and other enzymes. These compounds include drugs, insecticides, polycyclic hydrocarbons, and many others; the only obvious common denominator is that they are organic and lipophilic. It has also become apparent that although all inducers do not have the same effects, the effects tend to be nonspecific to the extent that any single inducer induces more than one enzymatic activity. Other enzymes often coinduced by P450 inducers include glutathione S-transferase, epoxide hydrolases and others; perhaps as a result of general induction of cellular processes including proliferative responses of endoplasmic reticulum, peroxisomes, and mitochondria.

***Specificity of Monooxygenase Induction.*** The majority of studies involving mono-oxygenase induction have been conducted in mammals. Mammals have at least 17 distinct CYP families, coding for as many as 50 to 60 individual CYP genes in any given species. Many of these CYP families are fairly specific for endogenous metabolic pathways and are not typically involved in metabolism of foreign chemicals. As discussed in Chapter 7, CYP families 1–4 are the predominant families involved in xenobiotic metabolism. These CYP families are also known for their ability to respond to xenobiotic challenges by increasing their protein levels. Many of the genes within families 1–4 are transcriptionally activated through one of four receptor-dependent mechanisms. Others, such as CYP2E1, are regulated at the level of mRNA stabilization and/or protein stabilization. These mechanisms of regulation are discussed later in this section.

Inducers of monooxygenase activity fall into four principle classes, exemplified by TCDD (inducer of CYP1A1), phenobarbital (inducer of the CYP2B and 3A families), rifampicin (inducer of CYP3A and 2C families), and ethanol (inducer of 2E1). Inducers of the phenobarbital-type tend to share few structural features other than lipophilicity, while TCDD-like inducers are primarily polycyclic hydrocarbons. Other inducers, such as ethanol, dexamethasone, and clofibrate are more specific. Many inducers require either fairly high dose levels or repeated dosing to be effective, frequently $>10$ mg/kg and some as high as 100 to 200 mg/kg. Some insecticides, however, such as mirex, can induce at dose levels as low as 1 mg/kg, while the most potent inducer known, 2,3,7,8-tetrachlorodibenzo-$p$-dioxin (TCDD), is effective at $1\mu$g/kg in some species.

In the liver, phenobarbital-type inducers cause a marked proliferation of the smooth endoplasmic reticulum as well as an increase in the amount of CYP content. Often these changes are sufficient to result in significant liver weight increases. Phenobarbital induction induces a wide range of oxidative activities including $O$-demethylation of $p$-nitroanisole, $N$-demethylation of benzphetamine, pentobarbital hydroxylation, and aldrin hydroxylation. CYP families that are primarily induced by phenobarbital and phenobarbital-like inducers include CYP2B, CYP2C, and CYP3A subfamilies.

In contrast with phenobarbital, induction by TCDD and polycyclic hydrocarbons does not cause proliferation of the endoplasmic reticulum, although the CYP content is increased. CYP1A1 is the primary isoform induced, although other non-CYP proteins such as uridine diphosphoglucuronyl transferase may also be induced. Induction of CYP1A1 by polycyclic hydrocarbons results in the induction of a relatively narrow range of oxidative activities, consisting primarily of reactions involving aryl hydrocarbon hydroxylase, the best-known reaction being the hydroxylation of benzo(a)pyrene.

Rifampicin and pregnenolone-16$\alpha$-carbonitrile (PCN) induce members of the CYP3A family and represent a third type of inducer, in that the substrate specificity of the microsomes from treated animals differs from that of the microsomes from either phenobarbital-treated or TCDD-treated animals. Inducing substrates of this class include endogenous and synthetic glucocorticoids (e.g., dexamethasone), pregnane compounds (e.g., pregenenolone 16$\alpha$-carbonitrile, PCN), and macrolide antibiotics (e.g., rifampicin).

Ethanol and a number of other chemicals, including acetone and certain imidazoles, induce CYP2E1. Piperonyl butoxide, isosafrole, and other methylenedioxyphenyl compounds are known to induce CYP1A2 by a non-Ah receptor-dependent mechanism. Peroxisome proliferators, including the drug, clofibrate, and the herbicide synergist tridiphane induce a CYP4A isozyme that catalyzes the $\omega$-oxidation of lauric acid.

All inducers do not fall readily into one or the other of these classes. Some oxidative processes can be induced by either type of inducer, such as the hydroxylation of aniline and the $N$-demethylation of chlorcyclizine. Some inducers, such as the mixture of PCBs designated Arochlor 1254, can induce a broad spectrum of CYP isoforms. Many variations also exist in the relative stimulation of different oxidative activities within the same class of inducer, particularly of the phenobarbital type.

It appears reasonable that because several types of CYP are associated with the endoplasmic reticulum, various inducers may induce one or more of them. Because each of these types has a relatively broad substrate specificity, differences may be caused by variations in the extent of induction of different cytochromes. Now that methods are available for gel electrophoresis of microsomes and identification of specific isoforms by immunoblotting and isoforms-specific antibodies, the complex array of inductive phenomena is being more logically explained in terms of specific isozymes.

Although the bulk of published investigations of the induction of monooxygenase enzymes have dealt with the mammalian liver, induction has been observed in other mammalian tissues and in nonmammalian species, both vertebrate and invertebrate. Many induced CYPs have now been cloned and/or purified from a variety of species. It is clear that many of these induced CYPs represent only a small percentage of the total CYP in the uninduced animal. For this reason the "constitutive" isozymes, those already expressed in the uninduced animal, must be fully characterized because they represent the available xenobiotic-metabolizing capacity of the normal animal.

***Mechanism and Genetics of Induction in Mammals.*** Many different mechanisms may be involved in CYP induction. These include increased transcription of DNA, increased mRNA translation to protein, mRNA stabilization, and protein stabilization. Induction can only occur in intact cells and cannot be achieved by the addition of inducers directly to cell fractions such as microsomes. It has been known for some time that in most cases of increase in monooxygenase activity there is a true induction involving synthesis of new enzyme, and not the activation of enzyme already synthesized, since induction is generally prevented by inhibitors of protein synthesis. For example, the protein synthesis inhibitors such as puromycin, ethionine, and cycloheximide inhibit aryl hydrocarbon hydroxylase activity. A simplified scheme for gene expression and protein synthesis is shown in Figure 9.7.

Perhaps the best understood example of induction involves induction of the aromatic hydrocarbon receptor (AhR) by compounds such as TCDD and 3-methylcholanthrene. The use of suitable inhibitors of RNA and DNA polymerase activity has shown that inhibitors of RNA synthesis such as actinomycin D and mercapto(pyridethyl)benzimidazole block aryl hydrocarbon hydroxylase induction, whereas hydroxyurea, at levels that completely block the incorporation of thymidine into DNA, has no effect. Thus it appears that the inductive effect is at the level of transcription and that DNA synthesis is not required.

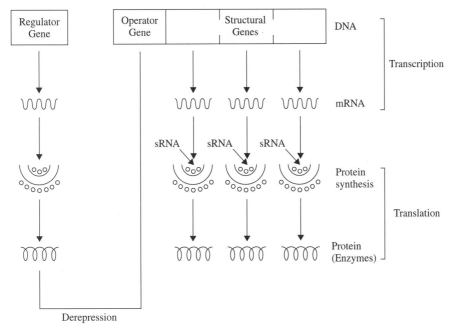

**Figure 9.7** Simplified scheme for gene expression in animals.

These findings imply that compounds that induce xenobiotic-metabolizing enzymes play a role as derepressors of regulator or other genes in a manner analogous to steroid hormones—namely by combining with a cytosolic receptor followed by movement into the nucleus and then derepression of the appropriate gene. In the case of the AhR, TCDD, or some other appropriate ligand enters the cell through the plasma membrane and binds to the cytosolic Ah receptor protein (Figure 9.8). After ligand binding, the receptor translocates to the nucleus where it forms a dimer with another protein known as ARNT. In the nucleus the transformed receptor interacts with specific sequences of DNA known as xenobiotic responsive elements (XREs). Two XREs are located approximately 1000 or more base pairs upstream from the transcriptional start site in the 5′ flanking region of the CYP1A1 gene. A third site is likely to be an inhibitory or suppressor site (Figure 9.9). The protein-DNA interaction that occurs at the XREs is thought to result in a bending of the DNA, which allows for increased transcription followed by increased protein synthesis. Another promoter region is located just upstream from the transcriptional start site. Although several transcription factors may interact with this binding site, including the TATA-binding protein, it has no binding sites for the AhR/Arnt proteins. Transfection experiments indicate that the TATA-binding site is essential for promoter function, while the other sites are less important. This promoter region is silent unless the upstream XREs have been appropriately activated by the AhR and Arnt proteins. The fact that several genes may be responsive to CYP1A inducers is indicative of the fact that similar XREs are found on many Ah-receptor inducible genes.

Although phenobarbital induction has been studied for many years, the mechanism for induction has only recently been established. In bacteria a key feature of phenobarbital induction was demonstrated to involve barbiturate-mediated removal of a

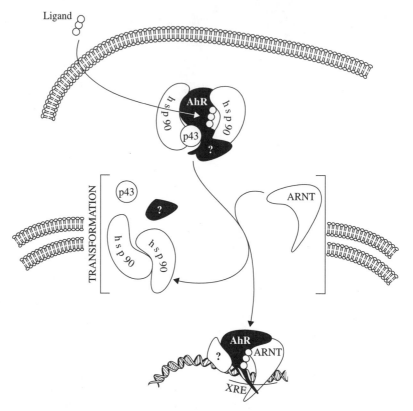

**Figure 9.8**   Proposed mechanism for ligand-activated AhR translocation and DNA binding (From J. C. Rowlands and J.-A. Gustafsson, *Crit. Rev. Toxicol.* **27**: 109, 1997.)

repressor protein from a 17-bp promoter regulatory sequence known as the "Barbie box." Although homologous promoter sequences have been observed in several phenobarbital responsive mammalian genes, ample evidence suggests that these sequences are not important in PB-induced transcription of mammalian CYP genes. Rather, in mammalian species the phenobarbital responsive sequences are found far upstream of the start codon.

The major advance in understanding phenobarbital induction came from a study using rat primary hepatocytes where phenobarbital responsiveness was demonstrated to be associated with a 163-bp DNA sequence at $-2318$ through $-2155$ bp of the CYP2B2 gene. Subsequent studies using in situ transfection of CYP2B2 promoter-luciferase constructs into rat livers confirmed this, as did similar studies involving mouse CYP2b10 and CYP2b9 genes. Additional deletion assays have narrowed phenobarbital responsiveness down to a minimum sequence of 51-bp from $-2339$ through $-2289$ of the Cyp2b10 gene; now known as the phenobarbital-responsive enhancer module (PBREM). The PBREM sequence has also been found in rat CYP2B1, CYP2B2, and human CYP2B6 genes. Multiple cis acting elements within this fragment cooperate to bring about increased DNA transcription. These include a nuclear factor 1 (NF1) binding site that is flanked by two nuclear receptor binding sites, designated NR1 and NR2. Inducibility requires at least one of the NR sites to be present to maintain phenobarbital inducibility. By

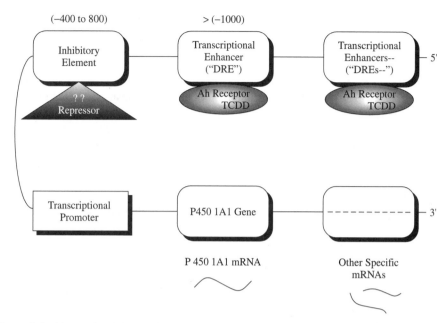

**Figure 9.9** Interaction of the Ah-receptor-ligand complex with the 5′ flanking region of the P450 1A1 gene. Two dioxin responsive elements (DREs) appear to lie approximately 1000 or more base pairs upstream from the 1A1 transcriptional s tart site. These elements appear to be transcriptional enhancers, whereas less direct evidence indicates an inhibitory element ("negative control element") between 400 and 800 bases upstream. The negative control element may inhibit the 1A1 promoted although the conditions for this inhibition are, as yet, undefined. (Adapted from A. B. Okey, *Pharmacol. Ther.* **45**: 241–298, 1990.)

contrast, although the NF1 site is necessary for the maximum phenobarbital response, it is nonessential for the basic phenobarbital response. This conclusion is supported by the fact that in rodent and human CYP2B genes the NR sites are highly conserved while the NF1 site is not.

The key factor that interacts with the PBREM is the orphan nuclear receptor known as a "constitutive active receptor" (CAR). CAR binds to each of the PBRE NR sites as a heterodimer with the retinoid X receptor (RXR), a common heterodimerization partner for many orphan nuclear receptors. Although the CAR-RXR binding does not require treatment with phenobarbital for activity, hence the term "constitutive," inclusion of phenobarbital substantially increases the activity of CYP2B and other PBRE related genes. It is thought that this is due to the displacement of two endogenous inhibitory androstane steroids that bind to the CAR-RXR heterodimer and inhibit its activity in the absence of phenobarbital like ligands. Thus, in the presence of phenobarbital, the binding of the inhibitory androstanes to CAR is abolished and the intrinsic activity of CAR becomes manifest, leading to the activation of PB responsive genes. Recent studies using CAR knockout mice indicate that many drug metabolizing genes are under CAR regulation, including isoforms of CYP2B, CYP3A, NADPH cytochrome P450 reductase, and enzymes involved in sulfotransferase metabolism.

In the early 1980s a distinct group of CYPs was described by several groups, which was characterized principally by its inducibility by steroidal chemicals. This particular group, belonging to the CYP3A subfamily, is well known for the diversity of substrates that it is capable of metabolizing. In humans the specific isoform CYP3A4 is responsible not only for the metabolism of endogenous compounds such as testosterone but also is credited for the metabolism of the largest number of currently used drugs. Many CYP3A substrates are further known for their ability to induce their own metabolism as well as the metabolism of other CYP3A substrates, resulting in the creation of potentially dangerous drug-drug type interactions. Regulation of the CYP3 family is likely to be primarily through enhanced transcription, although there are also some examples of post-translational regulation. For example, dexamethasone appears to increase CYP3A1 levels by stabilization of the mRNA while erythromycin acts by protein stabilization.

Several recent studies have begun to identify several elements on the 5′ upstream promoter region as well as receptors involved in CYP3 regulation (Figure 9.10). Deletion studies involving transfections of various chimeric reporter gene constructs into primary cultures of rat hepatocytes demonstrated the presence of a dexamethasone/PCN response element within the first 164 bp of the start of transcription. Subsequent studies demonstrated that for several CYP3A isoforms from different species contained nuclear receptor binding sites that are activated by DEX/PCN but exhibit low activation by rifampicin. Further work identified an additional 230-bp distal element called the xenobiotic-responsive enhancer module (XREM) located at −7836 through −7607 of the CYP3A4 that conferred responsiveness to both rifampicin and dexamethasone when combined with the proximal promoter region. XREM contains two nuclear receptor

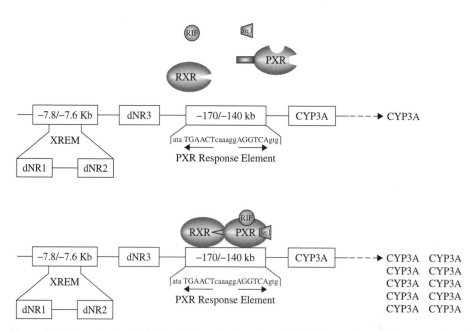

**Figure 9.10**   Illustration depicting DNA elements found in CYP3A genes and the activation of the human pregnane X receptor (PXR) by ligand (RIF) and subsequent transcriptional activation of CYP3A4 gene by the PXR/RXR heterodimer. dNR-1–3, nuclear receptors 1, 2, and 3, respectively; PXR, pregnane X receptor; RXR, retinoid X receptor; RIF, rifampicin; SRC-1, steroid receptor co-activator; XREM, xenobiotic responsive enhancer module.

binding sites (dNR1 and dNR2), neither of which is solely responsible for the activity of XREM. An additional nuclear receptor site, dNR3 located several hundred bases downstream of XREM also appears to have some importance in induction.

Recent work has demonstrated that the nuclear orphan receptor, pregnane X receptor (PXR) is the major determinant of CYP3A gene regulation by xenobiotics. Several lines of evidence support PXR involvement with CYP3A induction. First, both PXR and CYP3A isoforms are predominantly expressed in liver and intestine, with less expression found in lungs and kidneys. Second, PXR binds to human and rat CYP3A promoter regions and can activate expression of CYP3A4 promoter in transfection assays. Third, many of the same inducers of CYP3A isoforms also activate PXR. Fourth, interspecies differences in response to CYP3A inducers have been demonstrated to be due to the ability of these inducers to activate PXR in these species. Fifth, disruption of the mouse PXR gene eliminated induction of CYP3A by PCN and mice "humanized" with the PXR gene were able to respond to rifampicin induction. These observations suggest that many of the significant differences in CYP3A induction profiles between species may be due to differences in the PXR.

Peroxisome proliferators, including hypolipidemic drugs such as clofibrate, phthalate plasticizers, and herbicides bring about the induction of a CYP4A isoform that catalyzes the oxidation of many biologically important fatty acids, including arachidonic acid and other eicosanoids. CYP4A expression is part of a pleiotropic response in the rodent liver, which includes increased liver weight, proliferation of peroxisomes, and the elevation of several peroxisomal enzymes such as catalase. Peroxisome proliferators are often epigenetic carcinogens in rodents, but since the effect is primarily seen in rodents, its significance for other species such as humans is unclear. The receptor protein peroxisome proliferator-activated receptor-$\alpha$ PPAR$\alpha$ was first cloned in 1990. PPAR$\alpha$ knockout mice exposed to chemicals that normally induce CYP4A as well as peroxisome proliferation do not exhibit these characteristics, demonstrating the essential nature of PPAR$\alpha$ for these responses. Like PXR, PPAR$\alpha$, which is constitutively nuclear, also binds to DNA as a PPAR$\alpha$/RXR heterodimer in response to peroxisome proliferating chemicals.

CYP2E1 catalyzes metabolism of several low molecular weight xenobiotics including drugs (e.g., acetaminophen), solvents (e.g., ethanol and carbon tetrachloride), and procarcinogens (e.g., $N$-nitrosodimethylamine). Induction of CYP2E1 can occur as a result of exposure to several xenobiotics including ethanol, acetone, and imidazole, or alternatively, as a result of physiological conditions such as starvation and diabetes. Its induction by either fasting or diabetes is believed to be due to the high levels of ketones likely to be present in either of these conditions. It might also be noted that although CYP2E1 is in the same family as 2B1 and 2B2, it is not induced by phenobarbital-type inducers. In contrast to many other inducible CYPs, CYP2E1 induction is not accompanied by high levels of CYP2E1 mRNA, suggesting that regulation is by means of post-transcriptional mechanisms.

The regulation of CYP2E1 gene expression involves several mechanisms that do not primarily include increased transcription. Recent studies demonstrated that rapid increases in CYP2E1 protein levels following birth are due to stabilization of preexisting proteins by ketone bodies released at birth. Rats treated with ethanol or acetone can have three- to sixfold increases in CYP2E1 protein in the absence of increased CYP2E1 mRNA. Other studies have demonstrated that substrates including ethanol, imidazole, and acetone had little effects on CYP2E1 transcript content and that these substrates tend to prevent protein degradation. Thus increased protein expression in response to these substrates may be due to protein stabilization (e.g., decreased turnover), as a result of the inhibition of ubiquitin-mediated proteolysis. The ubiquitination process normally tags proteins with a chain of multiple ubiquitin moieties that can be detected as smears at the tops of SDS gels. The ubiquitin tags allow for the selective degradation of associated proteins by a cytosolic 26S protease, known as the proteasome. In recent studies an antibody prepared against a putative ubiquitination-target site on the CYP2E1 protein quenched ubiquitination in a concentration-dependent manner. The same antibody also prevented catalysis of chlorzoxazone. These results provide a plausible mechanistic explanation for the observation that substrate binding protects the CYP2E1 protein from ubiquitin-dependent proteolysis.

In other observations, diabetes is known to increase CYP2E1 expression at both the mRNA and protein levels in both chemically induced and spontaneous diabetic rats. Elevation of mRNA levels as a result of diabetes has been attributed to mRNA stabilization, which can be reversed by daily insulin treatment. Recent research has shown that insulin destabilizes CYP2E1 mRNA by binding to a 16-bp sequence within the 5' coding sequence of CYP2E1. The mechanism for regulation by this means is still uncertain, although other genes have also been reported with similar destabilizing sequences within their coding sequences. Many other possibilities for mRNA stabilization/destabilization exist within the 5' and 3' untranslated regions of the DNA that are being explored.

***Effects of Induction.*** The effects of inducers are usually the opposite of those of inhibitors; thus their effects can be demonstrated by much the same methods, that is, by their effects on pharmacological or toxicological properties in vivo or by the effects on enzymes in vitro following prior treatment of the animal with the inducer. In vivo effects are frequently reported; the most common ones are the reduction of the hexobarbital sleeping time or zoxazolamine paralysis time. These effects have been reported for numerous inducers and can be quite dramatic. For example, in the rat, the paralysis time resulting from a high dose of zoxazolamine can be reduced from 11 hours to 17 minutes by treatment of the animal with benzo(a)pyrene 24 hours before the administration of zoxazolamine.

The induction of monooxygenase activity may also protect an animal from the effect of carcinogens by increasing the rate of detoxication. This has been demonstrated in the rat with a number of carcinogens including benzo(a)pyrene, $N$-2-fluorenylacetamide, and aflatoxin $B_1$. Effects on carcinogenesis may be expected to be complex because some carcinogens are both activated and detoxified by monooxygenase enzymes, while epoxide hydrolase, which can also be involved in both activation and detoxication, may also be induced. For example, the toxicity of the carcinogen 2-naphthylamine, the hepatotoxic alkaloid monocrotaline, and the cytotoxin cyclophosphamide are all increased by phenobarbital induction—an effect mediated by the increased population of reactive intermediates.

Organochlorine insecticides are also well-known inducers. Treatment of rats with either DDT or chlordane, for example, will decrease hexobarbital sleeping time and offer protection from the toxic effect of warfarin. Persons exposed to DDT and lindane metabolized antipyrine twice as fast as a group not exposed, whereas those exposed to DDT alone had a reduced half-life for phenylbutazone and increased excretion of 6-hydroxycortisol.

Effects on xenobiotic metabolism in vivo are also widely known in both humans and animals. Cigarette smoke, as well as several of its constituent polycyclic hydrocarbons, is a potent inducer of aryl hydrocarbon hydroxylase in the placenta, liver, and other organs. The average content of CYP1A1 in liver biopsies from smokers was approximately fourfold higher than that from nonsmokers. Hepatic activity of CYP1A1 as measured by phenacetin $O$-deethylation, was also increased from 54 pmol/min/mg of protein in nonsmokers to 230 nmol/min/mg of protein in smokers. Examination of the term placentas of smoking human mothers revealed a marked stimulation of aryl hydrocarbon hydroxylase and related activities—remarkable in an organ that, in the uninduced state, is almost inactive toward foreign chemicals. These in vitro differences in metabolism are also observed in vivo, as smokers have been demonstrated to have increased clearance rates for several drugs metabolized principally by CYP1A1 including theophylline, caffeine, phenacetin, fluvoxamine, clozapine, and olanzapine.

**Induction of Xenobiotic-Metabolizing Enzymes Other Than Monooxygenases.** Although less well studied, xenobiotic-metabolizing enzymes other than those of the P450 system are also known to be induced, frequently by the same inducers that induce the oxidases. These include glutathione $S$-transferases, epoxide hydrolase, and UDP glucuronyltransferase. The selective induction of one pathway over another can greatly affect the metabolism of a xenobiotic.

### 9.5.3  Biphasic Effects: Inhibition and Induction

Many inhibitors of mammalian monooxygenase activity can also act as inducers. Inhibition of microsomal monooxygenase activity is fairly rapid and involves a direct interaction with the cytochrome, whereas induction is a slower process. Therefore, following a single injection of a suitable compound, an initial decrease due to inhibition would be followed by an inductive phase. As the compound and its metabolites are eliminated, the levels would be expected to return to control values. Some of the best examples of such compounds are the methylenedioxyphenyl synergists, such as piperonyl butoxide. Because P450 combined with methylenedioxyphenyl compounds in an inhibitory complex cannot interact with CO, the cytochrome P450 titer, as determined by the method of Omura and Sato (dependent on CO-binding to reduced cytochrome), would appear to follow the same curve.

It is apparent from extensive reviews of the induction of monooxygenase activity by xenobiotics that many compounds other than methylenedioxyphenyl compounds have the same effect. It may be that any synergist that functions by inhibiting microsomal monooxygenase activity could also induce this activity on longer exposure, resulting in a biphasic curve as described previously for methylenedioxyphenyl compounds. This curve has been demonstrated for NIA 16824 (2-methylpropyl-2-propynyl phenylphosphonate) and WL 19255 (5,6-dichloro-1,2,3-benzothiadiazole), although the results were less marked with R05-8019 [2,(2,4,5-trichlorophenyl)-propynyl ether] and MGK 264 [$N$-(2-ethylhexyl)-5-norbornene-2,3-dicarboximide].

## 9.6  ENVIRONMENTAL EFFECTS

Because light, temperature, and other in vitro effects on xenobiotic metabolizing enzymes are not different from their effects on other enzymes or enzyme systems, we are not concerned with them at present. This section deals with the effects of environmental factors on the intact animal as they relate to in vivo metabolism of foreign compounds.

*Temperature.* Although it might be expected that variations in ambient temperature would not affect the metabolism of xenobiotics in animals with homeothermic control, this is not the case. Temperature variations can be a form of stress and thereby produce changes mediated by hormonal interactions. Such effects of stress require an intact pituitary-adrenal axis and are eliminated by either hypothysectomy or adrenalectomy. There appear to be two basic types of temperature effects on toxicity: either an increase in toxicity at both high and low temperature or an increase in toxicity with an increase in temperature. For example, both warming and cooling increases the toxicity of caffeine to mice, whereas the toxicity of D-amphetamine is lower at reduced temperatures and shows a regular increase with increases in temperature.

In many studies it is unclear whether the effects of temperature are mediated through metabolism of the toxicant or via some other physiological mechanism. In other cases, however, temperature clearly affects metabolism. For example, in cold-stressed rats there is an increase in the metabolism of 2-naphthylamine to 2-amino-1-naphthol.

*Ionizing Radiation.* In general, ionizing radiation reduces the rate of metabolism of xenobiotics both in vivo and in enzyme preparations subsequently isolated. This has occurred in hydroxylation of steroids, in the development of desulfuration activity toward azinphosmethyl in young rats, and in glucuronide formation in mice. Pseudocholinesterase activity is reduced by ionizing radiation in the ileum of both rats and mice.

*Light.* Because many enzymes, including some of those involved with xenobiotic metabolism, show a diurnal pattern that can be keyed to the light cycle, light cycles rather than light intensity would be expected to affect these enzymes. In the case of hydroxyindole-$O$-methyltransferase in the pineal gland, there is a diurnal rhythm with greatest activity at night; continuous darkness causes maintenance of the high level. Cytochrome P450 and the microsomal monooxygenase system show a diurnal rhythm in both the rat and the mouse, with greatest activity occurring at the beginning of the dark phase.

*Moisture.* No moisture effect has been shown in vertebrates, but in insects it was noted that housefly larvae reared on diets containing 40% moisture had four times more activity for the epoxidation of heptachlor than did larvae reared in a similar medium saturated with water.

*Altitude.* Altitude can either increase or decrease toxicity. It has been suggested that these effects are related to the metabolism of toxicants rather than to physiological mechanisms involving the receptor system, but in most examples this has not been demonstrated clearly. Examples of altitude effects include the observations that at altitudes of $\geq 5000$ ft, the lethality of digitalis or strychnine to mice is decreased, whereas that of $D$-amphetamine is increased.

*Other Stress Factors.* Noise has been shown to affect the rate of metabolism of 2-napthylamine, causing a slight increase in the rat. This increase is additive with that caused by cold stress.

## 9.7  GENERAL SUMMARY AND CONCLUSIONS

It is apparent from the material presented in this chapter and the previous chapters related to metabolism that the metabolism of xenobiotics is complex, involving many enzymes; that it is susceptible to a large number of modifying factors, both physiological and exogenous; and that the toxicological implications of metabolism are important. Despite the complexity, summary statements of considerable importance can be abstracted:

- Phase I metabolism generally introduces a functional group into a xenobiotic, which enables conjugation to an endogenous metabolite to occur during phase II metabolism.
- The conjugates produced by phase II metabolism are considerably more water soluble than either the parent compound or the phase I metabolite(s) and hence are more excretable.
- During the course of metabolism, and particularly during phase I reactions, reactive intermediates that are much more toxic than the parent compound may be produced. Thus xenobiotic metabolism may be either a detoxication or an activation process.
- Because the number of enzymes involved in phase I and phase II reactions is large and many different sites on organic molecules are susceptible to metabolic attack, the number of potential metabolites and intermediates that can be derived from a single substrate is frequently very large.
- Because both qualitative and quantitative differences exist among species, strains, individual organs, and cell types, a particular toxicant may have different effects in different circumstances.
- Because exogenous chemicals can be inducers and/or inhibitors of the xenobiotic-metabolizing enzymes of which they are substrates; such chemicals may interact to bring about toxic sequelae different from those that might be expected from any of them administered alone.
- Because endogenous factors also affect the enzymes of xenobiotic metabolism, the toxic sequelae to be expected from a particular toxicant will vary with developmental stage, nutritional statue, health or physiological status, stress or environment.
- It has become increasingly clear that most enzymes involved in xenobiotic metabolism occur as several isozymes, which coexist within the same individual and, frequently, within the same subcellular organelle. An understanding of the biochemistry and molecular genetics of these isozymes may lead to an understanding of the variation among species, individuals, organs, sexes, developmental stages, and so on.

## SUGGESTED READING

Anderson, K. E., and A. Kappas. Dietary regulation of cytochrome P450. *An. Rev. Nutr.* **11**: 141–167, 1991.

Banerjee, A., T. A. Kocarek, and R. F. Novak. Identification of a ubiquitination-target/substrate-interaction domain of cytochrome P-450 (CYP) 2E1. *Drug Metabol. Disp.* **28**: 118–124, 2000.

Conney, A. H. Induction of drug-metabolizing enzymes: A path to the discovery of multiple cytochromes P450. *An. Rev. Pharmacol. Toxicol.* **43**: 1–30, 2003.

Denison, M. S., and J. P. Whitlock, Jr. Minireview: Xenobiotic-inducible transcription of cytochrome P450 genes. *J. Biol. Chem.* **270**: 18175–18178, 1995.

Goodwin, B., M. R. Redinbo, and S. A. Kliewer. Regulation of CYP3A gene transcription by the pregnane X receptor. *An. Rev. Pharmacol. Toxicol.* **42**: 1–23, 2002.

Guengerich, F. P. Cytochrome P-450 3A4: Regulation and role in drug metabolism. *An. Rev. Pharmacol. Toxicol.* **39**: 1–17, 1999.

Hodgson, E., and J. A. Goldstein. Metabolism of toxicants: Phase I reactions and pharmacogenetics. In *Introduction to Biochemical Toxicology*, 3rd ed., E. Hodgson, R. C. Smart, eds., New York: Wiley, 2001.

Kakizaki, S., Y. Yamamoto, A. Ueda, R. Moore, T. Sueyoshi, and M. Negishi. Phenobarbital induction of drug/steroid-metabolizing enzymes and nuclear receptor CAR. *Biochim. Biophys. Acta* **1619**: 239–242, 2003.

Kocarek, T. A., R. C. Zanger, and R. F. Novak. Post-transcriptional regulation of rat CYP2E1 expression: Role of CYP2E1 mRNA untranslated regions in control of translational efficiency and message stability. *Arch. Biochem. Biophys.* **376**: 180–190, 2000.

Lieber, C. S. Cytochrome P-4502E1: Its physiological and pathological role. *Physiol. Rev.* **77**: 517–544, 1997.

Lin, J. H., and A. Y. H. Lu. Interindividual variability in inhibition and induction of cytochrome P450 enzymes. *An. Rev. Pharmacol. Toxicol.* **41**: 535–567, 2001.

Newton, D. J., R. W. Wang, and A. Y. H. Lu. Cytochrome P450 inhibitors: Evaluation of specificities in the in vitro metabolism of therapeutic agents by human liver microsomes. *Drug Metab. Disp.* **23**: 154–158, 1995.

Okey A. B. Enzyme induction in the cytochrome P-450 system. *Pharmac. Ther.* **45**: 241–298, 1990.

Ronis, M. J. J., and H. C. Cunny. Physiological factors affecting xenobiotic metabolism. In *Introduction to Biochemical Toxicology*, 3rd ed., E. Hodgson, and R. C. Smart, eds. New York: Wiley, 2001.

Rose, R. L., and E. Hodgson. Adaptation to toxicants. In *Introduction to Biochemical Toxicology*, 3rd ed. E. Hodgson, and R. C. Smart eds. New York: Wiley, 2001.

Sueyoshi, T., and M. Negishi. Phenobarbital response elements of cytochrome P450 genes and nuclear receptors. *An. Rev. Pharmacol. Toxicol.* **41**: 123–143, 2001.

Smith, C. A. D., G. Smith, and C. R. Wolf. Genetic polymorphisms in xenobiotic metabolism. *Eur. J. Cancer* **30A**: 1935–1941, 1994.

Whitlock, J. P., Jr. Induction of cytochrome P4501A1. *An. Rev. Pharmacol. Toxicol.* **39**: 103–125, 1999.

# CHAPTER 10

# Elimination of Toxicants

GERALD A. LEBLANC

## 10.1  INTRODUCTION

The ability to efficiently eliminate toxic materials is critical to the survival of a species. The complexity of toxicant elimination processes has increased commensurate with the increased complexity associated with animal form. For unicellular organisms, passive diffusion can suffice for the elimination of toxic metabolic wastes produced by the organism. Similarly, as exogenous toxic materials derived from the environment diffuse into a unicellular organism, they can also readily diffuse out of the organism. The large surface area to mass ratio of these organisms ensures that a toxic chemical within the cell is never significantly distanced from a surface membrane across which it can diffuse.

As organisms evolved in complexity, several consequences of increased complexity compromised the efficiency of the passive diffusion of toxic chemicals:

1. They increased in size.
2. Their surface area to body mass decreased.
3. Their bodies compartmentalized (i.e., cells, tissues, organs).
4. They generally increased in lipid content.
5. They developed barriers to the external environment.

*Size.* With increased size of an organism, a toxic chemical has greater distance to traverse before reaching a membrane across which it can diffuse to the external environment. Thus overall retention of the chemical will increase as will propensity for the chemical to elicit toxicity.

*Surface Area to Body Mass Ratio.* Increased size of an organism is associated with a decrease in the surface area to body mass ratio. Accordingly, the availability of surface membranes across which a chemical can passively diffuse to the external environment decreases and propensity for retention of the chemical increases.

*A Textbook of Modern Toxicology, Third Edition,* edited by Ernest Hodgson
ISBN 0-471-26508-X  Copyright © 2004 John Wiley & Sons, Inc.

*Compartmentalization.* With increased complexity comes increased compartmentalization. Cells associate to form tissues and tissues associate to form organs. Compartmentalization increases the number of barriers across which chemicals must traverse before sites of elimination are reached. As different compartments often have different physicochemical characteristics (i.e., adipose tissue is largely fat while blood is largely aqueous), chemicals are faced with the challenge to be mobile in these various environments.

*Lipid Content.* As a general but not universal rule, the ability of organisms to store energy as fat increases with increased size of the organism. Thus large organisms tend to have significant lipid stores into which lipophilic chemicals can be stored for extended periods of time. These stored chemicals tend to be largely immobile and difficult to release from the adipose tissue.

*Barriers to the Environment.* Through evolution, increased complexity of organisms led to increased exploitation of various environments. In order to survive in these environments, organisms developed barriers such as skin and scales that protect from harsh conditions on the outside and minimize loss of vital constituents such as water on the inside. Likewise these barriers impede the elimination of toxic constituents by the organisms, requiring the development of specialized membranes and organs through which toxic materials can be eliminated.

A consequence of this hindrance to elimination of toxic materials by complex organisms was the development of specialized routes of elimination. These routes generally evolved in concert (i.e., co-evolved) with biotransformation processes that render chemicals amenable to these modes of elimination (see Chapters 7–9).

Three major routes of elimination culminate in the specialized organs of elimination, the liver, kidneys, and lungs. The liver serves as a major organ at which lipophilic materials are collected from the blood, biotransformed to generally less toxic and more polar derivatives, then eliminated into the bile. The kidneys complement the liver in that these organs collect wastes and other chemicals in the blood through a filtration process and eliminate these wastes in the urine. The respiratory membranes of the lungs are ideal for the removal of volatile materials from the blood into expired air. In addition to these major routes of elimination, several quantitatively minor routes exist through which toxic materials can be eliminated from the body. These include the following:

1. *Skin.* Skin constitutes the largest organ in the human body, and it spans the interface between the body and the external environment. While the skin's epidermis constitutes a relatively impervious membrane across which chemical elimination is difficult, the shear surface area involved requires consideration of this organ as a route of elimination. Volatile chemicals are particularly adept at traversing the skin and exiting the body through this route.

2. *Sweat.* Humans lose an average of 0.7 L of water per day due to sweating. This loss of fluid provides a route for the elimination of water-soluble chemicals.

3. *Milk.* Mother's milk is rich in lipids and lipoproteins. Milk thus serves as an ideal route for the elimination of both water-soluble and fat-soluble chemicals from the mother's body. For example, the DDT metabolite DDE, the flame retardant mirex, and the polychlorinated biphenyls (PCBs) often have been detected in mother's

milk. While lactation may provide a benefit to the mother by the elimination of toxic chemicals, transfer of these toxicants to the suckling infant can have dire consequences.

4. *Hair.* Growing hair can serve as a limited route through which chemicals can escape the body. Pollutants such as mercury and drugs such as cocaine have been measured in human hair, and hair analyses is often used as a marker of exposure to such materials.

## 10.2 TRANSPORT

For a chemical to be eliminated from the body at a site of elimination (i.e., kidney) that is distant from the site of storage (i.e., adipose tissue) or toxicity (i.e., brain), the chemical must be transported from the site of origin to the site of elimination. Chemicals are transported to the site of elimination largely via the circulatory system. Sufficiently water-soluble chemicals can freely dissolve into the aqueous component of blood and be transported by both diffusion and blood circulation to sites of elimination. With decreasing water solubility and increasing lipid solubility, chemicals are less likely to freely diffuse into blood and extraction of these chemicals from sites of toxicity or storage can be more challenging. These materials generally associate with transport proteins in the blood, which either contain binding sites for chemical attachment or lipophilic cores (lipoproteins) into which lipophilic chemicals can diffuse. The blood contains various transport proteins that are typically suited for the transport of specific endogenous chemicals. These include albumin, sex steroid-binding globulin, and lipoproteins. Often xenobiotics can utilize these proteins, particularly the nonspecific transporters, to facilitate mobilization and transport in the aqueous environment of the blood. At the site of elimination, xenobiotics may diffuse from the transport protein to the membranes of the excretory organ, or the transport protein may bind to surface receptors on the excretory organ, undergo endocytosis and intracellular processing, where the xenobiotic is released and undergoes processing leading to elimination.

## 10.3 RENAL ELIMINATION

The kidneys are the sites of elimination of water-soluble chemicals that are removed from the blood by the process of reverse filtration. Two characteristics are primarily responsible for determining whether a chemical will be eliminated by the kidneys: size and water solubility.

*Size.* The reverse filtration process requires that chemicals to be removed from the blood are able to pass through 70 to 100 A pores. As a rule, chemicals having a molecular mass of less than 65,000 are sufficiently small to be subject to reverse filtration.

*Water Solubility.* Non-water-soluble chemicals will be transported to the kidneys in association with transport proteins. Thus, in association with these proteins, the chemicals will not be able to pass through the pores during reverse filtration. Lipophilic chemicals are generally subject to renal elimination after they have undergone hydroxylation or conjugation reactions (Chapter 7) in the liver or elsewhere.

Blood is delivered to the human kidney by the renal artery. Blood flows to the kidneys of the adult human at a rate of roughly 1 L/min. The adult human kidney contains approximately 1 million functional units, called nephrons, to which the blood is delivered for removal of solutes. Collected materials are excreted from the body in the urine.

Blood entering the nephron passes through a network of specialized capillaries called the glomerulus (Figure 10.1). These capillaries contain the pores through which materials to be eliminated from the blood pass. Blood in the capillaries is maintained under high positive pressure from the heart coupled with the small diameter of the vessels. As a result these sufficiently small solutes and water are forced through the pores of the glomerulus. This filtrate is collected in the glomerular (or Bowman's) capsule in which the glomerulus is located (Figure 10.1). Included in this filtrate are water, ions, small molecules such as glucose, amino acids, urate, and foreign chemicals. Large molecules such as proteins and cells are not filtered and are retained in the blood.

Following glomerular filtration, molecules important to the body are re-absorbed from the filtrate and returned to the blood. Much of this re-absorption occurs in the proximal tubules (Figure 10.1). Cells lining the proximal tubules contain fingerlike projections that extent into the lumen of the tubule. This provides an expanse of cell surface area across which water and ions can diffuse back into the cells and, ultimately, be returned to the blood. The proximal tubules also contain active transport proteins that recover small molecules such as glucose and amino acids from the filtrate. From the proximal tubules, the filtrate passes through the Loop of Henle. Significant water

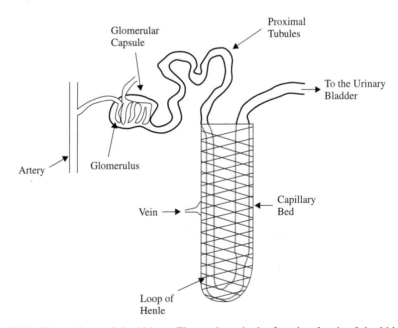

**Figure 10.1** The nephron of the kidney. The nephron is the functional unit of the kidney that is responsible for the removal of water-soluble wastes and foreign compounds from the blood.

re-absorption occurs in the descending portion of the loop, resulting in concentration of the filtrate. Water re-absorption does not occur in the ascending portion of the loop. Rather, the remaining, concentrated ions such as sodium, chloride, and potassium are re-absorbed. Those materials retained in the filtrate during passage through the nephron constitute the urine. The urine is transported through the ureters to the bladder and retained until excretion occurs.

The kidneys are a common site of chemical toxicity since the nephron functions to concentrate the toxicant and thus increase levels of exposure to the materials. This increased exposure can result from the concentration of the toxicant in the tubules. It also can occur by concentration within the cells of the nephrons when a chemical is capable of utilizing one of the active transport proteins and is shuttled from the lumen of the tubules into the renal cells.

## 10.4   HEPATIC ELIMINATION

The liver serves many vital functions to the body. It has a large capacity to hold blood and thus serves as a blood storage site. The liver synthesizes and secretes many substances that are necessary for normal bodily function. It cleanses the blood of various endogenous and foreign molecules. It biotransforms both endogenous and exogenous materials, reducing their bioreactivity and preparing them for elimination. It eliminates wastes and foreign chemicals through biliary excretion. Three of these functions occur coordinately in a manner that makes the liver a major organ of chemical elimination: chemical uptake from blood, chemical biotransformation, and biliary elimination of chemicals.

Blood is delivered to the liver from two sources. Oxygen-rich blood is delivered through the hepatic artery. In addition blood is shunted from the capillaries that service the intestines and spleen to the liver by the hepatic portal vein. These two vessels converge, and the entire hepatic blood supply is passaged through sinusoids (Figure 10.2). Sinusoids are cavernous spaces among the hepatocytes that are the functional units of the liver. Hepatocytes are bathed in blood as the blood passes through the sinusoids, as 70% of the hepatocyte surface membrane contacts the blood in the sinusoid. This provides for a tremendous surface area across which chemicals can diffuse to gain entry into the hepatocytes. Chemicals may passively diffuse across the sinusoidal

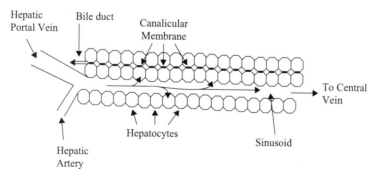

**Figure 10.2**  Diagrammatic representation of the basic architecture of the liver.

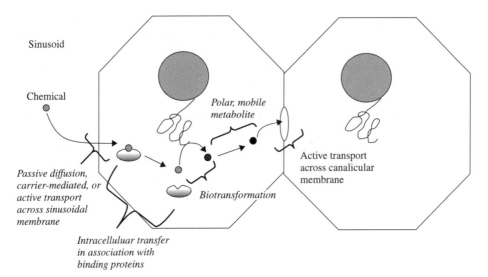

**Figure 10.3** Vectorial transport of a chemical from the liver sinusoid, through the hepatocyte, to the canalicular space.

membrane of the hepatocytes, they may be exchanged between blood transport proteins and the sinusoidal membranes, or their carrier proteins may bind to sinusoidal membrane receptors and then undergo endocytosis (Figure 10.3).

Lipophilic materials require intracellular carrier proteins to be optimally mobilized, just as they required transport proteins in the blood (Figure 10.3). Several intracellular carrier proteins that mobilize specific endogenous chemical have been characterized, although less is known of which proteins typically mobilize xenobiotics. Some of the cytosolic glutathione *S*-transferase proteins have been shown to noncatalytically bind xenobiotics and to be coordinately induced along with xenobiotic biotransformation enzymes and efflux transporters, suggesting that these proteins may function to mobilize xenobiotics.

Once mobilized in the hepatocyte, chemicals can contact and interact with biotransformation enzymes (Chapter 7). These enzymes generally increase the polarity of the chemical, thus reducing its ability to passively diffuse across the sinusoidal membrane back into the blood. Biotransformation reactions also typically render the xenobiotics susceptible to active transport across the canalicular membrane into the bile canaliculus and, ultimately, the bile duct (Figure 10.3). The bile duct delivers the chemicals, along with other constituents of bile, to the gall bladder that excretes the bile into the intestines for fecal elimination.

### 10.4.1 Entero-hepatic Circulation

Once in the gastrointestinal tract, chemicals that have undergone conjugation reactions in the liver may be subject to the action of hydrolytic enzymes that de-conjugate the molecule. De-conjugation results in increased lipophilicity of the molecule and renders them once again subject to passive uptake. Re-absorbed chemicals enter the circulation via the hepatic portal vein, which shunts the chemical back to the liver where

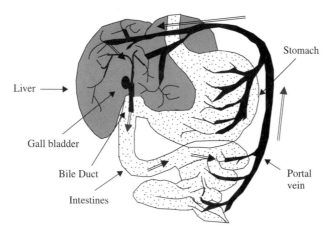

Stomach

Liver

Gall bladder

Bile Duct

Intestines

Portal
vein

**Figure 10.4**   Enterohepatic circulation (as indicated by ⟹). Polar xenobiotic conjugates are
secreted into the intestine via the bile duct and gall bladder. Conjugates are hydrolyzed in
the intestines, released xenobiotics are reabsorbed, and transported back to the liver via the
portal vein.

the chemical can be reprocessed (i.e., biotransformed) and eliminated. This process is
called entero-hepatic circulation (Figure 10.4). A chemical may undergo several cycles
of entero-hepatic circulation resulting in a significant increase in the retention time for
the chemical in the body and increased toxicity.

The liver functions to collect chemicals and other wastes from the body. Accordingly, high levels of chemicals may be attained in the liver, resulting in toxicity to
this organ. Biotransformation of chemicals that occur in the liver sometimes results in
the generation of reactive compounds that are more toxic than the parent compound
resulting in damage to the liver. Chemical toxicity to the liver is discussed elsewhere
(Chapter 14).

### 10.4.2   Active Transporters of the Bile Canaliculus

The bile canaliculus constitutes only about 13% of the contiguous surface membrane
of the hepatocyte but must function in the efficient transfer of chemical from the hepatocyte to the bile duct. Active transport proteins located on the canalicular membrane
are responsible for the efficient shuttling of chemicals across this membrane. These
active transporters are members of a multi-gene superfamily of proteins known as the
ATP-binding cassette transporters. Two subfamilies are currently recognized as having
major roles in the hepatic elimination of xenobiotics, as well as endogenous materials.
The *P*-glycoprotein (ABC B) subfamily is responsible for the elimination of a variety of
structurally diverse compounds. *P*-glycoprotein substrates typically have one or more
cyclic structures, a molecular weight of 400 or greater, moderate to low lipophilicity
($\log K_{ow} < 2$), and high hydrogen (donor)-bonding potential. Parent xenobiotics that
meet these criteria and hydroxylated derivatives of more lipophilic compounds are
typically transported by *P*-glycoproteins.

The multidrug-resistance associated protein (ABC C) subfamily of proteins largely
recognizes anionic chemicals. ABC C substrates are commonly conjugates of xenobiotics (i.e., glutathione, glucuronic acid, and sulfate conjugates). Thus conjugation

not only restricts passive diffusion of a lipophilic chemical but actually targets the xenobiotic for active transport across the canalicular membrane.

## 10.5   RESPIRATORY ELIMINATION

The lungs are highly specialized organs that function in the uptake and elimination of volatile materials (i.e., gasses). Accordingly, the lungs can serve as a primary site for the elimination of chemicals that have a high vapor pressure. The functional unit of the lung is the alveolus. These small, highly vascularized, membraneous sacs serve to exchange oxygen from the air to the blood (uptake), and conversely, exchange carbon dioxide from the blood to the air (elimination). This exchange occurs through passive diffusion. Chemicals that are sufficiently volatile also may diffuse across the alveolar membrane, resulting in removal of the chemical from the blood and elimination into the air.

## 10.6   CONCLUSION

Many processes function coordinately to ensure that chemicals distributed throughout the body are efficiently eliminated at distinct and highly specialized locations. This unidirectional transfer of chemicals from the site of origin (storage, toxicity, etc.) to the site of elimination is a form of vectorial transport (Figure 10.5). The coordinate action of blood binding proteins, active transport proteins, blood filtration units, intracellular binding proteins, and biotransformation enzymes ensures the unidirectional flow of chemicals, ultimately resulting in their elimination. The evolution of this complex interplay of processes results in the efficient clearance of toxicants and has provided the way for the co-evolution of complexity in form from unicellular to multi-organ organisms.

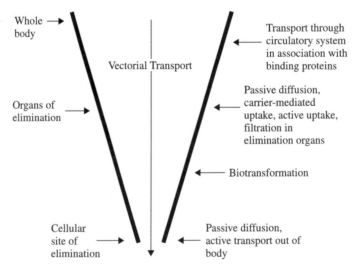

**Figure 10.5**   Processes involved in the vectorial transport of xenobiotics from the whole body point of origin to the specific site of elimination.

## SUGGESTED READING

LeBlanc, G. A., and W. C. Dauterman. Conjugation and elimination of toxicants. In *Introduction to Biochemical Toxicology*, E. Hodgson and R. C. Smart, eds. New York: Wiley-Interscience, 2001, pp. 115–136.

Kester, J. E. Liver. In *Encyclopedia of Toxicology*, vol. 2, P. Wexler, ed. New York: Academic Press, 1998, pp. 253–261.

Rankin, G. O. Kidney. In *Encyclopedia of Toxicology*, vol. 2, P. Wexler, ed. New York: Academic Press, 1998, pp. 198–225.

Klaassen, C. D. ed. *Casarett and Doull's Toxicology: The Basic Science of Poisons*, 6th ed. New York: McGraw-Hill, 2001.

# PART IV

# TOXIC ACTION

# CHAPTER 11

# Acute Toxicity

GERALD A. LEBLANC

## 11.1 INTRODUCTION

Acute toxicity of a chemical can be viewed from two perspectives. Acute toxicity may be the descriptor used as a qualitative indicator of an incident of poisoning. Consider the following statement: "methyl isocyanate gas, accidentally released from a chemical manufacturing facility in 1984, was *acutely toxic* to the residents of Bhopal, India." This statement implies that the residents of Bhopal were exposed to sufficiently high levels of methyl isocyanate over a relatively short time to result in immediate harm. High-level, short-term exposure resulting in immediate toxicity are all characteristics of acute toxicity. Alternatively, acute toxicity may represent a quantifiable characteristic of a material. For example, the statement: "the *acute toxicity* of methyl isocyanate, as measured by its LD50 in rats, is 140 mg/kg" defines the acute toxicity of the chemical. Again, the characterization of the quantified effects of methyl isocyanate as being acute toxicity implies that this quantification occurred during or following short-term dosing and that the effect measured occurred within a short time period following dosing. In terms of these qualitative and quantitative aspects, acute toxicity can be defined as *toxicity elicited immediately following short-term exposure to a chemical.* By this definition, two components comprise acute toxicity: acute exposure and acute effect.

## 11.2 ACUTE EXPOSURE AND EFFECT

In contrast to acute toxicity, chronic toxicity is characterized by prolonged exposure and sublethal effects elicited through mechanisms that are distinct from those that cause acute toxicity. Typically acute and chronic toxicity of a chemical are easily distinguished. For example, mortality occurring within two days of a single dose of a chemical would be a prime example of acute toxicity (Figure 11.1a). Similarly, reduced litter size following continuous (i.e., daily) dosing of the parental organisms would be indicative of chronic toxicity (Figure 11.1b). However, defining toxicity as being acute or chronic is sometimes challenging. For example, chronic exposure to a persistent, lipophilic chemical may result in sequestration of significant levels of the chemical

*A Textbook of Modern Toxicology, Third Edition,* edited by Ernest Hodgson
ISBN 0-471-26508-X Copyright © 2004 John Wiley & Sons, Inc.

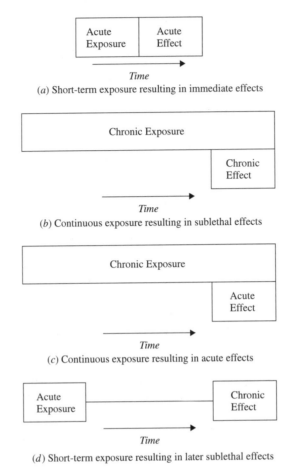

**Figure 11.1**  Examples of exposure/effect scenarios that result in either acute toxicity (*a*), chronic toxicity (*b*), or mixed acute/chronic toxicity (*c,d*). Examples for each scenario are provided in the text.

in adipose tissue of the organism with no resulting overt toxicity. Upon entering the reproductive phase, organisms may mobilize fatty stores, releasing the chemical into the blood stream resulting in overt toxicity including death (Figure 11.1*c*). One could argue under this scenario that chronic exposure ultimately resulted in acute effects. Lastly, acute exposure during a susceptible window of exposure (i.e., embryo development) may result in reproductive abnormalities and reduced fecundity once the organism has attained reproductive maturity (Figure 11.1*d*). Thus acute exposure may result in chronic toxicity.

An additional consideration is noteworthy when comparing acute and chronic toxicity. All chemicals elicit acute toxicity at a sufficiently high dose, whereas all chemicals do not elicit chronic toxicity. Paracelsus' often cited phrase "all things are poison ... the dose determines ... a poison" is clearly in reference to acute toxicity. Even the most benign substances will elicit acute toxicity if administered at a sufficiently high dose. However, raising the dose of a chemical does not ensure that chronic toxicity will ultimately be attained. Since chronic toxicity typically occurs at dosages below those

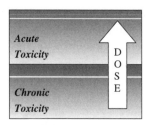

**Figure 11.2**  Relationships among chemical dose, acute toxicity and chronic toxicity. All chemicals elicit acute toxicity at a sufficiently high dose. However, chronic toxicity may not occur since dosage elevation may simply lead to acute toxicity.

that elicit acute toxicity, toxicity observed at the higher dosage may simply reflect acute, and not chronic, toxicity (Figure 11.2).

Effects encountered with acute toxicity commonly consist of mortality or morbidity. From a quantitative standpoint these effects are measured as the LD50, ED50, LC50, or EC50. The LD50 and ED50 represent the dose of the material that causes mortality (LD50) or some other defined effect (ED50) in 50% of a treated population. The LC50 and EC50 represent the concentration of the material to which the organisms were exposed that causes mortality (LC50) or some other defined effect (EC50) in 50% of an exposed population. LD50 and ED50 are normalize to the weight of the animal (i.e., mg chemical/kg body weight); whereas LC50 and EC50 are normalized to the environment in which the organisms were exposed (i.e., mg chemical/L water).

## 11.3  DOSE-RESPONSE RELATIONSHIPS

Acute toxicity of a chemical is quantified by its dose-response curve. This relationship between dose of the chemical administered and the resulting response is established by exposing groups of organisms to various concentrations of the chemical. Ideally doses are selected that will elicit >0% effect but <100% effect during the course of the experiment. At defined time periods following dosing, effects (e.g., mortality) are recorded. Results are plotted in order to define the dose-response curve (Figure 11.3*a*). A well-defined dose-response curve generated with a population of organisms whose susceptibility to the chemical is normally distributed will be sigmoidal in shape. The various segments (see Figure 11.3*a*) of the curve are represented as follows:

*Segment I*. This portion of the line has no slope and is represented by those doses of the toxicant that elicited no mortality to the treated population of organisms.

*Segment II*. This segment represents those dosages of the toxicant that affected only the most susceptible members of the exposed population. Accordingly, these effects are elicited at low doses and only a small percentage of the dosed organisms are affected.

*Segment III*. This portion of the line encompasses those dosages at which most of the groups of organisms elicit some response to the toxicant. Because most of the groups of exposed organisms respond to the toxicant within this range of dosages, segment III exhibits the steepest slope among the segments.

*Segment IV*. This portion of the line encompasses those dosages of the toxicant that are toxic to even the most tolerant organisms in the populations. Accordingly, high dosages of the toxicant are required to affect these organisms.

*Segment V*. Segment V has no slope and represents those dosages at which 100% of the organisms exposed to the toxicant have been affected.

A well-defined dose-response curve can then be used to calculate the LD50 for the toxicant. However, in order to provide the best estimate of the LD50, the curve is typically linearized through appropriate transformations of the data. A common transformation involves converting concentrations to logarithms and percentage effect to probit units (Figure 11.3*b*). Zero percent and 100% responses cannot be converted to probits; therefore data within segments I and V are not used in the linearization. A 95% confidence interval also can be determined for the linearize dose-response relationship (Figure 11.3*b*). As depicted in Figure 11.3*b*, the greatest level of confidence (i.e., the smallest 95% confidence interval) exists at the 50% response level, which is why LD50 values are favored over some other measure of acute toxicity (eg., LD05). This high level of confidence in the LD50 exists when ample data exist between the 51% and 99% response as well as between the 1% and 49% response.

Additional important information can be derived from a dose-response curve. The slope of the linearized data set provides information on the specificity of the toxicant. Steep slopes to the dose-response line are characteristic of toxicants that elicit toxicity by interacting with a specific target, while shallow slopes to the dose-response line are characteristic of toxicant that elicit more nonspecific toxicity such as narcosis.

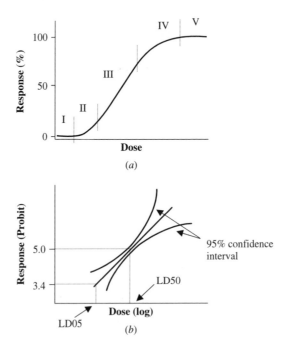

**Figure 11.3**   The dose-response relationship. (*a*) Five segments of the sigmoidal dose-response curve as described in the text. (*b*) Linearized dose-response relationship through log (dose)-probit (effect) transformations. Locations of the LD50 and LD05 are depicted.

The dose-response line also can be used to estimate the threshold dose. The threshold dose is defined as the lowest dose of the chemical that would be expected to elicit a response under conditions at which the assay was performed. The threshold dose is often empirically estimated as being a dose less that the lowest dose at which an effect was measured but higher than the greatest dose at which no effect was detected. Conceptually, the threshold dose is defined as the intercept of segments I and II of the dose-response curve (Figure 11.3*a*). Statistically, the threshold dose can be estimated from the linearized dose-response curve as the LC05. This value will closely approximate the threshold dose and can be statistically derived from the entire data set (i.e., the dose-response line). However, confidence in this value is greatly compromised, since it is derived from one end of the line (Figure 11.3*b*).

## 11.4  NONCONVENTIONAL DOSE-RESPONSE RELATIONSHIPS

The low-level effects of chemicals have received attention among pharmacologists for over 100 years. A current resurgence in interest among pharmacologists in low-level effects stems from use of homeopathic approaches to treating disease. Proponents of homeopathy maintain that low levels of toxic materials stimulate physiological responses that can target disease without eliciting adverse effects in the individual undergoing treatment. Homeopathic principles may have application in toxicology based on the premise that exposure to some chemicals at subthreshold levels, as defined by standard acute toxicity evaluations, can elicit toxicological as well as pharmacological effects. Both pharmacological and toxicological homeopathy may be the consequence of hormesis.

Hormesis is defined as an overcompensatory response to some disruption in homeostasis. Thus hormesis is typically evident at low doses of a chemical at which gross disruptions in homeostasis do not mask the hormetic response. Further, hormesis typically presents as an effect opposite to that elicited at higher levels of the chemical. For example, a chemical that stimulates corticosteroid secretion at high dosages resulting in hyperadrenocorticism might elicit a hormetic response at low dosages resulting in corticosteroid deficiency. A hypothetical nonconventional dose-response relationship resulting from such interactions is depicted in Figure 11.4. At the true threshold dose,

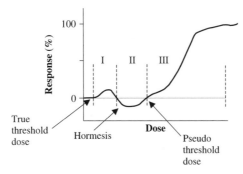

**Figure 11.4**  Nonconventional dose-response relationship involving low-dose effects and compensation. (I) True initiation of the response followed by a compensatory response that returns the effects to the 0% level. (II) A negative response due to overcompensation (hormesis) followed by recovery to the 0% effect level. (III) The standard sigmoidal dose-response relationship.

the organisms begin to exhibit increased stimulation in corticosteroid secretion. However, at slightly higher doses, a compensatory response occurs whereby corticosteroid secretion is decreased in order to maintain homeostasis within the organism. Overcompensation may actually result in a decrease in corticosteroid secretion at certain toxicant dosages. Finally the compensatory abilities of the organism are overcome by the high doses of the toxicant at the "pseudo" threshold dose, above which the standard dose-response relationship occurs. Nonconventional dose-response relationships have been observed with respect to both acute and chronic toxicity and are particularly relevant to the risk assessment process when establishing levels of exposure that are anticipated to pose no harm.

## 11.5   MECHANISMS OF ACUTE TOXICITY

An exhaustive review of the mechanisms by which chemicals cause acute toxicity is beyond the scope of this chapter. However, certain mechanisms of toxicity are relevant since they are common to many important classes of toxicants. Some of these mechanisms of acute toxicity are discussed.

### 11.5.1   Narcosis

Narcosis in toxicology is defined as toxicity resulting from chemicals associating with and disrupting the lipid bilayer of membranes. Narcotics are classified as either nonpolar (class 1) or polar (class 2) compounds. Members of both classes of compounds are lipid soluble. However, class 2 compounds possess constituents that confer some charge distribution to the compound (i.e., aliphatic and aromatic amines, nitroaromatics, alcohols). The aliphatic hydrocarbon (C5 through C8) are examples of powerful class 1 narcotics, whereas, ethanol is an example of a class 2 narcotic. The affinity of narcotics to partition into the nonpolar core of membranes (class 1 narcotics) or to distribute in both the polar and nonpolar components of membranes (class 2 narcotics) alters the fluidity of the membrane. This effect compromises the ability of proteins and other constituents of the membranes to function properly leading to various manifestation of narcosis. The central nervous system is the prime target of chemical narcosis and symptoms initially include disorientation, euphoria, giddiness, and progress to unconsciousness, convulsion, and death.

### 11.5.2   Acetylcholinesterase Inhibition

Acetylcholine is a neurotransmitter that functions in conveying nerve impulses across synaptic clefts within the central and autonomic nervous systems and at junctures of nerves and muscles. Following transmission of an impulse across the synapse by the release of acetylcholine, acetylcholinesterase is released into the synaptic cleft. This enzyme hydrolyzes acetylcholine to choline and acetate and transmission of the nerve impulse is terminated. The inhibition of acetylcholineasterase results in prolonged, uncoordinated nerve or muscle stimulation. Organophosphorus and carbamate pesticides (Chapter 5) along with some nerve gases (i.e., sarin) elicit toxicity via this mechanism.

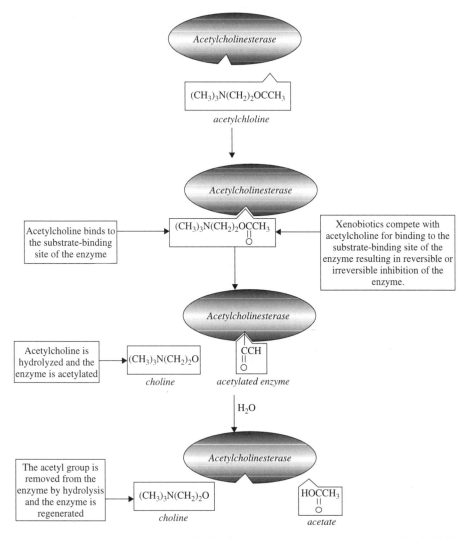

**Figure 11.5** Hydrolysis of acetylcholine by the enzyme acetylcholinesterase and its inhibition by toxicants such as organophosphorus and carbamate insecticides.

Inhibitors of acetylcholinesterase function by binding to the substrate-binding site of the enzyme (Figure 11.5). Typically the inhibitor or a biotransformation derivative of the inhibitor (i.e., the phosphodiester component of organophosphorus compounds) covalently binds to the enzyme resulting in its inhibition. Inhibition persists until the bound inhibitor is hydrolytically cleaved from the enzyme. This inhibition may be range from minutes in duration to permanent. Toxic effects of cholinesterase inhibition typically are evident when the enzyme activity is inhibited by about 50%. Symptoms include nausea and vomiting, increased salivation and sweating, blurred vision, weakness, and chest pains. Convulsions typically occur between 50% and 80% enzyme inhibition with death at 80–90% inhibition. Death is most commonly due to respiratory failure.

### 11.5.3 Ion Channel Modulators

Ion transport is central to nerve impulse transmission both along the axon and at the synapse and many neurotoxicants elicit effects by interfering with the normal transport of these ions (Figure 11.6). The action potential of an axon is maintained by the high concentration of sodium on the outside of the cell as compared to the low concentration inside. Active transporters of sodium ($Na^+K^+$ ATPases) that actively transport sodium out of the cell establish this action potential. One action of the insecticide DDT resulting in its acute toxicity is the inhibition of these $Na^+K^+$ ATPases resulting in the inability of the nerve to establish an action potential. Pyrethroid insecticides also elicit neurotoxicity through this mechanism. DDT also inhibits $Ca^{2+}Mg^{2+}$ ATPases, which are important to neuronal repolarization and the cessation of impulse transmission across synapses.

The $GABA_A$ receptor is associated with chloride channels on the postsynaptic region of the neuron and binding of gamma-aminobutyric acid (GABA) to the receptor causes opening of the chloride channel. This occurs after transmission of the nerve impulse across the synaptic cleft and postsynaptic depolarization. Thus activation of $GABA_A$ serves to prevent excessive excitation of the postsynaptic neuron. Many neurotoxicants function by inhibiting the $GABA_A$ receptor, resulting in prolonged closure of the chloride channel and excess nerve excitation. Cyclodiene insecticides (i.e., dieldrin), the organochlorine insecticide lindane, and some pyrethroid insecticides all elicit acute neurotoxicity, at least in part, through this mechanism. Symptoms of $GABA_A$ inhibition include dizziness, headache, nausea, vomiting, fatigue, tremors, convulsions, and death. Avermectins constitute a class of pesticides that are used extensively in veterinary medicine to treat a variety of parasitic conditions. While the mode of toxicity of these compounds is not precisely known, they appear to bind a distinct subset of chloride channels (GABA-insensitive chloride channels) resulting in disruptions in normal chloride transport across nerve cell membranes. Barbituates (i.e., phenobarbital) and ethanol elicit central nervous system effects, at least in part, by binding to $GABA_A$ receptors. However, unlike the previously discussed chemicals, these compounds enhance the ability of gamma-aminobutyric acid to bind the receptor and open the chloride channel. Accordingly, these compounds suppress nerve transmission which contributes to the sedative action of the chemicals.

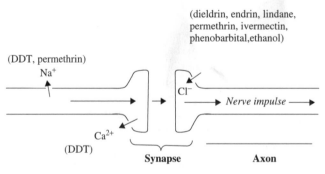

**Figure 11.6** Ion channels that facilitate nerve impulse transmission and that are susceptible to perturbation by various toxicants and drugs. Ion transport inhibitors are indicated in parentheses.

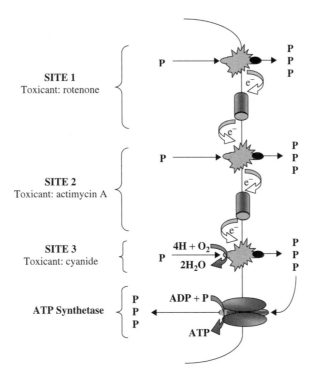

**SITE 1**
Toxicant: rotenone

**SITE 2**
Toxicant: actimycin A

**SITE 3**
Toxicant: cyanide

**ATP Synthetase**

$4H + O_2$

$2H_2O$

$ADP + P$

$ATP$

**Figure 11.7** Electron ($e^-$) transport along the inner mitochondrial membrane resulting in the pumping of protons ($P$) out of the mitochondrial matrix. Protons are shuttled back into the matrix through the ATP synthetase complex where ATP is generated. Sites of toxicant action are indicated.

## 11.5.4 Inhibitors of Cellular Respiration

Cellular respiration is the process whereby energy, in the form of ATP, is generated in the cell while molecular oxygen is consumed. The process occurs along respiratory assemblies that are located in the inner mitochondrial membrane. Electrons derived from NADH or $FADH_2$ are transferred along a chain of electron carrier proteins. This step-by-step transfer leads to the pumping of protons out of the mitochondrial matrix, resulting in the generation of a membrane potential across the inner mitochrondrial membrane. Protons are pumped out of the mitochrondrial matrix at three locations along the respiratory chain. Site 1 consists of the NADH-Q reductase complex, site 2 consists of the $QH_2$-cytochrome c reductase complex, and site 3 is the cytochrome c oxidase complex. ATP is generated from ADP when protons flow back across the membrane through an ATP synthetase complex to the mitochrondrial matrix. The transfer of electrons culminates with the reduction of molecular oxygen to water.

Many chemicals can interfere with cellular respiration by binding to the cytochromes that constitute the electron transport chain and inhibiting the flow of electrons along this protein complex. The pesticide rotenone specifically inhibits electron transfer early in the chain with inhibition of proton transport beginning at site 1. Actimycin A inhibits electron transfer and proton pumping at site 2. Cyanide, hydrogen sulfide, and azide inhibit electron flow between the cytochrome oxidase complex and $O_2$ preventing the generation of a proton gradient at site 3. Symptoms of toxicity from the inhibition of respiratory chain include excess salivation, giddiness, headache, palpitations, respiratory distress, and loss of consciousness. Potent inhibitors such as cyanide can cause death due to respiratory arrest immediately following poisoning.

Some chemicals do not interfere with electron transport leading to the consumption of molecular oxygen but rather interfere with the conversion of ADP to ATP. These uncouplers of oxidative phosphorylation function by leaking protons across the inner membrane back to the mitochondrial matrix. As a result a membrane potential is not generated, and energy required for the phosphorylation of ADP to ATP is lost. The uncoupling of oxidative phosphorylation results in increased electron transport, increased oxygen consumption, and heat production. The controlled uncoupling of oxidative phosphorylation is a physiologically relevant means of maintaining body temperature by hibernating animals, some newborn animals, and in some animals that inhabit cold environments. Chemicals known to cause uncoupling of oxidative phosphorylation include 2,4-dinitrophenol, pentachlorophenol, and dicumarol. Symptoms of intoxication include accelerated respiration and pulse, flushed skin, elevated temperature, sweating, nausea, coma, and death.

## SUGGESTED READING

Joy, R. M. Neurotoxicology: Central and peripheral. In *Encyclopedia of Toxicology*, vol. 2, P. Wexler, ed. New York: Academic Press, 1998, pp. 389–413.

Stryer, L. *Biochemistry*, 4th ed. San Francisco: W. H. Freeman, 1999.

Eaton, D. L., and C. D. Klaassen. Principles of toxicology In *Casarrett and Doull's Toxicology: The Basic Science of Poisons*, 6th ed. C. D. Klaassen, ed. New York: McGraw-Hill, 2001, pp. 11–34.

Calabrese, E. J., and L. A. Baldwin. U-shaped dose-responses in biology, toxicology, and public health. *An. Rev. Public Health* **22**: 15–33, 2001.

CHAPTER 12

# Chemical Carcinogenesis

ROBERT C. SMART

## 12.1 GENERAL ASPECTS OF CANCER

Carcinogenesis is the process through which cancer develops. Chemical carcinogenesis is the study of the mechanisms through which chemical carcinogens induce cancer and also involves the development/utilization of experimental systems aimed at determining whether a substance is a potential human carcinogen. An important aspect of toxicology is the identification of potential human carcinogens. To begin to appreciate the complexity of this subject, it is important to first have some understanding of cancer and its etiologies.

Cancer is not a single disease but a large group of diseases, all of which can be characterized by the uncontrolled growth of an abnormal cell to produce a population of cells that have acquired the ability to multiply and invade surrounding and distant tissues. It is this invasive characteristic that imparts its lethality on the host. Epidemiology studies have revealed that the incidence of most cancers increase exponentially with age (Figure 12.1). Epidemiologist have interpreted this exponential increase in cancer incidence to denote that three to seven critical mutations or "hits" within a single cell are required for cancer development. Molecular analyses of human tumors have confirmed the accumulation of mutations in critical genes in the development of cancer. These mutations can be the result of imperfect DNA replication/repair, oxidative DNA damage, and/or DNA damage caused by environmental carcinogens. Most cancers are monoclonal in origin (derived from a single cell) and do not arise from a single critical mutation but from the accumulation of sequential critical mutations in relevant target genes within a single cell (Figure 12.2). Initially a somatic mutation occurs in a critical gene, and this provides a growth advantage to the cell and results in the expansion of the mutant clone. Each additional critical mutation provides a further selective growth advantage resulting in clonal expansion of cells with mutations in multiple critical genes. It often requires decades for a cell clone to accumulate multiple critical mutations and for the progeny of this cell to clonally expand to produce a clinically detectable cancer. Thus the time required for accumulation of mutations in critical genes within a cell is likely related to the observation that cancer incidence increases exponentially with age.

*A Textbook of Modern Toxicology, Third Edition*, edited by Ernest Hodgson
ISBN 0-471-26508-X Copyright © 2004 John Wiley & Sons, Inc.

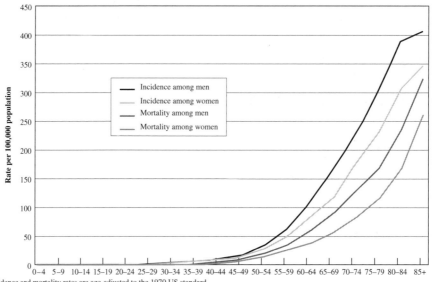

*Incidence and mortality rates are age-adjusted to the 1970 US standard.
**Source:** SEER Cancer Statistics Review, 1973–1998, Surveillance, Epidemiology,
and End Results Program, Division of Cancer Control and Population Sciences,       American Cancer Society, Surveillance Research, 2002
National Cancer Institute, 2001.

**Figure 12.1**   Colon/rectum cancer incidence and mortality rates (1994–1998) in the United States as related to age. (From *American Cancer Society's Facts and Figures—2002*, reprinted with permission of the American Cancer Society, Inc.)

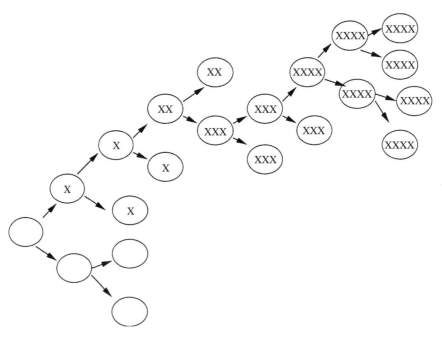

**Figure 12.2**   Monoclonal origin of cancer with the selection of cells with multiple mutations in critical genes. X designates the occurrence of a mutation in a critical gene.

Specific genes found in normal cells, termed proto-oncogenes, are involved in the positive regulation of cell growth and are frequently mutated in cancer. Mutational alteration of these proto-oncogenes can result in a gain of function, for example, the altered gene product can continually stimulate cell proliferation. Proto-oncogenes with gain-of-function mutations are now referred to as oncogenes. Another family of genes, known as tumor suppressor genes can be mutationally inactivated during carcinogenesis resulting in a loss of function. Tumor suppressor genes and the proteins they encode often function as negative regulators of cell growth. Tumor suppressor genes containing loss-of-function mutations encode proteins that are by and large inactive. Activation of oncogenes and inactivation of tumor suppressor genes within a single cell are important mutational events in carcinogenesis. A simple analogy can be made to the automobile; tumor suppressor genes are analogous to the brakes on the car while the proto-oncogenes are analogous to the accelerator pedal. Mutations within tumor suppressor genes inactivate the braking system while mutations in proto-oncogenes activate the acceleration system. Altering both the cellular brakes and cellular accelerator results in uncontrolled cell growth. In addition to the regulation in cell growth, some oncogenes and tumor suppressor genes can also impair the cells ability to undergo apoptosis or programmed cell death. Mutations in oncogenes and tumor suppressor genes provide a selective growth advantage to the cancer cell through enhanced cell growth and decreased apoptosis (Figure 12.3).

Cancer is a type of a neoplasm or tumor. While technically a tumor is defined as only a tissue swelling, the term is now used as a synonym for a neoplasm. A neoplasm or tumor is an abnormal mass of tissue, the growth of which exceeds and is uncoordinated with the normal tissue, and persists after cessation of the stimuli that evoked it. There are two basic types of neoplasms, termed benign and malignant. The general characteristics of these tumors are defined in Table 12.1. Cancer is the general name for a malignant neoplasm. In terms of cancer nomenclature, most adult cancers are carcinomas that are derived from epithelial cells (colon, lung, breast, skin, etc). Sarcomas are derived from mesenchymal tissues, while leukemias and lymphomas

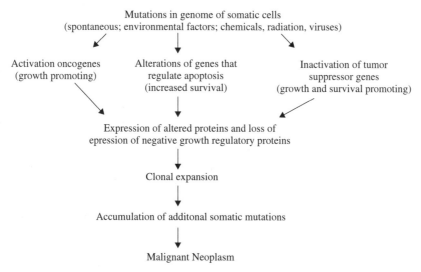

**Figure 12.3**   General overview of the cancer process.

**Table 12.1    Some General Characteristics of Malignant and Benign Neoplasms**

| Benign | Malignant |
| --- | --- |
| Generally slow growing | May be slow to rapid growing |
| Few mitotic figures | Numerous mitotic figures |
| Well-differentiated architecture, resembles that of parent tissue | Some lack differentiation, disorganized; loss of parent tissue architecture |
| Sharply demarcated mass that does not invade surrounding tissue | Locally invasive, infiltrating into surrounding normal tissue |
| No metastases | Metastases |

are derived from blood-forming cells and lymphoid tissue. Melanoma is derived from melanocytes and retinoblastoma, glioblastoma, and neuroblastoma are derived from the stem cells of the retina, glia, and neurons, respectively. According to the American Cancer Society, (1) the lifetime risk for developing cancer in the United States is about 1 in 3 for women and 1 in 2 for men, (2) in 2003 about 1.3 million new cancer cases are expected to be diagnosed not including carcinoma in situ or basal or squamous cell skin cancer, and (3) cancer is a leading cause of death in the United States and approximately 25% of all deaths are due to cancer.

## 12.2    HUMAN CANCER

Although cancer is known to occur in many groups of animals, the primary interest and the focus of most research is in human cancer. Nevertheless, much of the mechanistic research and the hazard assessment is carried out in experimental animals. A consideration of the general aspects of human carcinogenesis follows.

### 12.2.1    Causes, Incidence, and Mortality Rates of Human Cancer

Cancer cases and cancer deaths by sites and sex for the United States are shown in Figure 12.4. Breast, lung, and colon and rectum cancers are the major cancers in females while prostate, lung, and colon and rectum are the major cancer sites in males. A comparison of cancer deaths versus incidence for a given site reveals that prognosis for lung cancer cases is poor while that for breast or prostate cancer cases is much better. Age-adjusted cancer mortality rates (1930–1998) for selected sites in males are shown in Figure 12.5 and for females is shown in Figure 12.6. The increase in the mortality rate associated with lung cancer in both females and males is striking and is due to cigarette smoking. It is estimated that 87% of lung cancers are due to smoking. Lung cancer death rates in males and females began to increase in the mid-1930s and mid-1960s, respectively. These time differences are due to the fact that cigarette smoking among females did not become popular until the 1940s while smoking among males was popular in the early 1900s. Taking into account these differences along with a 20 to 25 year lag period for the cancer to develop explains the differences in the temporal increase in lung cancer death rates in males and female. Another disturbing statistic is that lung cancer, a theoretically preventable cancer, has recently surpassed breast cancer as the cancer responsible for the greatest number of cancer deaths in

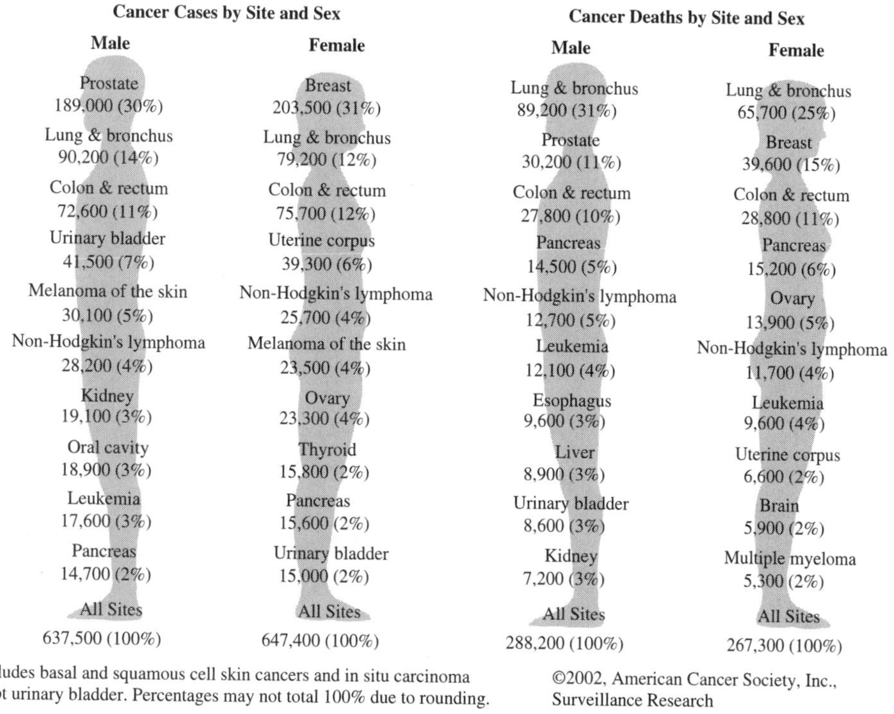

**Figure 12.4**  Cancer cases and cancer deaths by sites and sex: 2002 estimates. (From *American Cancer Society's Facts and Figures—2002*, reprinted with permission of the American Cancer Society, Inc.)

women. In addition to lung cancer, smoking also plays a significant role in cancer of the mouth, esophagus, pancreas, pharynx, larynx, bladder, kidney, and uterine cervix. Overall, the age-adjusted national total cancer death rate is increasing. In 1930 the number of cancer deaths per 100,000 people was 143. In 1940, 1950, 1970, 1984, and 1992 the rate had increased to 152, 158, 163, 170, and 172, respectively. According to the American Cancer Society, when lung cancer deaths due to smoking are excluded, the total age-adjusted cancer mortality rate had actually decreased by 16% between 1950 and 1993. However, it is important to realize that death and incidence rates for some types of cancers are increasing while the rates for others are decreasing or remaining constant.

Major insights into the etiologies of cancer have been attained through epidemiological studies that relate the roles of hereditary, environmental, and cultural influences on cancer incidence as well as through laboratory studies using rodent/cellular systems. Cancer susceptibility is determined by complex interactions between age, environment, and an individual's genetic makeup. It is estimated from epidemiological studies that 35–80% of all cancers are associated with the environment in which we live and work. The geographic migration of immigrant populations and differences in cancer incidence among communities has provided a great deal of information regarding the role of the environment and specific cancer incidences. For example, Japanese immigrants and the sons of Japanese immigrants living in California begin to assume a cancer death rate similar to the California white population (Figure 12.7). These results implicate

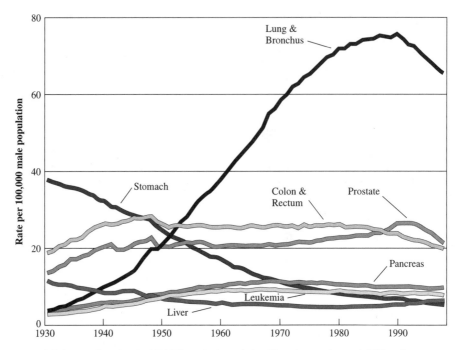

*Per 100,000, age-adjusted to the 1970 US standard population. **Note:** Due to changes in ICD coding, numerator information has changed over time. Rates for cancers of the liver, lung & bronchus, and colon & rectum are affected by these coding changes.

**Source:** US Mortality Public Use Data Tapes 1960-1998, US Mortality Volumes 1930-1959, National Center for Health Statistics, Centers for Disease Control and Prevention, 2001.

American Cancer Society, Surveillance Research, 2002

**Figure 12.5**   Age-adjusted mortality rates (1030–1998) for selected sites in males. (From *American Cancer Society's Facts and Figures—2002*, reprinted with permission of the American Cancer Society, Inc.)

a role of the environment in the etiology of cancer. It should be noted that the term environment is not restricted to exposure to human-made chemicals in the environment but applies to all aspects of our lifestyle including smoking, diet, cultural and sexual behavior, occupation, natural and medical radiation, and exposure to substances in air, water, and soil. The major factors associated with cancer and their estimated contribution to human cancer incidence are listed in Table 12.2. Only a small percentage of total cancer occurs in individuals with a hereditary mutation/hereditary cancer syndrome (ca. 5%). However, an individual's genetic background is the "stage" in which the cancer develops and susceptibility genes have been identified in humans. For example, genetic polymorphisms in enzymes responsible for the activation of chemical carcinogens may represent a risk factor as is the case for polymorphisms in the $N$-acetyl-transferase gene and the risk of bladder cancer. These types of genetic risk factors are of low penetrance (low to moderate increased risk); however, increased risk is usually associated with environmental exposure. While the values presented in Table 12.2 are a best estimate, it is clear that smoking and diet constitute the major factors associated with human cancer incidence. If one considers all of the categories that pertain to human-made chemicals, it is estimated that their contribution to human

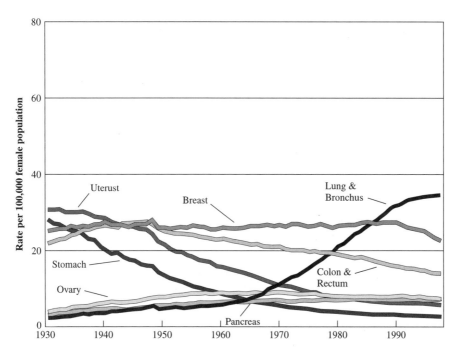

*Per 100,000, age-adjusted to the 1970 US standard population. †Uterus cancer death rates are for uterine cervix and uterine corpus combined.
**Note:** Due to changes in ICD coding, numerator information has changed over time. Rates for cancers of the liver, lung & bronchus, and colon & rectum are affected by these coding changes.
**Source:** US Mortality Public Use Data Tapes 1960–1998, US Mortality Volumes 1930–1959, National Center for Health Statistics, Centers for Disease Control and Prevention, 2001.

American Cancer Society, Surveillance Research, 2002

**Figure 12.6** Age-adjusted mortality rates (1030–1998) for selected sites in females. (From *American Cancer Society's Facts and Figures—2002*, reprinted with permission of the American Cancer Society, Inc.)

cancer incidence is approximately 10%. However, the factors listed in Table 12.2 are not mutually exclusive since there is likely to be interaction between these factors in the multi-step process of carcinogenesis.

### 12.2.2  Known Human Carcinogens

Two of the earliest observations that exposure of humans to certain chemicals or substances is related to an increased incidence of cancer were made independently by two English physicians, John Hill in 1771 and Sir Percival Pott in 1776. Hill observed an increased incidence of nasal cancer among snuff users, while Pott observed that chimney sweeps had an increased incidence of scrotal cancer. Pott attributed this to topical exposure to soot and coal tar. It was not until nearly a century and a half later in 1915 when two Japanese scientists, K. Yamagiwa and K. J. Itchikawa, substantiated Pott's observation by demonstrating that multiple topical applications of coal tar to rabbit skin produced skin carcinomas. This experiment is important for two major

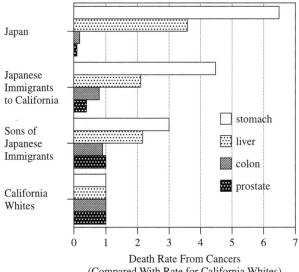

**Figure 12.7** Change in incidence of various cancers with migration from Japan to the United States provides evidence that the cancers are caused by components of the environment that differ in the two countries. The incidence of each kind of cancer is expressed as the ratio of the death rate in the populations being considered to that in a hypothetical population of California whites with the same age distribution; the death rates for whites are thus defined as 1. (Adapted from J. Cairns, in *Readings from Scientific American-Cancer Biology*, W. H. Freeman, 1986, p. 13.)

**Table 12.2   Proportions of Cancer Deaths Attributed to Various Different Factors**

| Major Factors | Best Estimate (%) | Range of Acceptable Estimates (%) |
|---|---|---|
| Diet | 35 | 10–70 |
| Tobacco | 30 | 25–40 |
| Infection | 10 | 1–? |
| Reproductive and sexual behavior | 7 | 1–13 |
| Occupation | 4 | 2–8 |
| Geophysical factors | 3 | 2–4 |
| Alcohol | 3 | 2–4 |
| Pollution | 2 | <1–5 |
| Food additives | 1 | −5–2 |
| Medicines | 1 | 0.5–3 |
| Industrial products | 1 | <1–2 |
| Unknown | ? | ? |

*Source*: Adapted from R. Doll and R. Peto, *The Causes of Cancer*, Oxford Medical Publications, 1981.

reasons: (1) it was the first demonstration that a chemical or substance could produce cancer in animals, and (2) it confirmed Pott's initial observation and established a relationship between human epidemiology studies and animal carcinogenicity. Because of these important findings, Yamagiwa and Itchikawa are considered the fathers of experimental chemical carcinogenesis. In the 1930s Kennaway and coworkers isolated a single active carcinogenic chemical from coal tar and identified it as benzo[a]pyrene, a polycyclic aromatic hydrocarbon that results from the incomplete combustion of organic molecules. Benzo[a]pyrene has also been identified as one of the carcinogens in cigarette smoke. The p53 tumor suppressor gene can be mutationally inactivated by numerous carcinogens, including the carcinogenic metabolite of benzo[a]pyrene.

Epidemiological studies have provided sufficient evidence that exposure to a variety of chemicals, agents, or processes are associated with human cancer. For example, the following causal associations have emerged between exposure and the development of specific cancers: vinyl chloride and hepatic cancer, amine dyes and bladder cancer, benzene and leukemia, diethylstilbestrol and clear cell carcinoma of the vagina, and cigarette smoking and lung cancer. Naturally occurring chemicals or agents such as asbestos, aflatoxin $B_1$, betel nut, nickel, and certain arsenic compounds are also associated with an increased incidence of certain human cancers. Both epidemiological studies and rodent carcinogenicity studies are important in the identification and classification of potential human carcinogens. The strongest evidence for establishing whether exposure to a given chemical is carcinogenic in humans comes from epidemiological studies. However, these studies are complicated by the fact that it often takes 20 to 30 years after carcinogen exposure for a clinically detectable cancer to develop. This delay is problematic and can result in inaccurate historical exposure information and additional complexity due to the interference of a large number of confounding variables. This lag period can also prevent the timely identification of a putative carcinogen and result in unnecessary exposure. Therefore methods to identify potential human carcinogens have been developed. The long-term rodent bioassay also known as the two-year rodent carcinogenesis bioassay (see Chapter 21) is currently used in an attempt to identify potential human carcinogens. It is clear that almost all human carcinogens identified to date are rodent carcinogens; however, it is not known if all rodent carcinogens are human carcinogens. Indeed, identification of possible human carcinogens based on rodent carcinogenicity can be extremely complicated (see below). Table 12.3 contains the list of the known human carcinogens as listed by the International Agency for Research on Cancer (IARC). In addition Table 12.3 includes information on carcinogenic complex mixtures and occupations associated with increased cancer incidence. In vitro mutagenicity assays are also used to identify mutagenic agents that may have carcinogenic activity (see Chapter 21).

### 12.2.3  Classification of Human Carcinogens

Identification and classification of potential human carcinogens through the two-year rodent carcinogenesis bioassay is complicated by species differences, use of high doses (MTD, maximum tolerated dose), the short life span of the rodents, high background tumor incidence in some organs, sample size, and the need to extrapolate from high to low doses for human risk assessment. Although these problems are by no means trivial, the rodent two-year bioassay remains the "gold standard" for the classification

**Table 12.3   List of Agents, Substances, Mixtures, and Exposure Circumstances Known to be Human Carcinogens**

Aflatoxins
4-Aminobiphenyl
Arsenic and certain arsenic compounds
Asbestos
Azathioprine
Benzene
Benzidine
Beryllium and certain beryllium compounds
*N,N-bis*-(2-Chloroethyl)-2-naphthylamine (chlornaphazine)
Bis(chloromethyl) ether and chloromethyl methyl ether
1,4-Butanediol dimethylsulfonate (Myleran®)
Cadmium and certain cadmium compounds
Chlorambucil
1-(2-Chloroethyl)-3-(4-methylcyclohexyl)-1-nitrosourea (MeCCNU)
Chromium and certain chromium compounds
Cyclophosphamide
Cyclosporin A (cyclosporin)
Diethylstilbestrol
Epstein-Barr virus
Erionite
Estrogen therapy
Estrogens, nonsteroidal
Estrogens, steroidal
Ethylene oxide
Etoposide in combination with cisplatin and bleomycin
*Helicobacter pylori*
Hepatitis B virus (chronic infection)
Hepatitis C virus (chronic infection)
Herbal remedies containing plant species of the genus *Aristolochia*
Human immunodeficiency virus, type 1
Human papillomavirus, type 16
Human papillomavirus, type 18
Human T-cell lymphotropic virus, type 1
Melphalan
Methoxsalen with ultraviolet A therapy (PUVA)
MOPP and other combined chemotherapy including alkylating agents
Mustard gas
2-Naphthylamine
Neutrons
Nickel compounds
*Opisthorchis viverrini*
Oral contraceptives
Radionuclides $\alpha$-particle emitting
Radionuclides $\beta$-particle emitting
Radon
*Schistosoma haematobium*
Silica
Solar radiation
Talc containing asbestiform fibers

**Table 12.3** (*continued*)

Tamoxifen
2,3,7,8-Tetrachlorodibenzo-*para*-dioxin
Thiotepa [*tris*(1-aziridinyl)phosphine sulfide]
Thorium dioxide
Treosulfan
Vinyl chloride
X and gamma($\gamma$) Radiation

*Mixtures*

Alcoholic beverages
Analgesic mixtures containing phenacetin
Betel quid with tobacco
Coal tar and coal pitches
Mineral oils
Salted fish
Shale oils
Soots
Tobacco smoke and tobacco smokeless products
Wood dust

*Exposure circumstances*

Aluminium production
Auramine manufacture
Boot and show manufacture and repair
Coal gasification
Coke gasification
Furniture and cabinet making
Haematite mining with exposure to radon
Iron and steel founding
Isopropanol manufacture
Manufacture of magenta
Painter
Rubber industry
Strong inorganic acid mists containing sulfuric acid

of potential human carcinogens. Criteria for the classification of carcinogens used by the National Toxicology Program's *Tenth Report on Carcinogens*, 2002, are shown in Table 12.4; the criteria used by Environmental Protection Agency (EPA) and the International Agency for Research on Cancer (IARC) are shown in Table 12.5. Carcinogens are classified by the weight of evidence for carcinogenicity referred to as sufficient, limited, or inadequate based on both epidemiological studies and animal data. EPA is planning to change their guidelines for carcinogen risk assessment and their carcinogen classification scheme. New guidelines will emphasize the incorporation of biological mechanistic data in the analysis, and will not rely solely on rodent tumor data. In addition the six alphanumeric categories listed in Table 12.5 will be replaced by three descriptors for classifying human carcinogenic potential. Carcinogens will be classified by the EPA as (1) known/likely to be a human carcinogen, (2) cannot be determined to be a human carcinogen, and (3) not likely to be a human carcinogen.

**Table 12.4   Carcinogen Classification System of the National Toxicology Program**

*Known to be a human carcinogen*

There is sufficient evidence of carcinogenicity from studies in humans which indicates a causal relationship between exposure to the agent, substance, or mixture and human cancer.

*Reasonably anticipated to be a human carcinogen*

There is limited evidence of carcinogenicity from studies in humans which indicates a causal interpretation is credible, but alternate explanations, such as chance, bias, or confounding factors, cannot adequately be excluded;

*or*

There is sufficient evidence of carcinogenicity from studies in experimental animals which indicates there is an increased incidence of malignant and/or a combination of malignant and benign tumors: (1) in multiple species or at multiple tissue sites, or (2) by multiple routes of exposure, or (3) to an unusual degree with regard to incidence, site, or type of tumor, and age at onset;

*or*

There is less than sufficient evidence of carcinogenicity in humans or laboratory animals, however the agent, substance or mixture belongs to a well defined, structurally-related class of substances whose members are listed in a previous *Report on Carcinogens* as either a known to be human carcinogen or reasonably anticipated to be human carcinogen, or there is convincing relevant information that the agent acts through mechanisms indicating that it would likely cause cancer in humans.

Conclusions regarding carcinogenicity in humans or experimental animals are based on scientific judgment, with consideration given to all relevant information. Relevant information includes, but is not limited to, dose response, route of exposure, chemical structure, metabolism, pharmacokinetics, sensitive subpopulations, genetic effects, and other data relating to mechanism of action or factors that may be unique to a given substance. For example, there may be a substance for which there is evidence of carcinogenicity in laboratory animals but there are compelling data indicating that the agent acts through mechanisms that do not operate in humans and it would therefore not reasonably be anticipated to cause cancer in humans.

*Source*: From the *Tenth Report on Carcinogens*, US Department of Health and Human Services, Public Health Service, National Toxicology Program.

## 12.3   CLASSES OF AGENTS ASSOCIATED WITH CARCINOGENESIS

Chemical agents that influence cancer development can be divided into two major categories based on whether or not they are mutagenic in in vitro mutagenicity assay. DNA-damaging agents (genotoxic) are mutagenic in in vitro mutagenicity assays and are considered to produce permanent alterations in the genetic material of the host in vivo, and epigenetic agents (nongenotoxic) are not mutagenic in in vitro assays. These agents are not believed to alter the primary sequence of DNA but are considered to alter the expression or repression of certain genes and/or to produce perturbations in signal transduction pathways that influence cellular events related to proliferation, differentiation, or apoptosis. Many epigenetic/nongenotoxic agents contribute to the clonal expansion of cells containing an altered genotype (DNA alterations) to form tumors, however in the absence of such DNA alterations these epigenetic agents have no effect on tumor formation.

**Table 12.5 IARC and EPA Classification of Carcinogens**

| IARC | EPA | |
|------|-----|---|
| 1 | Group A | Human carcinogens |
| | | Sufficient evidence from epidemiological studies to support a causal association between exposure to the agents and cancer |
| 2A | Group B | Probable human carcinogens |
| | Group B1 | Limited epidemiological evidence that the agent causes cancer regardless of animal data |
| | Group B2 | Inadequate epidemiological evidence or no human data on the carcinogenicity of the agent and sufficient evidence in animal studies that the agent is carcinogenic |
| 2B | Group C | Possible human carcinogens |
| | | Absence of human data with limited evidence of carcinogenicity in animals |
| 3 | Group D | Not classifiable as to human carcinogenicity |
| | | Agents with inadequate human and animal evidence of carcinogenicity or for which no data are available |
| 4 | Group E | Evidence of noncarcinogenicity for humans |
| | | Agents that show no evidence for carcinogenicity in at least two adequate animal tests in different species or in both adequate epidemiologic and animal studies |

## 12.3.1  DNA-Damaging Agents

DNA-damaging agents can be divided into four major categories. (1) Direct-acting carcinogens are intrinsically reactive compounds that do not require metabolic activation by cellular enzymes to covalently interact with DNA. Examples include $N$-methyl-$N$-nitrosourea and $N$-methyl-$N'$-nitro-$N$-nitrosoguanidine; the alkyl alkanesulfonates such as methyl methanesulfonate; the lactones such as beta propiolactone and the nitrogen and sulfur mustards. (2) Indirect-acting carcinogens require metabolic activation by cellular enzymes to form the ultimate carcinogenic species that covalently binds to DNA. Examples include dimethylnitrosamine, benzo[a]pyrene, 7,12-dimethylbenz[a]anthracene, aflatoxin B1 and 2-acetylaminofluorene (Figure 12.8). (3) Radiation and oxidative DNA damage can occur directly or indirectly. Ionizing radiation produces DNA damage through direct ionization of DNA to produce DNA strand breaks or indirectly via the ionization of water to reactive oxygen species that damage DNA bases. Ultraviolet radiation (UVR) from the sun is responsible for approximately 1 million new cases of human basal and squamous cell skin cancer each year. Reactive oxygen species can also be produced by various chemicals and cellular process including respiration and lipid peroxidation. (4) Inorganic agents such as arsenic, chromium and nickel are considered DNA-damaging agents although in many cases the definitive mechanism is unknown. DNA-damaging agents can produce three general types of genetic alterations: (1) gene mutations, which include point mutations involving single base pair substitutions that can result in amino acid substitutions in the encoded protein, and frame shift mutations

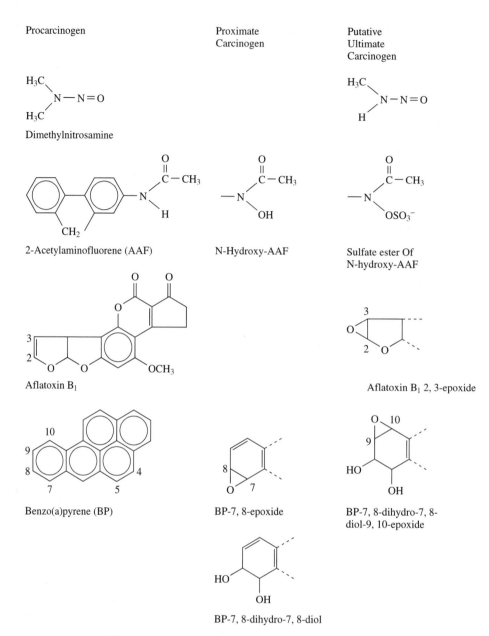

Procarcinogen

Proximate Carcinogen

Putative Ultimate Carcinogen

Dimethylnitrosamine

2-Acetylaminofluorene (AAF)

N-Hydroxy-AAF

Sulfate ester Of N-hydroxy-AAF

Aflatoxin B$_1$

Aflatoxin B$_1$ 2, 3-epoxide

Benzo(a)pyrene (BP)

BP-7, 8-epoxide

BP-7, 8-dihydro-7, 8-diol-9, 10-epoxide

BP-7, 8-dihydro-7, 8-diol

**Figure 12.8**  Examples of DNA-damaging carcinogens.

involving the loss or gain of one or two base pairs, resulting in an altered reading frame and gross alterations in the encoded protein; (2) chromosome aberrations, including gross chromosomal rearrangement such as deletions, duplications, inversions, and translocations; and (3) aneuploidy and polyploidy, which involve the gain or loss of one or more chromosomes.

## 12.3.2 Epigenetic Agents

Epigenetic agents that influence carcinogenesis are not thought to alter the primary sequence of DNA, but rather they are considered to alter the expression or repression of certain genes and/or produce perturbations in signal transduction pathways that influence cellular events related to proliferation, differentiation, or apoptosis. Many epigenetic agents favor the proliferation of cells with an altered genotype (cells containing a mutated oncogene(s) and/or tumor suppressor gene(s)) and allow the clonal expansion of these altered or "initiated" cells. Epigenetic agents can be divided into four major categories: (1) hormones such as conjugated estrogens and diethylstilbestrol; (2) immunosuppressive xenobiotics such as azathioprine and cyclosporin A; (3) solid state agents, which include plastic implants and asbestos; and (4) tumor promoters in rodent models, which include 12-*O*-tetradecanoylphorbol-13-acetate, peroxisome proliferators, TCDD and phenobarbital (Figure 12.9). In humans, diet (including caloric,

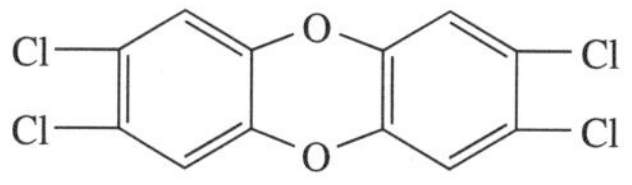

2, 3, 7, 8-Tetrachlorodibenzo-*p*-dioxin

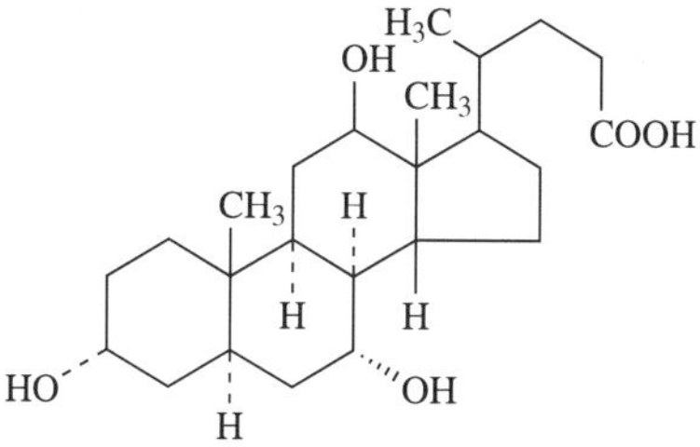

Cholic acid

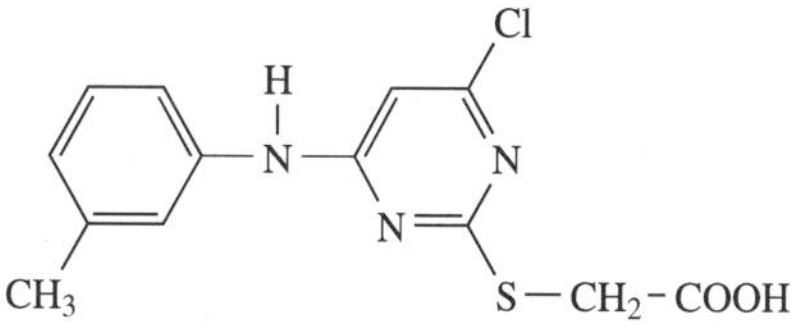

Wy-14,643

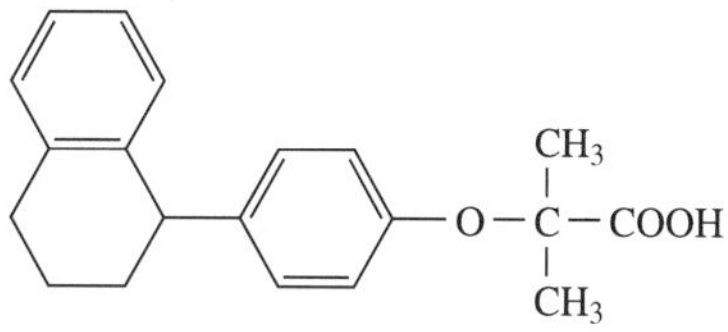

Nafenopin

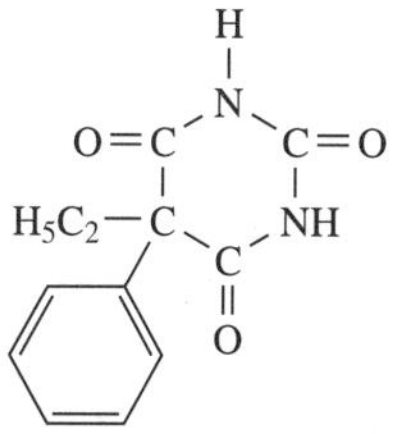

Phenobarbital

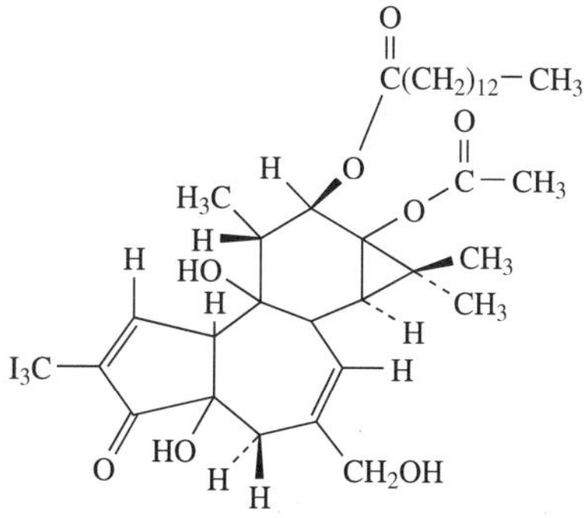

Tetradecanoyl phorbol acetate (TPA)

**Figure 12.9** Examples of tumor promoters.

fat, and protein intake), excess alcohol, and late age of pregnancy are considered to function through a promotion mechanism. While smoking and UVR have initiating activity, both are also considered to have tumor-promoting activity. By definition, tumor promoters are not classified as carcinogens since they are considered inactive in the absence of initiated cells. However, an altered genotype or an initiated cell can arise from spontaneous mutations resulting from imperfect DNA replication/repair oxidative DNA damage, or can result from environmental carcinogens. Theoretically, in the presence of a tumor promoter these mutant cells would clonally expand to form a tumor. Therefore the nomenclature becomes somewhat a matter of semantics as to whether the tumor promoter should or should not be classified as a carcinogen. Certain hormones and immunosuppressive agents are classified as human carcinogens, although it is generally considered that these agents are not carcinogenic in the absence of initiated cells. Rather, like tumor promoters, they may only allow for the clonal expansion of cells with an altered genotype.

Some nongenotoxic/epigenetic agents have been shown to induce DNA damage in vivo, for example, some "nongenotoxic/epigenetic agents" can induce oxidative DNA damage in vivo through the direct or indirect production of reactive oxygen species. For example, certain estrogens may possess this ability and such a characteristic may contribute to their carcinogenicity. Thus, as we gain a better understanding of chemical carcinogenesis, we find that there is functional and mechanistic overlap and interaction between these two major categories of chemical carcinogens.

## 12.4  GENERAL ASPECTS OF CHEMICAL CARCINOGENESIS

A great deal of evidence has accumulated in support of the somatic mutation theory of carcinogenesis, which simply states that mutations within somatic cells is necessary for neoplasia. As stated earlier, cancer development (carcinogenesis) involves the accumulation of mutations in multiple critical genes. These mutations can be the result of imperfect DNA replication/repair, oxidative DNA damage, and/or DNA damage caused by environmental carcinogens. Many chemical carcinogens can alter DNA through covalent interaction (DNA adducts) or direct and/or indirect oxidative DNA damage. Some chemical carcinogens are intrinsically reactive and can directly covalently bind to DNA, while others require metabolic via cytochromes P450 to produce reactive electrophilic intermediates capable of covalently binding to DNA (Figure 12.10). In the 1950s Elizabeth and James Miller observed that a diverse array of chemicals could produce cancer in rodents. In an attempt to explain this, they hypothesized that many carcinogens are metabolically activated to electrophilic metabolites that are capable of interacting with nucleophilic sites in DNA. The Millers termed this the electrophilic theory of chemical carcinogenesis. From this concept of metabolic activation, the important terms parent, proximate, and ultimate carcinogen were developed. A parent carcinogen is a compound that must be metabolized in order to have carcinogenic activity; a proximate carcinogen is an intermediate metabolite requiring further metabolism and resulting in the ultimate carcinogen, which is the actual metabolite that covalently binds to the DNA. The cell has many defense systems to detoxify the carcinogenic species, including cellular antioxidants and nucleophiles as well as a whole host of phase I and phase II enzymes. In addition reactive carcinogenic species may bind to noncritical sites in the cell, resulting in detoxification, or they can undergo spontaneous decomposition. If the carcinogenic species binds to DNA, the adducted DNA

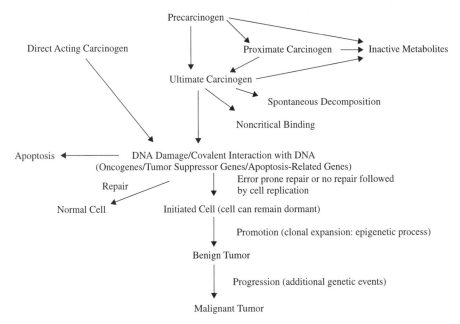

**Figure 12.10**  General aspects of chemically induced carcinogenesis.

can be repaired and produce a normal cell. If there is error in the repair of the DNA or the DNA adduct is not repaired before the cell replicates, an error in the newly synthesized DNA could occur, and if so, a mutation would occur in the DNA of the daughter cell. If this change has occurred in a critical gene, for example, in a proto-oncogene or tumor suppressor gene, it would represent an important mutagenic event(s) in carcinogenesis.

The mutationally altered cell or "initiated cell" has an altered genotype and may remain dormant (not undergo clonal expansion) for the lifetime of the animal. However, additional mutations or "hits" in critical genes followed by clonal expansion could lead to tumor development as described earlier in this chapter. In addition to this mechanism, chemical carcinogenesis in experimental models can be divided into at least three stages: termed initiation, promotion, and progression (Figure 12.10); this model is thus often referred to as the initiation/promotion model of chemical carcinogenesis. As mentioned above, the "initiated cell" may remain dormant (not undergo clonal expansion) for the lifetime of the animal. However, if the animal is repeatedly exposed to a tumor promoter, it will provide a selective growth advantage to the "initiated cell," which will clonally expand and eventually produce a benign tumor. This process is termed tumor promotion and is an epigenetic process favoring the growth of cells with an altered genotype. The development of a malignant tumor from a benign tumor encompasses a third step, termed progression and involves additional genetic changes.

## 12.5  INITIATION-PROMOTION MODEL FOR CHEMICAL CARCINOGENESIS

Experimentally, the initiation-promotion process has been demonstrated in several organs/tissues including skin, liver, lung, colon, mammary gland, prostate, and bladder

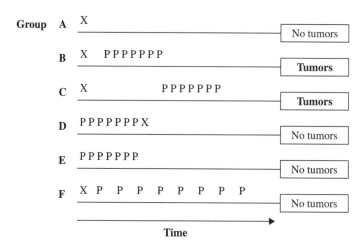

**Figure 12.11** Initiation/promotion model. $X$ = application of initiator, $P$ = application of promoter.

as well as in variety of cells in culture. While tumor promoters have different mechanisms of action and many are organ specific, all have common operational features (Figure 12.11). These features include (1) following a subthreshold dose of initiating carcinogen, chronic treatment with a tumor promoter will produce many tumors; (2) initiation at a subthreshold dose alone will produce very few if any tumors; (3) chronic treatment with a tumor promoter in the absence of initiation will produce very few if any tumors; (4) the order of treatment is critical as it must be first initiated and then promoted; (5) initiation produces an irreversible change; and (6) promotion is reversible in the early stages, for example, if an equal number of promoting doses are administered but the doses are spaced further apart in time, tumors would not develop or would be greatly diminished in number. Many tumor promoters are organ specific. For example, 12-$O$-tetradecanoylphorbol-13-acetate (TPA) also known as phorbol 12-myristate 13-acetate (PMA) belongs to a family of compounds known as phorbol esters. Phorbol esters are isolated from croton oil (derived from the seeds of the croton plant) and are almost exclusively active in skin. Phenobarbital, DDT, chlordane, TCDD and peroxisome proliferators Wy 24,643, clofibrate, and nafenopin are hepatic tumor promoters. TCDD is also a promoter in lung and skin. Some bile acids are colonic tumor promoters, while various estrogens are tumor promoters in the mammary gland and liver. There are multiple mechanisms of tumor promotion, and this may explain the organ specific nature of the many promoters. Under conditions in which the chemical produces tumors without tumor promoter treatment, the chemical agent is often referred to as a complete carcinogen.

It is generally accepted that tumor promoters allow for the clonal expansion of initiated cells by interfering with signal transduction pathways that are involved in the regulation of cell growth, differentiation, and/or apoptosis (Table 12.6). While the precise mechanisms of tumor promotion are not completely understood at the molecular/biochemical level, current research is providing new and promising mechanistic insights into how tumor promoters allow for the selective growth of initiated cells.

**Table 12.6  Some General Mechanisms of Tumor Promotion**

*Selective proliferation of initiated cells*

Increased responsiveness to and/or production of growth factors, hormones, and other active molecules

Decreased responsiveness to inhibitory growth signals

Perturbation of intracellular signaling pathways

*Altered differentiation*

Inhibition of terminal differentiation of initiated cells

Acceleration of differentiation of uninitiated cells

Inhibition of apoptosis in initiated cells

*Toxicity/compensatory hyperplasia*

Resistance to toxicity by initiated cells

## 12.6  METABOLIC ACTIVATION OF CHEMICAL CARCINOGENS AND DNA ADDUCT FORMATION

Having described the general aspects of chemical carcinogenesis including the initiation-promotion model, we now examine some aspects of chemical carcinogenesis in more detail. Metabolic activation of chemical carcinogens by cytochromes P450 is well documented. The metabolism of benzo[a]pyrene has been extensively studied and at least 15 major phase I metabolites have been identified. Many of these metabolites are further metabolized by phase II enzymes to produce numerous different metabolites. Extensive research has elucidated which of these metabolites and pathways are important in the carcinogenic process. As shown in Figure 12.12, benzo[a]pyrene is metabolized by cytochrome P450 to benzo[a]pyrene-7,8 epoxide, which is then hydrated by epoxide hydrolase to form benzo[a]pyrene-7,8-diol. Benzo[a]pyrene-7,8-diol is considered the proximate carcinogen since it must be further metabolized by cytochrome P450 to form the ultimate carcinogen, the bay region diol epoxide, (+)-benzo[a]pyrene-7,8-diol-9,10-epoxide-2. It is this reactive intermediate that binds covalently to DNA, forming DNA adducts. (+)-Benzo[a]pyrene-7,8-diol-9,10-epoxide-2 binds preferentially to deoxyguanine residues, forming *N*-2 adduct. (+)-Benzo[a]pyrene-7,8-diol-9,10-epoxide-2 is highly mutagenic in eukaryotic and prokaryotic cells and carcinogenic in rodents. It is important to note that not only is the chemical configuration of the metabolites of many polycyclic aromatic hydrocarbons important for their carcinogenic activity, but so is their chemical conformation/stereospecificity (Figure 12.12). For example, four different stereoisomers of benzo[a]pyrene-7,8-diol-9,10 epoxide are formed. Each one only differs with respect to whether the epoxide or hydroxyl groups are above or below the plane of the flat benzo[a]pyrene molecule, but only one, (+)-benzo[a]pyrene-7,8-diol-9,10-epoxide-2, has significant carcinogenic potential. Many polycyclic aromatic hydrocarbons are metabolized to bay-region diol epoxides. The bay-region theory suggests that the bay-region diol epoxides are the ultimate carcinogenic metabolites of polycyclic aromatic hydrocarbons.

DNA can be altered by strand breakage, oxidative damage, large bulky adducts, and alkylation. Carcinogens such as *N*-methyl-*N'*-nitro-*N*-nitrosoguanidine and methyl

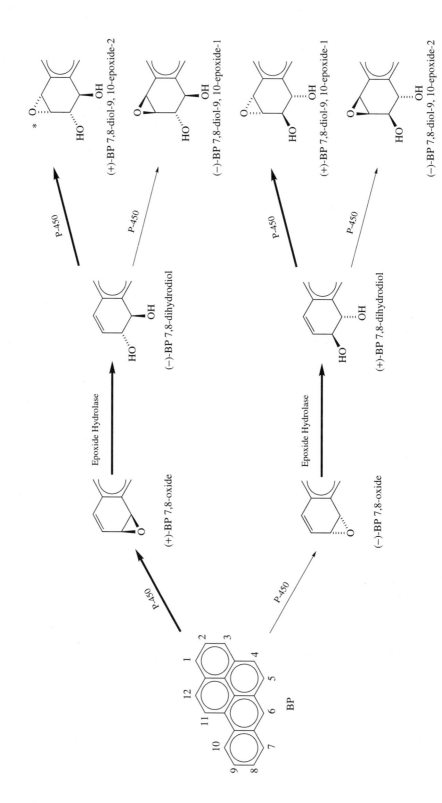

**Figure 12.12** Benzo[a]pyrene metabolism to the ultimate carcinogenic species. Heavy arrows indicate major metabolic pathways, * represents ultimate carcinogenic species. (Adapted from A. H. Conney, *Cancer Res.* **42**: 4875, 1982.)

methanesulfonate alkylate DNA to produce *N*-alkylated and *O*-alkylated purines and pyrimidines. Ionizing radiation and reactive oxygen species commonly oxidize guanine to produce 8-oxoguanine. Formation of DNA adducts may involve any of the bases, although the *N*-7 position of guanine is one the most nucleophilic sites in DNA. Of importance is how long the adduct is retained in the DNA. (+)-Benzo[a]pyrene-7,8-diol-9,10-epoxide-2 forms adducts mainly at guanine *N*-2, while aflatoxin B1 epoxide, another well-studied rodent and human carcinogen, binds preferentially to the *N*-7 position of guanine. For some carcinogens there is a strong correlation between the formation of very specific DNA-adducts and tumorigenicity. Quantitation and identification of specific carcinogen adducts may be useful as biomarkers of exposure. Importantly, the identification of specific DNA-adducts has allowed for the prediction of specific point mutations that would likely occur in the daughter cell provided that there was no repair of the DNA-adduct in the parent cell. As will be discussed in a later section, some of these expected mutations have been identified in specific oncogenes and tumor suppressor genes in chemically induced rodent tumors, providing support that the covalent carcinogen binding produced the observed mutation. In several cases, specific base pair changes in p53 tumor suppressor gene in human tumors are associated with a mutational spectrum that is consistent with exposure of the individual to a specific carcinogen. For example, the mutation spectra identified in p53 in human tumors thought to result from the exposure of the individual to ultraviolet radiation (UVR), aflatoxin, and benzo[a]pyrene (from cigarette smoke), are consistent with the observed specific mutational damage in p53 induced by these agents in experimental cellular systems.

## 12.7  ONCOGENES

### 12.7.1  Mutational Activation of Proto-oncogenes

Much evidence has accumulated for a role of covalent binding of reactive electrophilic carcinogens to DNA in chemical carcinogenesis. It is known that chemical mutagens and carcinogens can produce point mutations, frameshift mutations, strand breaks, and chromosome aberrations in mammalian cells. If the interaction of a chemical carcinogen with DNA leading to a permanent alteration in the DNA is a critical event in chemical carcinogenesis, then the identification of these altered genes and the function of their protein products is essential to our understanding of chemical carcinogenesis. While specific DNA-carcinogen adducts were isolated in the 1970s and 1980s, it was not until the early to mid-1980s that the identification of specific genes that were mutationally altered by chemical carcinogens became known. Certain normal cellular genes, termed proto-oncogenes, can be mutated by chemical carcinogens providing a selective growth advantage to the cell. The mutational activation of proto-oncogenes is strongly associated with tumor formation, carcinogenesis, and cell transformation. Proto-oncogenes are highly conserved in evolution and their expression is tightly regulated. Their protein products function in the control of normal cellular proliferation, differentiation, and apoptosis. However, when these genes are altered by a mutation, chromosome translocation, gene amplification, or promoter insertion, an abnormal protein product or an abnormal amount of product is produced. Under these circumstances these genes have the ability to transform cells in vitro, and they are termed oncogenes.

**Table 12.7  Oncogene Classification**

| Families | Genes |
|---|---|
| Growth factors | *sis, hst-1, int-2, wnt-1* |
| Growth factor receptor tyrosine kinases | *EGFR, fms, met/HGFR, ErbB2/neu/HER2, trk/NGFR* |
| Nonreceptor tyrosine kinases | *abl, src, fgr, fes, yes, lck* |
| Guanine nucleotide binding proteins | *H-ras, K-ras, N-ras, TC21, $GA_{12}$* |
| Serine/threonine kinases | *mos, raf, bcr, pim-1* |
| DNA-binding proteins | *myc, fos, myb, jun, E2F1, ets, rel* |

Over a 100 oncogenes have been identified with approximately 30 oncogenes having a major role in human cancer.

Most oncogene protein products appear to function in one way or another in cellular signal transduction pathways that are involved in regulating cell growth, differentiation or apoptosis. Signal transduction pathways are used by the cells to receive and process information to ultimately produce a biological cellular response. These pathways are the cellular circuitry conveying specific information from the outside of the cell to the nucleus. In the nucleus, specific genes are expressed, and their encoded proteins produce the evoked biological response. Oncogenes encode proteins that are components of this cellular circuitry (Table 12.7). If a component of the circuit is altered, then the entire cellular circuit of which the component is a part is altered. It is not difficult to imagine how an alteration in a pathway that regulates cellular growth, differentiation, or apoptosis could have very profound effects on cellular homeostasis. Indeed, this is the molecular basis of how oncogenes contribute to the cancer process.

### 12.7.2  *Ras* Oncogene

*Ras* genes are frequently mutated in chemically induced animal tumors and are the most frequently detected mutated oncogenes in human tumors. Approximately 20–30% of all human tumors contain mutated *ras*. The Ras subfamily includes H-*ras*, K-*ras*, and N-*ras*, and all have been found to be mutationally activated in numerous types of tumors from a large variety of species including humans.

Activated *ras* oncogenes have been detected in a large number of animal tumors induced by diverse agents including physical agents, such as radiation, and a large number of chemical carcinogens. Some chemical carcinogens bind covalently to DNA, forming specific adducts which upon DNA replication yields characteristic alterations in the primary sequence of the Ha-*ras* proto-oncogene. The study of the *ras* oncogene as a target for chemical carcinogens has revealed a correlation between specific carcinogen-DNA adducts and specific activating mutations of *ras* in chemically induced tumors. For example, 7,12-dimethylbenz[a]anthracene, a polycyclic aromatic hydrocarbon carcinogen, is metabolically activated to a bay-region diol epoxide that binds preferentially to adenine residues in DNA. Skin tumors isolated from mice treated with DMBA contain an activated H-*ras* oncogene with an A to T transversion of the middle base in the 61st codon of H-*ras*. Therefore the identified mutation in *ras* is consistent with the expected mutation based on the DMBA-DNA adducts which have been identified. Likewise rat mammary carcinomas induced by nitrosomethylurea contain a G to A transition in the 12th codon of H-*ras*, and this mutation is consistent with the modification of guanine residues by this carcinogen. Based on these events, the alteration of *ras* by specific chemical carcinogens appears to be an early event in carcinogenesis.

Ras proteins function as membrane-associated molecular switches operating downstream of a variety of membrane receptors. Ras is in the off position when it is bound to guanosine diphosphate (GDP). However, when stimulated by a growth factor receptor, Ras exchanges GTP guanosine triphosphate for GDP, and now Ras is in the on position. Ras communicates this "on" message to the next protein in the signaling circuitry, which through a kinase cascade ultimately results in the activation of several transcription factors. These transcription factors regulate the expression of genes involved in cell proliferation, for example. Once Ras has conveyed the "on" message, it turns itself off. Ras has intrinsic GTPase activity that hydrolyzes GTP to form GDP, and Ras is once again in the off position. Another protein, termed GAPp120 (GTPase activating protein), aids Ras in GTP hydrolysis. When *ras* is mutated in certain codons, including the 12th, 13th, or 61st codon, the intrinsic GTPase activity of Ras is greatly diminished as is its ability to interact with GAP. The net effect is that mutated Ras is essentially stuck in the "on" position continually sending a proliferative signal to the downstream circuitry.

## 12.8  TUMOR SUPPRESSOR GENES

### 12.8.1  Inactivation of Tumor Suppressor Genes

Activation of oncogenes results in a gain of function while inactivation of tumor suppressor genes results in a loss of function. Tumor suppressor genes have also been termed anti-oncogenes, recessive oncogenes, and growth suppressor genes. Tumor suppressor genes encode proteins that generally function as negative regulators of cell growth or regulators of cell death. In addition some tumor suppressor genes function in DNA repair and cell adhesion. The majority of tumor suppressor genes were first identified in rare familial cancer syndromes, and some are frequently mutated in sporadic cancers through somatic mutation. There are approximately 18 known tumor suppressor genes (e.g., p53, Rb, APC, p16, and BRCA1) that have been shown to have a role in cancer and another 12 putative tumor suppressors have been identified. When tumor suppressor genes are inactivated by allelic loss, point mutation, or chromosome deletion, they are no longer capable of negatively regulating cellular growth leading to specific forms of cancer predisposition. Generally, if one copy or allele of the tumor suppressor gene is inactivated, the cell is normal, and if both copies or alleles are inactivated, loss of growth control occurs. In some cases a single mutant allele of certain tumor suppressor genes, such as p53, can give rise to an altered intermediate phenotype. However, inactivation of both alleles is required for full loss of function and the transformed phenotype.

### 12.8.2  p53 Tumor Suppressor Gene

*p53* encodes a 53 kDa protein. *p53* is mutated in 50% of all human cancer and is the most frequently known mutated gene in human cancer. The majority (ca. 80%) of *p53* mutations are missense mutations and *p53* is mutated in approximately 70% of colon cancers, 50% of breast and lung cancers, and 97% of primary melanomas. In addition to point mutations, allelic loss, rearrangements, and deletions of *p53* occur in human tumors. p53 is a transcription factor and participates in many cellular functions, including cell cycle regulation, DNA repair, and apoptosis. The p53 protein is composed of 393 amino acids, and single missense mutations can inactivate the p53. Unlike *ras* genes, which have a few mutational codons that result in its activation, the p53 protein can be inactivated by hundreds of different single-point mutations in *p53*. It has been proposed that the mutation spectrum of *p53* in human cancer can aid in the identification of the specific carcinogen that is responsible for the genetic damage; that is to say, different carcinogens cause different characteristic mutations in *p53*. Some of the mutations in *p53* reflect endogenous oxidative damage, while others such as the mutational spectrum in *p53* in hepatocellular carcinomas from individuals exposed to aflatoxin demonstrate a mutation spectrum characteristic aflatoxin. In sun-exposed areas where skin tumors develop, the mutations found in *p53* in these tumors are characteristic of UV light induced pyrimidine dimers, and finally the mutation spectrum induced by (+)-benzo[a]pyrene-7,8-diol-9,10-epoxide-2 in cells in culture is similar to the mutational spectrum in *p53* in lung tumors form cigarette smokers. Thus certain carcinogens produce a molecular signature that may provide important information in understanding the etiology of tumor development.

p53 has been termed the "guardian of genome" because it controls a G1/S checkpoint, regulates DNA repair, and apoptosis. DNA damage results in the activation of p53 function and p53 prevents cells with damaged DNA from entering the S-phase of the cell cycle until the DNA damage is repaired. If the DNA damage is severe, p53 can cause the cell to undergo apoptosis. Mutation of *p53* disrupts these functions leading to the accumulation of mutations as cells enter S-phase with damaged DNA (mutator phenotype, genetic instability) and further development of malignant clones.

## 12.9 GENERAL ASPECTS OF MUTAGENICITY

Mutagens are chemical and physical agents that are capable of producing a mutation. Mutagens include agents such as radiation, chemotherapeutic agents, and many carcinogens. A mutation is a permanent alteration in the genetic information (DNA) of the cell. DNA-damaging agents/mutagens can produce (1) point mutations involving single base pair substitutions that can result in amino acid substitutions in the encoded protein and frame-shift mutations involving the loss or gain of one or two base pairs, resulting in an altered reading frame and gross alterations in the encoded protein, (2) chromosome aberrations including gross chromosomal rearrangement such as deletions, duplications, inversions, and translocations, and (3) aneuploidy and polyploidy, which involve the gain or loss of one or more chromosomes. Point mutations are classified as missense or nonsense mutations. A missense mutation produces an altered protein in which an incorrect amino acid has been substituted for the correct amino acid. A nonsense mutation is an alteration that produces a stop codon and results in a truncated protein. A point mutation can also be characterized based on the mutagen-induced substitution of one base for another within the DNA. When a point mutation produces a substitution of a purine for another purine (i.e., guanine for adenine) or a pyrimidine for another pyrimidine (i.e., thymine for cytosine), the mutation is referred as a transition. If a purine is substituted for a pyrimidine, and vice versa (i.e., thymine for adenine or guanine for cytosine), the mutation is referred to as a transversion.

## 12.10  USEFULNESS AND LIMITATIONS OF MUTAGENICITY ASSAYS FOR THE IDENTIFICATION OF CARCINOGENS

As mentioned earlier in this chapter, the two-year rodent carcinogenesis bioassay is considered the "gold standard" and is utilized to determine whether a test compound has carcinogenic potential. Identification and classification of potential human carcinogens through the two-year rodent carcinogenesis bioassay is complicated by species differences, use of high doses (MTD, maximum tolerated dose), the short life span of the rodents, sample size, and the need to extrapolate from high to low doses for human risk assessment. In addition the two-year rodent bioassay to is costly to conduct (>2 million dollars) and takes two to four years before complete results can be obtained. Since many carcinogens are mutagens, short-term test systems to evaluate the mutagenicity or genetic toxicity of compounds were developed with the idea that these tests could be used to quickly and inexpensively detect/identify chemical carcinogens. Short-term genotoxicity/mutagenicity assays were developed in a variety of organisms including bacteria, yeast, Drosophila, and human and rodent cells. These mutagenic assays or short-term genotoxicity tests directly or indirectly measure point mutations, frame-shift mutations, chromosomal damage, DNA damage and repair, and cell transformation.

In the 1970s it was reported that mutagenicity could predict rodent carcinogenicity 90% of the time. However, after extensive evaluation, it is now considered that mutagenicity can predict rodent carcinogenicity approximately 60% of the time. For certain classes of carcinogens such as the polycyclic aromatic hydrocarbons, short-term mutagenicity tests are generally highly accurate at predicting rodent carcinogenicity. For other classes of carcinogens such as the halogenated hydrocarbons, short-term genotoxicity tests often fail to detect these rodent carcinogens. Many of these halogenated hydrocarbons probably function through an epigenetic mechanism/tumor promoting mechanism.

In an important study published in 1987 by Tennant et al. (*Science* 236, pp. 933–941) 73 chemicals previously tested in the rodent two-year carcinogenesis bioassay were examined in four widely used short-term tests for genetic toxicity. The short-term assays measured mutagenesis in the Salmonella assay (Ames Assay) and mouse lymphoma assay, and chromosome aberrations and sister chromatid exchanges in Chinese hamster ovary cells. The concordance (% agreement between short-term genotoxicity test and rodent bioassay results) of each assay with the rodent bioassay data was approximately 60%. Within the limits of the study there was no evidence of complementarity among the four tests, and no battery of tests constructed from these assays improved substantially on the overall performance of the Salmonella assay. When interpreting the results of short-term test for genetic toxicity assays, it is important to consider (1) the structure and physical properties of the test compound, (2) the 60% concordance between the short-term test for genetic toxicity and rodent carcinogenicity, (3) epigenetic versus genetic mechanisms of carcinogenesis, and (4) the existence noncarcinogenic mutagens. It is also important to keep in mind that there is accumulating evidence that some compounds that are negative in short-term tests for mutagenicity can induce oxidative DNA damage in vivo through the direct or indirect production of reactive oxygen species. These compounds are in vivo mutagens but are negative in the short-term test of genetic toxicity.

Several bacterial and mammalian short-term tests for genetic toxicity as well as their biochemical and genetic rationale are described in Chapter 21 on toxicity testing. They include the salmonella assay, the Chinese hamster ovary cell/hypoxanthine-guanine phosphoribosyl transferase assay, the mouse lymphoma assay, the mammalian transformation assay, sister chromatid exchange, and the chromosome aberration assay.

## SUGGESTED READING

*10th Report on Carcinogens*, US Department of Health and Human Services, Public Health Service, National Toxicology Program. http://ehp.niehs.nih.gov/roc.toc10.html

Cancer Facts and Figures, American Cancer Society. www.cancer.org

International Agency for Research on Cancer. www.iarc.fr

Pitot, H. C., and Y. P. Dragan. Chemical Carcinogenesis. In *Casarett and Doull's Toxicology: The Basic Science of Poisons*, 6th ed., C. D. Klaassen, ed. New York: McGraw Hill, 2001, pp 241–319.

Smart, R. C., and J. K. Akunda. Carcinogenesis. In *An Introduction to Biochemical Toxicology*, E. Hodgson and R. C. Smart, eds. New York: Wiley, 2001, pp 343–396.

Tennant, R. W., et al. Prediction of chemical carcinogenicity in rodents from in vitro genetic toxicity tests. *Science* **236**: 933–941, 1987.

■■■■■■ **CHAPTER 13**

# Teratogenesis

STACY BRANCH

## 13.1 INTRODUCTION

Developmental toxicity is any morphological or functional alteration caused by chemical or physical insult that interferes with normal growth, homeostasis, development, differentiation, and/or behavior. Teratology is a specialized area of embryology. It is the study of the etiology of abnormal development (the study of birth defects). Teratogens therefore are xenobiotics and other factors that cause malformations in the developing conceptus. Examples of teratogens may include (Figure 13.1): pharmaceutic compounds, substances of abuse, hormones found in contraceptive agents, cigarette components, and heavy metals. Also included in this category are viral agents, altered metabolic states induced by stress, and nutrient deficiencies (e.g., folic acid deficiency).

## 13.2 PRINCIPLES OF TERATOLOGY

James Wilson (in 1959) proposed six principles of teratology. A simplified version of these is as follows:

1. Susceptibility to teratogenesis depends on the embryo's genotype that interacts with adverse environmental factors (G × E interaction).
2. The developmental stage of exposure to the conceptus determines the outcome.
3. Teratogenic agents have specific mechanisms through which they exert there pathogenic effects.
4. The nature of the teratogenic compound or factor determines its access to the developing conceptus/tissue.
5. The four major categories of manifestations of altered development are death, malformation, growth retardation, and functional deficits.
6. The manifestations of the altered development increase with increasing dose (i.e., no effect to lethality).

*A Textbook of Modern Toxicology, Third Edition,* edited by Ernest Hodgson
ISBN 0-471-26508-X  Copyright © 2004 John Wiley & Sons, Inc.

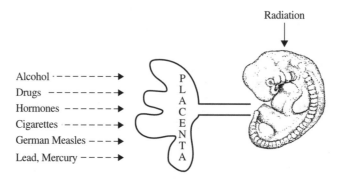

**Figure 13.1** Placenta and susceptibility to teratogens.

When describing teratogens, one may think of three basic characteristics of teratogens:

1. A given teratogen may be organ specific.
2. It may be species specific.
3. It can be dose specific.

Further discussion detailing these characteristics will be encountered later in the chapter when historical examples are addressed.

## 13.3 MAMMALIAN EMBRYOLOGY OVERVIEW

Prior to specific discussion of teratogenicity, an embryology review is provided to facilitate the understanding of the principles and descriptions associated with the effects of teratogen exposure. This section will address the development from the zygote state to the attainment of the three germ layers (gastrula). Figure 13.2 diagrams the events leading to the development of the three-layered embryo, the gastrula. Formation of the zygote marks the beginning of early embryonic development. The embryo proceeds from morula to the blastocyst while still within the zona pellucida. The aforementioned morula will give rise to the structure that attaches the early embryo to the uterus and feeds the embryo (trophoblast). Mammalian development is characterized by the formation of the blastocele-bearing embryo, the blastula (Figure 13.3). The blastula contains the mass of cells that will give rise to the actual embryo (conceptus). These cells, termed the inner cell mass (ICM), differentiate into ectoderm and endoderm prior to implantation. The ectoderm will eventually give rise to the epidermis and associated structures, the brain, and nervous system. The endoderm will give rise to glandular tissue such as the liver and pancreas and the linings of the gastrointestinal and respiratory tracts.

The inner cell mass gives rise to the epiblast (develops into ectoderm) and the hypoblast (develops into endoderm). Cells of the epiblast migrate toward the midline of the early embryo (Figures 13.4 and 13.5). The primitive streak is active proliferation of the cells with a loss of the basement membrane separating the epiblast and endoderm. The epiblast cells migrate and intermingle with the endoderm cells. The anterior end of

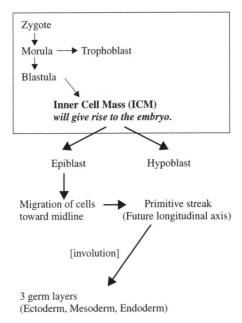

**Figure 13.2**  Development of the zygote to a three germ cell layered embryo.

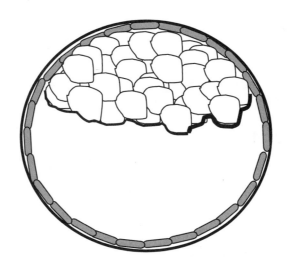

In the blastocyst stage (zona-intact) there is a
distinct inner cell mass and an outer layer of trophectoderm cells.

**Figure 13.3**  Blastula containing the inner cell mass that gives rise to the embryo proper.

the streak is defined by the Hensen's node. This node (also termed the primitive node) is associated with the organization of the developing embryo. The cellular migration (involution) leads to the creation of the third germ layer (the mesoderm). Somites are derived from the mesoderm. Figure 13.6 demonstrates the early stage embryo with visible somites and the succeeding embryonic to fetal stages. Somites are blocklike

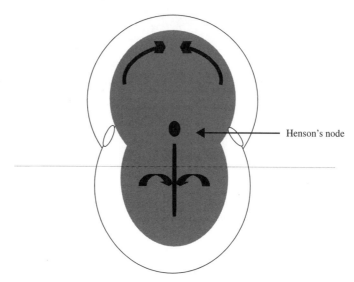

**Figure 13.4**   Dorsal view of the early embryo with the primitive evident. The horizontal dotted line represents sectioning that gives rise to the transverse view of the embryo in Figure 13.5. Curved arrows indicate migration of cells to the midline.

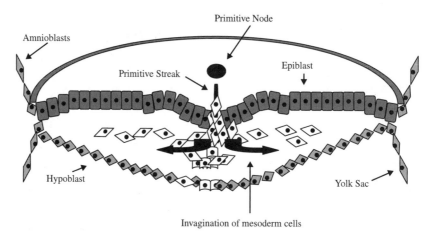

Invagination of these cells results in formation of the mesoderm and replacement of some of the hypoblast cells to produce the definitive endoderm.

**Figure 13.5**   View of the transverse section of embryo in Figure 13.6. Note the migration of cells to form the mesoderm (involution).

masses of mesoderm alongside the neural tube. They will form the vertebral column and segmental musculature. They will also develop into the excretory system, gonads, and the outer covering of internal organs. Also formed from mesoderm are mesenchymal cells. These are loose migratory cells forming the dermis (inner skin layer), bones and cartilage, and circulatory system.

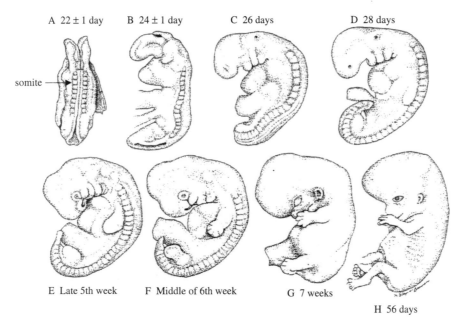

A  22 ± 1 day    B  24 ± 1 day    C  26 days    D  28 days

somite —

E  Late 5th week    F  Middle of 6th week    G  7 weeks

H  56 days

Somite: block-like mass of mesoderm alongside neural tube: forms vertebral column and segmental musculature

**Figure 13.6**  Early stage human embryo with visible somites (represents a 22-day-old human embryo or an 8-day-old mouse embryo).

## 13.4  CRITICAL PERIODS

Major fetal outcomes depend on the stage of pregnancy affected, as there are critical periods for the development of fetal processes and organs. Although embryogenesis is complex involving cell migrations, proliferation, differentiation, and organogenesis, one may divide the developmental stages in to three large categories: pre-implantation, implantation to organogenesis, and the fetal to neonatal stage. The outcomes associated with exposure during these periods vary. This is not to say there are exceptions based on the type of exposure. However, the primary outcomes are as follows:

| STAGE OF EXPOSURE | OUTCOME(S) |
| --- | --- |
| Pre-implantation | Embryonic lethality |
| Implantation to time of organogenesis | Morphological defects |
| Fetal → neonatal stage | Functional disorders, growth retardation, carcinogenesis |

The sensitivity of the embryo to the induction of morphological defects is increased during the period of organogenesis. This period is essentially the time of the origination and development of the organs. The critical period graph (Figure 13.7) demonstrates this point and defines the embryonic and fetal periods.

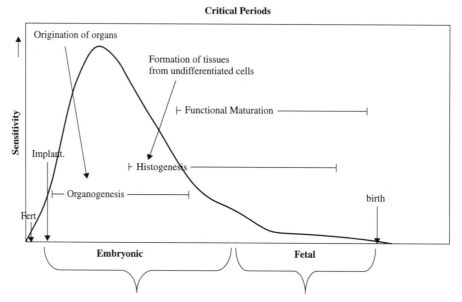

**Figure 13.7**    Critical periods graph including embryonic and fetal periods. Note the increased sensitivity to teratogenic events during organogenesis.

## 13.5   HISTORICAL TERATOGENS

### 13.5.1   Thalidomide

Thalidomide is a sedative-hypnotic drug used in Europe from 1957 to 1961. It was marketed for morning sickness, nausea, and insomnia. It went into general use and was widely prescribed in Europe, Australia, Asia, Africa, and the Americas. Women who had taken the drug from gestation days (GD) 35 to 50 gave birth to offspring suffering from a spectrum of different malformations, mainly amelia (absence of limbs) or phocomelia (severe shortening of limbs). Other malformations included: absence of the auricles with deafness, defects of the muscles of the eye and face, and malformations of the heart, bowel, uterus, and the gallbladder. The compound was withdrawn from the market in 1961 after about 10,000 cases had occurred.

### 13.5.2   Accutane (Isotetrinoin)

Accutane is a member of a family of drugs called retinoids, which are related to vitamin A. It is approved to treat serious forms of acne. These painful and disfiguring forms of acne do not respond to other acne treatments. Accutane is very effective, but its use is associated with a number of risks including birth defects. Exposure of pregnant women can lead to birth defects such as facial malformations, heart defects, and mental retardation.

### 13.5.3   Diethylstilbestrol (DES)

DES is a synthetic estrogen that inhibits ovulation by affecting release of pituitary gonadotropins. Some of its uses include treatment for hypogonadism, primary ovarian failure, and in some cases of prostate cancer. From 1940 to 1970, DES was used to help maintain pregnancy. In utero exposure to DES has been associated with abnormal development of the uterus. It has also been associated with certain types of tumors. Women who were exposed in utero often developed vaginal neoplasia, vaginal adenosis, and cervical erosion. Effects were not seen in offspring until they reached puberty. Clear cell carcinoma of the vagina is a type of adenocarcinoma found in young women who are exposed to diethylstilbestrol in utero. The reproductive organ of males can also be affected subsequent to in utero exposure. The outcomes include hypotrophic testes, poor semen volume and quality.

### 13.5.4   Alcohol

***Fetal Alcohol Syndrome.*** Fetal alcohol syndrome (FAS) is a pattern of mental and physical defects that develops in some offspring when exposed to alcohol in utero. The first trimester is the most susceptible period. Some babies with alcohol-related birth defects, such as lower birth weight and body size and neurological impairments, do not have all of the classic FAS symptoms. These outcomes are often referred to as fetal alcohol effects (FAE). Currently there is not total agreement among medical scientists concerning the precise differences between FAE and FAS. In addition to growth retardation, the most common outcomes of fetal alcohol syndrome include psychomotor dysfunction and craniofacial anomalies.

The observed growth deficiencies are associated with an inability of the baby to catch up due to a slower than normal rate of development. Other infrequent outcomes include skeletal malformations such as deformed ribs and sternum, scoliosis, malformed digits, and microcephaly. Distinctive facial anomalies have been associated with a diagnosis of fetal alcohol syndrome: small eye openings, epicanthal folds, failure of eyes to move in the same direction, short upturned nose, flat or absent groove between nose and upper lip, and thin upper lip. Visceral deformities may also be present: heart defects, genital malformations, kidney, and urinary defects.

A common concurrent manifestation of FAS include central nervous system defects. These include irregular arrangement of neurons and connective tissue. Mental retardation may also be present and associated with learning disabilities as well as difficulties in controlling body coordination.

### 13.5.5   "Non Chemical" Teratogens

Teratogens are not only xenobiotics. There may be other factors having the ability to cause malformations in the developing conceptus. Restraint stress in mice (12-hour restraint during early period of organogenesis) elicits axial skeletal defects (primarily supernumerary ribs). The Rubella virus (first reported in 1941, Austria) is associated with a number of fetal outcomes depending on the stage of development that the exposure occurs. Exposure during the first and second month of pregnancy was associated with heart and eye defects. Exposure during the third month was associated with hearing defects (and mental retardation in some cases).

## 13.6   TESTING PROTOCOLS

Formal testing guidelines were established after thalidomide disaster. In 1966 guidelines were established by the FDA: *Guidelines for Reproduction Studies for Safety Evaluation of Drugs for Human Use*. Since then (1994) new streamlined testing protocols have been developed with international acceptance. This newer approach, ICH (13.6.2), relies on the investigator to determine the model to access reproductive/developmental toxicity. However, many scientist currently conduct and publish the FDA version of testing (i.e., segment studies). Therefore both (FDA & ICH) approaches will be discussed. Further practical details of testing protocols may be found in Chapter 21 as well as some discussion of the regulatory implications for different agencies.

### 13.6.1   FDA Guidelines for Reproduction Studies for Safety Evaluation of Drugs for Human Use

*Multigenerational Study.* This approach involves the continuous exposure of a rodent species (usually mice). The parental animals are exposed shortly after weaning (30–40 days of age). At reproductive maturity, the animals are mated. The first generation is produced ($F_1$). From these an $F_2$ is produced and then subsequently an $F_3$ generation. The effects of the test is monitored through each generation. The measured parameters include fertility, litter size, and neonatal viability. This is a time-consuming effort that usually takes about two years to complete.

*Single-Generation Studies.* Single-generation studies are short-term studies conducted in three segments:

*Segment I: Evaluation of Fertility and Reproductive Performance.* Male rodents are treated for 70 days (to expose for one spermatogenic cycle), and nonpregnant females for 14 days (to exposure for several estrous cycles). Treatment is continued in the females during mating, pregnancy, and lactation. Fifty percent of the females are killed and the fetuses are examined for presence of malformations. The other 50% are allowed to give birth. After weaning, these offspring are killed and necropsied.

*Segment II: Assessment of Developmental Toxicity.* This involves the treatment of pregnant females only during the period covering implantation through organogenesis (typically from gestational days 6 to 15 in mice with 18-day gestational periods). One day prior to birth, the animals are killed and fetuses examined for viability, body weight, and presence of malformation.

*Segment II: Postnatal Evaluation.* Pregnant animals are treated from the last trimester of pregnancy until weaning. Evaluated are parturition process, late fetal development, neonatal survival, and growth as well as presence of any malformations.

### 13.6.2   International Conference of Harmonization (ICH) of Technical Requirements for Registration of Pharmaceuticals for Human Use (ICH) — US FDA, 1994

As previously mentioned, new streamlined testing protocols with international acceptance have been developed. Below is a description of these guidelines as it relates to similarity to a particular segment-type study.

*ICH 4.1.1: Fertility Assessment.* This study duration is typically shorter than segment I studies. Males are exposed for four weeks before mating and females two weeks before mating. Male reproductive organs are carefully evaluated: organ weights, histological analysis and sperm count, and mobility evaluation. For the females, fertility, litter size, and viability of conceptus are evaluated.

***ICH 4.1.2: Postnatal Evaluation and Pregnancy State Susceptibility.*** This study protocol is similar to the segment III study. Maternal toxicity is evaluated by comparing the degree of toxicity of the nonpregnant female to that of the pregnant female. Postnatal viability and growth are also evaluated. Offspring are also evaluated to assess functional development (i.e., presence of behavioral and reproductive deficits).

***ICH 4.1.3: Assessment of Developmental Toxicity.*** This is almost identical to the segment II study protocol. Pregnant animals are exposed from implantation through organogenesis. The parameters measured in the segment II study are similar. However, the study is usually conducted using at least two species. More specifically, at least one rodent and one nonrodent species.

### 13.6.3   Alternative Test Methods

A number of alternative test methods have been developed to reduce the number of whole animals used in studies and/or to obtain more rapid information concerning the potential of a compound to be a reproductive/developmental toxicant. Validation of many of the methods has been problematic, since they do not address the contribution of maternal factors or multiorgan contributions to outcomes. Some of these alternative methods include the use of cell or embryo culture. For example, the micromass culture involves the use of limb bud cells from rat embryos grown in micromass culture for five days. The processes of differentiation and cell proliferation are assessed. In the Chernoff/Kavlock Assay, pregnant rodents are exposed during organogenesis and allowed to deliver. Postnatal growth, viability, and gross morphology of litters are recorded (detailed skeletal evaluations are not performed). Other alternative tests involve the use of nontraditional test species such as *Xenopus* embryos (FETAX) and *Hydra*. *Xenopus* embryos are exposed for 96 hours and then evaluated for morphological defects, viability, and growth. The cells of *Hydra* aggregate to form artificial embryos. The dose response in these "embryos" is compared to that of the adult *Hydra*.

## 13.7   CONCLUSIONS

Understanding the mechanisms of the induction of birth defects is key to determine how to prevent these effects. Further, increasing the accuracy of experimental animal extrapolation will aid in the interpretation of experimental data in order to more accurately determine the risk of a given compound to elicit birth defects in humans.

### SUGGESTED READING

Ballantyne, B., Marrs, T. and P. Turner, eds. *General and Applied Toxicology*, college ed., New York: Macmillan, 1995.

Klaassen, C. D., ed. *Casarett and Doull's Toxicology: The Basic Science of Poisons*, 6th ed., New York: McGraw-Hill, 2001.

Korach, K. S. *Reproductive and Developmental Toxicology*. New York: Dekker, 1998.

Hood, R. D. *Handbook of Developmental Toxicology*. Boca Raton, FL: CRC Press, 1997.

# PART V

# ORGAN TOXICITY

# CHAPTER 14

# Hepatotoxicity

ERNEST HODGSON and PATRICIA E. LEVI

## 14.1 INTRODUCTION

### 14.1.1 Liver Structure

The basic structure of the liver consists of rows of hepatic cells (hepatocytes or parenchymal cells) perforated by specialized blood capillaries called sinusoids (see Chapter 10, Figures 10.2 and 10.3). The sinusoid walls contain phagocytic cells called Kupffer cells whose role is to engulf and destroy materials such as solid particles, bacteria, dead blood cells, and so on. The main blood supply comes to the liver from the intestinal vasculature. These vessels, along with those from the spleen and the stomach, merge with each other to form the portal vein (see Chapter 10, Figure 10.4). On entering the liver, the portal vein subdivides and drains into the sinusoids. The blood then perfuses the liver and exits by the hepatic veins, which merge into the inferior vena cava and return blood to the heart. The hepatic artery supplies the liver with oxygenated arterial blood.

Other materials, such as bile acids and many xenobiotics, move from the hepatocytes into the bile-carrying canaliculi, which merge into larger ducts that follow the portal vein branches. The ducts merge into the hepatic duct from which bile drains into the upper part of the small intestine, the duodenum. The gall bladder serves to hold bile until it is emptied into the intestine.

### 14.1.2 Liver Function

In the liver three main functions occur: storage, metabolism, and biosynthesis. Glucose is converted to glycogen and stored; when needed for energy, it is converted back to glucose. Fat, fat-soluble vitamins, and other nutrients are also stored in the liver. Fatty acids are metabolized and converted to lipids, which are then conjugated with proteins synthesized in the liver and released into the bloodstream as lipoproteins. The liver also synthesizes numerous functional proteins, such as enzymes and blood-coagulating factors. In addition the liver, which contains numerous xenobiotic metabolizing enzymes, is the main site of xenobiotic metabolism.

*A Textbook of Modern Toxicology, Third Edition,* edited by Ernest Hodgson
ISBN 0-471-26508-X  Copyright © 2004 John Wiley & Sons, Inc.

## 14.2  SUSCEPTIBILITY OF THE LIVER

The liver, the largest organ in the body, is often the target organ for chemically induced injuries. Several important factors are known to contribute to the liver's susceptibility. First, most xenobiotics enter the body through the gastrointestinal (GI) tract and, after absorption, are transported by the hepatic portal vein to the liver: thus the liver is the first organ perfused by chemicals that are absorbed in the gut. A second factor is the high concentration in the liver of xenobiotic metabolizing enzymes, primarily the cytochrome P450-dependent monooxygenase system. Although most biotransformations are detoxication reactions, many oxidative reactions produce reactive metabolites (Chapters 7 and 8) that can induce lesions within the liver. Often areas of damage are in the centrilobular region, and this localization has been attributed, in part, to the higher concentration of cytochrome P450 in that area of the liver.

## 14.3  TYPES OF LIVER INJURY

The types of injury to the liver depend on the type of toxic agent, the severity of intoxication, and the type of exposure, whether acute or chronic. The main types of liver damage are discussed briefly in this section. Whereas some types of damage—for example, cholestasis—are liver specific, others such as necrosis and carcinogenesis are more general phenomena.

### 14.3.1  Fatty Liver

Fatty liver refers to the abnormal accumulation of fat in hepatocytes. At the same time there is a decrease in plasma lipids and lipoproteins. Although many toxicants may cause lipid accumulation in the liver (Table 14.1), the mechanisms may be different. Basically lipid accumulation is related to disturbances in either the synthesis or the secretion of lipoproteins. Excess lipid can result from an oversupply of free fatty acids from adipose tissues or, more commonly, from impaired release of triglycerides from the liver into the plasma. Triglycerides are secreted from the liver as lipoproteins (very low density lipoprotein, VLDL). As might be expected, there are a number of points at which this process can be disrupted. Some of the more important ones are as follows (Figure 14.1):

- Interference with synthesis of the protein moiety
- Impaired conjugation of triglyceride with lipoprotein
- Interference with transfer of VLDL across cell membranes
- Decreased synthesis of phospholipids
- Impaired oxidation of lipids by mitochondria
- Inadequate energy (adenosine triphosphate [ATP] for lipid and protein synthesis

The role that fatty liver plays in liver injury is not clearly understood, and fatty liver in itself does not necessarily mean liver dysfunction. The onset of lipid accumulation in the liver is accompanied by changes in blood biochemistry, and for this reason blood chemistry analysis can be a useful diagnostic tool.

**Table 14.1   Examples of Hepatotoxic Agents and Associated Liver Injury**

*Necrosis and fatty liver*

| | | |
|---|---|---|
| Carbon tetrachloride | Dimethylnitrosamine | Phosphorous |
| Chloroform | Cyclohexamide | Beryllium |
| Trichloroethylene | Tetracycline | Allyl alcohol |
| Tetrachloroethylene | Acetaminophen | Galactosamine |
| Bromobenzene | Mitomycin | Azaserine |
| Thioacetamide | Puromycin | Aflatoxin |
| Ethionine | Tannic acid | Pyrrolizidine alkaloids |

*Cholestasis (drug-induced)*

| | | |
|---|---|---|
| Chlorpromazine | Imipramine | Carbarsone |
| Promazine | Diazepam | Chlorthiazide |
| Thioridazine | Methandrolone | Methimazole |
| Mepazine | Mestranol | Sulfanilamide |
| Amitriptline | Estradiol | Phenindione |

*Hepatitis (drug-induced)*

| | | |
|---|---|---|
| Iproniazid | Methoxyflurane | Halothane |
| Isoniazid | Papaverine | Zoxazolamine |
| Imipramine | Phenyl butazone | Indomethacin |
| 6-Mercaptopurine | Cholchicine | Methyldopa |

*Carcinogenesis (experimental animals)*

| | | |
|---|---|---|
| Aflatoxin B1 | Dimethylbenzanthracene | Acetylaminofluorene |
| Pyrrolizidine alkaloids | Dialkyl nitrosamines | Urethane |
| Cycasin | Polychlorinated biphenyls | |
| Safrole | Vinyl chloride | |

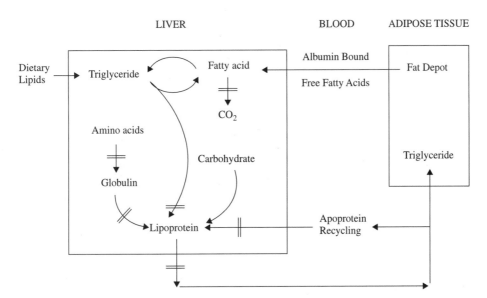

**Figure 14.1**   Triglyceride cycle in the pathogenesis of fatty liver. "=" are metabolic blocks. (From S. A. Meyer, *Introduction to Biochemical Toxicology*, 3rd ed., Wiley, 2001.)

## 14.3.2 Necrosis

Cell necrosis is a degenerative process leading to cell death. Necrosis, usually an acute injury, may be localized and affect only a few hepatocytes (focal necrosis), or it may involve an entire lobe (massive necrosis). Cell death occurs along with rupture of the plasma membrane, and is preceded by a number of morphologic changes such as cytoplasmic edema, dilation of the endoplasmic reticulum, disaggregation of polysomes, accumulation of triglycerides, swelling of mitochondria with disruption of cristae, and dissolution of organelles and nucleus. Biochemical events that may lead to these changes include binding of reactive metabolites to proteins and unsaturated lipids (inducing lipid peroxidation and subsequent membrane destruction) disturbance of cellular $Ca^{+2}$ homeostasis, inference with metabolic pathways, shifts in $Na+$ and $K+$ balance, and inhibition of protein synthesis. Changes in blood chemistry resemble those seen with fatty liver, except they are quantitatively larger. Because of the regenerating capability of the liver, necrotic lesions are not necessarily critical. Massive areas of necrosis, however, can lead to severe liver damage and failure.

## 14.3.3 Apoptosis

Apoptosis is a controlled form of cell death that serves as a regulation point for biologic processes and can be thought of as the counterpoint of cell division by mitosis. This selective mechanism is particularly active during development and senescence. Although apoptosis is a normal physiological process, it can also be induced by a number of exogenous factors, such as xenobiotic chemicals, oxidative stress, anoxia, and radiation. (A stimulus that induces a cell to undergo apoptosis is known as an *apogen*.) If, however, apoptosis is suppressed in some cell types, it can lead to accumulation of these cells. For example, in some instances, clonal expansion of malignant cells and subsequent tumor growth results primarily from inhibition of apoptosis.

Apoptosis can be distinguished from necrosis by morphologic criteria, using either light or electron microscopy. Toxicants, however, do not always act in a clear-cut fashion, and some toxicants can induce both apoptosis and necrosis either concurrently or sequentially.

## 14.3.4 Cholestasis

Cholestasis is the suppression or stoppage of bile flow, and may have either intrahepatic or extrahepatic causes. Inflammation or blockage of the bile ducts results in retention of bile salts as well as bilirubin accumulation, an event that leads to jaundice. Other mechanisms causing cholestasis include changes in membranes permeability of either hepatocytes or biliary canaliculi. Cholestasis is usually drug induced (Table 14.1) and is difficult to produce in experimental animals. Again, changes in blood chemistry can be a useful diagnostic tool.

## 14.3.5 Cirrhosis

Cirrhosis is a progressive disease that is characterized by the deposition of collagen throughout the liver. In most cases cirrhosis results from chronic chemical injury. The accumulation of fibrous material causes severe restriction in blood flow and in the liver's normal metabolic and detoxication processes. This situation can in turn cause further damage and eventually lead to liver failure. In humans, chronic use of ethanol is the single most important cause of cirrhosis, although there is some dispute as to whether the effect is due to ethanol alone or is also related to the nutritional deficiencies that usually accompany alcoholism.

### 14.3.6   Hepatitis

Hepatitis is an inflammation of the liver and is usually viral in origin; however, certain chemicals, usually drugs, can induce a hepatitis that closely resembles that produced by viral infections (Table 14.1). This type of liver injury is not usually demonstrable in laboratory animals and is often manifest only in susceptible individuals. Fortunately, the incidence of this type of disease is very low.

### 14.3.7   Oxidative Stress

Oxidative stress has been defined as an imbalance between the prooxidant/antioxidant steady state in the cell, with the excess of prooxidants being available to interact with cellular macromolecules to cause damage to the cell, often resulting in cell death. Although the occurrence of reactive oxygen species in normal metabolism and the concept of oxidative stress was derived from these studies, it is apparent that oxidative stress can occur in almost any tissue, producing a variety of deleterious effects. To date, a number of liver diseases, including alcoholic liver disease, metal storage diseases, and cholestatic liver disease, have been shown to have an oxidative stress component.

Reactive oxygen and reactive nitrogen radicals can be formed in a number of ways (Figure 14.2), the former primarily as a by-product of mitochondrial electron transport. Superoxide, hydrogen peroxide, singlet oxygen, and hydroxyl can all arise from this source. Other sources include monooxygenases and peroxisomes. If not detoxified, reactive oxygen species can interact with biological macromolecules such as DNA and protein or with lipids. Once lipid peroxidation of unsaturated fatty acids in phospholipids is initiated, it is propagated in such a way as to have a major damaging effect on cellular membranes. The formation, detoxication by superoxide dismutase and by glutathione-dependent mechanisms, and interaction at sites of toxic action are illustrated in Figure 14.2.

### 14.3.8   Carcinogenesis

The most common type of primary liver tumor is hepatocellular carcinoma; other types include cholangiocarcinoma, angiosarcoma, glandular carcinoma, and undifferentiated liver cell carcinoma. Although a wide variety of chemicals are known to induce liver cancer in laboratory animals (Table 14.1), the incidence of primary liver cancer in humans in the United States is very low.

Some naturally occurring liver carcinogens are aflatoxin, cycasin, and safrole. A number of synthetic chemicals have been shown to cause liver cancer in animals, including the dialkylnitrosamines, dimethylbenzanthracene, aromatic amines such as

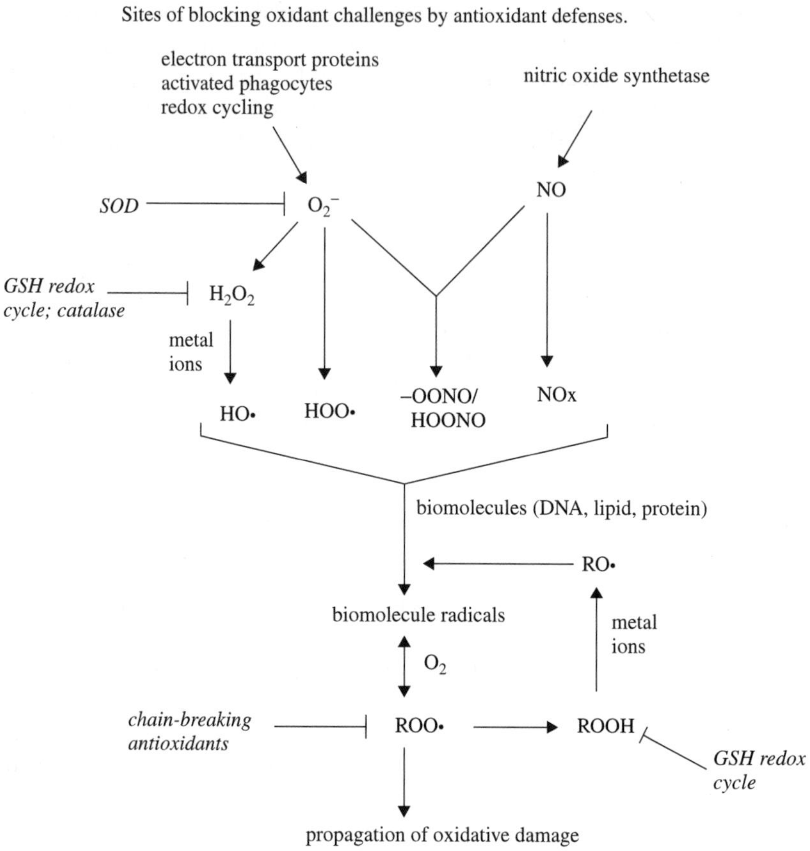

Sites of blocking oxidant challenges by antioxidant defenses.

**Figure 14.2** Molecular targets of oxidative injury. (From D. J., Reed, *Introduction to Biochemical Toxicology*, 3rd ed., Wiley, 2001.)

2-naphthylamine and acetylaminofluorene, and vinyl chloride. The structure and activation of these compounds can be found in Chapters 7 and 8. In humans, the most noted case of occupation-related liver cancer is the development of angiosarcoma, a rare malignancy of blood vessels, among workers exposed to high levels of vinyl chloride in manufacturing plants. For a discussion of chemical carcinogenesis, see Chapter 12.

## 14.4 MECHANISMS OF HEPATOTOXICITY

Chemically induced cell injury can be thought of as involving a series of events occurring in the affected animal and often in the target organ itself:

- The chemical agent is activated to form the initiating toxic agent.
- The initiating toxic agent is either detoxified or causes molecular changes in the cell.
- The cell recovers or there are irreversible changes.
- Irreversible changes may culminate in cell death.

Cell injury can be initiated by a number of mechanisms, such as inhibition of enzymes, depletion of cofactors or metabolites, depletion of energy (ATP) stores, interaction with receptors, and alteration of cell membranes. In recent years attention has focused on the role of biotransformation of chemicals to highly reactive metabolites that initiate cellular toxicity. Many compounds, including clinically useful drugs, can cause cellular damage through metabolic activation of the chemical to highly reactive compounds, such as free radicals, carbenes, and nitrenes (Chapters 7 and 8).

These reactive metabolites can bind covalently to cellular macromolecules such as nucleic acids, proteins, cofactors, lipids, and polysaccharides, thereby changing their biologic properties. The liver is particularly vulnerable to toxicity produced by reactive metabolites because it is the major site of xenobiotic metabolism. Most activation reactions are catalyzed by the cytochrome P450 enzymes, and agents that induce these enzymes, such as phenobarbital and 3-methylcholanthrene, often increase toxicity. Conversely, inhibitors of cytochrome P450, such as SKF-525A and piperonyl butoxide, frequently decrease toxicity.

Mechanisms such as conjugation of the reactive chemical with glutathione are protective mechanisms that exist within the cell for the rapid removal and inactivation of many potentially toxic compounds. Because of these interactions, cellular toxicity is a function of the balance between the rate of formation of reactive metabolites and the rate of their removal. Examples of these interactions are presented in the following discussions of specific hepatotoxicants.

## 14.5  EXAMPLES OF HEPATOTOXICANTS

### 14.5.1  Carbon Tetrachloride

Carbon tetrachloride has probably been studied more extensively, both biochemically and pathologically, than any other hepatotoxicant. It is a classic example of a chemical activated by cytochrome P450 to form a highly reactive free radical (Figure 14.3). First, $CCl_4$ is converted to the trichloromethyl radical ($CCl_3\cdot$) and then to the trichloromethylperoxy radical ($CCl_3O_2\cdot$). Such radicals are highly reactive and generally have a small radius of action. For this reason the necrosis induced by $CCl_4$ is most severe in the centrilobular liver cells that contain the highest concentration of the P450 isozyme responsible for $CCl_4$ activation.

Typically free radicals may participate in a number of events (Figure 14.4), such as covalent binding to lipids, proteins, or nucleotides as well as lipid peroxidation. It

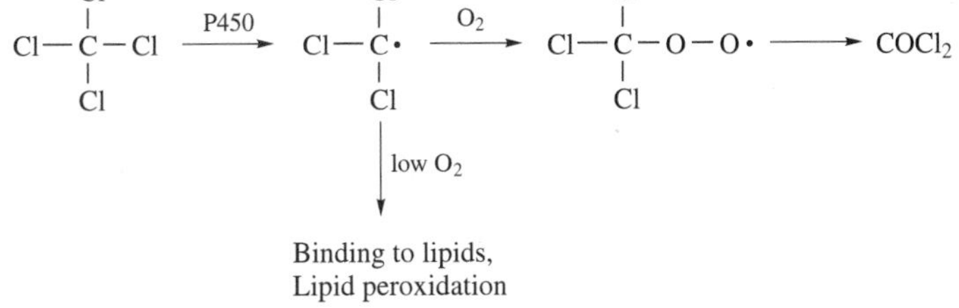

**Figure 14.3**  Metabolism of carbon tetrachloride and formation of reactive metabolites. (From P. E. Levi, *A Textbook of Modern Toxicology*, 2nd ed., Appleton and Lange, 1997.)

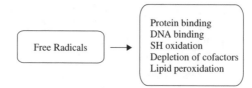

**Figure 14.4**   Summary of some toxic effects of free radicals. (From P. E. Levi, *A Textbook of Modern Toxicology*, 2nd ed., Appleton and Lange, 1997.)

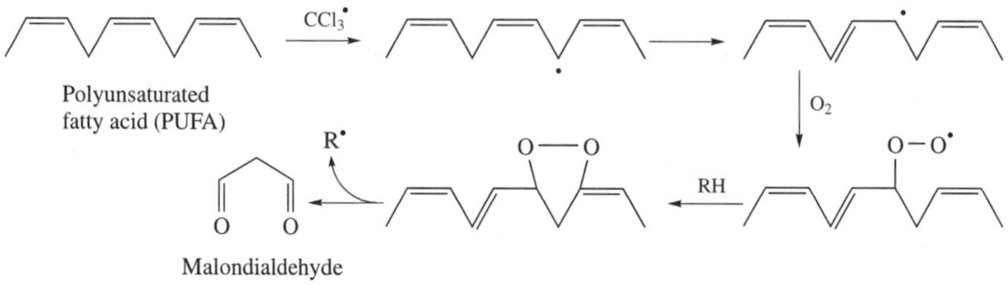

**Figure 14.5**   Schematic illustrating lipid peroxidation and destruction of membranes. (From P. E. Levi, *A Textbook of Modern Toxicology*, 2nd ed., Appleton and Lange, 1997.)

is now thought that $CCl_3\cdot$, which forms relatively stable adducts, is responsible for covalent binding to macromolecules, and the more reactive $CCl_3O_2\cdot$, which is formed when $CCl_3\cdot$ reacts with oxygen, is the prime initiator of lipid peroxidation.

Lipid peroxidation (Figure 14.5) is the initiating reaction in a cascade of events, starting with the oxidation of unsaturated fatty acids to form lipid hydroperoxides, which then break down to yield a variety of end products, mainly aldehydes, which can go on to produce toxicity in distal tissues. For this reason cellular damage results not only from the breakdown of membranes such as those of the endoplasmic reticulum, mitochondria, and lysosomes but also from the production of reactive aldehydes that can travel to other tissues. It is now thought that many types of tissue injury, including inflammation, may involve lipid peroxidation.

### 14.5.2   Ethanol

Alcohol-related liver diseases are complex, and ethanol has been shown to interact with a large number of molecular targets. Ethanol can interfere with hepatic lipid metabolism in a number of ways and is known to induce both inflammation and necrosis in the liver. Ethanol increases the formation of superoxide by Kupffer cells thus implicating oxidative stress in ethanol-induced liver disease. Similarly prooxidants (reactive oxygen species) are produced in the hepatocytes by partial reactions in the action of CYP2E1, an ethanol-induced CYP isoform. The formation of protein adducts in the microtubules by acetaldehyde, the metabolic product formed from ethanol by alcohol dehydrogenase, plays a role in the impairment of VLDL secretion associated with ethanol.

### 14.5.3   Bromobenzene

Bromobenzene is a toxic industrial solvent that is known to produce centrilobular hepatic necrosis through the formation of reactive epoxides. Figure 14.6 summarizes

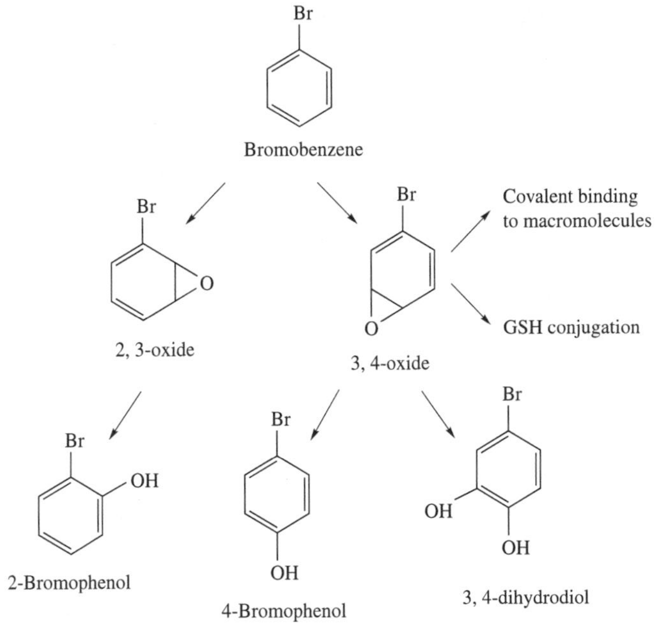

**Figure 14.6** Metabolism of bromobenzene. (From P. E. Levi, *A Textbook of Modern Toxicology*, 2nd ed., Appleton and Lange, 1997.)

the major pathways of bromobenzene metabolism. Both bromobenzene 2,3-epoxide and bromobenzene 3,4-epoxide are produced by P450 oxidations. The 2,3-epoxide, however, is the less toxic of the two species, reacting readily with cellular water to form the nontoxic 2-bromophenol. The more stable 3,4-epoxide is the form most responsible for covalent binding to cellular proteins. A number of pathways exist for detoxication of the 3,4-epoxide: rearrangement to the 4-bromophenol, hydration to the 3,4-dihydrodiol catalyzed by epoxide hydrolase, or conjugation with glutathione. When more 3,4-epoxide is produced than can readily be detoxified, cell injury increases.

Pretreatment of animals with inhibitors of cytochrome P450 is known to decrease tissue necrosis by slowing down the rate of formation of the reactive metabolite, whereas pretreatment of animals with certain P450 inducers can increase the toxicity of bromobenzene, (e.g., the P450 inducer phenobarbital increases hepatotoxicity by inducting a P450 isozyme that preferentially forms the 3,4-epoxide). However, pretreatment with another P450 inducer, 3-methylcholanthrene, decreases bromobenzene hepatotoxicity by inducing a form of P450 that produces primarily the less toxic 2,3-epoxide.

### 14.5.4 Acetaminophen

Acetaminophen is widely used analgesic that is normally safe when taken at therapeutic doses. Overdoses, however, may cause an acute centrilobular hepatic necrosis that can be fatal. Although acetaminophen is eliminated primarily by formation of glucuronide and sulfate conjugates, a small proportion is metabolized by cytochrome P450 to a reactive electrophilic intermediate believed to be a quinoneimine (see Chapter 8). This reactive intermediate is usually inactivated by conjugation with reduced glutathione and excreted. Higher doses of acetaminophen will progressively deplete hepatic glutathione levels, however, resulting in extensive covalent binding of the reactive metabolite to liver macromolecules with subsequent hepatic necrosis. The early administration of sulfhydryl compounds such as cysteamine, methionine, and N-acetylcysteine is very effective in preventing liver damage, renal failure, and death that would otherwise follow an acetaminophen overdose. These agents are thought to act primarily by stimulating glutathione synthesis.

In laboratory animals the formation of the acetaminophen-reactive metabolite, the extent of covalent binding, and the severity of hepatotoxicity can be influenced by altering the activity of various P450 isozymes. Induction of P450 isozymes with phenobarbital, 3-methylcholanthrene, or ethanol increases toxicity, whereas inhibition of P450 with piperonyl butoxide, cobalt chloride, or metyrapone decreases toxicity. Consistent with these effects in animals, it appears that the severity of liver damage after acetaminophen overdose is greater in chronic alcoholics and patients taking drugs that induce the levels of the P450 isozymes responsible for the activation of acetaminophen.

## 14.6   METABOLIC ACTIVATION OF HEPATOTOXICANTS

Studies of liver toxicity caused by bromobenzene, acetaminophen, and other compounds have led to some important observations concerning tissue damage:

- Toxicity may be correlated with the formation of a minor but highly reactive intermediate.
- A threshold tissue concentration of the reactive metabolite must be attained before tissue injury occurs.
- Endogenous substances, such as glutathione, play an essential role in protecting the cell from injury by removing chemically reactive intermediates and by keeping the sulfhydryl groups of proteins in the reduced state.
- Pathways such as those catalyzed by glutathione transferase and epoxide hydrolases play an important role in protecting the cell.
- Agents that selectively induce or inhibit the xenobiotic metabolizing enzymes may alter the toxicity of xenobiotic chemicals.

These same principles are applicable to the toxicity caused by reactive metabolites in other organs, such as kidney and lung as will be illustrated in the following sections.

## SUGGESTED READING

Hodgson, E., and S. A. Meyer. Pesticides. In *Comprehensive Toxicology: Hepatic and Gastrointestinal Toxicology*, vol. 9, I. G. Sipes, C. A. McQueen, and A. J. Gandolfi, eds. New York: Elsevier Science, 1997, p. 369.

Meyer, S. A. Hepatotoxicity. In *An Introduction to Biochemical Toxicology*, 3rd ed., E. Hodgson and R. C. Smart, eds. New York: Wiley, 2001, p. 487.

Reed, D. J. Mechanisms of chemically induced cell injury and cellular protection mechanisms. In *An Introduction to Biochemical Toxicology*, 3rd ed., E. Hodgson and R. C. Smart, eds. New York: Wiley, 2001, p. 221.

Treinen-Moslen, M. Toxic responses of the liver. In *Casarett and Doull's Toxicology: The Basic Sciences of Poisons*, 6th ed., C. D. Klaassen, ed. New York: McGraw-Hill, 2001, p. 471.

# CHAPTER 15

# Nephrotoxicity

ERNEST HODGSON and PATRICIA E. LEVI

## 15.1 INTRODUCTION

### 15.1.1 Structure of the Renal System

The renal system consists of the kidneys and their vasculature and innervation, the kidneys each draining through a ureter into a single median urinary bladder, and the latter draining to the exterior via a single duct, the urethra. The kidney has three major anatomical areas: the cortex, the medulla, and the papilla.

The renal cortex is the outermost region of the kidney and contains glomeruli, proximal and distal tubules, and peritubular capillaries. Cortical blood flow is high, the cortex receiving approximately 90% of the renal blood flow. Since blood-borne toxicants will be delivered preferentially to the cortex, they are more likely to affect cortical functions rather than those of medulla or papilla. The renal medulla is the middle portion and contains primarily loops of Henle, vasa recta, and collecting ducts. Although the medulla receives only about 6% of the renal blood flow, it may be exposed to high concentrations of toxicants within tubular structures. The papilla is the smallest anatomical portion of the kidney and receives only about 1% of the renal blood flow. Nevertheless, because the tubular fluid is maximally concentrated and luminal fluid is maximally reduced, the concentrations of potential toxicants in the papilla my be extremely high, leading to cellular injury in the papillary tubular and/or interstitial cells.

The nephron is the functional unit of the kidney. It is described in detail in Chapter 10 and illustrated in Figure 10.1.

### 15.1.2 Function of the Renal System

The primary function of the renal system is the elimination of waste products, derived either from endogenous metabolism or from the metabolism of xenobiotics. The latter function is discussed in detail in Chapter 10. The kidney also plays an important role in regulation of body homeostasis, regulating extracellular fluid volume, and electrolyte balance.

*A Textbook of Modern Toxicology, Third Edition,* edited by Ernest Hodgson
ISBN 0-471-26508-X  Copyright © 2004 John Wiley & Sons, Inc.

Other functions of the kidney include the synthesis of hormones that affect metabolism. For example, 25-hydroxy-vitamin $D_3$ is metabolized to the active form, 1,25-dihydroxy-vitamin $D_3$. Renin, a hormone involved in the formation of angiotensin and aldosterone, is formed in the kidney as are several prostaglandins. While kidney toxicity could affect any of these functions, the effects used clinically to diagnose kidney damage are related to excretory function damage, such as increases in urinary glucose, amino acids, or protein, changes in urine volume, osmolarity, or pH. Similarly changes in blood urea nitrogen (BUN), plasma creatinine, and serum enzymes can be indicative of kidney damage.

In animal studies of nephrotoxicity not only can histopathology be carried out but various biochemical parameters can be compared with those from untreated animals. They include lipid peroxidation and covalent binding to tissue macromolecules.

## 15.2  SUSCEPTIBILITY OF THE RENAL SYSTEM

Several factors are involved in the sensitivity of the kidney to a number of toxicants (Table 15.1), although the high renal blood flow and the increased concentration of excretory products following reabsorption of water from the tubular fluid are clearly of major importance. Although the kidneys comprise less than 1% of the body mass, they receive around 25% of the cardiac output. Thus significant amounts of exogenous chemicals and/or their metabolites are delivered to the kidney.

A second important factor affecting the kidneys sensitivity to chemicals is its ability to concentrate the tubular fluid and, as a consequence, as water and salts are removed, to concentrate any chemicals it contains. Thus a nontoxic concentration in the plasma may be converted to one that is toxic in the tubular fluid. The transport characteristics of the renal tubules also contribute to the delivery of potentially toxic concentrations of chemicals to the cells. If a chemical is actively secreted from the blood into the tubular fluid, it will accumulate initially within the cells of the proximal tubule or, if it is reabsorbed from the tubular fluid, it will pass into the cells in relatively high concentration.

The biotransformation of chemicals to reactive, and thus potentially toxic, metabolites is a key feature of nephrotoxicity. Many of the same activation reactions found in the liver are also found in the kidney and many toxicants can be activated in either organ, including acetaminophen, bromobenzene, chloroform, and carbon tetrachloride, thus having potential for either hepatotoxicity or nephrotoxicity. Some regions of the kidney have considerable levels of xenobiotic metabolizing enzymes, particularly cytochrome P450 in the pars recta of the proximal tubule, a region particularly susceptible to chemical damage. Since reactive metabolites are generally unstable, and therefore more or less transient, they are likely to interact with cellular macromolecular close to the site of generation. Thus, although the activity of activation enzymes such as

**Table 15.1  Factors Affecting the Susceptibility of the Kidney to Toxicants**

High renal blood flow
Concentration of chemicals in tubular fluid
Reabsorption and/or secretion of chemicals through tubular cells
Activation of protoxicants to reactive, and potentially toxic, metabolites

cytochrome P450 is lower in the kidney that in the liver, they are of greater importance in nephrotoxicity than those of the liver due to their proximity to site of action.

As with toxicity in other organs the ultimate expression of a toxic end point is the result of a balance between the generation of reactive metabolites and their detoxication. The high levels of glutathione found in the kidney doubtless play an important role in the detoxication process.

## 15.3 EXAMPLES OF NEPHROTOXICANTS

### 15.3.1 Metals

Many heavy metals are potent nephrotoxicants, and relatively low doses can produce toxicity characterized by glucosuria, aminoaciduria, and polyuria. As the dose increases, renal necrosis, anuria, increased BUN, and death will occur. Several mechanisms operate to protect the kidney from heavy metal toxicity. After low dose exposure and often before detectable signs of developing nephrotoxicity, significant concentrations of metal are found bound to renal lysosomes. This incorporation of metals into lysosomes may result from one or more of several mechanisms, including lysosomal endocytosis of metal-protein complexes, autophagy of metal-damaged organelles such as mitochondria, or binding of metals to lipoproteins within the lysosome. Exposure to high concentrations, however, may overwhelm these mechanisms, resulting in tissue damage.

***Cadmium.*** In humans, exposure to cadmium is primarily through food or industrial exposure to cadmium dust. In Japan, a disease called Itai-itai Byo is known to occur among women who eat rice grown in soils with a very high cadmium content. The disease is characterized by anemia, damage to proximal tubules, and severe bone and mineral loss. Cadmium is excreted in the urine mainly as a complex (CdMT) with the protein metallothionein (MT). MT is a low molecular weight protein synthesized in the liver. It contains a large number of sulfhydryl groups that bind certain metals, including cadmium. The binding of cadmium by MT appears to protect some organs such as the testes from cadmium toxicity. At the same time, however, the complex may enhance kidney toxicity because the complex is taken up more readily by the kidney than is the free metal ion. Once inside the cell, it is thought that the cadmium is released, presumably by decomposition of the complex within the lysosomes.

Cadmium has a long biological half-life, 10 to 12 years in humans; thus low-level chronic exposure will eventually result in accumulation to toxic concentrations.

***Lead.*** Lead, as $Pb^{2+}$, is taken up readily by proximal tubule cells, where it damages mitochondria and inhibits mitochondrial function, altering the normal absorptive functions of the cell. Complexes of lead with acidic proteins appear as inclusion bodies in the nuclei of tubular epithelium cells. These bodies, formed before signs of lead toxicity occur, appear to serve as a protective mechanism.

***Mercury.*** Mercury exerts its principle nephrotoxic effect on the membrane of the proximal tubule cell. In low concentrations, mercury binds to the sulfhydryl groups of membrane proteins and acts as a diuretic by inhibiting sodium reabsorption. Organomercurial diuretics were introduced into clinical practice in the 1920s and were used clinically into the 1960s. Despite their widespread acceptance as effective therapeutic diuretics, it was well known that problems related to severe kidney toxicity were possible. However, in the absence of other effective drugs, the organomercurials proved to be effective, sometimes life-saving, therapeutic agents. More recently organomercurial chemicals have been implicated as environmental pollutants, responsible for renal damage in humans and animals.

***Uranium.*** About 50% of plasma uranium is bound, as the uranyl ion, to bicarbonate, which is filtered by the glomerulus. As a result of acidification in the proximal tubule, the bicarbonate complex dissociates, followed by reabsorption of the bicarbonate ion; the released $UO_2^{2+}$ then becomes attached to the membrane of the proximal tubule cells. The resultant loss of cell function is evidenced by increased concentrations of glucose, amino acids, and proteins in the urine.

### 15.3.2   Aminoglycosides

Certain antibiotics, most notably the aminoglycosides, are known to be nephrotoxic in humans, especially in high doses or after prolonged therapy. The group of antibiotics includes streptomycin, neomycin, kanamycin, and gentamycin. Aminoglycosides are polar cations that are filtered by the glomerulus and excreted unchanged into the urine. In the proximal tubule, the aminoglycosides are reabsorbed by binding to anionic membrane phospholipids, followed by endocytosis and sequestration in lysosomes (Figure 15.1). It is thought that when a threshold concentration is reached, the lysosomes rupture, releasing hydrolytic enzymes that cause tissue necrosis.

### 15.3.3   Amphotericin B

With some drugs, renal damage may be related to the drugs' biochemical mechanism of action. For example, the polymycins, such as amphotericin B, are surface-active agents that bind to membrane phospholipids, disrupting the integrity of the membrane and resulting in leaky cells.

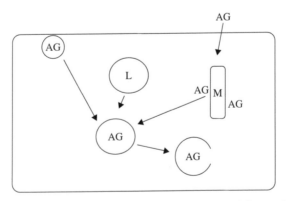

**Figure 15.1**   Possible cellular interactions of aminoglycosides. AG = aminoglycoside; M = mitochondrion; L = lysosome. (From E. Hodgson and P. E. Levi, eds., *A Textbook of Modern Toxicology.* 2nd ed., Appleton and Lange, Stamford, CT, 1997.)

## 15.3.4 Chloroform

Chloroform is a common industrial organic solvent that can be a hepatotoxicant or a nephrotoxicant in both humans and animals. As a nephrotoxicant it is both species and gender dependent. For example, following chloroform administration male mice develop primarily kidney necrosis whereas female develop liver necrosis.

As a nephrotoxicant, chloroform most probably undergoes metabolic activation in the kidney itself. Chloroform is metabolized to phosgene (Figure 15.2) by a cytochrome P450-dependent reaction, probably proceeding via an unstable hydroxylated product, trichloromethanol. Phosgene is capable of binding to cellular proteins to produce the cellular necrosis associated with chloroform toxicity to the kidney. Phosgene can also be further metabolized by a number of reactions (Figure 15.2), and as with most chemical-induced toxicity, the final expression of toxicity depends on a balance between activation and detoxication.

## 15.3.5 Hexachlorobutadiene

Hexachlorobutadiene is an industrial solvent and heat-transfer agent. It is a widespread environmental contaminant that is a potent and relatively specific nephrotoxicant. Hexachlorobutadiene first forms a glutathione conjugate, which is further metabolized by the mercapturic acid pathway to a cysteine conjugate (see Chapter 7 for details of glutathione conjugation and the mercapturic acid pathway). In the kidney, the cysteine conjugate is cleaved to a reactive intermediate by the enzyme, cysteine conjugate $\beta$-lyase.

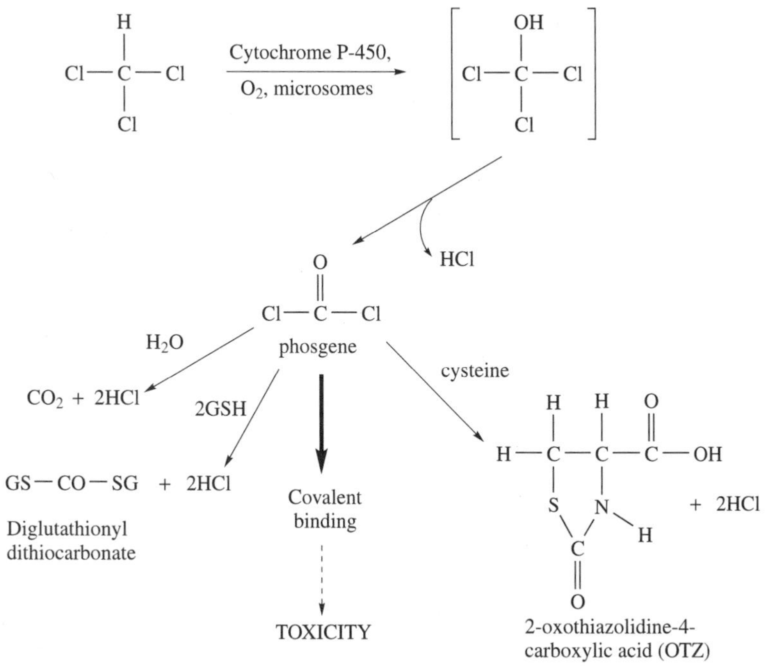

**Figure 15.2** Proposed mechanism of chloroform biotransformation. (From J. B. Tarloff in *An Introduction to Biochemical Toxicology*, 3rd ed., E. Hodgson and R. C. Smart eds., Wiley, 2001.)

### 15.3.6   Tetrafluoroethylene

The nephrotoxic mode of action of tetrafluoroethylene is similar to that of hexachloro-butadiene. It is first metabolized to a cysteine conjugate, which is metabolized by cysteine conjugate $\beta$-lyase to a reactive product that can bind to cellular macro-molecules.

## SUGGESTED READING

Tarloff, J. B. Biochemical mechanisms of renal toxicity. In *An Introduction to Biochemical Toxicology*, 3rd ed., E. Hodgson and R. C. Smart, eds. New York: Wiley, 2001, p. 641.

Schnellmann R. G, Toxic Responses of the kidney. In *Casarett and Doull's Toxicology: The Basic Sciences of Poisons*, 6th ed., C. D Klaassen, ed, New York: McGraw-Hill, 2001, p. 491.

■■■■■■ CHAPTER 16

# Toxicology of the Nervous System

BONITA L. BLAKE

## 16.1 INTRODUCTION

Many substances alter the normal activity of the nervous system. Sometimes these effects are immediate and transient, like the stimulatory effect of a cup of coffee or a headache from the fresh paint in your office. Other effects can be much more insidious, like the movement disorders suffered by miners after years of chronic manganese intoxication. Many agents are safe or even therapeutic at lower doses but become neurotoxic at higher levels. Trace metals and pyridoxine (vitamin B-6) fall into this category of dose-dependent effects. Since these agents affirm the maxim, "the dose makes the poison," it becomes necessary to have a meaningful definition of nervous system poisoning, or neurotoxicity. *Neurotoxicity refers to the ability of an agent to adversely affect the structural or functional integrity of the nervous system.* Structural damage to nervous system components usually results in altered functioning, although the reverse is not always true. Alterations in nervous system function may occur through toxicant interactions with the normal signaling mechanisms of neurotransmission, resulting in little or no structural damage. Nevertheless, it is easier to identify alterations, be they structural or functional, than it is to define adversity. For example, the stimulant effect of a morning cup of coffee may be too anxiety provoking for some individuals but a necessity to others.

In this chapter a brief introduction to the nervous system is presented and its functions are described. A discussion of some of the mechanisms of structural and functional neurotoxicant effects follows. These descriptions are not exhaustive, they are meant to illustrate the concepts of toxicant interaction with the nervous system. Finally some methods for testing toxicant effects in the nervous system are explored.

## 16.2 THE NERVOUS SYSTEM

Most multicellular animals possess a nervous system. In every case the function of the nervous system is to receive information about the external and internal environment, integrate the information, and then coordinate a response appropriate to the

*A Textbook of Modern Toxicology, Third Edition,* edited by Ernest Hodgson
ISBN 0-471-26508-X  Copyright © 2004 John Wiley & Sons, Inc.

environmental stimulus. In addition to these basic vital functions, the nervous systems of higher organisms are responsible for feeling, thinking, and learning. All of the other organ systems of the body are subject to control by the nervous system; thus damage to this "master" system by toxicants can have far-reaching and even devastating effects.

In vertebrates there are two major components of the nervous system. Although these two systems are anatomically separable, they are contiguous and they function interactively. The brain and spinal cord comprise the central nervous system (CNS), and the nervous tissue (ganglia and peripheral nerves) outside the brain and spinal cord comprise the peripheral nervous system (PNS). The PNS can be further divided into the somatic and autonomic nervous systems. Somatic afferents carry sensory information from the skin, muscle, and joints to the CNS, while motor efferents innervate skeletal muscle to cause contractive movement. The autonomic nervous system can be thought of as a motor system for visceral organs, since it projects to these organs to innervate and control the function of smooth muscle, cardiac muscle, and endocrine and exocrine glands. The autonomic nervous system is further divided anatomically and functionally into the sympathetic and parasympathetic subdivisions. Most organs are innervated by both subdivisions, and their influences generally oppose one another. For example, stimulation of sympathetic nerves increases heart rate, while stimulation of the vagus nerve, the primary parasympathetic innervation of the heart, slows its rate of contraction.

### 16.2.1 The Neuron

The basic unit of the nervous system is the neuron, a type of cell that is structurally and functionally specialized to receive, integrate, conduct, and transmit information. Although neurons are far more diverse than any other cell type in the body, they do have some common features. Neurons are polarized cells; that is, they have different characteristics on one end of the cell compared to the other (Figure 16.1). Typically the end of the neuron that receives information in the form of neurotransmitter stimulation from other neurons is highly branched into a region known as the dendritic tree. The branches are sometimes studded with projections, known as spines, which contain receptors that recognize and are activated by neurotransmitters. It is here that the neuron is in close contact with other neurons via specialized structures called synapses. Synapses are areas of close apposition where one neuron, the presynaptic neuron, releases neurotransmitter into the gap, or cleft, between the two neurons. The postsynaptic neuron then recognizes this chemical signal via receptors that are clustered in small densities opposing the presynaptic neuron. Once the neurotransmitter signal is recognized by its receptors, the dendritic region of the neuron transmits the information as intracellular and electrochemical signals to the regions of the neuron where signal integration takes place. In the typical neuron the arborizations of the dendritic tree converge on the soma, or cell body, where the nucleus and most of RNA- and protein-synthesizing machinery exist. The cell body usually then gives off a single axon, and it is in the region where the axon leaves the cell body (the axon hillock) that signals converge to be integrated into an all-or-none response.

Neurotransmission down an axon is in the form of electrochemical signals. In the resting state the interior of the axon is negatively charged with respect to the exterior. The membrane is then said to be polarized, and the charge difference across the membrane in this resting state is approximately $-70$ mV. Small depolarizing potentials

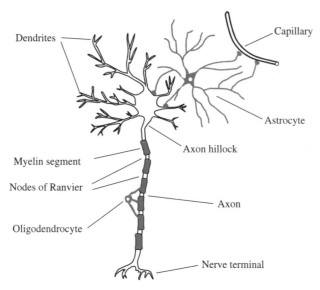

**Figure 16.1**   A neuron with accompanying astrocyte and myelinating oligodendrocyte.

arrive at the axon hillock from the dendritic regions where receptors have been stimulated, and this stimulation results in the opening and closing of ion channels. The depolarizing potentials occur primarily because of the opening of sodium channels, allowing sodium to transfer down its concentration gradient to the interior of the cell. Sodium brings with it a positive charge, and so the membrane in the region where the sodium channel opens becomes depolarized. The depolarization is then detected by voltage-sensing sodium channels, which allow further influx of sodium. When the spatial and temporal summation of these signals reaches a certain threshold (generally about +50 mV), the axon will generate an action potential at the axon hillock. Once this occurs, all of the sodium channels in the vicinity open, allowing a massive influx of sodium. Sodium channels stay open for only a short period of time, and once they close, they cannot reopen for a while. On the other hand, as sodium channels further down the axon sense the voltage change, they open, and thus a feedforward effect is created. The membrane is repolarized by the opening of potassium channels, which respond in a slightly delayed fashion, to the same signals that stimulate the sodium channels. As the potassium channels open, potassium rushes out of the cell down its own concentration gradient. This, combined with the closing of the sodium channels, produces a net efflux of positive charge, thereby repolarizing the membrane. The process of depolarization/repolarization continues down the length of the axon. In myelinated axons, the ion channels are clustered in regions between the segments of myelin in regions known as nodes of Ranvier. The myelin segments serve to insulate the axon, and they allow the action potential to jump from one node to the next, in a process called saltatory conduction (Figure 16.2). This results in much faster propagation of the action potential down the length of the axon.

Axons terminate at neuromuscular junctions, in effector organs (e.g., the heart), and in synapses with other neurons. When the action potential reaches the terminal of the axon, the depolarizing impulse stimulates the release of neurotransmitter from the terminal into the cleft between the presynaptic membrane and its effector or receiving

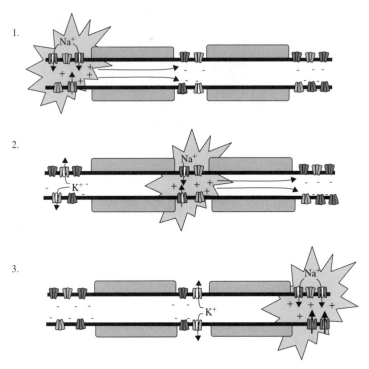

**Figure 16.2** Saltatory conduction. Myelin acts as an insulator to prevent current loss as the action potential travels down the axon. Sodium and potassium channels are clustered at the Nodes of Ranvier, where there is no myelin. Action potentials jump from one node to the next, reducing the overall membrane area involved in conduction, and speeding up electrical transmission.

neuron. The process of release usually involves the presence of packets of neurotransmitter called synaptic vesicles (Figure 16.3). These vesicles dock at the presynaptic membrane and, when stimulated to do so, fuse with the membrane to release their contents into the synaptic cleft. The signal to fuse is thought to be primarily an influx of calcium, mediated by calcium channels on the presynaptic membrane that are sensitive to changes in voltage. Proteins on the vesicle membrane and the presynaptic membrane form complexes with one another, and when stimulated by the localized increase in calcium ion concentration, mediate the fusion of the two membranes and release of neurotransmitter. Electrical signals at work within the neuron are thus converted to chemical signals at work between neurons in the form of neurotransmitters.

### 16.2.2 Neurotransmitters and their Receptors

Neurotransmitters are recognized by receiving neurons, neuromuscular junctions, or end effector organs via receptors that lie on the postsynaptic membrane. Receptors are generally selective for the neurotransmitter that they bind. The type of signaling that is characteristic of a given neurotransmitter is usually the result of the form of receptor to which it binds. For example, some receptors, like the nicotinic acetylcholine receptor found in neuromuscular junctions, are ion channels. The stimulation of the nicotinic

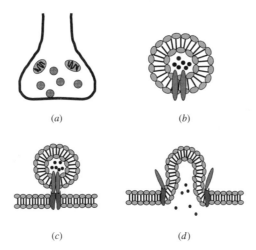

(a)

(b)

(c)

(d)

**Figure 16.3** Neurotransmitter release. (*a*) Presynaptic nerve terminal containing vesicles and other organelles. (*b*) Neurotransmitter-containing vesicles are made of lipid bilayers. Associated proteins participate in the release process. (*c*) The vesicle associates with the presynaptic membrane via protein complexes that mediate release. (*d*) Release of neurotransmitter into the synapse is by protein-mediated fusion of vesicle and presynaptic membranes.

receptor by acetylcholine results in the opening of its channel, which is permeable to sodium. The influx of sodium then serves to depolarize the muscle membrane that receives acetylcholinergic innervation. Neurotransmitter receptors that are ion channels thus mediate very fast and short-lived neurotransmission, particularly when compared to the other major type of neurotransmitter receptor, the G protein-coupled receptor. G protein-coupled neurotransmitter receptors activate intracellular signaling pathways that produce a more slow and sustained response to neurotransmitter stimulation. In general, these receptor-mediated pathways serve to modulate neurotransmission by ion channels, maintain and mediate changes in protein expression, and promote cell survival. Most neurotransmitters have both ion channel receptors and G protein-coupled receptors, although a few, like dopamine and norepinephrine, bind only to G protein-coupled receptors.

Neurotransmitters stimulate receptors on postsynaptic membranes, but the signals receptors send are not always excitatory to the receiving neuron. Receptors, directly or indirectly, modulate excitability of the postsynaptic neuron, so that it is more or less likely to fire an action potential. For example, the neurotransmitter glutamate binds to both ion channel receptors and G protein-coupled receptors, and in each case these receptors transmit a signal that enhances the excitability of the receiving neuron. On the other hand, the neurotransmitter GABA (for *g*amma-*a*mino *b*utyric *a*cid), while also binding to both types of receptors, is known for its ability to decrease the excitability of the postsynaptic neuron. Its message is therefore inhibitory to the propagation of signaling within a group of neurons.

### 16.2.3 Glial Cells

While neurons constitute the definitive unit of the nervous system, their function is critically dependent on the presence of glial cells. In fact there are far more glial cells in the nervous system than neurons. Glial cells perform many functions, including structural support, insulation, buffering, and guidance of migration during development. One class of glial cell, called microglia, is responsible for phagocytosis of cellular debris following injury and infection. The other types of glial cells, collectively known as macroglia, are the astrocytes and oligodendrocytes found in the CNS, and the Schwann cells found in the PNS. Oligodendrocytes and Schwann cells form myelin by wrapping multiple layers of plasma membrane around axons. Astrocytes are the most numerous of the glial cells, and they help form the blood-brain barrier, take up excess neurotransmitter and ions, and probably have some nutritive function. Metabolic enzymes expressed within astrocytes also regulate neuronal signaling by catabolizing excessive amounts of neurotransmitter. Monoamine oxidases, for example, catalyze the biotransformation of dopamine, norepinephrine, and serotonin into oxidation products that are substrates for further enzymatic reactions en route to excretion. The incidental bioactivation of the xenobiotic MPTP to its neurotoxic metabolite MPP+ by these enzymes will be discussed later in this chapter.

### 16.2.4   The Blood-Brain Barrier

The blood-brain barrier was conceptualized when it was noted that dyes injected into the bloodstream of animals stained nearly all tissues except the brain. It is thus this barrier and its PNS equivalent, the blood-nerve barrier, that prevents all but a select few molecules from entering the nervous system. The barrier itself is not a single unitary structure, but a combination of unique anatomical features that prevents the translocation of blood-borne agents from brain capillaries into the surrounding tissue. As mentioned above, astrocytes help form the barrier, surrounding capillary endothelial cells with extensions of their cytoplasm known as end-feet. There are also pericytes, whose function is not well known, that associate with the capillaries and may help induce a functional barrier. Another component of the barrier is the impermeable nature of the endothelial cells that line the interior of capillaries. Capillary endothelial cells in the nervous system are different from those in the periphery in at least three ways. First, brain capillaries form tight junctions of very high resistance between cells. In contrast, peripheral capillaries have low resistance tight junctions, and even openings, or fenestrations, that allow compounds to pass between cells. Second, compared to peripheral endothelial cells, brain endothelial cells are deficient in their ability to transport agents by endocytic mechanisms. Instead, only small lipophilic particles can be passed transcellularly. For larger molecules, carrier-mediated transport mechanisms are highly selective, and allow only one-way transport. Third, there is an enzymatic barrier that metabolizes nutrients and other compounds. Enzymes such as gamma-glutamyl transpeptidase, alkaline phosphatase, and aromatic acid decarboxylase are more prevalent in cerebral microvessels than in nonneuronal capillaries. Most of these enzymes are present at the lumenal side of the endothelium, Additionally the P-glycoprotein (P-gp) drug efflux transporter is presently thought to exist at the lumenal membrane surface, although some scientists argue that P-gp is actually associated with astrocytes. Finally the CNS endothelial cell displays a net negative charge at its lumenal side and at the basement membrane. This provides an additional selective mechanism by impeding passage of anionic molecules across the membrane.

Most of the toxicants that enter the nervous system do so by exploiting mechanisms designed to allow entry of essential molecules, such as nutrients, ions, and neurotransmitter precursors. Small, lipophilic molecules are able to cross the blood-brain barrier relatively easily. Some agents can be recognized by active transport systems and thereby traverse the blood-brain barrier along with endogenous ligands. The Parkinson's disease therapeutic agent levodopa enters the brain in this manner. In some cases the blood-brain barrier is itself subject to damage by neurotoxicants. For example, lead, cadmium, mercury, and manganese accumulate in endothelial cells and damage their membranes, leading to brain hemorrhage and edema.

### 16.2.5   The Energy-Dependent Nervous System

Nervous tissue has a high demand for energy, yet nerve cells can only synthesize ATP through glucose metabolism in the presence of oxygen. Critical ATP-dependent processes in the nervous system include regulation of ion gradients, release and uptake of neurotransmitters, anterograde and retrograde axonal transport, active transport of nutrients across the blood-brain barrier, P-gp function, phosphorylation reactions, assembly of mitochondria, and many others. The highest demand for energy (up to 70%) is created by the maintenance of resting potential in the form of sodium and potassium concentration gradients across the nerve cell membrane. These gradients are maintained primarily by the activity of the $Na^+/K^+$ ATPase pump. The pump uses the energy of hydrolyzing each ATP molecule to transport three sodium ions out of the cell, and two potassium ions into the cell. Maintenance of the resting potential is not the only benefit of this pump's activity, however. The gradients created by the pump are also important for maintaining osmotic balance, and for the activity of indirect pumps that make use of the sodium gradient to transport other molecules against their own concentration gradient. Neurotransmission is thus heavily dependent on the proper functioning of the $Na^+/K^+$ ATPase pump.

Another process dependent on energy metabolism is axonal transport. Axonal transport carries organelles, vesicles, viruses, and neurotrophins between the nerve nucleus

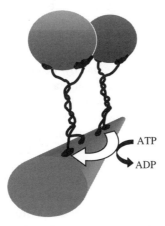

**Figure 16.4**   The microtubule motor protein kinesin. Kinesin consists of two ATP-hydrolyzing subunits that contact microtubules, a stem region, and regions that interact with vesicle and organelle proteins. One ATPase subunit binds and hydrolyses ATP, generating the force required to advance it forward. As this happens, the other subunit releases ADP, in preparation for binding another ATP, and its own advancement.

and the terminal. This distance can be quite long when one considers that the length of the sciatic nerve, for example, can be up to one meter. Anterograde transport (from cell body to terminal) is accomplished by two mechanisms defined by their rate, fast axonal transport and slow axonal transport. Fast axonal transport proceeds at rates of approximately 400 mm/day, and is mediated by the ATP-dependent motor protein kinesin. Kinesin forms cross-bridges between vesicles or organelles and microtubules, and dual projections of these cross-bridges shift back-to-front along microtubules in a coordinated, ATP-dependent manner, such that the entire molecule appears to be walking (Figure 16.4). Slow axonal transport is used to carry cytoskeletal elements such as tubulin and neurofilaments to the far ends of the axon, and it proceeds at approximately 0.2 to 5 mm/day. Traditionally slow transport has been regarded as passively dependent on axoplasmic flow; however, recent evidence suggests that the cytoskeletal elements actually move rather quickly but frequently stall in a stop-and-go fashion. Fast axonal transport also proceeds retrogradely, mediated by the ATP-dependent motor protein dynein. The rate of retrograde transport is about 200 mm/day. Neurons use retrograde transport for recycling membranes, vesicles, and their associated proteins. Neurotrophic factors, and some viruses and toxins (e.g., tetanus toxin) are also transported by this mechanism.

## 16.3   TOXICANT EFFECTS ON THE NERVOUS SYSTEM

Neurotoxicants affect the nervous system in a number of different ways. Some neurotoxicants damage the distal portions of axons without much effect on the remainder of the cell, while others produce outright cell death. Still others affect signaling processes in the nervous system, without causing structural damage. This wide variety of neurotoxicant effects is due in part to the unique nature of the different types of neurons and glia. Neurons may be differentially vulnerable to certain neurotoxicants because of their functional characteristics, as in the case of the targeting of substantia nigra neurons by the active metabolite of MPTP, an agent that causes Parkinson's disease. The substantia nigra, a brain region where neurons that synthesize dopamine are particularly abundant, sends out axons that project to other parts of the brain where the dopamine is released. After release, these neurons take back up the synaptic dopamine via selective transporters on the nerve endings. The damaging metabolite of MPTP, MPP+, is not distinguished from dopamine by the uptake transporter, so when present, MPP+ is taken up as well. MPP+ kills substantia nigra neurons by affecting mitochondrial energy production and promoting free radical formation. The death of these neurons results in a lack of dopamine release in an area of the brain called the stratum, which is responsible for the control of movement. The loss of dopamine in the stratum causes the hallmark symptoms of the neurodegenerative disease Parkinsonism: slowed movement, rigidity, and tremors. Because epidemiological studies have linked Parkinsonism in some patients with the agricultural use of pesticides, many of which are toxic to mitochondria, scientists believe that at least some cases of Parkinson's disease may be related to long-term exposure to environmental toxins.

MPTP is one example of a toxicant that causes direct structural damage to neurons, resulting in loss of function. In the following sections, other types of structural and functional effects of neurotoxicants are described. Structural effects are divided into three primary types: effects on myelin formation, primary damage to axons, and direct promotion of cell death. Neurons may also be secondarily affected by neurotoxicants that target other cells in the nervous system, disrupting normal homeostatic function and causing structural or functional damage. Another method by which toxicants may affect the function of the nervous system is by directly altering synaptic neurotransmission.

### 16.3.1 Structural Effects of Toxicants on Neurons

***Demyelination.*** The role of myelin in the nervous system is to aid in signal transduction. Myelin acts like an electrical insulator by preventing loss of ion current, and intact myelin is critical for the fast saltatory nerve conduction discussed above. Neurotoxicants that target the synthesis or integrity of PNS myelin may cause muscle weakness, poor coordination, and paralysis. In the brain, white matter tracts that connect neurons within and between hemispheres may be destroyed, in a syndrome known as toxic leukoencephalopathy. A multifocal distribution of brain lesions is reflected in mental deterioration, vision loss, speech disturbances, ataxia (inability to coordinate movements), and paralysis.

Demyelination occurs secondary to axonal degeneration, a topic covered in the section on axonopathy. Neurotoxicants that produce primary demyelination are uncommon, but may be divided into those that affect the integrity of the myelin sheath without or prior to damage to the myelinating cells, and those that directly injure myelin-producing cells. The former include agents like hexachlorophene, isoniazid, and the organotins. These compounds cause reversible edema between the layers of myelin by a mechanism that is yet unclear. The optic nerve is particularly susceptible to demyelination by hexachlorophene and organic solvents, whereas other cranial nerves, such as the trigeminal and vestibulocochlear, are vulnerable to styrene, xylene, and to trichloroethylene, an agent used in dry-cleaning. The metalloid tellurium damages myelin by inhibiting an enzyme involved in the synthesis of cholesterol, a major component of myelin.

In contrast to agents that target the integrity of the myelin sheath, chronic exposure to cyanide and carbon monoxide is thought to directly injure myelin-producing Schwann cell bodies in the PNS and oligodendrocytes in the CNS. Inorganic lead also causes direct damage myelinating cells. Oligodendrocytes appear more sensitive to lead toxicity than astrocytes or neurons. One mechanism for the devastating developmental effects of lead exposure may be the preferential inhibition of oligodendrocyte precursor cell differentiation.

***Axonopathy.*** Axonopathy is a specialized form of neuronal damage, involving degeneration of the axon, while leaving the cell body intact. When axons die and degenerate, myelin breaks down as well, yet Schwann cells may survive and guide regeneration of the axon in some cases. Some toxicants produce axonal injury by directly targeting the axon itself. Others are thought to cause degenerative changes in axons by compromising the metabolic systems of the neuron. In the latter case the distal portion of the axon degenerates first, because it is this region that is most heavily dependent on intact axonal transport mechanisms. Since axonal transport is energy-dependent, toxicants that interfere with ATP production, such as the nicotinamide analogue Vacor, may cause distal regions to degenerate initially. Agents that target tubulin, like the vinca alkaloids, also cause this type of injury, because the tubulin-derived microtubules are critically important for axonal transport. With continued exposure the degeneration progresses proximally, and eventually will affect the entire neuron. This distal-to-proximal degeneration is called "dying back neuropathy," and it affects the longest and largest diameter neurons most severely. If exposure to the toxicant is discontinued before death of the entire proximal axon and cell body, axons in the PNS will often regenerate, but axonal regeneration does not occur within the CNS. Regeneration in the PNS is dependent on Schwann cells proliferating and guiding growth of regenerating axon tips back to the target tissue.

In the 1850s Augustus Waller described the sequence of events that occurred following transection of a nerve fiber. These events have subsequently become known as Wallerian degeneration. The essential features of this type of degeneration include swelling of the axon at the distal end of the proximal segment of the transected axon, distal axonal dissolution and phagocytosis by inflammatory cells, and dissolution of myelin, with preservation and proliferation of Schwann cells along the length of the former axon. Certain neurotoxicants are capable of chemically transecting an axon, producing Wallerian degeneration similar to that occurring after slicing the nerve in half. Hexane, for example, forms covalent adducts with neurofilament proteins. This chemical crosslinking is thought to result in neurofilamentous axonal swellings that essentially block transport to regions of the axon distal to the swelling. The distal regions then die due to lack of communication with the neuron cell body, undergoing Wallerian degeneration.

Axonopathy can manifest as defects in sensory or motor functions, or a combination of the two. For most neurotoxicants, sensory changes are noticed first, followed by progressive involvement of motor neurons. One historically important case that illustrates these effects is that of the epidemic poisoning resulting from the consumption of "Ginger Jake" during Prohibition. Extracts of ginger used in tonics were legally required to contain 5 grams of ginger per milliliter of alcohol. The Department of Agriculture checked for compliance with this requirement by sampling the tonics, boiling off the ethanol, and weighing the solid content. Bootleggers soon discovered that a good deal of money could be saved by cutting back on the ginger and substituting it with adulterating agents like castor oil and molasses. It was such an attempt at adulterating Ginger Jake that led to the addition of Lyndol, a triorthocresyl phosphate (TOCP)-containing oil used in lacquers and varnishes, to tonic that was consumed by hundreds of thousands of people. The earliest signs that developed in people who had consumed the product were noted days to weeks later, and began with tingling and numbness in the hands and feet. In many, this progressed to leg cramps, weakness of the legs and arms, and ataxia. Those with minor symptoms improved, but perhaps thousands of people were left permanently paralyzed by the incident. Today TOCP is used to study the syndrome of delayed effects caused by some organophosphate compounds, commonly known as organophosphate-induced delayed neuropathy (OPIDN). The nature of OPIDN is still poorly understood. It appears not to be associated with organophosphate inhibition of acetylcholinesterase, but rather with another neuronal enzyme, the neuropathy target esterase (NTE). The physiological role of NTE is unknown.

***Neuronopathy.*** Neuronopathy refers to generalized damage to nerve cells, with the primary damage occurring at the nerve cell body. Axonal and dendritic processes die secondarily in response to loss of the cell body. Like other cells in the body, neurons die by one of two processes distinguished by their morphological and molecular features: apoptosis and necrosis. (This division is overly simplistic; there is much debate over the characteristics of the two categories, and whether there are more than two categories of cell death. Nevertheless, only these two will be considered here.) Often the same neurotoxicants can promote either form of cell death, depending on the intensity of the insult. For example, methylmercury given to rats at a high dose for one week causes widespread histopathological damage consistent with necrosis, whereas lower doses spread out over a longer time period results in apoptotic changes restricted primarily to cerebellar granule cells. It is thought that the severe and abrupt loss of cellular energy production by impairment of mitochondrial activity and plasma membrane disruption are responsible for necrotic death of neurons. This affects surrounding tissue more than apoptosis, since the dying cells release their contents and localized inflammatory responses ensue. On the other hand, apoptosis is death that is encoded within individual cells. Apoptosis is characterized by a process of cell shrinkage, pyknosis and fragmentation of nuclei, and membrane budding. The dying cell breaks apart into small membrane-bound apoptotic fragments that are phagocytosed, and thus collateral damage is reduced because only cells with activated death programs are affected. It is important to remember that apoptotic and necrotic mechanisms of cell death can occur concomitantly or sequentially, and thus are part of a continuum of effects associated with dose-dependent alterations in cellular energy production and the differential sensitivity of neuronal subtypes.

Many neurotoxicants produce their effects by promoting cell death in neurons. Neurotoxicant-induced cytotoxicity has been associated with the pathogenesis of a number of neurodegenerative disorders, including Alzheimer's disease, Parkinson's disease, and amyotrophic lateral sclerosis (ALS), and the weight of evidence suggests that toxicant exposure is a risk factor for these diseases. One area of intense research focus has been the toxic effects of excessive signaling by glutamate and other excitatory amino acids (EAAs), and the role that EAAs may play in neurodegenerative disorders. Glutamate activates ion channel receptors that open to allow influx of calcium and other ions into the neuron. This influx of ions, combined with other second messenger events that promote further intracellular release of calcium, contribute to calcium overload. Signaling cascades are then activated in response to the intracellular calcium, and these pathways eventually lead to activation of oxidative stress and programmed cell death (apoptotic) pathways. EAA-mediated neuronopathic injury has been extensively studied for its role in ischemic and seizure-induced brain damage. It is furthermore thought that the glutamate receptor agonist, domoic acid, a toxin produced by algae, is responsible for several outbreaks of shellfish poisoning. In one incident in 1987, several people died, and dozens became ill with dizziness, seizures, and memory loss after consuming blue mussels. The mussels were contaminated with domoic acid, found in high levels following an algae bloom near Prince Edward Island, Canada. More recently domoic acid produced by algae blooms has been blamed for the abnormal behavior and deaths of pelicans, cormorants, and sea lions on the California coast.

## 16.3.2   Effects of Toxicants on Other Cells

Toxicants may selectively target glial cells for a number of reasons. Myelinating glial cells constantly synthesize cholesterol and cerebroside for myelin production; thus toxicants that affect these synthetic pathways will preferentially affect myelination. The hydrophobic nature of myelin may serve as a reservoir for lipophilic toxicants as well. The specialized structures of cells that form myelin also present unique challenges to cellular homeostasis, increasing the vulnerability of these cells to toxicant action.

In general, a toxic or physical insult to either neurons or glial cells eventually leads to changes in the other cell type. Because of this, it is often difficult to determine whether the primary insult was neuronal or glial. Metals such as lead, cadmium, and aluminum are capable of inducing cell death in cultured astrocytes and endothelial cells. Methylmercury preferentially accumulates in astrocytes and to some extent in microglia, causing cellular swelling. The swelling is presumably the effect of methylmercury interfering with ion channels, because ion channel blockers can reverse this effect. Astrocytes are important reservoirs of excess glutamate, and swollen astrocytes release glutamate in and around synapses, potentially causing the excitotoxicity described above. Astrocyte swelling also has effects on brain blood flow, since astrocyte end-feet surround the blood vessels of the CNS. Not only does swelling result in reduced lumen size, but the distances substrates and waste products must diffuse to reach the bloodstream is increased.

Astrocytes and microglial cells become activated secondarily to brain injury, such as acute trauma or toxic lesioning. This activation, known as "reactive gliosis" is characterized by astrocyte hypertrophy and hyperplasia. Reactive astrocytes have greatly enlarged cytoplasmic processes, and produce increased amounts of a protein known as glial fibrillary acidic protein (GFAP). GFAP is often used as a quantitative histochemical marker for toxicant-mediated injury in the nervous system.

In addition to their function as phagocytic cells, glial cells produce neurotrophic factors to prevent neuronal death and promote axonal growth after injury. Glial cells also have xenobiotic biotransforming enzymes, but in some cases glial metabolism results in xenobiotic activation to a toxic metabolite. As discussed above, MPP+ is the toxic metabolite of MPTP. MPTP is taken up by astrocytes, where monoamine oxidase B converts MPTP to MPP+. While MPP+ seems to cause no damage to the astrocytes themselves, the astrocytes release this reactive metabolite into synapses, where it is selectively taken up by dopamine re-uptake transporters on the endings of neurons that normally release and recycle dopamine. MPP+ then kills the dopaminergic neurons, and this insult to the movement-controlling brain circuitry results in the classic motor symptoms of Parkinson's disease. The properties of MPTP have only become known since the 1980s, when its accidental ingestion by heroin addicts resulted in acute Parkinson-like symptoms. These incidents spurred multiple investigations, leading to much of what is known today about the pathogenesis of Parkinson's disease.

### 16.3.3 Toxicant-Mediated Alterations in Synaptic Function

Nervous system function may be adversely affected by neurotoxicants without necessarily causing structural damage to tissue. In many cases neurotoxicants interfere with signaling processes within the nervous system by activating or inhibiting receptors, or altering the amount of neurotransmitter available to activate receptors. This type of neurotoxicity is illustrated by the well-characterized actions of the organophosphates and carbamates on acetylcholine signaling.

Organophosphates inhibit acetylcholinesterase, the enzyme responsible for breaking down acetylcholine into acetic acid and choline. After acetylcholine has been released into the synapse or the neuromuscular junction, acetylcholinesterase terminates receptor-stimulating activity by binding acetylcholine in its active site. Separate sites within the binding pocket of acetylcholinesterase bind the quaternary nitrogen of the choline group, and the carbonyl of the ester group. A hydrolytic reaction results in the loss of choline, leaving an acylated serine residue, which is then rapidly hydrolyzed. The biologically active oxon forms of organophosphates also bind to the active site of acetylcholinesterase, covalently phosphorylating the serine residue in the catalytic site of the enzyme. The phosphorylation of acetylcholinesterase creates a relatively stable inactive enzyme that persists for hours to days before hydrolysis of the phosphate moiety occurs spontaneously, and acetylcholinesterase activity is restored. The rate of spontaneous hydrolysis is increased with larger alkyl groups attached to the phosphate moiety. When one or more of these alkyl groups is lost, in a process known as "aging," spontaneous reactivation of acetylcholinesterase by hydrolysis of the phosphate moiety is impossible, and the enzyme is permanently inactivated. Carbamates similarly inhibit acetylcholinesterase by carbamylating the enzyme active site. The stability of carbamylation is much less than phosphorylation, however, and spontaneous reactivation thus occurs faster than with organophosphates.

The effects of acetylcholinesterase inhibition can be seen throughout the nervous system. Acetylcholine and its receptors mediate neurotransmission in sympathetic and parasympathetic autonomic ganglia, in the effector organs where autonomic nerves terminate, in neuromuscular junctions, and in the brain and spinal cord. The signs of hypercholinergic activity are thus very diverse, and include effects mediated by both nicotinic and muscarinic types of acetylcholine receptor. Hyperstimulation of nicotinic receptors in neuromuscular junctions results in muscle weakness, in rapid, localized contractions called fasciculations, and in paralysis. Nicotinic receptors are also found in sympathetic and parasympathetic ganglia, and so stimulation of both divisions of the autonomic system is apparent as hypertension, increased heart rate, and papillary dilation. Muscarinic receptors in the PNS mediate postganglionic parasympathetic effects on the smooth muscle present in the end organs such as the lung, gastrointestinal tract, eye, bladder, and secretory glands. Hyperstimulation of these receptors results in a pattern of toxicity known by the mnemonic SLUDGE (salivation, lacrimation, urination, defecation, GI upset, emesis). Bronchospasm and bradycardia are also muscarinic effects. In the CNS, confusion, anxiety, restlessness, ataxia, seizures, and coma are effects of both muscarinic and nicotinic receptor overstimulation. Death generally occurs from respiratory paralysis.

Treatment for toxicity by organophosphates and carbamates is directed at counteracting hyperstimulation and regenerating acetylcholinesterase enzymatic activity. Atropine is a muscarinic receptor antagonist (it blocks acetylcholine from binding to the muscarinic receptor), and is used to counteract the effects of cholinergic overactivity. Atropine has no effect at the nicotinic receptor, however, so the skeletal muscular and some of the sympathetic effects of cholinergic hyperstimulation will remain after administration of atropine. Inhibition of acetylcholinesterase activity by organophosphates can be reversed by administration of oxime compounds (e.g., pralidoxime and 2-PAM). These compounds contain a quaternary nitrogen that binds to the choline binding site of acetylcholinesterase, positioning the oxime portion of the molecule near the esteratic site. Oximes are themselves reversible inhibitors of acetylcholinesterase, but their mechanism of organophosphate reversal is by attack of the covalent phosphoserine bond, releasing the phosphate group. Oximes are not effective on dealkylated or "aged" enzymes, so they must be administered soon after organophosphate intoxication in order to be effective. They are also ineffective against carbamate-mediated toxicity, and some researchers believe they actually worsen carbamate effects by stabilizing the carbamylation of the enzyme.

Many biological toxins produce hyperstimulation of receptors by directly binding and activating them (agonism), or reduce receptor stimulation by prohibiting the endogenous ligand from activating them (antagonism). A number of snake and spider venoms, mushroom and plant alkaloids, affect nervous system function by these mechanisms. As the binding of receptors by these agents is usually reversible, their effects are reversible as well (although some may still cause death by massively altering neuronal signaling). Beyond the receptor, an active area of current research is the role of intracellular signaling molecules in mediating the effects of neurotoxicants. The effects of a number of metals, in particular, may be related to their ability to act as cofactors for proteins involved in intracellular signaling. To date, however, few signal transduction molecules have been shown to be directly affected by neurotoxicants. Notable exceptions are cholera and pertussis toxins, which selectively target G proteins, but their primary effects are on the gastrointestinal and respiratory systems, respectively, rather than on the nervous system.

On the other hand, the Clostridium toxins, botulinum (causing botulism) and tetanospasmin (causing tetanus), block neurotransmission by inhibiting release of neurotransmitter into synapses and at motor end-plates in muscle. Both of these agents are structurally similar proteases, but the effects they cause are vastly different. Botulinum toxin enters presynaptic motor neurons in the PNS, where it cleaves proteins that are involved in the fusion of synaptic vesicles with membranes. This cleavage results in the inhibition of acetylcholine release from the presynaptic terminal, and thus muscles cannot be stimulated to contract. The clinical result of botulinum intoxication (usually by ingestion) is a flaccid paralysis. Recovery occurs when the presynaptic neuron sprouts new nerve endings that contact the muscle and create new motor end-plates. Tetanus toxin causes a completely different clinical picture, even though its substrate specificity for cleavage of proteins is very similar. Once taken up into the presynaptic nerve endings, tetanus associates with endosomes, and like endosomes, is transported retrogradely toward the neuron cell body. Tetanospasmin then continues its trek to the dendritic regions of the neurons, where it is released, again retrogradely, into synapses. Usually these synapses are within the spinal cord, where interneurons send an inhibitory signal (via the neurotransmitters glycine and GABA) to motor neurons to slow their activity and prevent massive muscular contraction. The presynaptic membranes of the interneurons take up tetanospasmin, and this is where most of its activity occurs. By cleaving release-regulating proteins in the interneuron terminal, tetanospasmin prevents the release of glycine and GABA onto the motor neurons. The motor neurons then become hyperactive, and this results in overstimulation of the motor end-plate with acetylcholine. Clinically this results in a spasms, stiffness, and whole-body paralysis. Again, the interneurons themselves do not die, but they must form new synapses with the motor neurons. Fortunately, in all but the most severe cases, recovery is complete. The reformation of new synapses by neurons, even in the CNS, is an example of the remarkable plasticity of the nervous system. The continual formation and reformation of synaptic connections allows the organism to change and adapt to an inconstant environment.

## 16.4  NEUROTOXICITY TESTING

A large number of the chemicals used in industry today remain poorly characterized with respect to their toxic effects on the nervous system. In order to determine potential risks to human and environmental well-being, existing neurotoxicants must be identified, and the approximately 2000 new chemicals introduced each year must be screened for their potential neurotoxic effects. Often, a tiered approach is used, with the first tier consisting of general screening tests to identify acute hazards. The Environmental Protection Agency (EPA) has proposed screening guidelines for tests in rodents that include a functional observational battery (FOB, see below) to evaluate sensory, motor, and autonomic effects, tests that identify changes in motor activity, and neuropathological assessment. Interpretation of the outcome of tier 1 screening may lead to more selective testing and examining the effects of repeated exposures in the second tier. Specialized tests for behavioral effects, developmental neurotoxicity, or delayed organophosphate effects may be required. If necessary, a third tier of testing characterizes mechanisms of neurotoxicant-induced injury. Complete and comprehensive evaluation of potential neurotoxicant effects requires that data from different types of sources be considered; this can range from molecular interactions to whole animal and human exposure analysis. Below are examples of techniques commonly used for testing neurotoxic effects.

### 16.4.1  In vivo Tests of Human Exposure

Historically the first indication of neurotoxic potential by a chemical has often followed accidental human exposure in the workplace. Case reports of incidents involving individuals, or clusters of individuals, are useful for documentation but generally provide a limited amount of information about the specific details of an exposure. Procedures included in most case reports include a patient medical history and clinical neurological exam, sometimes supplemented with psychiatric or neurophysiological tests, and/or neuroimaging. Although the specific tests involved vary depending on the clinician, most basic clinical neurological exams rely heavily on evaluation of mental status (level of consciousness, orientation, mood, etc.) and sensorimotor function (gait, coordination, muscle tone, sensitivity to touch, reflexes).

Human epidemiological studies generally represent a deeper investigation into the causal relationship between an exposure and neurotoxicological effects. Some of the methods used to identify neurotoxic effects in epidemiological studies include behavioral assessments, neurophysiological evaluations, and neuroimaging techniques. Neurobehavioral assessments examine a variety of psychological and cognitive functions such as mood, attention, memory, perceptual and visuospatial ability, and psychomotor performance. In an effort to standardize neurotoxicological testing of human behavioral effects, particularly for studies involving worksite exposure, the World Health Organization (WHO) and the US National Institute for Occupational Safety and Health (NIOSH) devised a the Neurobehavioral Core Test Battery (NCBT). The NCBT (Table 16.1) consists of seven tests that were shown previously to be sensitive indicators of neurotoxicant exposure. The battery is designed to be administered one on one by an examiner. Although this battery has a relatively narrow

**Table 16.1  The WHO Neurobehavioral Core Test Battery (NBCT)**

| Domain | Analysis | Test | Task |
|---|---|---|---|
| Psychomotor performance | Motor speed, motor steadiness | Pursuit aiming | Follow a pattern of small circles, placing a dot in each circle around a pattern; subject's score is number of taps in circle within one minute. |
| | Manual dexterity, hand–eye coordination | Santa Ana Dexterity Test | Perform skillful movements with hands and arms. |
| Perceptual coding and perceptual motor speed | | Wechsler Digit Symbol Test | Each number in a list is associated with a simple symbol. On a list of random digits with blank spaces below them, write the correct symbols in blank spaces as fast as possible. |
| Attention and short-term memory | Attention and response speed | Simple reaction time | Test reactions of hands or feet from visual and auditory signals. |
| | Visual perception and memory | Benton Visual Retention Test | Recall and reproduce figures. |
| | Auditory memory | Wechsler Digit Span Test | Recall digits in series forwards and backwards immediately after hearing them. |
| Mood and affect | | Profile of Mood States | Evaluate, by questionnaire, anger, tension, confusion, depression, etc. |

focus, primarily on the effects most commonly seen in CNS toxicity, it also provides suggestions for the selection of further testing depending on the exposure setting. The NCBT has been widely used because of its ease of administration, relatively low cost, and its large base of control data. A broader battery of cognitive and psychomotor tests that is often used is the Neurobehavioral Evaluation System (NES). The NES consists of a combination of automated (computerized) and hand-administered tests. The sensitivity of the NES to effects caused by neurotoxicants in industrial settings has been validated internationally.

Neurobehavioral examinations are useful for identifying neurotoxicant-mediated deficits, but it is often difficult to localize the site of toxic action from such tests. For example, sensorimotor tests of reaction time, manual dexterity, hand-eye coordination, and finger tapping can indicate either neuromuscular or psychomotor damage.

The results of these tests thus should be interpreted in the context of other experiments. For example, electrophysiological techniques can help to focus an investigation to the site of the lesion, and characterize electrical dysfunction within the damaged nerves. Electrophysiological nerve conduction studies can distinguish between proximal and distal axonal lesions in peripheral nerves and can be performed noninvasively (i.e., with skin surface electrodes). Characteristic changes in the velocity, duration, amplitude, waveform, or refractory period of peripheral nerves may be detected, depending on the agent. Evoked potentials represent another useful electrophysiological endpoint. These procedures measure the function of an entire system, such as the visual, auditory, or motor systems. The specific pathway is stimulated by an evoking stimulus, such as a flash of light or electrical nerve stimulation. The evoked potentials are read as changes in ongoing electroencephalograms (EEGs) in response to the stimulation. Thus the activity of the entire neural circuit is evaluated in the brain after peripheral stimulation. Evoked potentials can be very sensitive indicators of changes in neural activity when performed in a carefully controlled environment, and when interpreted in light of behavioral or other physiological findings.

An increasingly popular method of documenting brain pathology is the use of neuroimaging methods. Computerized axial tomography (CAT) and magnetic resonance imaging (MRI) can produce images of the brain that can show structural changes in the volume or density of a specific region or ventricle. Other techniques, such as positron emission tomography (PET) and single photon emission computerized tomography (SPECT), use radioactive tracer molecules to determine functional biochemical changes in processes like glucose utilization or receptor binding. The number of cases so far analyzed with neuroimaging techniques is still relatively small, and thus specific toxicant-mediated effects are not well characterized. Nevertheless, this growing field promises to contribute significantly to neurotoxicity studies in the future.

### 16.4.2   In vivo Tests of Animal Exposure

The primary approach currently used to detect and characterize potential neurotoxicants involves the use of animal models, particularly rodents. Behavioral and neurophysiological tests, often similar to the ones used in humans, are typically administered. The sensitivity of these measures to neurotoxicant exposure is widely accepted. Although it is often not possible to test toxicant effects on some higher behavioral functions in animals (e.g., verbal ability, cognitive flexibility), there are other neurobehavioral outcomes such as memory loss, motivational defects, somatosensory deficits, and motor dysfunction that can be successfully modeled in rodents. These behaviors are based on the ability of the nervous system to integrate multiple inputs and outputs, thus they cannot be modeled adequately in vitro. Although the bulk of neurotoxicity data has been collected in rodents, birds and primates are also used to model human behavioral outcomes.

As mentioned above, a useful screening tool for neurotoxicant exposure is a battery of observational tests of function known as an FOB. FOBs, like the one developed by the EPA, are used to detect overt changes in behavior and physiology of animals exposed to neurotoxicants. In the typical exam, an observer documents cageside observations regarding the appearance and activity of the animal. Then the animal is handled and examined for obvious signs such as lacrimation, salivation, or piloerection. Pupillary light responses and temperature are recorded, and the ease of handling the animal is also noted. The animal is then placed in an open field, such as the top of a laboratory cart, and observed for a set period of time, during which the observer records exploratory behaviors, excretion rate, and whether or not there are any motor abnormalities. A number of manipulations are then performed to assess hearing, sensitivity to touch, righting reflex, coordination, and grip strength. The FOB used in the EPA guidelines has shown remarkable consistency in detecting chemical effects in diverse testing laboratories and situations. A test of motor activity can be administered along with the FOB, consisting of quantitative evaluations of the animal's movement in either an open field or a maze. Motor activity tests reflect integrative abilities of the nervous system to process sensory input, association, and motor output. A number of agents, such as toluene, triadimefon, and chlorinated hydrocarbons, increase or decrease motor activity in a toxicant-specific manner, unrelated to their general effects on the health of the animal.

More in-depth behavioral tests are required if dose-related toxicant effects are noted in screening tests. These tests may also be required as part of more selective toxicological screening, such as for developmental neurotoxicity. Focused tests of neuromotor function and activity, sensory functions, memory, attention, and motivation help to identify sites of toxicant-mediated lesioning, aid in the classification of neurotoxicants, and may suggest mechanisms of action. Some of these tests, like the schedule-controlled operant behavior tests for cognitive function, require animal training and extensive operator interaction with the animals.

### 16.4.3   In vitro Neurochemical and Histopathological End Points

In vitro methods for studying neurotoxicant effects are a valuable supplement to whole animal and human testing, allowing the researcher to supplement findings, test hypotheses, and reduce the number of animals used for toxicity testing. Much of the neurochemical and histopathological data on neurotoxicant effects in humans and animals is gathered concomitantly with, or immediately after, performing behavioral tests. This may involve collection of bodily fluids or samples from living subjects for the purpose of analyses such as red blood cell or plasma acetylcholinesterase or NTE activity, determination of hormone or neurotransmitter concentration, or detection of the presence of toxicant or metabolite in the cerebrospinal fluid.

Postmortem tissues can provide a wealth of information about the location, timing, extent, and mechanism of neurotoxicant-induced damage. For example, changes in the gross morphology and weight of brain or nerves may be seen at higher levels of toxicant exposure. Microscopically, fixed and stained tissues reveal the type of damage to target cells, such as axonopathic or demyelinating lesions. Degenerative changes in cells may be indicative of the injurious process, and whether cells are dying by necrosis or apoptosis. Typical stains such as cresyl violet (Nissl stain) and silver (Golgi stain) are useful for cell morphology and counting. Other stains are selective for damaged cells, like the specialized silver impregnation techniques that are frequently used to identify neurotoxicant-mediated degeneration of neurons.

Tissue sections may also be processed for immunohistochemical staining. A frequently used immunochemical marker for neuropathologic insult is glial fibrillary acidic protein (GFAP). GFAP is produced in large amounts by reactive astrocytes that proliferate in response to tissue injury. Stress proteins, apoptotic signals, and immediate early genes are also utilized as markers of neuronal activity and injury. Other protein markers can be used to quantitatively identify specific types of neurons, which may be reduced in numbers after selective neurotoxicant-induced cell death. For example, tyrosine hydroxylase (TH) is an enzyme involved in dopamine synthesis and, as such, is selectively expressed in dopamine-containing neurons. Loss of TH immunoreactivity is used to identify dopaminergic cell death after administration of the neurotoxicant MPTP.

In homogenized tissue preparations, mechanistic information can be obtained from analyzing tissue levels of neurotransmitter and metabolites, signaling proteins, and receptor-binding affinities. Protein and lipid peroxidation and oxygen radical formation are commonly seen with toxicants that target mitochondrial function. Neurotoxicants may alter the levels or activation state of many proteins, including kinases, phosphatases, and proteases, quantifiable with activity or immunological techniques.

Cell culture protocols are a useful adjunct to neurotoxicity testing. Individual cell lines are particularly well suited for identifying selective cellular and molecular toxicity and for studying the mechanistic aspects of neurotoxicant injury. Clonal cells lines, as well as primary cultures of neurons or glial cells, may be used, and the choice of cell type or particular clonal line depends on the particular end points under study. For example, if a researcher wished to study the effects of a given neurotoxicant on neurotransmitter release, she might choose the rat pheochromocytoma PC12 cell, which releases catecholamine neurotransmitter upon stimulation with a variety of agents. The relative inexpensive and ease of manipulating exposure make cellular techniques an attractive alternative for many types of studies. Cultured cell studies cannot, however, reproduce systemic metabolic and kinetic effects, or mimic the complex neuronal circuitry that is present in vivo. Thus, while cell studies provide a vehicle for in-depth examination of the nature of toxicant-cellular interactions, extrapolation to in vivo conditions is often not possible.

## 16.5  SUMMARY

The nervous system is at once unique in structure and staggeringly complex, exquisitely sensitive, yet capable of amazing adaptability. Because of these attributes, the neurotoxic potential of many agents, to say little of their underlying mechanisms, remains unknown. Particularly concerning are the possibilities that chronic low levels of chemical exposure are having an effect on the behavioral development of children, and contributing subtly to neurodegenerative diseases in the elderly. The huge task of testing natural and synthetic chemicals for neurotoxic effects has been facilitated in recent years with the development of behavioral testing batteries, advances in pathological and biochemical techniques, and a more focused attention of regulatory agencies on issues relating to neurotoxicology.

## SUGGESTED READING

Harry, J., B. Kulig, M. Lotti, H. Tilson, and G. Winneke, eds. *Neurotoxicity Risk Assessment for Human Health*. Environmental Health Criteria 223. Geneva: World Health Organization, 2001.

Massaro, E., ed. *Handbook of Neurotoxicology*. Totowa, NJ: Humana Press, 2002.

Neurotoxicity: Identifying and Controlling Poisons of the Nervous System. Washington, DC: Congress of the US Office of Technology Assessment, 1990.

Tilson, H., and J. Harry, eds. *Neurotoxicology*. Philadelphia: Taylor and Francis, 1999.

# CHAPTER 17

# Endocrine System

GERALD A. LEBLANC

## 17.1 INTRODUCTION

Among the various organ systems of the body, the endocrine system is somewhat unique. While most systems are associated with a specific physiological task (respiration, reproduction, excretion, etc.), the endocrine system functions to regulate many of the activities associated with these other systems. Accordingly the endocrine system is integral to the maintenance of total normal bodily function (homeostasis), and disruption of normal endocrine function by exogenous chemicals can result in multiple, diverse, and dire consequences. Toxicity to the endocrine system is most commonly associated with altered development, growth, maturation, and reproduction (Table 17.1). However, endocrine toxicity also can present as gastro-intestinal dysfunction, malaise, neurological and other disorders (Table 17.1). This is why endocrine toxicity often can be misconstrued as toxicity to some other endocrine-regulated system of the body.

The endocrine system, as an authentic target of chemical toxicity, tragically entered the limelight as a consequence of the widespread use of the drug diethylstilbestrol (DES). DES, a nonsteroidal synthetic estrogen, was prescribed to pregnant woman from the 1940s to the 1960s as a prophylactic against miscarriage (see Section 17.4.1). Following the discovery of the endocrine toxicity of this drug, many additional drugs and environmental chemicals have been shown to mimic the action of hormones or interfere with their hormonal function. These activities often have been clearly shown, in laboratory studies, to result in endocrine-related toxicity. In some instances drug use or exposure to ambient environmental chemicals has been shown to result in endocrine toxicity. Such examples will be presented at the end of this chapter.

## 17.2 ENDOCRINE SYSTEM

The endocrine system can be broadly described as *an assemblage of organs (glands) that produce chemical messengers (hormones) that regulate various bodily functions.* The bodily functions regulated by the endocrine system can be categorized as those

*A Textbook of Modern Toxicology, Third Edition,* edited by Ernest Hodgson
ISBN 0-471-26508-X Copyright © 2004 John Wiley & Sons, Inc.

**Table 17.1  Processes Regulated by Some Hormones of the Endocrine System Susceptible to Disruption by Endocrine Toxicants**

| Hormone Group | Example | Origin | Regulated Process |
|---|---|---|---|
| Androgens | Testosterone | Testes, adrenals | Sexual differentiation, fertility, secondary sex characteristics, sexual function, libido |
| Estrogens | 17β-Estradiol | Ovaries, testes | Sexual differentiation, fertility, secondary sex characteristics, bone density maintenance, blood coagulation |
| Glucocorticoids | Cortisol | Adrenals | Bone formation, wound healing, growth, development |
| Thyroid hormones | Thyroxine | Thyroid gland | Fetal brain and bone development, oxygen consumption, gut motility |

involved in the maintenance of homeostasis and those involved in physiological progression. Functions regulated by the endocrine system resulting in homeostasis include maintenance of the reproductive system, energy production, and metabolism. Functions regulated by the endocrine system resulting in physiological progression include fetal development, growth, and maturation. Endocrine processes related to physiological progression historically have received the greatest attention in endocrine toxicology and will be emphasized in this chapter.

Both the maintenance of homeostasis and the regulation of physiological progression require that the endocrine system detect signals, either external or internal, and transduce these signals to the appropriate target sites within the body. These target sites then respond in the appropriate manner to maintain homeostasis or institute change related to development, maturation, and so on. In many species these initial signals are of external origin. For example, many species initiate reproductive maturation in response to changes in environmental temperature and day length. Reproductively mature organisms often respond to external visual or olfactory stimuli produced by sexually receptive individual to initiate sexual behavior.

The signal to be transduced by the endocrine system initiates in the central nervous system. In mammals, the hypothalamus commonly initiates the endocrine signaling pathway by secreting peptide hormones. These neuro-endocrine hormones can be rapidly synthesized, secreted, and degraded to allow near-instantaneous, short-lived responses to the stimulatory signal. Accordingly they can be present in the body in pulses and secretory rhythms that often contribute to their signaling function. For example, the hypothalamic peptide hormones "growth hormone releasing hormone" (GHRH) and somatostatin are secreted in an alternating pulsatile fashion. Both hormones target the pituitary gland, though GHRH stimulates and somatostatin inhibits growth hormone secretion by the pituitary. As a result the secretory pattern of the secondary hormone messenger in this cascade, growth hormone, is highly controlled. Disruption of this rhythm in rodent models can alter hepatic enzyme expression and other dynamic processes. Disruption of the growth hormone secretory rhythm associated with sleep has been shown to interfere with normal growth in children. Hormone secretory rhythms have been associated with other physiological processes including sleep, sexual behavior, and ovulation.

Endocrine signaling pathways from the central nervous system to the target organ typically occur along axes (Figure 17.1). An axis is defined by the endocrine glands that produce signaling hormones along the cascade (i.e., hypothalamic–pituitary–gonadal axis), and sometimes, a terminal target organ of the signaling pathway (i.e., hypothalamic–pituitary–gonadal–hepatic axis).

Endocrine signaling cascades offer several advantages over a single hormone signaling strategy. Cascades provide several sites at which the signal can be regulated thus ensuring maintenance of the appropriate endocrine signal (Figure 17.2). For example, testosterone is secreted by the testis but regulates its own secretion by acting upstream in the axis at the pituitary gland and hypothalamic gland. Signaling cascades also

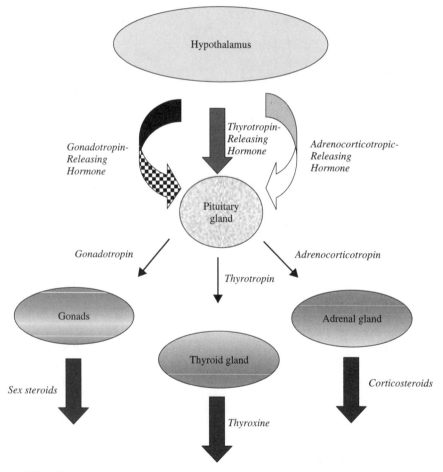

**Figure 17.1**  Some major neuro–endocrine axes that transduce endocrine signals to target organs. Neuro–endocrine signaling is initiated by the secretion of releasing hormones, or in some instances inhibiting hormones, that regulate secretion of the secondary hormone signal by the pituitary. Pituitary hormones then regulate secretion of the tertiary hormone, often a steroid hormone, by the appropriate endocrine gland. The tertiary hormones then stimulate gene transcription at target organs.

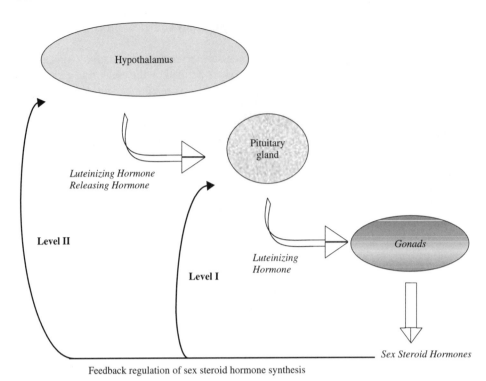

Feedback regulation of sex steroid hormone synthesis

**Figure 17.2** The hypothalamic–pituitary–gonadal axis. Endocrine signaling cascades provide multiple sites for regulation and ensure optimum signaling.

utilize multiple hormones with differing properties to contribute to the process. Peptide hormones are commonly the intermediate messengers along a signaling cascade, while the terminal hormone is often of nonpeptide origin (i.e., steroids). Peptide hormones offer advantages as intermediate messengers in that they can be rapidly synthesized and degraded (i.e., turned "on" and "off"). Peptide hormones also do not require cell entry to elicit activity but rather bind to cell surface receptors. This facilitates a rapid physiological response to the hormone. Steroid and other nonpeptide hormones are typically more stable, they are maintained in circulation at a relatively constant, physiologically appropriate level, they can be stored as precursor molecules or apolar conjugates, they can be mobilized as polar conjugates, and most often, they require cell entry to interact with its receptor and elicit a response. Accordingly the nonpeptide terminal hormones offer the advantages of constant availability but lack the advantages of rapid modulation.

### 17.2.1 Nuclear Receptors

Toxicologically the function of the terminal hormones of endocrine cascades (i.e., steroid, retinoid, thyroid hormones) appear to be most susceptible to disruption by chemicals. This is because many foreign molecules share sufficient characteristics with these hormone molecules to allow binding to the nuclear receptors of these hormones in either an agonistic or antagonistic fashion. The binding of the xenobiotic to the receptor results in aberrant receptor function with associated toxicological outcome. The nuclear receptors are so called since these receptors initiate their classical physiological responses within the cell nucleus. Cell surface receptors to peptide hormones, on the other hand, can likely discriminate between peptide molecules and nonpeptide xenobiotics thus minimizing the likelihood of interaction and associated disruption of function.

The nuclear receptor superfamily consists of members of the steroid receptor family and the thyroid receptor family (Figure 17.3). Members of these two receptor families are distinct in many structural and functional attributes. Steroid receptor family members typically exist in the extranuclear matrix of the cell in association with various accessory proteins (*hsp*90, *hsp*70, *hsp*56). These accessory proteins stabilize the receptor molecule and help maintain the molecule's integrity. Binding of hormone ligand to the receptor protein stimulates dissociation with the accessory proteins, homodimerization of two receptor molecules, and nuclear localization (Figure 17.4). Here the receptor complex interacts with hormone response elements (Table 17.2) that are associated with hormone responsive genes, and transcription of these genes is regulated.

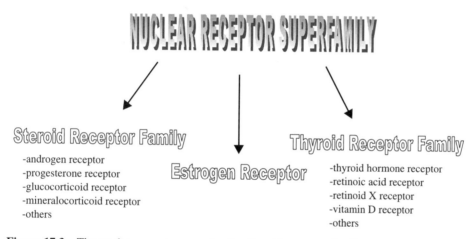

NUCLEAR RECEPTOR SUPERFAMILY

Steroid Receptor Family

-androgen receptor
-progesterone receptor
-glucocorticoid receptor
-mineralocorticoid receptor
-others

Estrogen Receptor

Thyroid Receptor Family

-thyroid hormone receptor
-retinoic acid receptor
-retinoid X receptor
-vitamin D receptor
-others

**Figure 17.3** The nuclear receptor superfamily. Steroid receptor family members and thyroid receptor family members differ in several structural and functional properties. The estrogen receptors share properties with both steroid and thyroid receptor families and are likely an evolutionary precursor to both families.

**Table 17.2  DNA Recognition Sequences Utilized by Steroid, Estrogen, and Thyroid Receptor Family Members**

| Receptor Family | Recognition Sequence |
|---|---|
| Steroid | AGAACA ... TGTTCT |
| Estrogen | AGGTCA ... TGACCT |
| Thyroid | AGGTCA ... AGGTCA |

*Note*: Recognition sequences are split by spacer nucleotides, which are denoted by the dots.

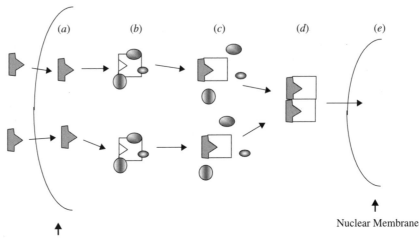

Cell Surface Membrane

Nuclear Membrane

**Figure 17.4**   Intracellular steroid receptor activation by hormone ligands. (*a*) Steroid hormones diffuse across the cell membrane into the cell. (*b*) Steroid hormone receptors in the basal state bound to accessory proteins. (*c*) Steroid hormones bind to receptors and accessory proteins are dissociated from the receptors. (*d*) Hormone : receptor complexes dimerize. (*e*) Dimer complexes enter the nucleus and initiate transcription of responsive genes.

In contrast to the steroid receptor family, members of the thyroid receptor family typically do not associate with accessory proteins and are not localized to the extranucleus matrix. Rather, these receptors exist in the basal state associated with chromatin in the cell nucleus. When bound by hormone ligand, thyroid receptor family members dissociate from the chromatin and typically form heterodimeric combinations with the retinoid-X receptor (RXR). RXR also is capable of homodimerization in association with its ligand 9-*cis* retinoic acid. Thus high 9-*cis* retinoic acid levels apparently promote homodimerization, and low levels are permissive of heterodimerization of RXR with activation by the partner ligand.

The estrogen receptor shares structural and functional attributes of both nuclear receptor families. For example, the estrogen receptor resides in the cell nucleus, but it associates with accessory proteins rather than chromatin. The estrogen receptor shares a high degree of sequence homology with the thyroid receptor family members but does not heterodimerize with RXR. Rather, active estrogen receptor exists as a homodimer. The observation that the estrogen receptor shares attributes of both nuclear receptor families supports nucleotide sequence evidence that the estrogen receptor is an ancestral precursor to both receptor families.

### 17.2.2  Membrane-Bound Steroid Hormone Receptors

Some cellular responses occur too rapidly following steroid hormone exposure to involve the multi-step process of nuclear receptor activation. For example, 17$\beta$-estradiol can rapidly stimulate adenylate cyclase and cause a near-instantaneous increase in intracellular cAMP in cultured prostate cells. These effects are mediated by the interaction of steroid hormones with cell surface proteins.

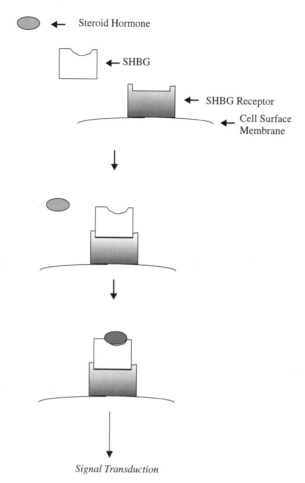

**Figure 17.5** Endocrine signaling pathway involving steroid hormone, sex hormone binding globulin (SHBG), and the SHBG receptor.

Due to their lipophilic nature, steroid hormones are mobilized in the circulatory system by transfer proteins. Sex hormone-binding globulin (SHBG) is one such transfer protein that binds testosterone, $17\beta$-estradiol, and other sex steroids. Roughly half of circulating testosterone and $17\beta$-estradiol is bound to SHBG. Receptors exist on the surface of some cells that are capable of binding unliganded SHBG (Figure 17.5). Unliganded SHBG, which is bound to the cell surface receptor, can subsequently bind steroid hormone. Binding of an appropriate hormone to the SHBG then stimulates a signal-transduction pathway within the cell. Some steroid hormones ($17\beta$-estradiol, $5\alpha$-androstan-$3\alpha$,$17\beta$-diol) function as SHBG : SHBG-receptor agonists, while others (testosterone, $5\alpha$-dihydrotestosterone) function as antagonists. Interestingly $5\alpha$-androstan-$3\alpha$,$17\beta$-diol had previously been considered an inactivation product of the potent androgen $5\alpha$-dihydrotestosterone (DHT). Studies in human prostate cells have shown that activation of this SHBG-dependent pathway stimulates DNA synthesis and cell growth. These observations, in combination with studies in dogs that have shown $5\alpha$-androstan-$3\alpha$,$17\beta$-diol to stimulate benign prostatic hyperplasia, have led to suggestions that the SHBG-receptor pathway is involved in this disease condition. The susceptibility of these membrane-bound receptor pathways in endocrine toxicity has received little attention, though conceivably toxicants could perturb these pathways by competing with endogenous hormone for binding to SHBG, resulting in the loss of stimulatory activity (antagonists) or inappropriate stimulation of activity (agonists).

## 17.3   ENDOCRINE DISRUPTION

Xenobiotics have the ability to disrupt hormone activity through a variety of mechanisms, though the predominant mechanisms appear to involve binding to the hormone receptor, either as an agonist or antagonist, or by modulating endogenous steroid hormone levels.

### 17.3.1   Hormone Receptor Agonists

A hormone receptor agonist is defined as *a compound that binds to and activates a hormone receptor.* Endogenous hormones function as agonists to their respective receptors. Xenobiotics can act as receptor agonists and stimulate receptor-dependent physiological processes in the absence of the endogenous receptor ligand (hormone). Such inappropriate stimulation can result in the errant expression of hormone-dependent processes such as breast development in males (gynecomastia).

***Estrogen Receptor.*** Among the steroid hormone receptors, the estrogen receptor appears most susceptible to the agonistic action of xenobiotics. Estrogen receptor agonists are quite diverse in molecular structure (Figure 17.6). Several steric considerations, associated with the steroid structure, in conjunction with electrostatic (charge) properties of the outer surface of the molecule seem to dictate whether a xenobiotic can fit into the binding-pocket of the receptor and function as a receptor agonist. It is not clear why the estrogen receptor would be more susceptible to the agonistic action of xenobiotics as compared to other steroid hormone receptors. High ligand specificity may be an evolutionary trait of most "modern" receptors that is not associated with this presumed ancestral precursor to other nuclear receptor superfamily members. The estrogen receptor is often referred to as a promiscuous receptor because of this susceptibility to agonistic interactions with xenobiotics.

Some drugs are rather potent estrogens (i.e., DES); however, environmental chemicals with estrogenic activity are typically weak agonists with activity several orders-of-magnitude less than that of $17\beta$-estradiol (Table 17.3). Because of this weak activity, xenoestrogens are typically not associated with endocrine toxicity to adult females owing to the large amount of $17\beta$-estradiol in these individuals. However, adult males, immature individuals, and embryos all have been shown to exhibit endocrine toxicity resulting from xeno-estrogen exposure. For example, in utero exposure of male or female rodents and humans to DES causes proliferation of epithelial cells associated with the reproductive system resulting in abnormalities of this system. Gynecomastia is a common side effect of estrogenic drugs such as DES and fosfestrol when administered to adult males. The physiological consequences of xeno-estrogenic activity is typically characteristic of feminization, that is, the acquisition of female characteristics.

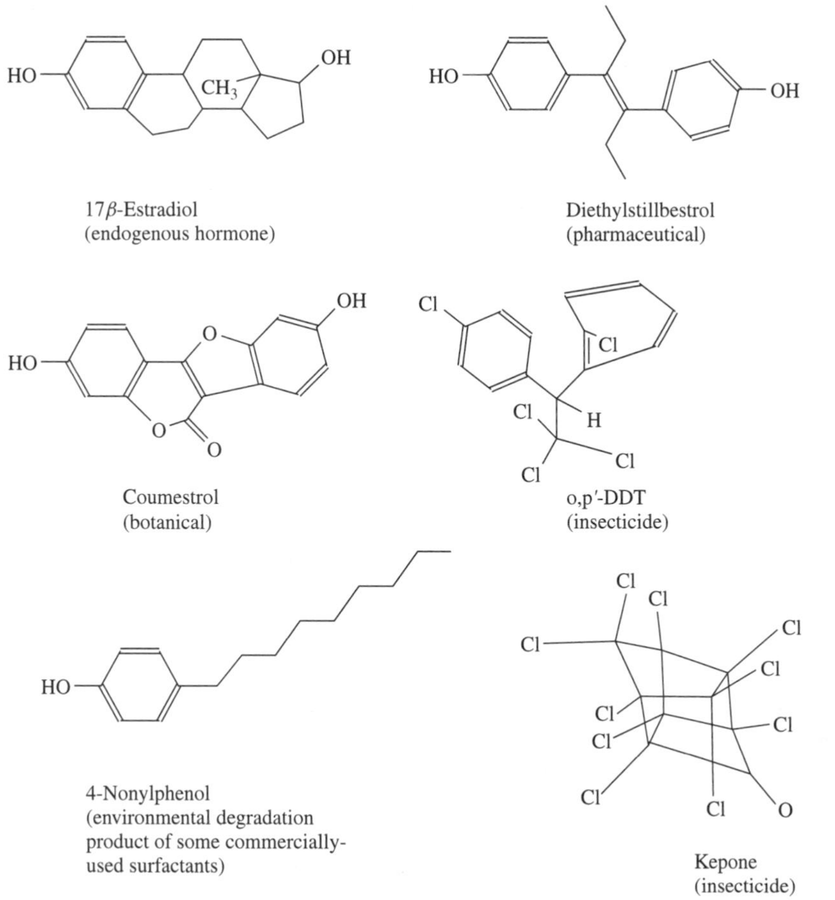

17β-Estradiol
(endogenous hormone)

Diethylstillbestrol
(pharmaceutical)

Coumestrol
(botanical)

o,p′-DDT
(insecticide)

4-Nonylphenol
(environmental degradation
product of some commercially-
used surfactants)

Kepone
(insecticide)

**Figure 17.6** Estrogen receptor agonists. Molecules are presented in a low-energy state that may represent their natural three-dimensional conformation.

***Ecdysone Receptor.*** Ecdysteroids are a class of steroid hormones that regulate a variety of processes related to development, growth, and reproduction in insects and other arthropods but are not utilized by vertebrates. Many compounds of plant origin, or derivations thereof, have been identified that are ecdysteroid receptor agonists (i.e., cucurbitacins, withasteroids). The ecdysteroid agonists are presumed to have evolved in plants as a means of protection against insect predation. Some environmental chemicals of anthropogenic origin also have been shown to exhibit ecdysteroid receptor agonistic activity (i.e., tebufenozide) and have been exploited as insecticides due to their ability to interfere with insect development and growth.

***Retinoic Acid Receptor.*** Most of the biological effects of retinoids are mediated through the retinoic acid receptor (RAR) and the retinoid X receptor (RXR). Both all-*trans*-retinoic acid and 9-*cis*-retinoic acid serve as agonists of RAR, while only 9-*cis*-retinoic acid functions as an agonist of RXR. The functional RAR exists as a heterodimer with RXR, while functional RXR exists as a homodimer. Methoprene is a juvenile hormone III analogue that mimics the activity of this insect hormone.

**Table 17.3   Potency of Some Xenoestrogens Relative to 17$\beta$-Estradiol**

| Chemical | Potency |
|---|---|
| 17$\beta$-Estradiol | 100 |
| Diethylstilbestrol | 74 |
| 4-Nonylphenol | 0.005 |
| 4-Octylphenol | 0.003 |
| 4-*tert*-Octylphenol | 0.00036 |
| $o'$, $p'$-DDT | 0.00011 |
| $o'$, $p'$-DDE | 0.00004 |
| 2', 5'-Dichloro-4-biphenylol | 0.62 |
| 2', 4', 6-Trichloro-4-biphenylol | 1.0 |
| 2', 3', 4', 5'-Tetrachloro-4-biphenylol | 0.82 |
| Bisphenol A | 0.005 |
| Butylbenzylphthalate | 0.0004 |

*Source*: Coldham, N. G. et al., 1997, *Environ. Health Perspect.* **105**(7): 734–742.

*Note*: Estrogenic potency of the compounds was measured using a recombinant yeast cell bioassay.

Exposure of juvenile insects to methoprene results in various abnormalities associated with development and ultimately death. The environmental degradation product of methoprene, methoprenic acid was found to serve as an RXR agonist and specifically activate genes responsive to RXR homodimers. In addition exposure of frog larvae to methoprenic acid caused developmental deformities consistent with those that have been observed in recent years in wild populations and consistent with those caused by exposure to retinoic acid under laboratory conditions. These observations indicate that methoprenic acid functions as an RXR agonist, and that this activity could contribute to the occurrence of amphibians deformities documented in the environment.

### 17.3.2   Hormone Receptor Antagonists

While the estrogen receptor appears somewhat unique among vertebrate nuclear hormone receptors in its promiscuity toward receptor agonists, many nuclear hormone receptors have been shown to be susceptible to chemical antagonism. Receptor antagonists are defined as *chemicals that bind to a hormone receptor but do not activate the receptor*. Rather, these chemicals inhibit receptor activity by preventing the endogenous hormone from binding to and activating the receptor.

***Estrogen Receptor.*** Chemicals often bind to the estrogen receptor and function as mixed agonists/antagonists (discussed below). For example, the drug tamoxifen functions as an estrogen receptor antagonists in reproductive tissue but functions as an agonist with respect to the preservation of bone mineral density and reducing serum cholesterol concentrations. Accordingly tamoxifen can function as a prophylactic against the growth of estrogen-responsive breast cancers and osteoporosis via two different mechanisms (estrogen receptor antagonism and agonism, respectively). Other drugs that bind to the estrogen receptor as an antagonist or mixed agonist/antagonist include raloxifene,

ICI 164,384, and toremifene. Environmental estrogen receptor antagonists include some phytochemicals (i.e., flavonoids) and PCBs (i.e., 3, 3′, 4, 4′-tetrachlorobiphenyl). Consequences of estrogen receptor antagonism are typically considered de-feminization (loss of female traits). In laboratory animal studies, estrogen receptor antagonists have been shown in females to disrupt estrous cycles, impair fertility, increase preimplantation loss, and cause embryolethality.

***Androgen Receptor.*** Chemicals that bind to the androgen receptor in an antagonistic fashion include the pharmaceuticals spironolactone, cimetidine, cyperoterone acetate, and hydroxyflutamide. Environmental chemicals that have been shown to act as androgen receptor antagonists include the metabolites of the agricultural fungicide vinclozolin, the DDT metabolite $p$, $p'$-DDE, some hydroxylated PCBs, and the organophosphate insecticide fenitrothion. The consequence of androgen receptor antagonism is typically considered demasculinization (loss of male traits). Demasculinizing effects of antiandrogens in laboratory animal studies have included reductions in the size of the ventral prostate and seminal vesicle weights along with deformities of the penis.

***Glucocorticoid Receptor.*** Some drugs (i.e., mifepristone) elicit antagonistic activity toward the glucocorticoid receptor. This property has been associated with adverse side effects of some drugs and also has been capitalized upon therapeutically for the modulation of the glucocorticoid receptor. Antiglucocorticoids typically are steroidal compounds that are capable of binding to the receptors but are relatively ineffective in activating the receptor. As such, these compounds are typically mixed agonists/antagonists (see below). Glucocorticoid receptor antagonists can adversely affect growth, development, and other glucocorticoid-regulated processes (Table 17.1). Little is known of the ability of environmental chemicals to function as glucocorticoid receptor antagonists.

***Mixed Agonists/Antagonists.*** Chemicals often can function as either a receptor agonist or antagonist depending on the level of endogenous hormone. A weak agonist may bind to a receptor and stimulate some low-level receptor-mediated activity in the absence of the endogenous hormone. However, in the presence of the hormone, binding of the xenobiotic to the receptor may prevent binding of the endogenous hormone, and if the xenobiotic is a much weaker activator of receptor-mediated activity, then the net effect is loss of activity. Thus, in the presence of the endogenous hormone, the xenobiotic functions as a receptor antagonist. Whether a weak xeno-agonist functions as an agonist or antagonist depends on (1) the concentration of the xeno-agonist, (2) the binding affinity of the xeno-agonist to the receptor, (3) the concentration of the endogenous hormone to the receptor, and (4) the binding affinity of the endogenous hormone to the receptor. These compounds are classified mixed agonists/antagonists.

## 17.3.3  Organizational versus Activational Effects of Endocrine Toxicants

Effects of receptor agonists or antagonists on endocrine related processes are often described as being either organizational or activational. An organizational effect of an endocrine toxicant is one that typically results from neonatal or prenatal exposure during which time hormones are directing various irreversible aspects of development.

Accordingly the disrupting effect of the toxicant also is irreversible. These organizational effects may be evident only later in life during maturation or reproduction. Neonatal exposure to DES resulting in proliferation of epithelial cells of the reproductive tract at reproductive maturity is an example of an organizational effect of an endocrine toxicant. Organizational effects of endocrine toxicants have been of great concern to toxicologists and are the most difficult type of toxicity to diagnose owing to the temporal separation between exposure and effect.

An activational effect of an endocrine toxicant occurs in the same time frame as the exposure and is the consequence of the toxicant disrupting the immediate role of a hormone in some physiological process. Activational effects are reversible following cessation of exposure to the toxicant. For example, androgens contribute to maintenance of the prostate gland in the adult male. Exposure of adult males to an antiandrogen can result in a decrease in prostate size. Cessation of exposure to the antiandrogen then results in restoration of the prostate gland to its normal size.

### 17.3.4 Inhibitors of Hormone Synthesis

Endocrine toxicants can elicit antihormone activity by lowering levels of endogenous hormone in the body. With steroid hormones, chemicals typically elicit this effect by inhibiting enzymes necessary for synthesis of the hormone. For example, the cytochrome P450 enzyme CYP19 is responsible for the aromatization of testosterone to form $17\beta$-estradiol. CYP19 inhibitors such as fadrozol, anastrozole, and letrozole, can lower endogenous $17\beta$-estradiol levels resulting in de-feminization. Cytochrome P450s enzymes also are critical to various hydroxylation reactions that contribute to the synthesis of androgens and other steroid hormones and inhibition of these enzymes can result in a variety of antisteroid hormone effects. For example, the agricultural and medicinal fungicides propiconazole, ketoconazole, and fenarimol are capable of inhibiting P450 enzymes and reducing synthesis and circulating levels of testosterone and other steroid hormones. Toxicological consequences of the lowering of endogenous steroid hormone levels are typically comparable to those effects elicited by antagonists of the hormone's receptor.

### 17.3.5 Inducers of Hormone Clearance

In most species, steroid and thyroid hormones are inactivated and cleared from the body by the same biotransformation processes that are involved in chemical detoxification (see Chapter 7). Predominant among the hormone biotransformation processes in vertebrates are hydroxylation, glucuronic acid conjugation, and sulfate conjugation. The thyroid hormones $T_3$ and $T_4$ are inactivated and cleared following sulfate and glucuronic acid conjugation, respectively. The glucuronosyl transferase enzymes that are responsible for the elimination of $T_4$ are induced following exposure to phenobarbital-type inducers and Ah receptor ligands (see Chapter 9). Thus exposure to chemicals such as some dioxins and PCBs can result in enhanced clearance of thyroid hormone resulting in low circulating thyroid hormone levels. The resulting hypothyroid state can result in a variety of pathological conditions. In newborn infants, hypothyroidism is associated with cretinism. This organizational syndrome is characterized by mental retardation, short stature, and various neurological abnormalities. In children, hypothyroidism can cause delayed growth and mental development while advancing the onset of puberty in adolescents. Hypothyroidism in adults results in various activational abnormalities including impaired cardiovascular, pulmonary, intestinal, and renal function. Chronic fatigue, lethargy, and difficulty in concentration are also associated with thypothyroidism in adults.

Increased clearance of steroid hormones due to induction of hepatic biotransformation enzymes following chemical exposure often has been cited as a possible mechanism by which toxicants could lower circulating testosterone or $17\beta$-estradiol levels. While enhanced clearance of sex steroids has been demonstrated following chemical exposure and induction of hepatic biotransformation enzymes, elegant feedback control mechanisms tend to ensure that more hormone is produced and homeostasis is maintained (Figure 17.2). Enhanced clearance of sex steroids can contribute to endocrine disruption if the toxicity also results in impaired hormone synthesis (i.e., gonadal toxicity or interference with the feedback control of hormone synthesis). 2,3,7,8-Tetrachlorodibenzodioxin appears to lower circulating sex steroid levels via this dual effect.

### 17.3.6   Hormone Displacement from Binding Proteins

Steroid and thyroid hormones are typically distributed throughout the body while bound to serum-binding proteins such as sex hormone-binding globulin, corticosteroid-binding globulin, thyroxine-binding globulin (transthyretin), and albumin. Most steroid and thyroid hormones (>95%) are present in the blood reversibly bound to proteins. This bound hormone is not available for cell entry where it may interact with nuclear receptors or undergo inactivation/elimination reactions. Rather, the bound hormone serves as a reservoir from which hormone can be liberated (free hormone) for cell entry.

Some xenobiotics can compete with hormones for binding to the blood proteins. As a result the circulating hormone reservoir can be depleted and free hormone becomes limited. A variety of phenolic compounds, including hydroxylated metabolites of polychlorinated biphenyls (PCBs), chlorophenols, chlorophenoxy acids, and nitrophenols, have been shown to interfere with thyroxine binding to thyroxine-binding globulin during in vitro experiments. In some instances compounds that displace thyroxine from the binding protein also have been shown to decrease circulating thyroxine levels in exposed animal models or in humans. In vitro experiments also have revealed that testosterone and $17\beta$-estradiol can be displaced from sex hormone-binding globulin by some chemicals such as 4-nonylphenol, 4-tert-octylphenol, bisphenol A, $O$-hydroxybiphenyl, and pyrethroid insecticides. However, it is not clear whether these chemicals would significantly displace sex steroids from the binding globulin at concentrations typically measured in human blood.

## 17.4   INCIDENTS OF ENDOCRINE TOXICITY

### 17.4.1   Organizational Toxicity

In utero exposure to estrogens or antiandrogens has been shown, in animal models, to elicit a variety of organizational effects associated with development of the reproductive system. The best-described example of the organizational effects of a drug administered to humans involves the synthetic estrogen DES. As discussed previously, DES was prescribed to over two million pregnant women in the United States between the 1940s and 1960s to prevent miscarriage. Offspring exposed to DES during fetal development experienced a variety of problems upon attainment of sexual maturity. DES daughters experience a significantly increased risk of clear cell adenocarcinoma of the vagina and cervix. DES daughters have increased risk of a variety of reproductive disorders including structural abnormalities of the reproductive tract, infertility, ectopic pregnancy, miscarriage, and pre-term delivery.

Less is known of the risks faced by males exposed to DES during fetal development. Animal studies have revealed that male rodents exposed to DES have increased incidence of prostatic metaplasia. Epidemiological studies of DES sons have suggested increased risk of various testicular abnormalities including epididymal cysts, testicular varicoceles, and undescended testis. Hyperplasia and metaplasia of the prostatic ducts in DES sons also have been reported.

The effects elicited by fetal exposure to DES appear to be largely the consequence of the estrogenic activity of this drug. Estrogens orchestrate organizational events during fetal development that promote female reproductive tract development. Excess estrogen exposure resulting from DES treatment of either female or male fetuses resulted in permanent alterations, many of which became evident only upon attainment of reproductive maturity.

Organizational effects on reproductive development resulting from perinatal exposure to endocrine toxicants of environmental origin also have been reported to occur. In 1973 a fire retardant containing polybrominated biphenyls (PBBs) was mistakenly added to cattle feed in Michigan. An estimated 4000 people subsequently were exposed to the PBBs by consuming dairy products derived from these cattle. PBBs are long-lived chemicals that are stored in the fat of exposed individuals. PBBs have been reported to elicit endocrine toxicity-like symptoms in animal models consistent with hypothyroidism. For example, offspring from maternal rats provided PBBs during gestation and lactation showed signs of neurological deficit and growth retardation. Daughters of mothers that were exposed to PBBs during the Michigan incident were monitored for possible adverse effects on the female reproductive system. The initiation of menarche (menstruation) among these daughters correlated with the likely severity of PBB exposure. The most highly exposed daughters began menstruating approximately 1 year ahead of females that were less severely exposed. Early initiation of menarche is consistent with precocious puberty associated with hypothyroidism. The initiation of menarche also is under the regulation of $17\beta$-estradiol and early initiation of menarche may reflect an estrogen-type organizational effect of the PBBs during perinatal exposure.

### 17.4.2  Activational Toxicity

***Estrogenic Pharmaceuticals.*** Administration of estrogenic pharmaceuticals to children or adults can result in a variety of abnormalities associated largely with secondary sex characteristics that are reversible upon cessation of drug treatment.

Gynecomastia, the development of breast tissue in males, is often the consequence of perturbations in the normal androgen/estrogen ratio. As discussed earlier in this chapter, prolonged administration of drugs with estrogenic or antiandrogenic activity can cause gynecomastia. Gynecomastia had been reported in the medical literature to occur as a result of frequent intercourse when an estrogen-containing cream was used as a vaginal lubricant and among morticans who applied estrogen-containing skin creams to corpses without the use of gloves.

Similar to gynecomastia in adult males, activational toxicity from estrogenic drugs has been reported to cause pseudoprecocious puberty in children. Pseudoprecocious puberty is characterized by the development of some indicators of puberty (pubic or facial hair, morphological changes in sex organs, breast development, etc.) in preadolescent individuals. An outbreak of pseudoprecocious puberty was reported among a group of children ranging in age from 4 months to 2 years of age following application of a skin cream to treat dermatitis. Symptoms included pigmentation of the nipples, breast development, the presence of pubic hair, and vaginal discharge and bleeding among the females. Breast development also was reported in prepubertal boys following use of an estrogen-containing hair cream. These reports highlight the fact that dermal exposure can be adequate to attain a sufficient dose of endocrine-active compound to elicit adverse responses. In all of these cases the symptoms of endocrine toxicity resolved following cessation of exposure to the causative agent.

***Environmental Estrogens.*** Thelarche is defined as the development of breast tissue in preadolescent females (typically <8 years of age). Since 1979 physicians have monitored an epidemic level of thelarche on the island of Puerto Rico. The cause of thelarche in Puerto Rico is not known; however, evidence strongly implicates exposure to endocrine-disrupting agents. Analyses of blood samples from thelarche and nonthelarche children for environmental chemicals with known estrogenic activity revealed that 68% of the thelarche children contained significantly high levels of several types of phthalate esters. Only a single nonthelarche child contained a significant amount of phthalate ester and only one type of phthalate ester was found in this individual. Phthalate esters are used as plasticizers and are ubiquitous environmental contaminants. They have been shown to cause a variety of endocrine-related effects in animal models and some phthalate esters have been shown to be estrogenic in vitro. The association between phthalate ester exposure and the high incidence of thelarche in Puerto Rico does not establish causality but has generated concern that environmental agents are responsible for this condition.

Kepone (chlordecone) is an organochlorine insecticide (Figure 17.5) that was manufactured in Hopewell, Virginia, from the mid-1960s to 1975. In 1975 the Center for Disease Control determined that employees of the manufacturing facility and other residents of Hopewell, totaling over 200 individuals, had been significantly contaminated with this insecticide. Exposed individuals reported a variety of symptoms. Foremost, among the symptoms of "Kepone sickness" were neurological disorders presenting as tremors, weight loss, and nervousness. However, subsequent evaluations revealed that males exposed to Kepone also experienced testicular dysfunction that was characteristic of estrogen exposure. Later laboratory studies demonstrated that Kepone was an estrogen receptor agonist, which could explain its adverse effects on the male reproductive system.

## 17.4.3  Hypothyroidism

Hypothyroidism describes the clinical state arising from a deficiency in thyroid hormone. Toxicity resulting in hypothyroidism is manifested at several organ systems as

**Table 17.4   Clinical Manifestations of Hypothyroidism**

| Organ System | Manifestation |
|---|---|
| Skin | Puffy appearance, dry, course, yellow-tinted skin brittle nails, wound healing slowed, hair loss |
| Cardiovascular | Enlarged heart, changes in electrocardiographs |
| Respiratory | Maximal breathing capacity reduced, obstructive sleep apnea, fluid accumulation in the pleural cavity |
| Digestive | Reduced appetite with modest weight gain |
| Muscle | Stiffness, aching |
| Nervous | Slowing of intellectual functions, lethargy, headaches |

described in Table 17.4, and individual effects may be misdiagnosed as organ-specific toxicity. Hypothyroidism can result from various causes other than chemical toxicity including diseases of the hypothalamic–pituitary–thyroidal axis, iodine deficiency, and heritable defects in thyroid hormone production. Chemical agents that have historically been recognized for their ability to cause hypothyroidism include phenylbutazone, resorcinol, lithium, and para-aminosalicylic acid.

Disruptions in thyroid hormone levels can occur through chemical-induced increases in the metabolic inactivation and elimination of the hormone. Chemicals that are capable of increasing the metabolic clearance of thyroid hormone include the polycyclic halogenated hydrocarbons (i.e., dioxins, furans, polychlorinated biphenyls, polybrominated biphenyls). A study reported in the *New England Journal of Medicine* suggested that environmental or occupational exposure to such chemicals can result in hypothyroidism in humans. The study consisted of a comparison of thyroid status in workers who were occupationally exposed to polybrominated biphenyls as compared to workers who were not exposed to any polyhalogenated hydrocarbons. Four of 35 exposed workers and none of 89 unexposed workers exhibited signs of hypothyroidism that included increased plasma levels of thyrotropin and decreased plasma levels of thyroxine. Thyrotropin is secreted by the pituitary gland and stimulates the thyroid gland to produce thyroxine (see Figure 17.1). The increase in thyrotropin and decrease in thyroxine is consistent with hypothyroidism caused by increased clearance of the thyroxine. As discussed earlier in this chapter, perinatal exposure to PBBs during the Michigan milk contamination also produced symptoms characteristics of hypothyroidism.

## 17.5   CONCLUSION

The endocrine system possesses many targets at which toxicants can elicit either reversible or permanent effects on an individual. Effects of chemicals on endocrine-regulated processes such as development, maturation, growth, and reproduction have been well documented in both laboratory and epidemiological studies. Less is know of the potential effects of endocrine toxicants on more generalized endocrine-regulated processes such as bone maintenance, general organ function, and metabolism. The US Environmental Protection Agency has been mandated by the US Congress to develop and implement a program for the screening and testing of chemicals for endocrine-disrupting toxicity. At this writing, the EPA is in the process of developing such a program that will focus on the effects of chemicals on the androgen/estrogen and thyroid hormone regulated processes. Once implemented, this required testing will greatly expand our knowledge of the extent to which humans are exposed to chemicals that interfere with processes regulated by these hormones. However, it is important to recognize that chemicals have the potential ability to interfere with other hormone cascades, including those involving mineralcorticoids, glucocorticoids, retinoids, and perhaps some peptide hormones. Research is needed to increase our understanding of the susceptibility of endocrine signaling pathways involving these hormones to chemical toxicity and, ultimately, to our establishing chemical exposure limits that include these considerations.

## SUGGESTED READING

Bahn, A. K., J. L. Mills, P. J. Snyder, et al. Hypothyroidism in workers exposed to polybrominated biphenyls. *N. Engl. J. Med.* **302**: 31–33, 1980.

Beas, F., L. Vargas, R. P. Spada, and N. Merchack. Pseudoprecocious puberty in infants caused by dermal ointment containing estrogens. *J. Pediatr.* **75**: 127–130, 1962.

Blanck, H. M., M. Marcus, P. E. Tolbert, C. Rubin, A. K. Henderson, V. S. Hertzberg, R. H. Zhang, and L. Cameron. Age at menarche and tanner stage in girls exposed in utero and postnatally to polybrominated biphenyls. *Epidemiology* **11**: 641–647, 2000.

Colborn, T., and C. Clement, eds. Chemically-induced alterations in sexual and functional development: The Wildlife/Human Connection. In *Advances in Modern Environmental Toxicology*, M. A. Mehlman, ed. Princeton, NJ: Princeton Scientific Publishing, 1992.

DiRaimondo, C. V., A. L. Roach, and C. K. Meador. Gynecomastia from exposure to vagina estrogen cream. *N. Engl. J. Med.* **302**: 1089–1090, 1980.

Greenspan, F. S., and G. J. Strewler, eds. *Basic and Clinical Endocrinology.* Stamford, CT: Appleton and Lange, 1997.

Guillette, L, Jr. and D. A. Crain, eds. *Environmental endocrine disrupters.* New York: Taylor and Francis, 2000.

McLachlan, J. A. Environmental signaling: What embryos and evolution teach us about endocrine disrupting chemicals. *Endocrine Rev.* **22**: 319–341, 2001.

Wilson, J. D., and D. W. Foster, eds. *Textbook of Endocrinology.* Philadelphia: Saunders, 1992.

# CHAPTER 18

# Respiratory Toxicity

ERNEST HODGSON, PATRICIA E. LEVI, and JAMES C. BONNER

## 18.1 INTRODUCTION

Pulmonary diseases caused by agents in the environment have been known for centuries and have been associated with occupations such as stone quarrying, coal mining, and textiles. The problem is more complex and widespread today because new agents are constantly being added to the environment. They include all types of inhalant toxicants, gases, vapors, fumes, aerosols, organic and inorganic particulates, and mixtures of any or all of these. Gasoline additives and exhaust particles, pesticides, plastics, solvents, deodorant and cosmetic sprays, and construction materials are all included. Table 18.1 lists some of the more important industrial lung toxicants, the exposure sources, and associated injuries.

### 18.1.1 Anatomy

Air enters the respiratory systems of mammals through the nose or mouth, some, including humans, utilizing both while others being obligatory nose breathers. Inhaled air then passes into the proximal airways, the trachea, and the main bronchi to each lung. The main bronchi then subdivide several times into numerous bronchi, finally into terminal bronchioles and respiratory bronchioles, ultimately ending in alveolar ducts and alveoli (Figure 18.1).

### 18.1.2 Cell Types

As shown in Table 18.2, there are many different cell types in the respiratory system with considerable variation in both structure and function from the nasal epithelium to the alveoli. The various cell types of the airway epithelium are shown in Figure 18.2.

### 18.1.3 Function

The nasal passages have an olfactory function, but with regard to inhaled toxicants they have primarily a defensive function and form the initial defensive barrier against inhaled

*A Textbook of Modern Toxicology, Third Edition,* edited by Ernest Hodgson
ISBN 0-471-26508-X  Copyright © 2004 John Wiley & Sons, Inc.

**Table 18.1   Some Important Industrial Lung Toxicants and Associated Injury**

| Toxicant | Source | Damage |
| --- | --- | --- |
| Aluminum dust | Ceramics, paints, fireworks, electrical goods | Fibrosis |
| Ammonia | Manufacture of fertilizers, explosives, ammonia | Irritation |
| Arsenic | Manufacture of pesticides, glass, pigments, alloys | Lung cancer, bronchitis |
| Asbestos | Mining, construction, shipbuilding | Asbestosis, lung cancer |
| Beryllium | Ore extraction, ceramics, alloys | Fibrosis, lung cancer |
| Cadmium oxide | Welding, smelting, manufacture of electronics, alloys, pigments | Emphysema |
| Chlorine | Manufacture of pulp and paper, plastics, chlorinated chemicals | Irritation |
| Chromium | Manufacture of Cr compounds, paint pigments | Lung cancer |
| Coal dust | Coal mining | Fibrosis |
| Hydrogen fluoride | Manufacture of chemicals, plastics, photographic film, solvents | Irritation, edema |
| Iron oxides | Welding, steel manufacturing, mining, foundry work | Fibrosis |
| Nickel | Nickel extraction and smelting, electroplating | Nasal cancer lung cancer, edema |
| Nitrogen oxides | Welding, explosive manufacturing | Emphysema |
| Ozone | Welding, bleaching, deodorizing | Emphysema |
| Phosgene | Production of pesticides, plastics | Edema |
| Silica | Mining, quarrying, farming | Fibrosis (silicosis) |
| Sulfur dioxide | Bleaching, refrigeration, fumigation coal combustion | Irritation |
| Talc | Rubber industry, cosmetics | Fibrosis |
| Tetrachloroethylene | Dry cleaning, metal degreasing | Edema |

toxicants. Since the nasal epithelium contains relatively high levels of xenobiotic-metabolizing enzymes such as CYP and FMO isoforms, it can function as a detoxification site. However, as these enzymes, particularly CYP, may also activate toxicants, the nose is also a site for toxicant-induced lesions.

The trachea and bronchi likewise have a protective function. Mucous and serous cells secrete fluids that together comprise the mucus, which is moved toward the pharynx by the cilia of the ciliated cells. The movement of mucus serves to move entrapped particles toward the pharynx where they are eliminated by swallowing or expectoration. The mucus may also have other protective functions, protecting the epithelial cells by free radical scavenging and antioxidant properties. The Clara cells are known to contain high concentrations of xenobiotic metabolizing enzymes.

The principal function of the lungs is gas exchange, providing $O_2$ to the tissues and removing $CO_2$. This gas exchange takes place in the alveoli. Because the lung has a large surface area and exchanges a significant volume of air (100,000–20,000 L/day

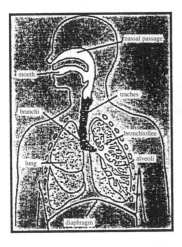

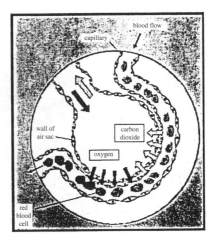

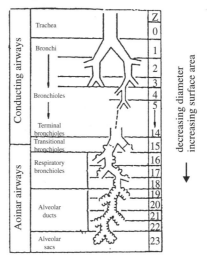

Conducting airways
• Delivery and structural support
• Defense:
1. (mucous and ciliated airway epithelial cells,
   "the mucociliary escalator".
2. inflammatory cells.

Alveoli
• *Gas exchange* (type I epithelium
  /cap. endothelium)

• Surfactant production (type II
  epithelial cells)

• defense (macrophages)

**Figure 18.1**  Structure and function of the respiratory system.

**Table 18.2  Cell Types of the Respiratory System**

| | |
|---|---|
| Nasal | Stratified squamous to mucociliated epithelium with olfactory cells |
| Tracheo-bronchial region | Mucociliated epithelium (ciliated, mucous cells, basal cells); smooth muscle cells; fibroblasts; neuroendocrine cells; immune cells |
| Bronchioles | Mucociliated epithelium with Clara cells in distal bronchioles and alveolar ducts |
| Alveoli | Type I and type II epithelium, alveolar macrophages |

for the average adult), the lung is the major interface between an organism the environment and any toxicants present in the air. It is also significant, from an efficiency point of view, that the entire cardiac output goes to the lungs.

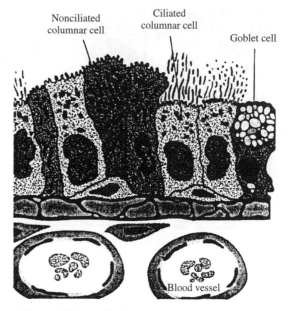

Nonciliated columnar cell

Ciliated columnar cell

Goblet cell

Blood vessel

**Figure 18.2**   Cell types of the airway epithelium.

## 18.2   SUSCEPTIBILITY OF THE RESPIRATORY SYSTEM

### 18.2.1   Nasal

The nasal epithelia are the first point of contact for respiratory toxicants. Because they contain xenobiotic metabolizing enzymes, they are susceptible to toxic effects caused by reactive intermediates.

### 18.2.2   Lung

In addition to being in direct contact with airborne toxicants, the entire body blood volume passes through the lung one to five times a minute, exposing the lung to toxicants and drugs within the systemic circulation. Thus the possibility of damage from both inhaled and circulating agents is enormous. As with the liver and kidney, the lungs possess significant levels of many xenobiotic metabolizing enzymes and thus can play a large role in the activation and detoxication of exogenous chemicals.

## 18.3   TYPES OF TOXIC RESPONSE

Although many different agents may damage the lung, the patterns of cellular injury and repair are relatively constant, and most fall into one or more of the categories described below.

### 18.3.1   Irritation

Perhaps one of the most obvious and familiar chemical effects is irritation caused by volatile compounds such as ammonia or chlorine gas. Such irritation, especially if severe or persistent, results in constriction of the airways. Edema and secondary infection frequently follow severe or prolonged irritation. Such damage is known to result from exposure to agents such as ozone, nitrogen oxides, and phosgene.

### 18.3.2  Cell Necrosis

Severe damage to the cells lining the airways can result in increased cell permeability, followed by cell death.

### 18.3.3  Fibrosis

Fibrosis, or formation of collagenous tissue, was perhaps one of the earliest recognized forms of occupational diseases. *Silicosis*, resulting from inhalation of silica ($SiO_2$), is thought to involve first the uptake of the particles by macrophages and lysosomal incorporation, followed by rupture of the lysosomal membrane and release of lysosomal enzymes into the cytoplasm of the macrophages. Thus the macrophage is digested by its own enzymes. After lysis, the free silica is released to be ingested by fresh macrophages, and the cycle continues. It is also thought that the damaged macrophages release chemicals that are instrumental in initiating the collagen formation in the lung. Fibrosis may become massive and impair the respiratory function of the lung significantly. *Asbestosis* was recognized as long ago as 1907; however, the magnitude of the risk has become apparent only recently, primarily due to the increased incidence of lung cancer among asbestosis sufferers, especially those who are also cigarette smokers. Both silicosis and asbestosis are thought to be premalignant conditions.

### 18.3.4  Emphysema

Emphysema is characterized by an enlargement of the airspaces with the destruction of the gas-exchange surface area. The loss of tissue and air-trapping capacity results in a distended lung that no longer effectively exchanges $O_2$ and $CO_2$. Although cigarette smoking is the major cause of emphysema, other toxicants can also cause this condition.

### 18.3.5  Allergic Responses

Numerous agents, including microorganisms, spores, dust, and chemicals, are known to elicit allergic responses resulting in constriction of the airways. Several diverse examples are farmer's lung from the spores of a mold that grows on damp hay, maple bark stripper's disease from spores of a fungus growing on maple trees, cheese washer's lung from penicillin spores, and mushroom picker's lung from the mushroom spores. Byssinosis comes from the inhalation of cotton, flax, or hemp dusts. This condition, however, does not seem to result from bacterial or fungal exposure but from an apparent toxicant or allergen associated with the plant dusts.

### 18.3.6  Cancer

Perhaps the most severe response of the lung to injury is cancer, with the primary cause of lung cancer being cigarette smoking. Cigarette smoke contains many known carcinogens as well as lung irritants. Many of the polycyclic aromatic hydrocarbons, such as benzo(*a*)pyrene, can be metabolized in the lung by pulmonary P450 enzymes to reactive metabolites capable of initiating cancer. In addition cigarette smoke contains numerous compounds that can act as tumor promoters. Asbestos is associated with two forms of cancer—lung cancer and malignant mesothelioma, a tumor of the cells covering the surface of the lung and the adjacent body wall.

### 18.3.7 Mediators of Toxic Responses

Most of the toxic responses summarized above involve a relatively small number of biochemical events related to oxidative injury, signaling pathways, and genotoxicity.

Reactive oxygen species (ROS) such as superoxide anion ($\cdot O_2$) and hydroxyl radical ($\cdot OH$) are balanced by antioxidant enzymes and radical scavengers. Imbalance favoring ROS generation leads to lipid peroxidation, with resultant membrane damage, and DNA damage (genotoxicity).

Changes in the concentration of growth factor and/or cytokine signaling molecules may lead to inflammation, proliferative responses, or apoptosis.

## 18.4 EXAMPLES OF LUNG TOXICANTS REQUIRING ACTIVATION

### 18.4.1 Introduction

The activation of pulmonary toxicants falls into three main categories or mechanisms, depending either on the site of formation of the activated compound or on the nature of the reactive intermediate.

1. The parent compound may be activated in the liver, with the reactive metabolite then transported by the circulation to the lung. As would be expected, the activated compounds may lead to covalent binding and damage to both liver and lung tissue.
2. A toxicant entering the lung, either from inhaled air or the circulatory system, may be metabolized to the ultimate toxic compound directly within the lung itself. Although the total concentration of P450 is less in the lung than in the liver, the concentration varies considerably in the different cell types, with the highest concentration being found in the nonciliated bronchiolar epithelial (Clara) cells of the terminal bronchioles. Because of this, the Clara cells are often a primary target for the effects of activated chemicals.
3. Another means of metabolic activation is the cyclic reduction/oxidation of the parent compound, resulting in high rates of consumption of NADPH and production of superoxide anion. Either the depletion of NADPH and/or the formation of reactive oxygen radicals could lead to cellular injury.

The following three chemicals serve to illustrate these three mechanisms of activation.

### 18.4.2 Monocrotaline

The pyrrolizidine alkaloids, found in the genus *Senecio* and a number of other plant genera, are plant toxins of environmental interest that have been implicated in a number

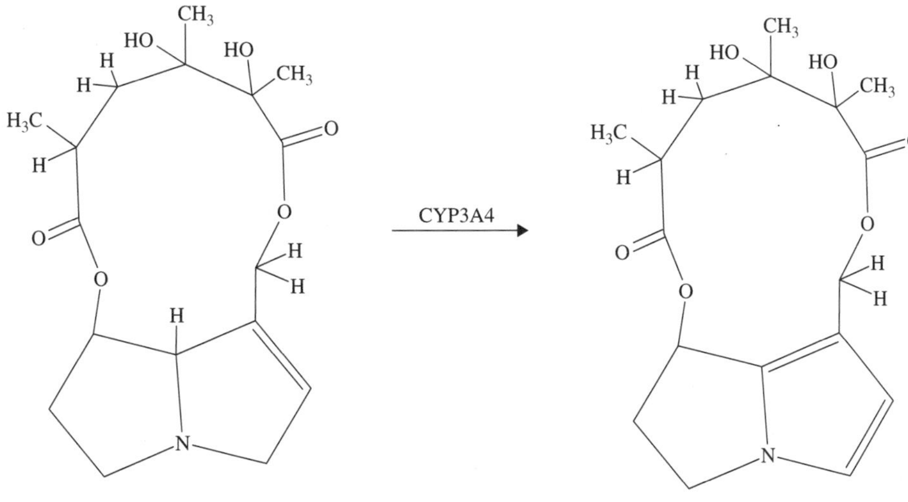

**Figure 18.3**   Structure and activation of monocrotaline.

of livestock and human poisonings. Grazing animals may be poisoned by feeding on pyrrolizidine alkaloid-containing pastures, and human exposure may occur through consumption of herbal teas and contaminated grains and milk. The chemical structure for monocrotaline is shown in Figure 18.3.

Monocrotaline, found in the leaves and seeds of the plant *Crotalaria spectabilis*, has been the most extensively studied of the pyrrolizidine alkaloids. When monocrotaline is given to rats and other animals at high doses, a pronounced liver injury occurs, and animals usually die of acute effects, presumably liver failure. Lower doses, however, that are only mildly hepatotoxic, result in lung injury that is associated with pulmonary hypertension and usually death in several weeks. It is thought that activation of monocrotaline to its dehydro metabolite occurs in the liver and is a reductive reaction mediated by cytochrome P450 3A4 (Figure 18.3). Even though monocrotaline acts as a pneumotoxicant, several lines of evidence indicate that the lung is incapable of activating monocrotaline or can only do so to a limited extent. Furthermore the main site of pulmonary injury occurs in the endothelial cells, a target site consistent with a reactive metabolite being absorbed from the circulatory system.

## 18.4.3  Ipomeanol

One of the best-known examples of a toxic compound being activated in the lung is 4-ipomeanol (Figure 18.4). This naturally occurring furan is produced by the mold *Fusarium solani* that infects sweet potatoes. Lung edema in cattle is known to be associated with the ingestion of mold-damaged sweet potatoes. A similar pulmonary lesion can be produced in a number of species regardless of the route of administration. Pulmonary injury by 4-ipomeanol is caused, not by the parent compound, but by a highly reactive alkylating metabolite produced in the lung by lung-specific P450 isozymes. In addition these isozymes are highly concentrated in the Clara cells, which are most affected by 4-ipomeanol toxicity. Although the reactive metabolite has not been identified unambiguously, considerable data suggest a reactive epoxide.

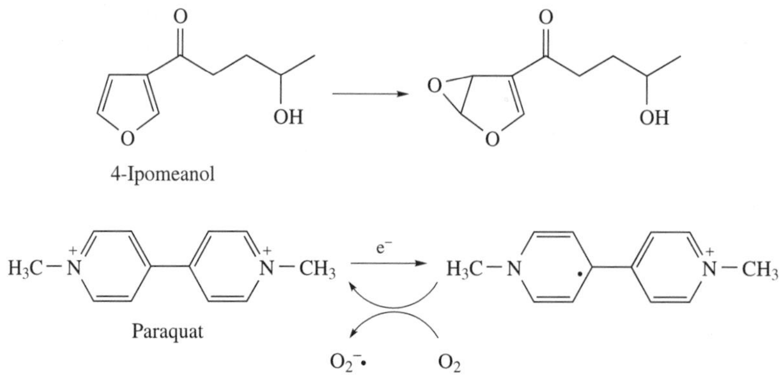

**Figure 18.4** Structure and activation of ipomeanol and paraquat.

Other toxic lung furans, such as the atmospheric contaminants 2-methylfuran and 3-methylfuran, may exert their toxicity through the formation of reactive metabolites, probably reactive aldehydes.

### 18.4.4 Paraquat

Systemic administration of compounds such as the herbicide paraquat (Figure 18.4), bleomycin (a cancer therapeutic agent), and nitroflurantoin (an antibiotic used for urinary tract infections) initiate a progression of degenerative and potentially lethal lesions in the lung by a mechanism known as *redox cycling*. These compounds are reduced by cytochrome P450 reductase and NADPH, forming a free radical. Although the free radical could potentially react with tissue macromolecules, one molecule of oxygen is reduced to superoxide that can then be converted to other toxic oxygen species. These reactive compounds may cause peroxidation of cellular membranes. The specific toxicity of paraquat to the lung results from the uptake of this compound by the polyamine transport system in the lung as well as from the high pulmonary oxygen tension. Nitroflurantoin, however, is not actively accumulated in the lung, and its tissue specificity probably results from the high pulmonary oxygen tension.

### 18.5 DEFENSE MECHANISMS

There are two important defense mechanisms against inhaled particles. The first of these involves the mucociliary escalator and consists of the trapping of particles in mucus followed by the upward movement of the mucus brought about by the upward beating of cilia on the airway epithelial airway cells. The material is then either swallowed or expectorated. The second mechanism is macrophage mediated. Macrophages engulf particles and either deposit them on the mucociliary escalator or enter the lymphatic system.

Toxicants may also be detoxified by xenobiotic-metabolizing enzymes such as cytochrome P450 or the flavin-containing monooxygenase. It should be borne in mind, however, that these enzymes may also activate toxicants to more reactive, and potentially more toxic, metabolites.

## SUGGESTED READING

Bond, J. A. Metabolism and elimination of inhaled drugs and airborne chemicals from the lungs. *Pharmacol. Toxicol.* **72**: 36–47, 1993.

Cho, M., C. Chichester, C. Plopper, and A. Buckpitt. Biochemical factors important in Clara cell selective toxicity in the lung. *Drug Metabol. Rev.* **27**: 369–386, 1995.

Dahl, A. R., and J. L. Lewis. Respiratory tract uptake of inhalants and metabolism of xenobiotics. *An. Rev. Pharmacol. Toxicol.* **32**: 383–407, 1993.

Foth, H. Role of the lung in accumulation and metabolism of xenobiotic compounds-implications for chemically induced toxicity. *Crit. Rev. Toxicol.* **25**: 165–205, 1995.

Henderson, R. J., and K. J. Nikula. Respiratory tract toxicity. In *Introduction to Biochemical Toxicology*, 3rd ed., E. Hodgson and R. C. Smart, eds. New York: Wiley-Interscience, 2001.

Wheeler, C. W., T. M. Guenthner. Cytochrome P450-dependent metabolism of xenobiotics in human lung. *J. Biochem. Toxicol.* **6**: 163–169, 1991.

Witschi, H. R., and J. A. Last. Toxic responses of the respiratory system. In *Casarett and Doull's Toxicology: The Science of Poisons*, 6th ed., C. D. Klaassen, ed. New York: McGraw-Hill, 2001, pp. 515–534.

# CHAPTER 19

# Immunotoxicity

MARYJANE K. SELGRADE

## 19.1 INTRODUCTION

A properly functioning immune system is essential to good health. It defends the body against infectious agents and in some cases tumor cells. Individuals with immune deficiencies resulting from genetic defects, diseases (e.g., AIDS, leukemia), or drug therapies are more susceptible to infections and certain types of cancer, the consequences of which can be life-threatening. On the other hand, the immune system may react to foreign substances that would otherwise be relatively innocuous, such as certain chemicals, pollens, and house dust. The resulting allergic reactions can produce an array of pathologies, ranging from skin rashes and rhinitis to more life-threatening asthmatic and anaphylactic reactions. A crucial part of immune function is the ability to distinguish endogenous components (*self*) from potentially harmful exogenous components (*non-self*). Failure to make this distinction results in autoimmune disease.

Immunotoxicology is the study of undesired effects resulting from the interactions of xenobiotics with the immune system (Figure 19.1). There is evidence that some xenobiotics can cause immune suppression. Xenobiotics can also interact with the immune system to either cause or exacerbate allergic disease. Finally there is growing concern that xenobiotics could have some involvement in autoimmune disease. This chapter provides a brief overview of the immune system, chemicals associated with immune suppression and immune pathologies, and approaches to testing for these effects.

## 19.2 THE IMMUNE SYSTEM

Cells of the immune system include several types of leukocytes (white blood cells) (Table 19.1), which are derived from bone marrow. T lymphocytes, a subset of immune

Disclaimer: This chapter has been reviewed by the National Health and Environmental Effects Research Laboratory, US Environmental Protection Agency and approved for publication. Approval does not signify that the contents necessarily reflects the views and policies of the Agency, nor does mention of trade names or commercial products constitute endorsement or recommendation for use.

*A Textbook of Modern Toxicology, Third Edition,* edited by Ernest Hodgson
ISBN 0-471-26508-X Copyright © 2004 John Wiley & Sons, Inc.

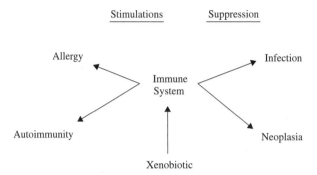

**Figure 19.1** Potential consequences of immunotoxicity.

**Table 19.1 Leukocytes**

| |
|---|
| *Granulocytes (polymorphonuclear leukocytes)* |
| Neutrophils |
| Eosinophils |
| Basophils/mast cells[a] |
| *Monocytes* |
| Lymphocytes |
| Monocytes/macrophages[a] |
| Natural killer cells |

[a]Found in blood/more activated form found in tissues.

cells, undergo differentiation and maturation in the thymus. Leukocytes circulate throughout the body in blood and lymph and populate other lymphoid tissues including the spleen, lymph nodes (scattered throughout the body), tonsils, and adenoids, as well as aggregates of lymphoid tissue in the lung, gut, and skin, which are referred to as bronchus-, gut- and skin-associated lymphoid tissue (BALT, GALT, and SALT). Also immune cells can be recruited to almost any tissue in the body where there is injury or infection. Accumulation of leukocytes in tissues in response to injury is known as inflammation. Cytokines (e.g., interleukins, interferons, and chemokines), soluble mediators produced by immune cells as well as cells outside the immune system, control the maturation, differentiation, and mobilization of immune cells. Immune responses are divided into innate responses directed nonspecifically against foreign substances, and acquired responses directed against specific antigens. There is considerable interaction between these two types of immunity.

Innate immunity provides a rapid, although usually incomplete, antimicrobial defense. Granulocytes, natural killer cells, and macrophages are important mediators of innate immunity. Granulocytes have the capacity to phagocytize (engulf) infectious agents or other types of particles and to destroy or remove them from the tissue. They release a variety of soluble mediators that can kill invading organisms, increase vascular permeability, and recruit more leukocytes to the tissue. Natural killer cells are large granular lymphocytes that nonspecifically kill tumor and virus-infected cells. Macrophages are also phagocytic, can release chemotactic and cytotoxic cytokines, and, when activated, can kill tumor or virus-infected cells. Mediators released from all of these cells during the acute inflammatory response influence the development of acquired immune responses.

Acquired immunity *specifically* recognizes foreign substances (called antigens) and *selectively* eliminates them. On re-encountering the same antigen there is an enhanced response providing protection against reinfection. Vaccination against infectious agents is based on this principle. T lymphocytes and B lymphocytes (T cells and B cells) are the major players in acquired immunity (Figure 19.2). In both cases there are millions of different clones, groups of immune cells that have specific receptors for a particular antigen. When a cell encounters that specific antigen, clonal expansion occurs; that is, B and T cells with that particular specificity divide and differentiate and are thus activated to respond to the current crisis (e.g., infection). Memory cells develop that represent an enlarged clone of long-lived cells that are committed to respond rapidly, by clonal expansion, upon re-exposure to the same antigen.

B cells recognize native or denatured forms of proteins or carbohydrates in soluble, particulate, or cell-bound form. Activated B cells differentiate into plasma cells and produce antibodies, soluble proteins known as immunoglobulins (Ig), that circulate freely and react specifically with the invoking antigen. There are several classes (called isotypes) of Ig molecules—IgM, IgG, IgA, IgE, and IgD. IgM is the predominant antibody in the primary immune response (following initial exposure to an antigen). IgG usually appears later, following a primary infection, but is the predominant antibody

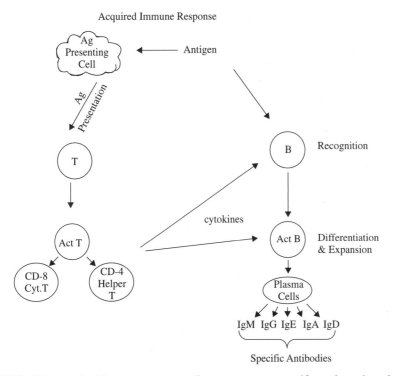

**Figure 19.2** The acquired immune response. In response to a specific antigen there is clonal expansion of B cells and subsequent production of antibodies (Ig) specific for that antigen. Antigen presenting cells process and present antigen to T cells. Again there is clonal expansion of cells specific for that antigen.

in the response to subsequent exposures. IgE acts as a mediator of allergy and parasitic immunity. IgA is found in secretions such as mucous, tears, saliva, and milk, as well as serum, and acts locally to block entrance of pathogens through mucous membranes. IgD is mainly membrane bound on B cells. Little is known about the function of this isotype. It does not appear to have a unique role that affects host immunity.

A given B cell will form antibody against just one single antigen; however, during the lifetime of this cell, it can switch to make a different class of antibody. Isotype switching is mediated by T helper cells. B cells recognize two types of antigen: T-independent antigens, which activate the cell without T cell help (predominantly an IgM response), and T-dependent antigens, which required T cell help in order to activate B cells. Most antigens belong to this latter category. Antibodies that specifically recognize microbial antigens can, in combination with plasma proteins known as complement, lyse bacterial cells or neutralize virus. Also microbes complexed with antibody are more readily phagocytized.

T cells recognize antigen that is presented via an antigen-presenting cell (APC) such as macrophages or dendritic cells. APCs process and present short peptide fragments complexed with major histocompatibility (MHC) molecules on the surface of the APC. This processing and presentation is required for T cell activation. There are two major divisions of T cells that are distinguished by expression of different cell surface markers (CD4 and CD8). CD-4 cells are also know as T-helper cells because they provide help for B cell activation. CD-8 cells are also known as cytotoxic T cells because they lyse cells expressing specific viral or tumor antigens.

As indicated above the thymus plays a key role in T cell differentiation. Pre-T cells migrate from the bone marrow to the thymus. As relatively immature cells, T cells express both CD4 and CD8 molecules. As maturation progresses these cells undergo both positive and negative selection. During positive selection only cells that bind to MHC with a certain affinity survive. As a result of this process T cells become MHC restricted; that is, they will only respond to antigen presented in association with MHC. Cells that survive positive selection are potentially able to respond to self proteins. However, before T cells leave the thymus negative selection occurs during which self-reactive cells are removed or functionally inactivated. During the course of positive and negative selection CD4+ CD8+ cells down-regulate the expression of one of these molecules such that mature T cells express only CD4 or CD8. Mature T cells leave the thymus and populate secondary lymphoid organs.

## 19.3 IMMUNE SUPPRESSION

Experimental studies in laboratory rodents have demonstrated that a diverse array of chemical exposures suppress immune function (Table 19.2). In addition a limited number of clinical and epidemiologic studies have reported suppression of immune function and/or increased frequency of infectious and/or neoplastic disease following exposure of humans to some of these agents. From the description above it is clear there are a number of cellular and molecular targets for chemicals that act as immunosuppressants. Clearly, a chemical that disrupts cell proliferation would affect clonal expansion. Disruption of T cell maturation in the thymus is another potential mechanism for immune suppression. Chemicals may also interfere with receptor ligand binding at the cell

**Table 19.2   Selected Examples of Immunosuppressive Agents**

*Drugs*

Cyclosporin A, cyclophosphamide, glucocorticoids (Dexamethazone), azothioprine

*Metals*

Lead, cadmium, methylmercury, organotins[a]

*Pesticides*

Chlorodane[a], DDT[a], Dieldrin[a]

*Industrial compounds*

2,3,7,8-Tetrachlorodibenzo-p-dioxin (TCDD), polychlorinated and polybrominated biphenyls (PCBs and PBBs), benzene, poly aromatic hydrocarbons[a]

*Addictive substances*

Cocaine, ethanol, opiates, cannabinoids, nicotine

*Air pollutants*

Environmental tobacco smoke, ozone, nitrogen dioxide

*Microbial toxins*

Aflatoxin,[b] ochratoxin A,[b] trichothecenes T-2 toxin[b]

*Radiation*

Ionizing, UV

*Other*

Asbestos, diethylstilbestrol (DES), dimethylnitrosamine

[a]Effects in humans are unknown; for all other compound without superscripts changes have been demonstrated in both rodents and humans.
[b]Effects in humans unknown, but veterinary clinicians have noted immunosuppression in livestock ingesting mycotoxins at levels below those that cause overt toxicity.

surface and/or the cascade of signals that lead to transcription of genes responsible for generating and regulating the appropriate immune responses.

Because of the complexity of the immune system, tiered approaches to testing chemicals for immunosuppressive potential have been developed. Like other types of toxicity testing, the first level of the tier (Table 19.3) frequently relies solely on structural end points, including changes in the weight of thymus and other lymphoid organs, histopathology of these organs, or differential blood cell counts. This type of evaluation is convenient because it can be carried out along with an evaluation for other organ systems during routine toxicity testing using one set of animals. However, although these nonfunctional endpoints may be effective in identifying gross (high dose) immunotoxic effects, they are not very accurate in predicting changes in immune function or alterations in susceptibility to challenge with infectious agents or tumor cells at lower chemical doses. Hence the first testing tier (Table 19.3) often includes functional end points designed to assess (1) antibody-mediated responses, (2) T-cell-mediated responses, and (3) NK cell activity. The most commonly used immune function assay in laboratory animals assesses the ability of a mouse or rat to respond to challenge with an antigen, usually sheep red blood cells (SRBC) (Figure 19.3). The response is assessed by determining the number of antigen specific antibody (IgM)

**Table 19.3    Tier I Tests (Screen) for Immune Suppression Using Laboratory Rodents**

| | |
|---|---|
| Immunopathology | Hematology: Complete blood count and differential |
| | Weights: body, spleen, thymus |
| | Histology: Spleen, thymus, lymph node |
| Antibody-mediated immunity | IgM plaque-forming cell (PFC) response to T cell-dependent antigen (e.g., SRBC) |
| Cell-mediated immunity | Lymphoproliferative response: T cell mitogens (Con A and PHA) Allogeneic mixed leukocyte response (MLR) |
| Nonspecific immunity | Natural killer (NK) cell activity |

Note: For details on specific assays see M. I. Luster et al., *Fund Appl. Tox.* **10**: 2–19, 1988.

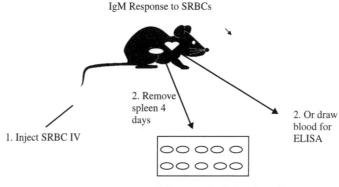

**Figure 19.3**   Assessing chemicals for immunosuppressive effects. The most common approach to accomplish this goal is to inject chemical and vehicle treated mice or rats with antigen and assess the antibody response. Most often the antigen injected is sheep red blood cells (SRBC); four days later slides are made with a single cell suspension of spleen cells, sheep red blood cells, and complement immobilized in agar. Slides are incubated and spleen cells making antibody against SRBC lyse the surrounding RBCs generating plaques. Plaques are counted to determine the number of antibody forming cells. Alternatively, serum can be obtained and an ELISA assay performed to detect SRBC specific antibody.

forming cells (AFC) in the spleen (Jerne assay) or by assessing antigen specific antibodies in serum using an enzyme-linked immunosorbent assay (ELISA). Because the SRBC is a T-dependent antigen, T and B cells, as well as antigen presenting cells, must be functional to have a successful immunization. Suppression of this response is highly predictive of suppression of other immune function tests and also correlates well with tests that assess resistance to challenge with an infectious agent or tumor cells. The disadvantage to this test is that it usually requires a dedicated set of animals because of the antigen challenge. The most common approach has been to treat the animals for 14 to 28 days with the xenobiotic of interest, inject the antigen at the end of that exposure, and collect spleen or serum 4 to 5 days later. Unlike the tests for antibody-mediated immunity, tier 1 tests for cell-mediated immunity, and natural killer cell activity can be done ex vivo and do not require a dedicated set of animals. However, these tests focus on one cell type and are not as predictive of overall immunocompetence as the antibody assays.

**Table 19.4    Tier II More Indepth Evaluation of Immunosuppressive Chemicals**

| | |
|---|---|
| Immunopathology | Quantitation of B and T cell numbers using flow cytometry |
| Antibody-mediated immunity | IgG PFC to SRBC |
| | IgM PFC to T cell-independent antigen (e.g., TNP-LPS) |
| Cell-mediated immunity | Cytotoxic T lymphocyte (CTL) cytolysis |
| | Delayed hypersensitivity response (DHR) |
| Nonspecific immunity | Macrophage: phagocytosis, bactericidal/tumoricidal activity) |
| | Neutrophil: function (phagocytosis and bactericidal activity) |
| Host resistance models | Response to challenge with infectious agent or tumor cells |

Note: For details on specific assays see M. I. Luster et al., *Fund Appl. Tox.* **10**: 2–19, 1988.

When immunosuppressive effects are noted in tier 1, an in-depth evaluation using more sophisticated tests may be carried out (tier 2, Table 19.4). This might include enumeration of lymphocyte subsets (B cells, total T cells, and CD4+ and CD8+) using flow cytometry or assessment of the IgM response to a T-independent antigen in an effort to determine what portion of the immune response is the actual target. Unlike tier 1, tests of cell-mediated immunity in tier 2 require administration of an antigen and subsequent test for cytotoxic T cells (e.g., against an immunizing tumor cell) or a delayed type hypersensitivity response (similar to the response to a tuberculin test). In order to understand the mechanism's underlying immune suppression, a host of other tests can be carried out, including expression of an assortment of cytokines.

Tier 2 also include host resistance models, tests in which an animal is exposed to a xenobiotic and then challenged with an infectious agent or tumor cells. This is considered the ultimate test for an adverse effect on the immune system. However, it should be noted that the amount of immune suppression that can be tolerated is greatly dependent on the dose and virulence of the challenging agent, as well as the genetics of the host. Manipulation of these variables can affect greatly results obtained in host resistance tests.

As in animal studies, human clinical data obtained from routine hematology (differential cell counts) and clinical chemistry (serum immunoglobulin levels) may provide general information on the status of the immune system in humans. However, as with the animal studies, these may not be as sensitive nor as informative as assays that target specific components of the immune system and/or assess function. The assessment of certain lymphocyte surface antigens has been successfully used in the clinic to detect and monitor the progression or regression of leukemias, lymphomas, and HIV infections, all diseases associated with severe immunosuppression. However, there is considerable variability in the "normal" human population, such that the clinical significance of slight to moderate quantitative changes in the numbers of immune cell populations is difficult to interpret. There is consensus within the immunotoxicology community that tests that measure the response to an actual antigen challenge are likely to be more reliable predictors of immunotoxicity than flow cytometric assays for cell surface markers because the latter generally only assesses the state of the immune system at rest. For ethical reasons it is not possible to immunize humans with SRBC. One approach under consideration is assessing responses to vaccines in chemically exposed populations. This approach has been used successfully to demonstrate a link between mild, stress-induced suppression of the antibody response to influenza vaccine and enhanced risk of infectious disease.

There is some debate over how to interpret immunotoxicity data with respect to adversity. The most conservative interpretation is that any significant suppression of an immune response is adverse because a linear relationship between immune suppression and susceptibility is assumed. Supporting this notion is the fact that apparently immunocompetent individuals suffer from infections, suggesting that adverse effects can occur even when known immune suppression is zero. Others argue that there is clearly redundancy and reserve capacity in the immune response and that some suppression should be tolerable. It is impossible to establish a quantitative relationship between immune suppression and increased risk of infection because both the genetics of the host and the virulence and dose of the infectious agent will influence this relationship. Immunocompetence in a population can probably be represented as a bell-shaped curve, such that a portion of the population is highly susceptible to infection, a portion is highly resistant, and the remaining population falls somewhere in between (Figure 19.4). Genetics, age, and preexisting disease all contribute to the risk represented by this curve. In addition the portion of the population at risk is determined by the dose and virulence of any infectious agent that might be encountered. The higher the dose and the virulence, the more people are at risk. Exposure to an immunosuppressive agent shifts the whole bell-shaped curve to the left, thus increasing the population at risk. Unfortunately, it is difficult to determine more quantitatively the relationship between small decrements in immune responsiveness and the degree of change in the population at risk.

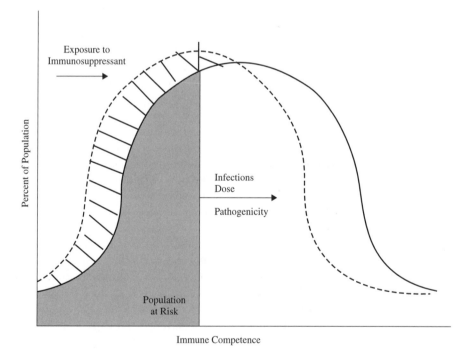

**Figure 19.4** Adverse effect of immune suppression. Immune competence is represented by a bell-shape curve. The shaded area represents the population at risk of infection, which increases or decreases depending on the dose and virulence of infectious agents that are encountered. Exposure to an immunosuppressant shifts the whole curve to the left, such that a larger population is at risk for any given infectious challenge.

## 19.4  CLASSIFICATION OF IMMUNE-MEDIATED INJURY (HYPERSENSITIVITY)

Under certain circumstances immune responses can produce tissue damage. These deleterious reactions are collectively known as hypersensitivity or allergy. Hypersensitivity reactions have been divided into four types (originally proposed by Gell and Coombs) based on mechanism (Table 19.5). In all cases the adverse effects of hypersensitivity develop in two stages: (1) Induction (sensitization) requires a sufficient or cumulative exposure dose of the sensitizing agent to induce immune responses that cause no obvious symptoms. (2) Elicitation occurs in sensitized individuals upon subsequent exposure to the antigen and results in adverse antigen-specific responses that include inflammation.

Type I hypersensitivity (sometimes referred to as atopy) is mediated by an antigen-specific cytophilic antibody (usually IgE) that binds to mast cells and basophils. On subsequent exposure, the allergen binds to these cell-bound antibodies and cross-links IgE molecules, causing the release of mediators such as histamine and slow-reacting substance of anaphylaxis (SRS-A). These mediators cause vasodilation and leakage of fluid into the tissues, plus sensory nerve stimulation (leading to itching, sneezing, and cough). Type I is also called immediate-type hypersensitivity because reactions occur within minutes after exposure of a previously sensitized individual to the offending

**Table 19.5   Classification of Hypersensitivity Reactions**

| Type | Mechanisms | | Example |
|---|---|---|---|
| | Induction (Initial Exposure to Antigen) | Elicitation (Re-exposure to Antigen) | |
| I (immediate) | Clonal expansion of B cells; Cytophilic antibody (IgE) generated; binds to mast cells | Antigen binds to cell bound antibody, cross-links receptors, causing release of mediators | Anaphylactic response to bee sting |
| II (cytolytic) | Clonal expansion of B cells; IgM, IgG generated. Antigen binds to cell membrane | Anamnestic[a] Ig response; binds to cell bound antigen and in the presence of complement or activated macrophages cell lysis occurs | Rh factor incompatibility, Hemolytic anemia in reaction to drug treatment |
| III (Arthus) | Clonal expansion of B cells; IgM, IgG generated | Anamnestic Ig response; antigen antibody complexes form in some tissues leading to inflammation | glomerular nephritis, rheumatic heart disease, farmers lung |
| IV (delayed) | Clonal expansion of antigen-specific T cells occurs | T cells activated, release cytokines, activate macrophages, inflammation | contact dermatitis |

[a]Heightened response on re-exposure to antigen.

antigen. Type I reactions include immediate asthmatic responses to allergen, allergic rhinitis (hay fever), atopic dermatitis (eczema), and acute urticaria (hives). The most severe form is systemic anaphylaxis (e.g., in response to a bee sting), which results in anaphylactic shock, and potentially death.

Type II hypersensitivity is the result of antibody-mediated cytotoxicity that occurs when antibodies respond to cell surface antigens. Antibodies bound to antigen on the cell surface activate the complement system and/or macrophages, leading to lysis of the target cell. Frequently blood cells are the targets, as in the case of an incompatible blood transfusion or Rh blood incompatibility between mother and child. The basement membrane of the kidney or lung may also be a target. Autoimmune diseases can result from drug treatments with penicillin, quinidine, quinine, or acetaminophen. Apparently these drugs interact with the cell membrane such that the immune system detects "foreign" antigens on the cell surface. This type of autoimmune disease may also have unknown etiologies.

Type III reactions are the result of antigen-antibody (IgG) complexes that accumulate in tissues or the circulation, activate macrophages and the complement system, and trigger the influx of granulocytes and lymphocytes (inflammation). This is sometimes referred to as the Arthrus reaction and includes postinfection sequelae such as rheumatic heart disease. Farmer's lung, a pneumonitis caused by molds has been attributed to both type III and type IV, and some of the late phase response (4–6 hours after exposure) in asthmatics may be the result of Arthrus-type reactions.

Unlike the preceding three types, type IV, or delayed-type hypersensitivity (DTH), involves T cells and macrophages, not antibodies. Activated T cells release cytokines that cause accumulation and activation of macrophages, which in turn cause local damage. This type of reaction is very important in defense against intracellular infections such as tuberculosis, but is also responsible for contact hypersensitivity responses (allergic contact dermatitis) such as the response to poison ivy. Inhalation of beryllium can result in a range of pathologies, including acute pneumonitis, tracheobronchitis, and chronic beryllium disease, all of which appear to be due to type IV beryllium-specific immune responses. The expression of type IV responses following challenge is delayed, occurring 24 to 48 hours after exposure.

The different types of immune-mediated injury are not mutually exclusive. More than one hypersensitivity mechanism may be involved in the response to a particular antigen. Also the resulting pathology, particularly that caused by type III and IV, reactions may appear very similar, although the mechanisms leading to the effect are different.

## 19.5 EFFECTS OF CHEMICALS ON ALLERGIC DISEASE

Xenobiotics can affect allergic disease in one of two ways. They can themselves act as antigens and elicit hypersensitivity responses, or they can enhance the development or expression of allergic responses to commonly encountered allergens, such as dust mite. Chemicals that act as allergens include certain proteins that can by themselves induce an immune response and low molecular weight chemicals (known as haptens) that are too small to induce a specific immune response but may react with a protein to induce an immune response that is then hapten specific. Haptens have been associated with both allergic contact dermatitis (ACD), sometimes called contact hypersensitivity

(CHS), and respiratory hypersensitivity. Proteins have been associated with respiratory hypersensitivity and food allergies. When a chemical is an allergen or a hapten, there are two doses of concern, the sensitizing dose and the elicitation dose. In general, the dose required for sensitization is greater than that required to elicit a response in a sensitized individual. Chemicals that enhance the development of allergic sensitization are referred to as adjuvants. Air pollutants have been associated both with enhanced sensitization and exacerbation of allergic respiratory symptoms.

### 19.5.1 Allergic Contact Dermatitis

Allergic contact dermatitis (ACD) or contact hypersensitivity (CHS) is one of the most common occupational health problems and hence is one of the most common problems associated with immunotoxicity. It is a type IV response that occurs as a result of dermal exposure to chemicals that are haptens. Following dermal exposure, the chemical reacts with host cell protein at the surface of the skin and is picked up by epidermal dendritic cells, known as Langerhans cells. Cytokines released from the epidermal keratinocytes and from Langerhans cells cause maturation and mobilization of the Langerhans cells, which travel to the draining local lymph node and present antigen to lymphocytes. Clonal expansion occurs, enlarging the number of T lymphocytes specific for that allergen and generating memory cells that, in addition to specificity for the allergen, have the propensity to home to the skin. On re-exposure to the chemical, these specific T cells are activated, proliferate, home rapidly to the site of exposure, and produce erythema and edema typical of a type IV response. The reaction to poison ivy is the classic example.

Methods to assess chemicals (drugs, pesticides, dyes, cosmetics, and household products, etc.) for potential to induce CHS are well established, and several protocols using guinea pigs have been in use since the 1950s. These protocols assess the actual disease end point, skin erythema, and edema, following sensitization and challenge with the test agent. Two commonly used tests are the guinea pig maximization test and the Buehler occluded patch test. The sensitization procedure for the maximization tests includes intradermal injection of the test chemical with an adjuvant (intended to enhance the sensitization process) as well as topical application. The Buehler test relies on topical sensitization alone. In both cases, after approximately 2 weeks, animals are challenged at a different site on the skin and erythema and edema are assessed 24 to 48 hours later. This assessment is somewhat subjective and these tests are fairly expensive.

A chemical is considered to be a sensitizer if 30% (maximization) or 15% (Buehler) of the animals respond. Recently a more economical, less subjective, test for CHS has been developed using mice. This test, the local lymph node assay (LLNA), assesses the proliferative response of lymphocytes in the draining lymph node following application of the agent to the ear and is based on our understanding of the immunologic mechanisms underlying CHS; that is, clonal expansion has to occur in the draining lymph node if there is to be allergic sensitization (Figure 19.5). The LLNA is gaining acceptance as a stand-alone alternative to the guinea pig tests and is likely to become the assay of choice.

Finally structure activity approaches have recently been developed to identify contact sensitizers. This approach is based on the concept that the biologic mechanisms that determine a chemical's effect are related to its structure and hence chemicals with similar structures will have similar effects. Computer models have been developed to compare the structure of an unknown chemical to structures in a database for known

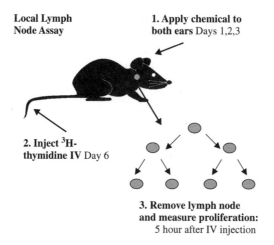

**Local Lymph
Node Assay**

**1. Apply chemical to
both ears** Days 1,2,3

**2. Inject** $^3$**H-
thymidine IV** Day 6

**3. Remove lymph node
and measure proliferation:**
5 hour after IV injection

**Figure 19.5**  Assessing chemicals for potential contact sensitivity. In the local lymph node assay the chemical in question is applied to both ears on three consecutive days. Control mice are treated with vehicle. Radioisotope is injected intravenously on day 6. The draining lymph nodes are removed 5 hours later and the proliferative response is measured by the incorporation of radio isotope. Results are frequently presented as a stimulation index (counts per min (cpm) for the test chemical/cpm for control). (Picture adapted from D. Sailstad, *Lab Animal* **31**: 36, 2002.)

contact sensitizers. CHS lends itself to this approach because there is a large database of chemicals known to cause it, and there is a reasonable understanding of chemical characteristics that facilitate skin penetration, chemical reactivity with host proteins, and immune reactivity.

Because nonspecific inflammatory responses also can occur following chemical exposure to the skin, a distinction must be made between an irritant and a sensitizer. An irritant is an agent that causes local inflammatory effects but induces no immunological memory. Therefore, on subsequent exposures, local inflammation will again result, but there is no enhancement of the magnitude of the response and no change in the dose required to induce the response. In immunologically mediated inflammation (hypersensitivity) there may be no response to a sensitizer during the induction stage, but responses to subsequent exposures are exacerbated. The dose required for elicitation is usually less than that required to achieve sensitization.

### 19.5.2  Respiratory Allergens

There is evidence that both occupational and environmental exposures to chemicals (both proteins and haptens) can result in the induction or exacerbation of respiratory allergies (Table 19.6). Of particular concern is the induction of allergic asthma. In sensitized asthmatic individuals the antigen challenge generally causes a type I (IgE-mediated) immediate hypersensitivity response with release of mediators responsible for bronchoconstriction. Between 2 and 8 hours after the immediate response, asthmatics experience a more severe and prolonged (late phase) reaction that is characterized by mucus hypersecretion, bronchoconstriction, airway hyperresponsiveness to a variety of nonspecific stimuli (e.g., histamine, methacholine), and airway inflammation characterized by eosinophils. This later response is not mediated by IgE.

**Table 19.6 Example of Chemicals Associated with Respiratory Allergy**

| *Proteins* |
| --- |
| Enzymes |
| Latex |
| Animal dander |
| Dust mite |
| Molds |
| Cockroach |
| Microbial pesticides |

*Low molecular weight (<3000)-haptens*

Toluene diisocyanate
Diphenylmethane diisocyanate
Phthalic anhydride
Trimellitic anhydride
Platinum salts
Reactive dyes

*Adjuvants*

Ozone
Nitrogen dioxide
Diesel exhaust
Residual oil fly ash

Although proteins are generally immunogens, not all proteins are allergens and there is a range of potencies for those that are. There is also a strong genetic component associated with susceptibility to develop allergic reactions to proteins. Susceptible individuals are called atopic. There is at present no structural motif that can be used to characterize a protein as an allergen for hazard identification. Examples of occupational protein exposures associated with respiratory allergy and asthma include enzymes, latex, flour (both the grain itself and fungal contaminants), and animal dander. Environmental (mostly indoor) exposure including molds, spores, dust mite, animal dander, and cockroach have also been associated with this type of respiratory disease. Because this is a type 1 response, cytophilic antibodies (IgE) specific for the allergen are frequently used to identify proteins that may cause this effect. For example, in order to determine the etiology of occupational asthma in human subjects, the skin prick test is often used. Different proteins are injected under the skin to test for the presence of cytophilic antibodies in order to identify which proteins are causing a response in an individual. Serum may also be tested for protein specific IgE. Because IgE can sometimes be detected in the absence of respiratory responses, a positive IgE test may be followed by an assessment of respiratory responses.

Under very controlled situations patients may be exposed via the respiratory route to suspect allergens (broncho- provocation test) and respiratory function monitored to pinpoint the offending allergen. Guinea pigs and mice have been used to test proteins for potential allergenicity. Animals are usually sensitized by the respiratory route and monitored for the development of cytophilic antibody (IgG1 in guinea pigs, IgE in mice) as well as increased respiratory rate and other changes in pulmonary function.

The guinea pig intratracheal test has been used to establish the relative potency of different detergent enzymes and establish safe occupational exposure levels. As the name implies, guinea pigs are sensitized by intratracheal exposure and induction of cytophilic antibodies are assessed. Dose responses obtained for new enzymes are compared to a reference enzyme for which safe exposure levels have been established. The relative potency of the new enzyme to this reference is used to establish a safe exposure level for the new enzyme.

Exposure to certain low (<3000) molecular weight compounds (haptens) has also been associated with the development of occupational asthma. Highly reactive compounds such as the diisocyanates or acid anhydrides have the capacity to react with protein and induce an immune response. Toluene diisocyanate (TDI) and trimellitic anhydride are the compounds that have been most extensively studied in this regard. There is a great deal of interest in developing a test to screen chemicals for this type of effect in order to avoid induction of immune responses that could lead to occupational asthma. Although specific IgE antibodies have been detected in some individuals with TDI asthma, it has not been uniformly present and some of these individuals exhibit the late phase but not the immediate response. Hence, unlike proteins, there is less certainty about the mechanisms underlying respiratory allergic responses to low molecular weight compounds.

Structure activity approaches similar to those described for contact sensitizers have been developed, but this approach has limitations because the database of known respiratory sensitizers is small compared to contact sensitizers and the underlying mechanisms are less well defined. At the other extreme guinea pigs have been exposed by inhalation for a number of days, rested, and then challenged at a later date by inhalation with subsequent monitoring of respiratory responses. Although this approach has produced a good model of TDI asthma, it is too cumbersome and expensive for routine testing. Because the capacity to interact with protein is a pre-requisite to allergenicity, it has been suggested that testing for protein reactivity in vitro could provide an initial screening test for chemicals. Also, because it appears that respiratory sensitizers are a subset of chemicals that produce positive results in a contact sensitivity test, it has been suggested that the LLNA test be used as the first tier in screening chemicals for this effect. The problem then becomes separating chemicals that are strictly contact sensitizers from those that have the capacity to cause respiratory sensitization. Efforts have been made to determine whether differences in responses to dermal application of these chemicals could provide a means for making this distinction. One proposal is to assess total serum IgE following dermal exposure, assuming that respiratory sensitizers would produce a bigger IgE signal. Another approach has been to assess cytokine profiles in the draining lymph node following dermal exposure. Different subsets of T helper (Th) cells, have been associated with type I immediate (Th2) and type IV delayed (Th1) responses. These different populations of T cells are distinguished by different cytokine profile and efforts are underway to use these differing profiles to distinguish respiratory from contact sensitizers. However, there is as yet no well-validated, well-accepted test to assess low molecular weight chemicals for the capacity to induce respiratory allergy. This remains a subject of intense research.

### 19.5.3  Adjuvants

An adjuvant is a compound administered in conjunction with an antigen that non-specifically enhances the immune response to that antigen. Adjuvants are used in vaccines to promote immunogenicity. There is now growing concern that chemicals in our environment (particularly, air pollutants) might act as adjuvants for allergic sensitization to common allergens such as dust mite and pollen. Laboratory rodents have been used to show that nitrogen dioxide, residual oil fly ash, and diesel exhaust enhance allergic sensitization and disease. Enhanced sensitization to an allergen has also been demonstrated in rhesus monkeys exposed to ozone and humans exposed to diesel exhaust. The significance of these findings in terms of enhanced burden of respiratory allergies in the human population is unclear. As in other areas of toxicology, simultaneous environmental exposures to agents that are not the agent of immediate concern can certainly influence outcomes. Adjuvancy is a concern that likely extends beyond air pollution and type 1 responses.

## 19.6   EMERGING ISSUES: FOOD ALLERGIES, AUTOIMMUNITY, AND THE DEVELOPING IMMUNE SYSTEM

There are several emerging issues in immunotoxicology. These active areas of research will be only briefly described here because there are currently more questions than answers.

Toxicologists have recently been drawn into the area of food allergy by advances in biotechnology and the need to assess the safety of genetically modified foods in terms of potential allergenicity. There is concern that insertion of a novel gene into a food crop (e.g., to increase yield or pest resistance) might inadvertently introduce a new allergen into the food supply. Food allergies are relatively rare, affecting approximately 5% of children and 2–3% of adults, and even in these individuals, most proteins are not food allergens. However, when food allergy does occur, the consequences can be severe. Anaphylactic (life-threatening) reactions to peanuts provide the best example. Unfortunately, the mechanisms underlying food allergies (or the mechanisms that protect most of people from developing reactions to the foreign proteins they eat), the characteristics that make a protein a food allergen, and the characteristics that make an individual susceptible to food allergies are poorly understood at this time. These are some of the issues that need to be resolved in order to develop appropriate safety assessment tools.

Autoimmune diseases affect about 3% of the population and comprise a diverse array of both organ specific (e.g., type I diabetes, thyroiditis) and systemic (systemic lupus erythematosis) diseases. Susceptibility includes a strong genetic component, and in some cases women appear to be more vulnerable than men. Xenobiotics might affect the development or progression of autoimmune disease. A variety of mechanisms could contribute to xenobiotic effects on the development and maintenance of immune tolerance or unmasking or modification of self proteins. There is evidence that exposure to certain drugs, heavy metals, silica, and endocrine disruptors are a concern in this regard. Current research includes both human and animals studies to determine the extent of risk and ways to assess and control it.

Finally there is growing concern that the developing immune system may be particularly vulnerable to xenobiotic exposures and that perinatal and/or in utero exposures may have a lifelong impact on susceptibility to infectious, allergic, or autoimmune disease. As in other areas of toxicology, tests designed to assess the risk of immunotoxicity for adults may not be sufficient to protect children and research is currently underway to determine how best to meet this need.

Clearly, exposure to xenobiotics can have a number of effects on the immune system that in turn can affect an array of health outcomes. In some areas of immunotoxicology significant progress has been made in terms of identifying and understanding the risks associated with xenobiotic exposure. In other areas more research is needed.

## SUGGESTED READING

Benjamini, E., G. Sunshine, and S. Leskowitz. *Immunology A Short Course*, 4th ed. New York: Wiley, 2000.

Gilmour, M. I., M. J. K. Selgrade, and A. L. Lambert. Enhanced allergic sensitization in animals exposed to particulate air pollution. *Inhalation Toxicol.* **12** (suppl. 3): 373–380, 2000.

Kimber, I., I. L. Bernstein, M. H. Karol, M. K. Robinson, K. Sarlo, and M. J. K. Selgrade. Workshop overview: Identification of respiratory allergens. *Fundam. Appl. Toxicol.* **33**: 1–10, 1996.

Luster, M. I., A. E. Munson, P. T. Thomas, M. P. Holsapple, J. D. Fenters, K. L. White, L. E. Lauer, D. R. Germolec, G. J. Rosenthal, and J. H. Dean. Development of a testing battery to assess chemical-induced immunotoxicity: National Toxicology Program's guidelines for immunotoxicity evaluation in mice. *Fundam. Appl. Toxicol.* **10**: 2–19, 1988.

Luster, M. I., et al. Risk assessment in immunotoxicology I. Sensitivity and predictability of immune tests. *Fundam. Appl. Toxicol.* **18**: 200–210, 1992, and II. Relationships between immune and host resistance tests. *Fundam. Appl. Toxicol.* **21**: 71–82, 1993.

Metcalfe, D. E., et al. Assessment of the allergenic potential of foods derived from genetically engineered crop plants. *Crit. Rev. Food Sci. Nutr.* **36S**: S165–S186, 1996.

Roitt, I. *Roitt's Essentials in Immunology*. London: Blackwell, 2001.

Sailstad, D. M. Murine local lymph node assay: An alternative test method for skin hypersensitivity testing. *Lab. Animal* **31**: 36–41, 2002.

Sarlo, K. Human Health Risk Assessment: Focus on Enzymes. In *Proceeding of the 3rd World Conference on Detergents*, A. Cahn, ed. Champaign, I: Am. Oil Chem. Soc. Press, 1994, pp. 54–57.

# CHAPTER 20

# Reproductive System

STACY BRANCH

## 20.1 INTRODUCTION

What is reproductive toxicity? Reproductive toxicity refers to any adverse effect on any aspect of male or female sexual structure, function, and lactation including effects on the reproductive potential and viability of the offspring. This concept may also include the following:

*Organ toxicity*. Can interfere with normal system function.

*Teratogenicity*. Ability to cause dysmorphogenesis in the developing fetus.

*Behavioral teratogenicity*. Ability to adversely affect the mental development of the fetus.

*Developmental toxicity*. Includes both teratogenicity terms above and abnormal postnatal development.

This chapter discusses both male and female reproductive toxicity and provides an overview of reproductive physiology.

## 20.2 MALE REPRODUCTIVE PHYSIOLOGY

The anterior pituitary is stimulated by the hypothalamus (via the gonadotrophic hormone releasing hormone) to release gonadotrophic hormones (leutenizing hormone—LH; and follicle stimulating hormone—FSH). In addition to LH and FSH, prolactin is released by the anterior pituitary. The target of LH and FSH in the male is the testis. While LH stimulates steroidogenesis, FSH has its primary effects on the sertoli cells. The role of prolactin (which is inhibited by dopamine) is to modulate the effects of LH in the testicular tissue. Critical points within the hypothalamic–pituitary–gonadal axis (Figure 20.1) may be susceptible to alterations by xenobiotics, leading to altered reproductive function and pathology.

*A Textbook of Modern Toxicology, Third Edition,* edited by Ernest Hodgson
ISBN 0-471-26508-X  Copyright © 2004 John Wiley & Sons, Inc.

**Physiology of Reproductive System**

*Hypothalamic-pituitary-Gonadal Axis*

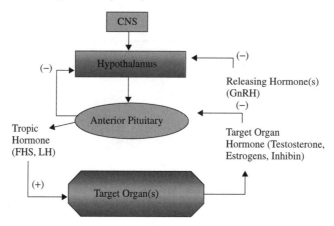

**Figure 20.1** Hypothalamic–pituitary–gonadal axis. Negative feedback is designated by (−), and positive feedback is designated by (+).

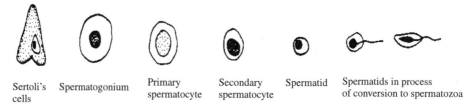

| Sertoli's cells | Spermatogonium | Primary spermatocyte | Secondary spermatocyte | Spermatid | Spermatids in process of conversion to spermatozoa |

**Figure 20.2** Spermatogenesis. (Adapted from J. A. Thomas, in *Casarett and Doull's Toxicology: The Basic Science of Poisons*, 6th ed., C. D. Klassen, ed., McGraw-Hill, 2001.)

Two major components of the testis are the seminiferous tubules (site of spermatogenesis) and the interstitial compartment. The interstitial compartment contains Leydig cells, which produce testosterone under the influence of LH. Androgens control spermatogenesis (Figure 20.2), growth and activity of accessory sex glands, masculinization, male behavior, and various metabolic functions. Secretion of androgen by the developing fetal testis is essential for differentiation of the gonads, which includes regression of Müllerian ducts and the development of Wolffian ducts.

## 20.3 MECHANISMS AND TARGETS OF MALE REPRODUCTIVE TOXICANTS

### 20.3.1 General Mechanisms

Toxicants may mimic endogenous compounds (i.e., hormones), thus acting as agonists or antagonists. Toxicants may be directly cytotoxic or may be activated to toxic compounds. Some toxicants may have indirect effects by inhibiting key enzymes involved in steroid synthesis. Below are examples of how selected toxicants affect various susceptible targets of the male reproductive tract.

### 20.3.2 Effects on Germ Cells

Epidemiological data have indicated that wives of men exposed to the compound vinyl chloride experience spontaneous abortion. Vinyl chloride is used to manufacture polyvinyl chloride (PVC). PVC is a component of various plastic products, including pipes, furniture, and automobile upholstery. Ethylnitrosourea (ENU) is a mutagenic agent that has been used extensively in mouse mutagenesis. ENU acts on male spermatogonial stem cells, introducing mutations. It is also capable of inducing reversible sterility in mice. Actinomycin D is an older chemotherapeutic drug that has been used in cancer therapy for many years. Actinomycin D is commonly used in the treatment of gestational trophoblastic cancers, testicular cancer, Wilm's tumor, and rhabdomyosarcoma. Actinomycin D intercalates with DNA and disrupts the structure and function of the DNA of the spermatozoa.

### 20.3.3 Effects on Spermatogenesis and Sperm Quality

The antibiotics and antimetabolites used in cancer treatment (i.e., vinblastine) are spermatotoxic and affect semen quality. Ionizing radiation is also spermatotoxic (with spermatogonial cells being the most sensitive). Prolonged scrotal heating is a factor that affects the earlier states of spermatogenesis.

### 20.3.4 Effects on Sexual Behavior

Anabolic steroids, antidepressants and drugs of abuse affect libido, potency, and ejaculatory function. Anabolic steroids are derivatives of testosterone, and have strong genitotropic effects. There is published evidence indicating that anabolic steroids increases sexual desire; however, the frequency of erectile dysfunction is also increased. Treatment with the antidepressant fluoxetine has been associated with sexual side effects including delayed or nonexistent ejaculation and hyposexuality. Mice treated in utero with the anitleukemic agent 5-aza-2'-deoxycytidine exhibit abnormal reproductive behavior and low reproductive capacity.

### 20.3.5 Effects on Endocrine Function

Cimetidine (for treatment of peptic ulcers) competes with dihydrotestosterone for receptors in the testis and accessory sex glands. The more common sequelae are low sperm count and gynacomastia. Epidemiological evidence has shown that occupation exposure to oral contraceptives can induce gynacomastia in exposed males. Diethylstilbestrol (DES) antagonizes the activity of fetal testosterone. In the male offspring, testicular hypoplasia, abnormal semen parameters, and infertility result. Ketoconazole has be shown to be transported to the seminal fluid and to immobilize the sperm.

## 20.4 FEMALE REPRODUCTIVE PHYSIOLOGY

As described previously for the male, the female hormonal signaling is composed of four primary levels: CNS, hypothalamus, anterior pituitary, and gonads. The gonadotropin-releasing hormones of the hypothalamus stimulates the anterior pituitary to release LH and FSH. Subsequently LH and FSHS stimulate the release of estrogen and progesterone from the ovaries (Figure 20.3). Estrogen is secreted in the growing follicle and has effects on the uterus. The oocytes are formed before birth, then develop into the primary oocytes after meiosis. At the time of puberty, the release of gonadotropin stimulates the oocytes to develop into graafian follicles (Figure 20.3).

***Ovarian Cycles.*** Estrus is the period when the female mammal is most receptive to the male (coincides with high levels of circulating estrogen). Rodents are considered to be polyestrous and have a succession of estrus cycles. Cats are seasonally (spring, early fall) polyestrus, while dogs are monestrous. Humans and higher primates cycle at monthly intervals. Although most mammals ovulate spontaneously, some mammals (cats and minks) undergo provoked or induced ovulation (i.e., stimulated by mating). The estrus cycle and the resulting differences in circulating hormone concentrations at different stages of the cycle are depicted in Figure 20.4. The changes in circulating hormone and the stage of follicle development during an adverse toxicant insult results in a variety of toxicological manifestations.

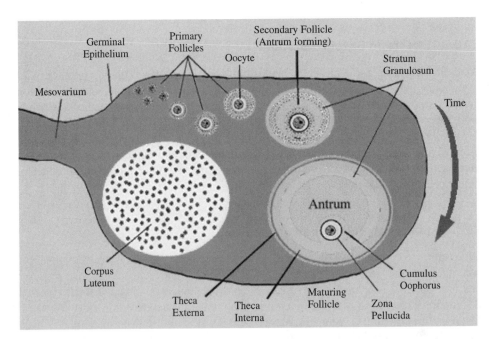

**Figure 20.3** The arrow follows the ovarian follicles (time course) from their maturation from primary follicles to the corpus luteum. (Adapted from Web site of Dr. Steven Scadding and Dr. Sandra K. Ackerley, http://www.uoguelph.ca/zoology/devobio/210labs/ovary4.html.)

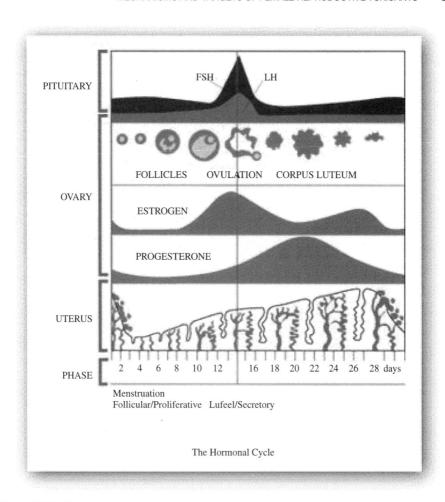

**Figure 20.4**   Women's menstrual cycle. (Courtesy of Women's Health Interactive, http://www. womens-health.com/health_center/gynecology/gyn_repro_menstrual.html.)

## 20.5   MECHANISMS AND TARGETS OF FEMALE REPRODUCTIVE TOXICANTS

Xenobiotics can adversely affect the normal functions of the cells/organs of the reproductive system. These agents may induce a variety of outcomes, including prevention of ovulation and impairment of ovum transport, fertilization, or implantation. Endocrine disruptors may mimic endogenous hormones as well as directly destroy cellular components, leading to cell death. More indirect effects may include inhibition of key enzymes involved in steroid synthesis.

### 20.5.1   Tranquilizers, Narcotics, and Social Drugs

Compounds within this class of substances can inhibit hypothalamic–pituitary–ovarian axis function by inhibiting gonadotropin secretion. Subsequently ovulation and estrus is suppress leading to infertility or reduced fertility.

## 20.5.2 Endocrine Disruptors (EDs)

Endocrine disruptors are compounds (synthetic and naturally occurring) that can alter the normal hormonal balance and function in animals. The historical ED diethylstilbestrol (DES) is a classic example of an endocrine disrupter affecting female reproductive health. In utero exposure of females to DES is associated with the induction of vaginal carcinomas apparent after puberty. In experimental mice, estrogenic substances cause accelerated sexual maturation and irregular estrous cycles and prolonged estrous. In rats, xenoestrogens such as kepone and methoxychlor cause masculinization of the exposed female rats. These rats do not ovulate, lack stimulation of the LH surge, and exhibit male sexual behavior. In humans, estrogen mimicking compounds can alter natural hormonal cycles and have been associated with breast cancer induction. Certain environmental EDs may function as promoters or inducers of carcinogenesis. Polychlorinated biphenyls (PCBs) and a trichloroethane compound (DDT) are persistent in the environment. Serum DDE (a DDT metabolite) levels have been found to correlate with breast cancer incidence.

## 20.5.3 Effects on Germ Cells

As previously described for the male reproductive toxicity, the class of toxicants affecting germ cells can alter the structure of genetic material (chromosomal aberrations, alterations in meiosis, DNA synthesis, and replication). Mature oocytes have a DNA repair capacity different from that of mature sperm, but this capacity decreases at the period of meiotic maturation.

## 20.5.4 Effects on the Ovaries and Uterus

Cyclophosphamide and vincristine are examples of alkylating agents capable of inducing gonadal dysfunction. Premature menopause is a primary outcome of exposure to these agents. Amenorrhea and abnormal hormonal levels are characteristics of the ovarian dysfunction induced by cyclophosphamide.

Premature ovarian failure can be induced in offspring exposed in utero by active metabolites such as 6-mercaptopurine. Tamoxifen (treatments for breast cancer) and clomiphene (to induce ovulation) are antiestrogens that can inhibit uterine decidual induction in pseudopregnant rats.

## 20.5.5 Effects on Sexual Behavior

Normal sexual activity is associated with ovulation in most female mammals. Compounds affecting this process can adversely affect female libido. Ovarian failure induced by xenobiotic compounds has been associated with a decrease in libido in women. Certain types of oral contraceptives as well as drugs of abuse (methadone, cannabis, alcohol) cause decreases in female libido. The treatment for hirsutism, excessive growth of hair in both normal and abnormal locations, is the compound cyproterone acetate. It is an antiandrogen that has the side effect of severely decreasing libido in women.

## SUGGESTED READING

Klaassen, C. D. *Cassarett and Doull's Toxicology: The Basic Science of Poisons*, 6th ed. New York: McGraw-Hill, 2001.

Korach, K. S. *Reproductive and Developmental Toxicology*. New York: Dekker, 1998.

Naz, R. K. *Endocrine Disruptors*. Boca Raton, FL: CRC Press, 1997.

Ballantyne, B., T. Marrs, and P. Turner, eds. *General and Applied Toxicology*, college ed., New York: Macmillan, 1995.

# PART VI

# APPLIED TOXICOLOGY

# CHAPTER 21

# Toxicity Testing

HELEN CUNNY and ERNEST HODGSON

## 21.1 INTRODUCTION

Although testing for toxicity, usually for the purposes of human health risk assessment, might be expected to be one of the more routine aspects of toxicology, it is actually one of the more controversial. Among the many areas of controversy are the use of animals for testing and the welfare of the animals, extrapolation of animal data to humans, extrapolation from high-dose to low-dose effects, and the increasing cost and complexity of testing protocols relative to the benefits expected. New tests are constantly being devised and are often added to testing requirements already in existence.

Most testing can be subdivided into in vivo tests for acute, subchronic, or chronic effects and in vitro tests for genotoxicity or cell transformation, although other tests are used and are described in this chapter. Any chemical that has been introduced into commerce or that is being developed for possible introduction into commerce is subject to toxicity testing to satisfy the regulations of one or more regulatory agencies. Furthermore compounds produced as waste products of industrial processes (e.g., combustion products) are also subject to testing.

*Toxicity assessment* is the determination of the potential of any substance to act as a poison, the conditions under which this potential will be realized, and the characterization of its action. *Risk assessment*, however, is a quantitative assessment of the probability of deleterious effects under given exposure conditions. Both are involved in the regulation of toxic chemicals. *Regulation* is the control, by statute, of the manufacture, transportation, sale, or disposal of chemicals deemed to be toxic after testing procedures or according to criteria laid down in applicable laws.

Testing in the United States is carried out by many groups: industrial, governmental, academic, and others. Regulation, however, is carried out by a narrow range of governmental agencies, each charged with the formulation of regulations under a particular law or laws and with the administration of those regulations. The principal regulatory agencies for the United States are shown in Table 21.1. Other industrialized countries have counterpart laws and agencies for the regulation of toxic chemicals.

*A Textbook of Modern Toxicology, Third Edition,* edited by Ernest Hodgson
ISBN 0-471-26508-X  Copyright © 2004 John Wiley & Sons, Inc.

**Table 21.1   Some Agencies and Statutes Involved in Regulation of Toxic Chemicals in the United States**

*Food and Drug Administration (FDA)*

Food, Drug, and Cosmetic Act

*Labor Department*

Occupational Safety and Health Act

*Consumer Product Safety Commission*

Consumer Product Safety Act

*Environmental Protection Agency (EPA)*

Federal Insecticide, Fungicide, and Rodenticide Act
Clean Air Act
Federal Water Pollution Control Act
Safe Drinking Water Act
Toxic Substances Control Act
Resource Conservation and Recovery Act

*State governments*

Various state and local laws
Enforcement of certain aspects of federal law delegated to states

Although the objective of much, but by no means all, toxicity testing is the elimination of potential risks to humans, most of the testing is carried out on experimental animals. This is necessary because our current knowledge of quantitative structure activity relationships (QSAR) does not permit accurate extrapolation to new compounds. Human data are difficult to obtain experimentally for ethical reasons but necessary for such deleterious effects as irritation, nausea, allergies, odor evaluation, and some higher nervous system functions. Some insight may be obtained in certain cases from occupational exposure data, although this often tends to be irregular in time and not clearly defined as to the composition of the toxicant or the exposure levels, because multiple exposure is common. Clearly, any experiments involving humans must be carried out under carefully defined conditions after other testing is complete.

Although for a variety of reasons extrapolation from experimental animals to humans presents problems, including differences in metabolic pathways, dermal penetration, mode of action, and others, experimental animals present numerous advantages in testing procedures. These advantages include the possibility of clearly defined genetic constitution and their amenity to controlled exposure, controlled duration of exposure, and the possibility of detailed examination of all tissues following necropsy.

Although not all tests are required for all potentially toxic chemicals, any of the tests shown in Table 21.2 may be required by the regulations imposed under a particular law. The particular set of tests required depends on the predicted or actual use of the chemical, the predicted or actual route of exposure, and the chemical and physical properties of the chemical.

**Table 21.2   Summary of Toxicity Tests and Related End Points**

I Chemical and physical properties
>   For the chemical in question, probable contaminants from synthesis as well as intermediates and waste products from the synthetic process

II Exposure and environmental fate
>   A. Degradation studies—hydrolysis, photodegradation, etc.
>   B. Degradation in soil, water, under various conditions
>   C. Mobility and dissipation in soil, water, and air
>   D. Accumulation in plants, aquatic animals, wild terrestrial animals, food plants, and animals, etc.

III In vivo tests
>   A. Acute
>     1. LD50 and LC50—oral, dermal or inhaled
>     2. Eye irritation
>     3. Dermal irritation
>     4. Dermal sensitization
>   B. Subchronic
>     1. 30- to 90-day feeding
>     2. 30- to 90-day dermal or inhalation exposure
>   C. Chronic/reproduction
>     1. Chronic feeding (including oncogenicity tests)
>     2. Teratogenicity
>     3. Reproduction (multi-generation)
>   D. Special tests
>     1. Neurotoxicity
>     2. Potentiation
>     3. Metabolism
>     4. Pharmacodynamics
>     5. Behavior

IV In vitro tests
>   A. Mutagenicity—prokaryote (Ames test)
>   B. Mutagenicity—eukaryote (*Drosophila*, mouse, etc.)
>   C. Chromosome aberration (*Drosophila*, sister chromatid exchange, etc.)

V Effects on wildlife
>   Selected species of wild mammals, birds, fish, and invertebrates: acute toxicity, accumulation, and reproduction in laboratory-simulated field conditions

## 21.2   EXPERIMENTAL ADMINISTRATION OF TOXICANTS

### 21.2.1   Introduction

Regardless of the chemical tested and whether the test is for acute or chronic toxicity, all in vivo testing requires the reproducible administration of a known dose of the chemical under test, applied in a reproducible manner, that is generally related to the expected route of humans exposure. The nature and degree of the toxic effect can be

**Table 21.3  Variation in Toxicity by Route of Exposure**

| Chemical | Species/Gender | Route | LD50 (mg/kg) |
|---|---|---|---|
| N-Methyl-N-(1-naphthyl) fluoroacetamide[a] | Mouse/M | Oral | 371 |
| | | Dermal | 402 |
| | | Subcutaneous | 250 |
| | Rat/M | Oral | 115 |
| | | Dermal | 300 |
| | | Subcutaneous | 78 |
| Chlordane[b] | Rat/M | Oral | 335 |
| | | Dermal | 840 |
| Endrin[b] | Rat/M | Oral | 18 |
| | | Dermal | 18 |

[a]Data from Y. Hashimoto, et al., *Tox. Appl. Pharmacol.* **12**: 536–547, 1968.
[b]Data from J. R. Allen, et al., *Pharmacol. Therap.* **7**: 513–547, 1979.

affected by the route of administration (Table 21.3). This may be related to differences at the portals of entry or to effects on pharmacokinetic processes. In the latter case, one route (e.g., intravenous) may give rise to a concentration high enough to saturate some rate-limiting process, whereas another (e.g., subcutaneous) may distribute the dose over a longer time and avoid such saturation. Another key question is that of appropriate controls. To identify effects of handling and other stresses as well as the effects of the solvents or other carriers, it is usually better to compare treated animals with both solvent-treated and untreated or possibly sham-treated controls.

### 21.2.2  Routes of Administration

*Oral.* Oral administration is often referred to as administration per os (PO). Compounds can be administered mixed in the diet, dissolved in drinking water, by gastric gavage, by controlled-release capsules, or by gelatin capsules. In the first two cases, either a measured amount can be provided or access can be ad libitum (available 24 hours per day), with the dose estimated from consumption measurements. For certain tests pair-feeding of controls should be considered; that is, controls are permitted only the amount of food consumed by treated animals. In any case it is essential to consider possible nutritional effects caused by reduction of food intake due to distasteful or repellent test materials. In gastric gavage, the test material is administered through a stomach tube or gavage needle; if a solvent is necessary for preparation of dosing solutions or suspensions, the vehicle is administered also to control animals.

*Dermal.* Dermal administration is required for estimation of toxicity of chemicals that may be absorbed through the skin, as well as for estimation of skin irritation and photosensitization. Compounds are applied, either directly or in a suitable solvent, to the skin of experimental animals after hair has been removed by clipping. Often dry materials are mixed with water to make a thick paste that can be applied in a manner that ensures adequate contact with the skin. Frequently the animals must be restrained to prevent licking and hence oral uptake of the material. Solvent and restraint controls should be considered when stress is involved. Skin irritancy tests may be conducted on either animals or humans, using volunteer test panels for human tests.

*Inhalation.* The respiratory system is an important portal of entry, and for evaluation purposes animals must be exposed to atmospheres containing potential toxicants. The generation and control of the physical characteristics of such contaminated atmospheres is technically complex and expensive in practice. The alternative—direct instillation into the lung through the trachea—presents problems of reproducibility as well as stress and for these reasons is generally unsatisfactory.

Inhalation toxicity studies are conducted in inhalation chambers. The complete system contains an apparatus for the generation of aerosol particles, dusts, or gas mixtures of defined composition and particle size, a chamber for the exposure of experimental animals, and a sampling apparatus for the determination of the actual concentration within the chamber. All these devices present technical problems that are difficult to resolve. For rat studies, a particle size of 4 microns is usually targeted.

Animals are normally exposed for a fixed number of hours each day and a fixed number of days each week. Exposure may be nose only, in which the nose of the animal is inserted into the chamber through an airtight ring; or whole body, in which animals are placed inside the chamber. In the latter case, variations due to unequal distribution in the test atmosphere are minimized by rotation of the position of the cages in the chamber during subsequent exposures. Whole body exposure results are usually less satisfactory due to test material accumulation on the fur of the animals and subsequent ingestion during grooming. Figure 21.1 shows a typical inhalation system and supporting equipment.

*Injection.* Except for certain pharmaceuticals and drugs of abuse, injection (parenteral administration) does not correspond to any of the expected modes of exposure. Injection may be useful, however, in mechanistic studies or in QSAR studies in order to bypass absorption and/or permit rapid action. Methods of injection include intravenous (IV), intramuscular (IM), intraperitoneal (IP), and subcutaneous (SC). Infusion of test

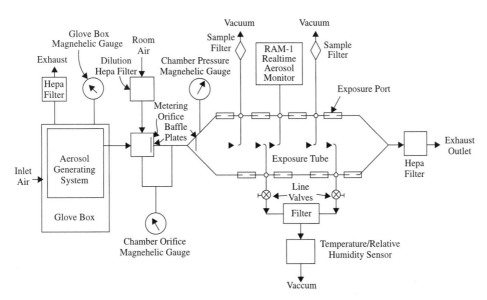

**Figure 21.1** An inhalation exposure system (Modified from Adkins, B. et al. *Am. Int. Hyg. Assoc. J.* **41**: 494, 1980.)

materials over an extended period is also possible. Again, both solvent controls and untreated controls are necessary for proper interpretation of the results.

## 21.3   CHEMICAL AND PHYSICAL PROPERTIES

Although the determination of chemical and physical properties of known or potential toxicants does not constitute a test for toxicity, it is an essential preliminary for such tests.

The information obtained can be used in many ways. Structure activity comparisons with other known toxicants may indicate the most probable hazards. These comparisons may also aid identification in subsequent poisoning episodes. Determination of stability to light, heat, freezing, and oxidizing or reducing agents may enable preliminary estimates of persistence in the environment as well as indicate the most likely breakdown products that may also require testing for toxicity. Establishing such properties as the lipid solubility or octanol/water partition coefficient may enable preliminary estimates of rate of uptake and persistence in living organisms. Vapor pressure may indicate whether the respiratory system is a probable route of entry. Knowledge of the chemical and physical properties must be acquired in order to develop analytical methods for the measurement of the compound and its degradation products. If the chemical is being produced for commercial use, similar information is needed on intermediates in the synthesis or by-products of the process because both are possible contaminants in the final product.

## 21.4   EXPOSURE AND ENVIRONMENTAL FATE

Data on exposure and environmental fate are needed, not to determine toxicity, but to provide information that may be useful in the prediction of possible exposure in the event that the chemical is toxic. These tests are primarily useful for chemicals released into the environment such as pesticides, and they include the rate of breakdown under aerobic and anaerobic conditions in soils of various types, the rates of leaching into surface water from soils of various types, and the rate of movement toward groundwater. The effects of physical factors on degradation through photolysis and hydrolysis studies and the identification of the product formed can indicate the rate of loss of the hazardous chemical or the possible formation of hazardous degradation products. Tests for accumulation in plants and animals and movement within the ecosystem are considered in Section 21.7.

## 21.5   IN VIVO TESTS

Traditionally the basis for the determination of toxicity has been administration of the test compound, in vivo, to one or more species of experimental animal, followed by examination for clinical signs of toxicity and/or mortality in acute tests. In addition pathological examination for tissue abnormalities is also performed, especially in tests of longer duration. The results of these tests are then used by a variety of extrapolation techniques to estimate hazard to humans. These tests are summarized in the remainder of this section. While these tests offer many advantages and are widely used, they suffer from a number of disadvantages: they require the use of animals, whose numbers are often deemed unnecessary by both animal rights and animal welfare advocates; they are extremely expensive to conduct, and they are time-consuming. As a result they have been supplemented by many specialized in vitro tests, some of which are summarized in Section 21.6, and research is ongoing to further develop tests with fewer disadvantages.

### 21.5.1  Acute and Subchronic Toxicity Tests

Acute toxicity test methods measure the adverse effects that occur within a short time after administration of a single dose of a test substance. This testing is performed principally in rodents and is usually done early in the development of a new chemical or product to provide information on its potential toxicity. This information is used to protect individuals who are working with the new material and to develop safe handling procedures for transport and disposal. The information gained also serves as the basis for hazard classification and labeling of chemicals in commerce. Acute toxicity data can help identify the mode of toxic action of a substance and may provide information on doses associated with target-organ toxicity and lethality that can be used in setting dose levels for repeated-dose studies. This information may also be extrapolated for use in the diagnosis and treatment of toxic reactions in humans. The results from acute toxicity tests can provide information for comparison of toxicity and dose-response among members of chemical classes and help in the selection of candidate materials for further work. They are further used to standardize certain biological products such as vaccines.

The results of acute toxicity tests have a wide variety of regulatory applications. These include determination of the need for childproof packaging, determination of reentry intervals after pesticide application, establishment of the requirement and basis for training workers in chemical use, determination of requirements for protective equipment and clothing, and decision making about general registration of pesticides or their restriction for use by certified applicators. Acute oral toxicity may be used in risk assessments of chemicals for humans and nontarget environmental organisms.

The various national and international regulatory authorities have used different hazard classification systems in the past. In light of the importance of hazard classification, the Organisation for Economic Cooperation and Development (OECD) recently harmonized criteria for hazard classification for global use. For example, the five harmonized categories for acute oral toxicity (in mg/kg body weight) are 0–5, 5–50, 50–300, 300–2000, and 2000–5000.

***Acute Oral Testing.*** Traditionally acute oral toxicity testing has focused on determining the dose that kills half of the animals (i.e., the median lethal dose or LD50), the timing of lethality following acute chemical exposure, as well as observing the onset, nature, severity, and reversibility of toxicity. The LD50 concept was developed by Trevan in 1927. Original testing methods were designed to characterize the dose-response curve by using several animals (usually at least 5/sex) at each of several test doses. Data from a minimum of three doses is required. The LD50 values are presented as estimated doses (mg/kg) with confidence limits. The simplest method for the determination of the LD50 is a graphic one and is based on the assumption that the effect is a quantal one (all or none), that the percentage responding in an experimental group is dose related, and that the cumulative effect follows a normal distribution. Data from a typical example (see Figure 11.3), their analysis, and implications are discussed in Chapter 11 on acute toxicity.

As a result of much recent controversy, the LD50 test has been the subject of considerable regulatory attention and as a result changes in requirements have been promulgated. These changes are intended to obtain more information but, at the same time, use fewer animals.

*Criticism of the LD50 Test.* The criticisms of the test include:

- Used uncritically, it is an expression of lethality only, not reflecting other acute effects.
- It requires large numbers of experimental animals to obtain statistically acceptable values. Moreover the results of LD50 tests are known to vary with species, strain, sex, age, and so on (Table 21.4); thus the values are seldom closely similar from one laboratory to another despite the numbers used.
- Because, for regulatory purposes, the most important information needed concerns chronic toxicity, little useful information is derived from the LD50 test. The small amount of information that is acquired could be obtained as well from an approximation requiring only a small number of animals.
- Extrapolation to humans is difficult.

*Support of the LD50 Test.* Continued use of the test has been advocated, however, on the grounds that it is of use in the following ways:

- Properly conducted, acute toxicity tests yield not only the LD50 but also information on other acute effects such as cause of death, time of death, symptomatology, nonlethal acute effects, organs affected, and reversibility of nonlethal effects.
- Information concerning mode of action and metabolic detoxication can be inferred from the slope of the mortality curve.
- The results can form the basis for the design of subsequent subchronic studies.
- The test is useful as a first approximation of hazards to workers.
- The test is rapidly completed.

For the previously listed reasons, there has been a concerted effort in recent years to modify the concept of acute toxicity testing as it is embodied in the regulations of many countries and to substitute more meaningful methods that use fewer experimental animals. The article by Zbinden and Flury-Roversi is an excellent summary of the

**Table 21.4   Factors Causing Variation in LD50 Values**

| Species | Health | Temperature |
|---------|--------|-------------|
| Strain | Nutrition | Time of day |
| Age | Gut contents | Season |
| Weight | Route of administration | Human error |
| Gender | Housing | |

factors affecting LD50 determinations, the advantages and disadvantages of requiring such tests, and the nature and value of the information derived. It concludes that the acute toxicity test (single-dose toxicity) is still of considerable importance for the assessment of risk posed by new chemical substances, and for a better control of natural and synthetic agents in the human environment. It is not permissible, however, to regard a routine determination of the LD50 in various animal species as a valid substitute for an acute toxicity study.

*Current Status.* Recently attention has been focused on developing alternatives to the classical LD50 test to reduce the number of animals used or refine procedures to make exposures less stressful to animals. OECD adopted several alternative methods for determining acute oral toxicity: a limit test for materials with anticipated low toxicity, a fixed dose procedure, an acute toxic class method, and an up-and-down procedure. The fixed-dose procedure and the acute toxic class method estimate the LD50 within a dose range for use in classification and labeling. The up-and-down procedure generates point estimates and confidence intervals of the LD50 and therefore may be useful in a wider set of applications.

The fixed-dose procedure (Guideline OECD 420) aims to identify the appropriate hazard class for new chemicals; it does not provide a point estimate of the LD50. This method calls for testing animals sequentially at one of four doses: 5, 50, 300, or 2000 mg/kg body weight. The test begins with a sighting study in which animals are tested, one at a time, at doses selected from the set doses. Once clear signs of toxicity appear, additional animals (females, or the more sensitive sex) are dosed at that level for a total of five animals. Subsequent groups of animals may receive doses at higher or lower levels, if necessary, depending on the outcome of the previous group. Decision criteria based on the number of animals surviving or showing evident toxicity provide for classification decisions.

The acute toxic class method (guideline OECD 423) aims to identify the appropriate hazard and labeling classification and provides a range for lethality rather than a point estimate of the LD50. Groups of three animals (females, or the more sensitive sex) receive one of four or five doses: 5, 50, 300, 2000 and if necessary 5000 mg/kg body weight. Depending on the survival or mortality of the first group of animals, three or more animals may receive the same or a higher or lower dose. The number of animals that survive or die determines the classification decisions.

The up-and-down procedure (Guideline OECD 425) employs sequential dosing, using only a single animal at each step, the dosage depending on whether the previously dosed animal lives or dies. The test provides a point estimate of lethality and confidence intervals, and can be used to evaluate lethality up to 5000 mg/kg. The main test incorporates elements of range finding and uses a flexible stopping point. A sequential limit test uses up to five animals. Default dose spacing is 3.2 times the previous dose. The starting dose should be slightly below the estimated LD50. If no information is available to estimate the LD50, the starting dose is 175 mg/kg. A computer program was developed by the US EPA to simplify both the experimental phase of the test and the calculation of the LD50 and confidence intervals.

For all three guidelines, selection of a starting dose close to the actual LD50 should decrease the number of animals necessary, reduce study duration, and decrease the amount of test substance needed. Therefore it is desirable that all information on the test substance be made available to the testing laboratory for consideration prior to conducting the study. Such information includes the identity and chemical structure of the substance, its physicochemical properties, the results of any other toxicity test tests on the substance, toxicological data on structurally related substances, the anticipated uses of the substance, or cytotoxicity data on the substance. This information will aid the testing laboratory in selecting the most appropriate test to satisfy regulatory requirements and in choosing the starting dose.

As with the traditional acute oral toxicity methods, the alternative tests involve the administration of a single-bolus dose of a test substance to fasted healthy young adult rodents by oral gavage, observation for morbidity/mortality for up to 14 days after dosing, with recording of body weight (weekly) and clinical signs (daily), and a necropsy at study termination. At the time of dosing, each animal should be between 8 and 12 weeks old and its weight should fall in an interval within $+/-20\%$ of the mean weight of all previously dosed animals taken on their day of dosing. Observation of the postdosing effects on each animal should be for at least 48 hours or until it is clear whether the dosed animals will survive. However, depending on the characteristics of the test material, investigators can vary this time between dosing, so long as the interval is sufficient. Only when the results are clear can a decision be made about whether an additional dose is necessary, and if so, whether to dose the next animal or group of animals at the same, higher, or lower dose. The information from every animal, even those that die after the initial observation period, is used in the final determination of the test outcome.

These newer methods call for testing to be done in a single sex to reduce variability in the test population. This reduction in variability in turn minimizes the number of animals needed. Normally females are used. Although there is usually little difference in sensitivity between males and females, in those cases where there are observable differences, females are most commonly the more sensitive sex. Normally animal suppliers have an excess of female rats because many researchers order only male rats to avoid physiological changes associated with estrus cycling in females; therefore preferential use of female animals for acute testing should not result in excess male animals.

***Eye Irritation.*** Because of the prospect of permanent blindness, ocular toxicity has long been a subject of both interest and concern. Although all regions of the eye are subject to systemic toxicity, usually chronic but sometimes acute, the tests of concern in this section are tests for irritancy of compounds applied topically to the eye. The tests used are all variations of the Draize test, and the preferred experimental animal is the albino rabbit.

The test consists of placing the material to be tested directly into the conjunctival sac of one eye, with the other eye serving as the control. The lids are held together for a few seconds, and the material is left in the eye for at least 24 hours. After that time it may be rinsed out, but in any case, the eye is examined and graded after 1, 2, and 3 days. Grading is subjective and based on the appearance of the cornea, particularly as regards opacity; the iris, as regards both appearance and reaction to light; the conjunctiva, as regards redness and effects on blood vessels; and the eyelids, as regards swelling. Fluorescein dye may be used to assist visual examination because the dye is more readily absorbed by damaged tissues, which then fluoresce when the eye is illuminated. Each end point in the evaluation is scored on a numerical scale, and chemicals are compared on this basis. In addition to the "no-rinse" test, some protocols also investigate the effect of rinsing the eye one minute after exposure to determine if this reduces the potential for irritation. In addition eyes may be graded for up to 21 days after administration of an irritating test material to evaluate recovery.

The eye irritation test is probably the most criticized by advocates of animal rights and animal welfare, primarily because it is inhumane. It has also been criticized on narrower scientific grounds in that both concentration and volumes used are unrealistically high, and that the results, because of high variability and the greater sensitivity of the rabbit eye, may not be applicable to humans. It is clear, however, that because of great significance of visual impairment, tests for ocular toxicity will continue.

Attempts to solve the dilemma have taken two forms: to find substitute in vitro tests and to modify the Draize test so that it becomes not only more humane but also more predictive for humans. Substitute tests consist of attempts to use cultured cells or eyes from slaughtered food animals, but neither method is yet acceptable as a routine test. Modifications consist primarily of using fewer animals. Usually one animal is tested first and, if the material is severely irritating no further eye testing is conducted. EPA has reduced the required number of animals from 6 to 3. In addition eye irritation should never be carried out on materials with a pH of less than 2 or more than 10 as these materials can be assumed to be potential eye irritants.

### *Dermal Irritation and Sensitization.*

*Dermal Irritation and Sensitization.* There are tests for dermal irritation caused by topical application of chemicals. These fall into four general categories: primary irritation, cutaneous sensitization, phototoxicity, and photosensitization. Because many foreign chemicals come into direct contact with the skin, including cosmetics, detergents, bleaches, and many others, these tests are considered essential to the proper regulation of such products. Less commonly, dermal effects may be caused by systemic toxicants.

In the typical primary irritation test, the backs of albino rabbits are clipped free of hair and an area of about 5 $cm^2$ on each rabbit is used in the test. This area is then treated with either 0.5 ml or 0.5 g of the compound to be tested and then covered with a gauze pad. The entire trunk of the rabbit is wrapped to prevent ingestion. After 4 to 24 hours the tape and gauze are removed, the treated areas are evaluated for erythematous lesions (redness of the skin produced by congestion of the capillaries) and edematous lesions (accumulation of excess fluid in SC tissue), each of which is expressed on a numerical scale. After an additional 24 to 48 hours, the treated areas are again evaluated.

Skin sensitization tests are designed to test the ability of chemicals to affect the immune system in such a way that a subsequent contact causes a more severe reaction than the first contact. This reaction may be elicited at a much lower concentration and in areas beyond the area of initial contact. The antigen involved is presumed to be formed by the binding of the chemical to body proteins, the ligand-protein complex then being recognized as a foreign protein to which antibodies can be formed. Subsequent exposure may then give rise to an allergic reaction. Skin sensitization tests generally follow protocols that are modifications of the Buchler (dermal inductions) method or the Magnesson and Kligman (intradermal inductions) method. The test animal commonly used in skin sensitization tests is the guinea pig; animals are treated with the test compound in a suitable vehicle, with the vehicle alone, or with a positive control such as 2,4-dinitrochlorobenzene (a relatively strong sensitizer) or cinnamaldehyde (a relatively weak sensitizer) in the same vehicle. During the induction phase the animals are treated for each of 3 days evenly spaced during a 2-week period. This is followed by a 2-week rest period followed by the challenge phase of the test. This consists of a 24-hour topical treatment carried out as described for primary skin irritation tests. The lesions are scored on the basis of severity and the number of animals responding (incidence). If there is a greater skin reaction in the animals given induction doses compared to those given the test material for the first time, the compound is considered to be a dermal sensitizer.

Other test methods include those in which the induction phase is conducted by intradermal injection together with Freund's adjuvant (a chemical mixture that enhances the antigenic response) and the challenge by dermal application, or tests in which both induction and challenge doses are topical but the former is accompanied by intradermal injections of Freund's adjuvant. It is important that compounds that cause primary skin irritation be tested for skin sensitization at concentrations low enough that the two effects are not confused.

Phototoxicity tests are designed to evaluate the combined dermal effects of light (primarily ultraviolet [UV] light) and the chemical in question. Tests have been developed for both phototoxicity and photoallergy. In both cases the light energy is believed to cause a transient excitation of the toxicant molecule, which, on returning to the lower energy state, generates a reactive, free-radical intermediate. In phototoxicity these organic radicals act directly on the cells to cause lesions, whereas in photoallergy they bind to body proteins. These modified proteins then stimulate the immune system to produce antibodies, because the modifications cause them to be recognized as foreign or "nonself" proteins. These tests are basically modifications of the tests for primary irritation and sensitization except that, following application of the test chemical, the treated area is irradiated with UV light. The differences between the animals treated and irradiated and those treated and not irradiated is a measure of the phototoxic effect.

***Safety Pharmacology Studies.*** Safety pharmacology studies investigate the potential undesirable pharmacodynamic effects of a test article on physiological functions in relationship to exposure. These tests are typically conducted as part of the development of new drugs. The objectives of safety pharmacology studies are threefold: to identify undesirable effects of a test article that may have relevance to its use in humans, to evaluate a test article for possible effects observed in toxicology or clinical studies, and to investigate the mechanism underlying any undesirable effects of the test article. Safety pharmacology consists of a core battery of studies with follow-up studies as indicated by preliminary findings. The core studies are designed to target vital organ systems, particularly the central nervous system, cardiovascular system, and pulmonary system. These studies are typically conducted using small numbers of rats and dogs. In the study for pulmonary function, end points measured are respiratory rate, minute volume, and tidal volume. In the cardiovascular telemetry study, end points include heart rate, blood pressure, and electrocardiogram evaluation. In telemetry studies a radio transmitter is implanted in all animals to permit continuous monitoring for 24 hours pretest and 24 hours after dosing. A cardiopulmonary study can also be conducted in which respiratory rate, minute volume, tidal volume, blood pressure, heart rate, electrocardiogram, and body temperature are monitored in restrained animals for typically 2 hours after dosing.

***Subchronic Tests.*** Subchronic tests examine toxicity caused by repeated dosing over an extended period, but not one that is so long as to constitute a significant portion of the expected life span of the species tested. A 28- or 90-day oral study in the rat or dog would be typical of this type of study, as would a 21- to 28-day dermal application study or a 28- to 90-day inhalation study. Such tests provide information on essentially all types of chronic toxicity other than carcinogenicity and are usually believed to be essential for establishing the dose regimens for prolonged chronic studies. They are frequently used as the basis for the determination of the no observed effect level (NOEL). This value is often defined as the highest dose level at which no deleterious or abnormal effect can be measured, and is often used in risk assessment calculations. Subchronic tests are also useful in providing information on target organs and on the potential of the test chemical to accumulate in the organism.

**Twenty-Eight- to Ninety-Day Tests.** Chemicals are usually tested by administration in the diet, less commonly in the drinking water, and only when absolutely necessary by gavage, because the last process involves much handling and subsequent stress. Numerous experimental variables must be controlled and biologic variables evaluated. In addition the number of end points that can be measured is large, and as a consequence record keeping and data analysis must be carefully planned. If all is done with care, much may be learned from such tests.

*Experimental (Nonbiologic) Variables.* Several environmental variables may affect toxicity evaluations, some directly and others by their effects on animal health. Major deviations from the optimum temperature and humidity for the species in question can cause stress reactions. Stress can also be caused by housing more than one species of experimental animal in the same room. Many toxic or metabolic effects show diurnal variations that are related to photoperiod. Cage design and the nature of the bedding have also been shown to affect the toxic response. Thus the optimum housing conditions are clean rooms, each containing a single species, with the temperature, humidity, and photoperiod being constant and optimized for the species in question. Cages should be the optimum design for the species, bedding should be inert (not cause enzyme induction or other metabolic effect), and cages should not be overcrowded, with individual caging whenever possible.

Dose selection, preparation, and administration are all important variables. Subchronic studies are usually conducted using three (less often, four) dose levels. The highest should produce obvious toxicity but not high mortality and the lowest no toxicity (NOEL), whereas the intermediate dose should give effects clearly intermediate between these two extremes. Although the doses can be extrapolated from acute test, such extrapolation is difficult, particularly in the case of compounds that accumulate in the body, and frequently a 14-day range-finding study is made. Although the route of administration should ideally mimic the expected route of exposure in humans, in practice, the chemical is usually administered ad libitum in the diet because this is, on average, most appropriate. Diets containing known amounts of the test material are presented to the animals. Measurement of food consumption is recommended to provide an estimate of the test material consumed. In cases where a highly accurate measurement of dose is an important factor in the experimental design, the animals may be treated by gavage or by capsules containing the test material.

To avoid effects from nonspecific variations on the diet, enough feed from the same batch should be obtained for the entire study. Part is set aside for the controls, and the remainder is mixed with the test chemical at the various dose levels. Care should be taken to store all food in such a way that not only does the test chemical remain stable, but also the nutritional value is maintained. The identity and concentration of the test chemical should be checked periodically by chemical analysis. Treated diets may be prepared at set intervals, such as weekly, depending on the stability of the test material in the diet.

Subchronic studies are usually conducted with 10 to 20 males and 10 to 20 females of a rodent species at each dose level and 4 to 8 of each sex of a larger species, such as the dog, at each dose level. Animals should be drawn from a larger group and assigned to control or treatment groups by a random process, but the larger group should not vary so much that the mean weights and ages of the subgroups vary significantly at the beginning of the experiment.

*Biologic Variables.* Subchronic studies should be conducted on two species, ideally a rodent and a nonrodent. Although the species chosen should be those with the greatest pharmacokinetic and metabolic similarity to humans for the compound in question, this information is seldom available. In practice, the most common rodent used is the rat, and the most common nonrodent used is the dog. It has long been held that inbred rodent strains should be used to reduce variability. This and the search for strains that were sensitive to chemical carcinogenesis but did not have an unacceptably high spontaneous tumor rate led to widespread use of the F344 rat and B6C3F1 mouse. Other researchers believe that an outbred strain such as the Sprague Dawley rat is more robust and prefer to use them.

Although ideally the age should be matched to the expected exposure period in terms of the stage of human development, this is not often done. Young adult or adolescent animals that are still growing are preferred in almost all cases, and both sexes are routinely used.

Good animal care is critical at all times because toxicity has been shown to vary with diet, disease, and environmental factors. Animals should be quarantined for some time before being admitted to the test area, their diet should be optimum for the test species, and the facility should be kept clean at all times. Regular inspection by a veterinarian is essential, and any animals showing unusual symptoms not related to the treatment (e.g., in controls or in low dose but not high dose animals) should be removed from the test and autopsied.

*Results.* Although the information required from subchronic tests varies somewhat from one regulatory agency to another, the requirements are basically similar (Table 21.5). For explanatory purposes the data obtained from these tests can be described as two types: that obtained from living animals during the course of the test and that obtained from animals sacrificed either during or at the end of the test period. Many of the tests performed on living animals can be carried out first before the test period begins to provide a baseline for comparison to subsequent measurement. A satellite group of treated animals can be added to the test for evaluation of "recovery." For these animals the treated food would be removed at the end of the test period and they would be returned to the control diet for 21 to 28 days while the various end points are followed. This is done to establish whether any effects noted are reversible. Autopsies should be performed on all animals found dead or moribund during the course of the test. The following is a list of end points that may be measured during a 90-day oral toxicity study.

**Table 21.5  Summary of Subchronic Test Guidelines by Regulatory Agency**

| Character of tests | EPA Pesticide Assessment Guidelines | FDA "Red Book" | FDA IND/NDA Pharmacology Review Guidelines | OECD | EPA Health Effects Test Guidelines | NTP |
|---|---|---|---|---|---|---|
| Purpose | Pesticide registration support | Food and color additives; safety assessment | IND/NDR pharmacology review guidelines | Assessment and evaluation of toxic characteristics | Select chronic dose levels | Predict dose range for chronic study |
| | No-observed-effect level | No observed adverse effects, no-effect level | Characterize pharmacology, toxicology, pharmacokinetics, and metabolism of drugs for precautionary clinical decisions | Select chronic dose levels. Useful information and permissible human exposure | Establish safety criteria for human exposure, no-observed-effect level | |
| Species | Rat, dog, mouse | Rat, dog | Rat, mouse, other rodents, dog, monkey, other nonrodents | Rat, dog | Rat, dog | Fischer 344 rats B6C3F$_1$ mice |
| Doses | Three dose levels | Three dose levels | Three dose levels | Three dose levels | Three dose levels | Five dose levels |
| Endpoints | Clinical signs Ophthalmology Hematology Clinical chemistry Histopathology | Clinical signs Ophthalmology Hematology Clinical chemistry Histopathology | Ophthalmology Hematology Clinical chemistry Histopathology | Ophthalmology Hematology Clinical chemistry Histopathology | Ophthalmology Hematology Clinical chemistry Histology | Weight loss, histopathology |
| | Target organs | Target organs | Target organs, behavioral and pharmacological effects | Target organs | Target organs | Target organs |

*Source*: From the National Toxicology/Program. Washington, DC: Department of Health and Human Services. *Report of the NTP Ad hoc Panel on Chemical Carcinogenesis Testing and Evaluation.*
*Note*: EPA, Environmental Protection Agency; FDA, Food and Drug Administration; IND/NDA, investigative new drug/new drug assessment; OECD, Organization for Economic Cooperation and Development; NTP, National Toxicology Program.

A. In-life Data. Interim tests are carried out at intervals before the study to establish baselines, at intervals during the study, and at the end of the study.

1. *Appearance.* Mortality and morbidity as well as the condition of the skin, fur, mucous membranes, and orifices should be checked at least daily. Presence of palpable masses or external lesions should be noted.

2. *Eyes.* Ophthalmologic examination of both cornea and retina should be carried out at the beginning and at the end of the study.

3. *Food consumption.* Weekly.

4. *Body weight.* Weekly.

5. *Behavioral abnormalities.*

6. *Respiration rate.*

7. *ECG.* Particularly with the larger animals.

8. *EEG.* Particularly with the larger animals.

9. *Hematology.* Assessment should be made prior to chemical administration (pretest) and at least prior to termination. Hemoglobin, hematocrit, RBC, WBC, differential counts, platelets, reticulocytes, and clotting parameters should be assessed.

10. *Blood chemistry.* Pretest, and at least prior to termination, electrolytes and electrolyte balance, acid-base balance, glucose, urea nitrogen, serum lipids, serum proteins (albumin-globulin ratio), enzymes indicative of organ damage such as transaminases and phosphatases should be measured. Toxicant and metabolite levels should be assessed as needed.

11. *Urinalysis.* Pretest, and at least prior to termination, microscopic appearance (sediment, cells, stones, etc.), pH, specific gravity, chemical analysis for reducing sugars, proteins, ketones, and bilirubin should be measured. Toxicant and metabolite levels should be assessed as needed.

12. *Fecal analysis.* Occult blood, fluid content, and toxicant and metabolite levels should be assessed if needed.

B. Termination Tests. Because the number of tissues that may be sampled is large (Table 21.6) and the number of microscopic methods is also large, it is necessary to consider all previous results before carrying out the pathological examination. For example, clinical tests or blood chemistry analyses may implicate a particular target organ that can then be examined in greater detail. All control and high-dose animals are examined in detail. If lesions are found, the next lowest dose group is examined for these lesions, and this method continues until a no effect group is reached.

Because pathology is largely a descriptive science with a complex terminology that varies from one practitioner to another, it is critical that the terminology be defined at the beginning of the study and that the same pathologist examine the slides form both treated and control animals. Pathologist are not in agreement on the necessity or the wisdom of coding slides so that the assessor is not aware of the treatment given the animal from which a particular slide is derived. Such coding however, eliminates unintentional bias, a hazard in a procedure that depends on subjective evaluation. Other items of utmost importance are quality control, slide

**Table 21.6  Tissues and Organs to Be Examined Histologically in Chronic and Subchronic Toxicity Tests**

| | | |
|---|---|---|
| Adrenals | Larynx | Salivary gland |
| Bone and bone marrow | Liver | Sciatic nerve |
| Brain | Lungs and bronchi | Seminal vesicles |
| Cartilage | Lymph nodes | Skin |
| Cecum | Mammary glands | Spinal cord |
| Colon | Mandibular lymph node | Spleen |
| Duodenum | Mesenteric lymph node | Stomach |
| Esophagus | Nasal cavity | Testes |
| Eyes | Ovaries | Thigh muscle |
| Gall bladder | Parathyroids | Thymus |
| Ileum | Pituitary | Urinary bladder |
| Jejunum | Prostate | Uterus |
| Kidneys | Rectum | |

identification, and data recording. Many tissues may be examined; consequently an even larger number of tissue blocks must be prepared. Because each of these may yield many slides to be stained, comparable quality of staining and the accurate correlation of a particular slide with its parent block, tissue, and animal is critical.

1. *Necropsy.* This must be conducted with care to avoid postmortem damage to the specimens. Tissues are removed, weighed, and examined closely for gross lesions, masses, and so on. Tissues are then fixed in buffered formalin for subsequent histologic examination.

2. *Histology.* The tissues listed in Table 21.6 plus any lesions, masses, or abnormal tissues are embedded, sectioned, and strained for light microscopy. Parafin embedding and staining with hemotoxylin and eosin are the preferred routine methods, but special stains may be used for particular tissues or for a more specific examination of certain lesions. Electron microscopy may also be used for more specific examination of lesions or cellular changes after their initial localization by more routine methods.

**Repeated Dose Dermal Tests.** Twenty-one to 28-day dermal tests are particularly important when the expected route of human exposure is by contact with the skin, as is the case with many industrial chemicals or pesticides. Compounds to be tested are usually applied daily to clipped areas on the back of the animal, either undiluted or in a suitable vehicle. In the latter case, if a vehicle is used, it is also applied to the controls. Selection of a suitable solvent is difficult because many affect the skin, causing either drying or irritation, whereas others may markedly affect the rate of penetration of the test chemical. Corn oil, methanol, or carboxymethyl cellulose are preferred to dimethyl sulfoxide (DMSO) or acetone. It should also be considered that some of the test chemical may be ingested as a result of grooming by the animal, although this can be controlled to some extent by use of restraining collars and/or wrapping.

The criteria for environment, dose selection, and species selection, for example, are not greatly different from the criteria used for 90-day feeding tests, although the list of end points to be examined is often shorter (e.g., fewer organs may be examined).

It is necessary, however, to pay close attention to the skin at the point of application because local effects may be as important as systemic ones.

***Twenty-Eight- to Ninety-Day Inhalation Tests.*** Inhalation studies are indicated whenever the route of exposure is expected to be through the lungs. Animals are commonly exposed for 6 to 8 hours each day, 5 days each week, in chambers of the type previously discussed. Even in those cases where the animals are maintained in the inhalation chambers during nonexposure hours, food is always removed during exposure. Nevertheless, exposure tends to be in part dermal and, due to grooming of the fur, in part oral. Environmental and biologic parameters are the same as those for other subchronic tests, as are the routine end points to be measured before, during, and after the test period. Particular attention must be paid, however, to effects on the tissues of the nasal cavity and the lungs because these are the areas of maximum exposure.

If the test material is particulate, consideration must be given to the particle size and its inhalation potential. Particles of 4 microns in size are considered to be inhalable; larger particles will be cleared from the respiratory tract by ciliary action and subsequently swallowed (oral exposure) or expelled by sneezing or expectoration.

## 21.5.2 Chronic Tests

Chronic tests are those conducted over a significant part of the life span of the test animal. The duration of a chronic study is generally one year or more. Typically rat and dog are the preferred species; for carcinogenicity studies, rats and mice are used.

***Chronic Toxicity and Carcinogenicity.*** Descriptions of tests for both chronic toxicity and carcinogenicity are included here because the design is similar—so similar in fact that they can be combined into one test. Chronic toxicity tests are designed to discover any of numerous toxic effects and to define safety margins to be used in the regulation of chemicals. As with subchronic tests, two species are usually used, one of which is either a rat or a mouse strain, in which case the tests are run for 2 years or 1.5 to 2.0 years, respectively. Data are gathered after 1 year to determine chronic effects without potential confounding effects of aging. Data are gathered after 1.5 years (mouse) or 2 years (rat) to determine carcinogenic potential. The nonrodent species used may be the dog, a nonhuman primate, or, rarely, a small carnivore such as the ferret. Chronic toxicity tests may involve administration in the food, in the drinking water, by capsule, or by inhalation, the first being the most common. Gavage is rarely used. The dose used is the maximum tolerated dose (MTD) and usually two lower doses, perhaps 0.25 MTD and 0.125 MTD with the lowest dose being a predicted no effect level.

*MTD.* The MTD has been defined for testing purposes by the US Environmental Protection Agency (EPA) as:

> [T]he highest dose that causes no more than a 10% weight decrement, as compared to the appropriate control groups; and does not produce mortality, clinical signs of toxicity, or pathologic lesions (other than those that may be related to a neoplastic response) that would be predicted to shorten the animals' natural life span.

This dose is determined by extrapolation from subchronic studies.

The requirements for animal facilities, housing, and environmental conditions are as described for subchronic studies. Special attention must be paid to diet formulation because it is impractical to formulate all of the diets for 2-year study from a single batch. In general, semisynthetic diets of specified components should be formulated regularly and analyzed before use for test material content.

The end points used in these studies are those described for the subchronic study: appearance, ophthalmology, food consumption, body weight, clinical signs, behavioral signs, hematology, blood chemistry, urinalysis, organ weights, and pathology. Some animals may be killed at fixed intervals during the test (e.g., 6, 12, or 18 months) for histologic examination. Particular attention is paid to any organs or tests that showed compound related changes in the subchronic tests.

Carcinogenicity tests have many requirements in common (physical facilities, diets, etc.) with both chronic and subchronic toxicity tests as previously described. Because of the numbers and time required, these tests are usually carried out using rats and/or mice, but in some cases a nonrodent species may also be used. The chemical under test may be administered in the food, in the drinking water, by dermal application, by gavage, or by inhalation, the first two methods being the most common. Because the oncogenic potency of chemicals varies through extreme limits, the purity of the test chemical is of great concern. A 1% contaminant needs only to be 100 times as potent as the test chemical to have an equivalent effect, and differences of this magnitude and greater are not unheard of.

Dosing is carried out over the major part of the life span for rodents, beginning at or shortly after weaning. The highest dose used is the MTD. The principal end point is tumor incidence as determined by histologic examination. The statistical problem of distinguishing between spontaneous tumor occurrence in the controls and chemical-related tumor incidence in the treated animals is great; for that reason large numbers of animals are used. A typical test involves 50 or more rats or mice of each gender in each treatment group. Some animals are necropsied at intermediate stages of the test (e.g., at 12 months), as are all animals found dead or moribund. All surviving animals are necropsied at the end of the test. Tissues to be examined are listed in Table 21.6, with particular attention being paid to abnormal masses and lesions.

### 21.5.3   Reproductive Toxicity and Teratogenicity

The aim of developmental and reproductive testing is to examine the potential for a compound to interfere with the ability of an organism to reproduce. This includes testing to assess reproductive risk to mature adults as well as the developing individual at various stages of life, from conception to sexual maturity. Traditionally animal studies have been conducted in three "segments": (I) in adults, treatment during a pre-mating period and optionally continuation for the female through implantation or lactation; (II) in pregnant animals treatment during the major period of organogenesis; and (III) treatment of pregnant/lactating animals from the completion of organogenesis through lactation (peri- and postnatal study). Although guidelines addressing treatment regimens have been rather similar throughout the world, required end points measured in adults and developing organisms have varied. International harmonization of guidelines has demonstrated the need for flexibility in testing for reproductive and developmental toxicity, and toxicologists are now often challenged to design unique studies to examine potential effects on all the parameters considered in the classical segment I, II, and III studies. In adults, these include development of mature egg and

sperm, fertilization, implantation, delivery of offspring (parturition), and lactation. In the developing organism, these include early embryonic development, major organ formation, fetal development and growth, and postnatal growth including behavioral assessments and attainment of full reproductive function. These evaluations are usually best carried out in several separate studies.

*Some Definitions in Reproductive Biology.* Discussion of a bit of reproductive biology may be helpful in the understanding of study designs to evaluate reproductive and developmental toxicity. Tests to assess general reproductive performance and fertility are generally conducted using rats. In the rat, multiple eggs are ovulated from mature follicles in the ovary. The follicle that an egg leaves behind develops into glandular tissue known as a corpora lutea. The corpora lutea secretes progesterone, a hormone needed to maintain pregnancy (unlike humans in which progesterone is secreted by the placenta). Corpora lutea are visible as blisterlike protuberances on the ovary. A count of the corpora lutea in the ovary allows one to determine the maximum number of potential offspring for that pregnancy. Fertilized eggs develop into zygotes that may attach to the wall of the uterus (implantation). The discrete areas of implantation may be observed and counted upon examination of the uterus at C-section. Calculation of pre- and postimplantation loss are important end points in a reproductive toxicity study. Pre-implantation loss is the death of a fertilized ova prior to implantation in the uterine wall. Postimplantation loss (i.e., resorption and/or fetal death) is the death of the conceptus after implantation in the uterine wall and prior to parturition. Postimplantation loss can be broken down into early and late resorptions and fetal death. A late resorption has discernable features such as limbs, eyes, and nose, whereas an early resorption has none of these features.

*Single- and Multiple-Generation Tests.* Fertility and general reproductive performance can be evaluated in single and multiple generation tests. These tests are usually conducted using rats. Fertility is defined as the ability to produce a pregnancy, while the ability to produce live offspring is known as fecundity. An abbreviated protocol for a single-generation test is shown in Figure 21.2.

In typical tests 25 males per dose group are treated for 70 days prior to mating and 25 females per dose group are treated for 14 days pre-mating. The number of animals is chosen to yield at least 20 pregnant females per dose group including controls. The treatment durations are selected to coincide with critical times during which spermatogenesis and ovulation occur. It takes approximately 70 days in the rat for spermatogonial cells to become mature sperm capable of fertilization. In the female rat the estrus cycle length is 4 to 5 days and a 14-day dosing period is considered sufficient time to detect potential effects on hormonal or other systems that may effect ovulation. In some study designs both males and females are treated for 70 days pre-mating.

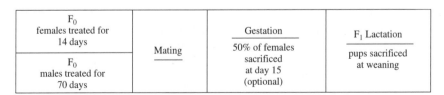

| $F_0$ females treated for 14 days | Mating | Gestation | $F_1$ Lactation |
|---|---|---|---|
| $F_0$ males treated for 70 days | | 50% of females sacrificed at day 15 (optional) | pups sacrificed at weaning |

**Figure 21.2**   Abbreviated protocol for a one generation reproductive toxicity test.

Treatment of the females is continued through pregnancy (21 days) and until the pups are weaned. Pups are usually 21 days of age. The test compound is administered at three dose levels either in the feed, in drinking water, or by gavage. The high dose is chosen to cause some, but not excessive, maternal toxicity (e.g., an approximate 10% decrease in body weight gain, or effects on target organs). Low doses are generally expected to be no-effect levels.

After the pre-mating period, the rats are placed in cohabitation, with one male and one female caged together. Mating is confirmed by the appearance of spermatozoa in a daily vaginal smear. Day 1 of gestation is the day insemination is confirmed. The females bear and nurse their pups. After birth, the pups are counted, weighed, and examined for external abnormalities. The litters are frequently culled to a constant number (usually 8–10) after 4 days. At weaning, the pups are killed and autopsied for gross and internal abnormalities. In a multigeneration study, approximately 25/sex/group are saved to produce the next generation. Brother-sister pairings are avoided. Treatment is continuous throughout the test, which can be carried out for two, sometimes, three generations. An abbreviated protocol for a multiple generation test can be seen in Figure 21.3. Note that the parental generation is known as the $F_0$ generation and the

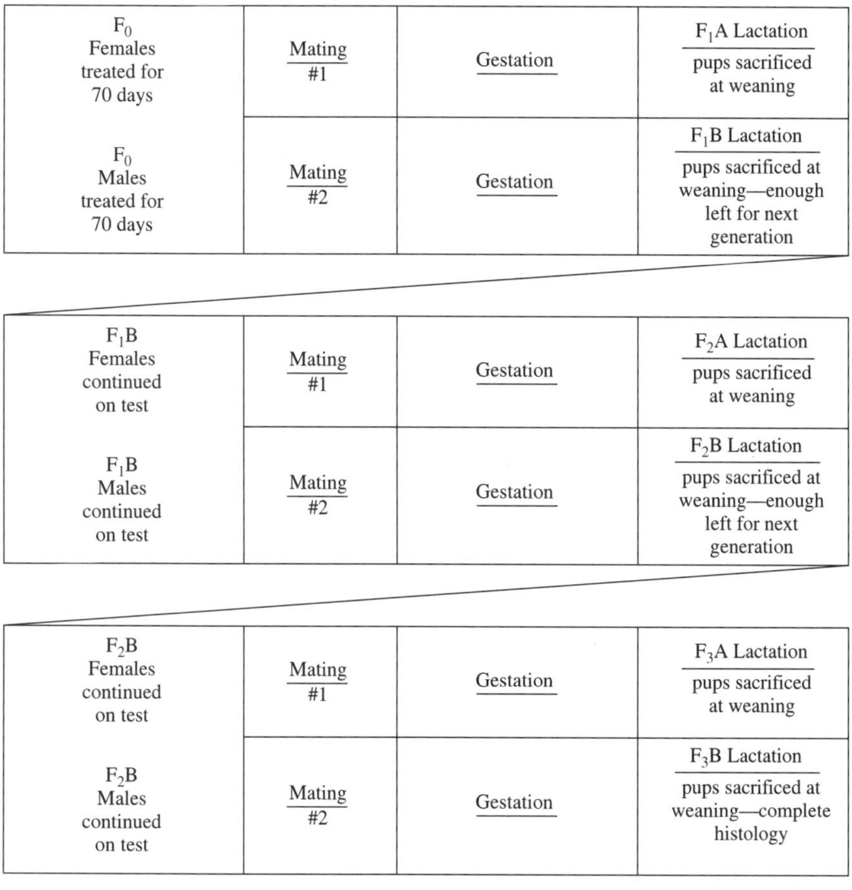

| $F_0$ Females treated for 70 days | Mating #1 | Gestation | $F_1A$ Lactation — pups sacrificed at weaning |
| $F_0$ Males treated for 70 days | Mating #2 | Gestation | $F_1B$ Lactation — pups sacrificed at weaning—enough left for next generation |

| $F_1B$ Females continued on test | Mating #1 | Gestation | $F_2A$ Lactation — pups sacrificed at weaning |
| $F_1B$ Males continued on test | Mating #2 | Gestation | $F_2B$ Lactation — pups sacrificed at weaning—enough left for next generation |

| $F_2B$ Females continued on test | Mating #1 | Gestation | $F_3A$ Lactation — pups sacrificed at weaning |
| $F_2B$ Males continued on test | Mating #2 | Gestation | $F_3B$ Lactation — pups sacrificed at weaning—complete histology |

**Figure 21.3** Abbreviated protocol for a multigeneration reproductive toxicity test.

offspring are known as the $F_1$s and $F_2$s. In some studies, parents produce two litters, for example the $F_1A$ and $F_1B$ litters.

Because both males and females are treated in this type of study design, it is not possible to distinguish between maternal and paternal effects in the reproductive performance. To permit this separation, it is necessary to dose additional animals to the stage of mating and then breed them to untreated members of the opposite sex. Similarly, if effects are seen postnatally, it may not be possible to distinguish between effects mediated in utero or mediated by lactation. This distinction can be made by "cross-fostering" the offspring of treated females to untreated females, and vice versa.

The end points observed in these types of tests, depending on study design, are as follows:

1. Fertility index, the number of pregnancies relative to the number of matings.
2. The number of live births, relative to the number of total births.
3. Pre-implantation loss, or number of corpora lutea in the ovaries relative to the number of implantation sites.
4. Postimplantation loss, or the number of resorption sites in the uterus relative to the number of implantation sites.
5. Duration of gestation.
6. Effects on male or female reproductive systems.
7. Litter size and condition, gross morphology of pups at birth, gender, and anogenital distance.
8. Survival of pups.
9. Weight gain and performance of adults and pups.
10. Time of occurrence of developmental landmarks, such as eye opening, tooth eruption, vaginal opening in females, preputial separation in males.
11. Morphological abnormalities in weanlings.

Results from single and multiple generation tests provide important information for assessment of test materials that may perturb a variety of systems including the endocrine system. A number of variations of the single and multiple generation tests exist. For example, a number of weanlings may be left to develop and be tested later for behavioral and/or physiological defects (e.g., developmental neurotoxicity testing).

*Teratology.* Teratology is the study of abnormal fetal development. For an agent to be labeled a teratogen, it must significantly increase the occurrence of adverse structural or functional abnormalities in offspring after its administration to the female during pregnancy or directly to the developing organism. In teratology testing, exposure to the test chemical may be from implantation to parturition, although it has also been restricted to the period of major organogenesis, the most sensitive period for inducing structural malformations. Observations may be extended throughout life, but usually they are made immediately prior to birth after a C-section. The end points observed are mainly morphologic (structural changes and malformations), although embryo-fetal mortality is also used as an end point. Figure 21.4 shows an outline of a typical teratology study.

Teratology studies are carried out in two species, a rodent species (usually the rat) and in another species such as the rabbit (rarely in the dog or primate). Enough females

Teratology

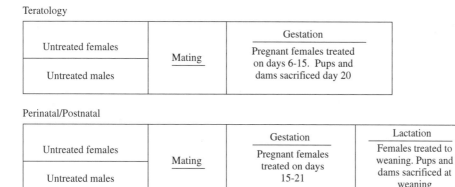

Perinatal/Postnatal

**Figure 21.4** Abbreviated protocol for a teratology test and for a perinatal/postnatal toxicity test in rats.

should be used so that, given normal fertility for the strain, there are 20 pregnant females in each dosage group. Traditionally the timing of compound administration has been such that the dam is exposed during the period of major organogenesis, that is days 6 through 15 of gestation in the rat or mouse and days 6 through 18 for the rabbit. Newer study designs call for dosing until C-section. Day 1 is the day spermatozoa appear in the vagina in the case of rats, or the day of mating in the rabbit.

The test chemical is typically administered directly into the stomach by gavage, which is a requirement of EPA and some other regulatory agencies. This method of dosing allows a precise calculation of the amount of test material received by the animal. Studies typically have three dose levels and a control group that receives the vehicle used for test material delivery. The high dose level is chosen to be one at which some maternal toxicity is known to occur, but never one that would cause more than 10% mortality. The low dose should be one at which no maternal toxicity is apparent, and the intermediate dose(s) should be chosen as a predicted low effect level.

The test is terminated by performing a C-section on the day before normal delivery is expected. The uterus is examined for implantation and resorption sites and for live and dead fetuses, and the ovaries are examined for corpora lutea. In rodent studies, half of the fetuses are examined for soft tissue malformations, and the remaining are examined for skeletal malformations. In nonrodents, all fetuses are examined for both soft tissue and skeletal malformations. The various end points that may be examined include maternal toxicity, embryo-fetal toxicity, external malformations, and soft tissue and skeletal malformations.

Careful evaluation of maternal toxicity is necessary in assessing the validity of the high-dose level, and the possibility that maternal toxicity is involved in subsequent events. The parameters evaluated include body weight, food consumption, clinical signs, and necropsy data such as organ weights. Because exposure starts after implantation, conception and implantation rates should be the same in controls and all treatment levels. If not, the test is suspect, with a possible error in the timing of the dose or use of animals from a source unsuitable for this type of testing.

Embryo-fetal toxicity is determined from the number of dead fetuses and resorption sites relative to the number of implantation sites. In addition to the possibility of lethal malformations, such toxicity can be due to maternal toxicity, stress, or direct toxicity to the embryo or fetus that is not related to developmental malformations. Fetal weight and fetal size may also be a measure of toxicity but should not be confused with the variations seen as a result of differences in the number of pups per litter. Smaller litters tend to have larger pups while larger litters have smaller pups.

Anomalies may be regarded either as variations that may not adversely affect the fetus and not have a fetal outcome or as malformations that have adverse effects on the fetus. For some findings there is disagreement as to which class it belongs, such as the number of ribs in the rabbit that inherently has a large amount of variability. Common external anomalies are listed in Table 21.7 and are determined by examination of

**Table 21.7   External Malformations Commonly Seen in Teratogenicity Tests**

*Brain, cranium, spinal cord*

Encephalocele—protrusion of brain through an opening of the skull.
Cerebrum is well formed and covered by transparent connective tissue.
Exencephaly—lack of skull with disorganized outward growth of the brain.
Microcephaly—small head on normal sized body.
Hydrocephaly—marked enlargement of the ventricles of the cerebrum.
Craniorachischisis—exposed brain and spinal cord.
Spina bifida—Nonfusion of spinal processes. Usually ectoderm covering is missing and spinal cord is evident.

*Nose*

Enlarged naris—enlarged nasal cavities
Single naris—a single naris, usually median

*Eye*

Microphthalmia—small eye
Anophthalmia—lack of eye
Open eye—no apparent eyelid, eye is open

*Ear*

Anotia—absence of the external ear
Microtia—small ear

*Jaw*

Micrognathia—small lower jaw
Agnathia—absence of lower jaw
Aglossia—absence of tongue
Astomia—lack of mouth opening
Bifid tongue—forked tongue
Cleft lip—either unilateral or bilateral cleft of upper lip

*Palate*

Cleft palate—a cleft or separation of the median portion of the palate

*Limbs*

Clubfoot—foot that has grown in a twisted manner, resulting in an abnormal shape or position. It is possible to have a malposition of the whole limb
Micromelia—abnormal shortness of the limb
Hemimelia—absence of any of the long bones, resulting in a shortened limb
Phomelia—absence of all of the long bones of a limb, the limb is attached directly to the body

**Table 21.8   Some Common Visceral Anomalies Seen in Teratogenicity Tests**

*Intestines*

Umbilical hernia—protrusion of the intestines into the umbilical cord
Ectopic intestines—extrusion of the intestines outside the body wall

*Heart*

Dextrocardia—rotation of the heart axis to the right
Enlarged heart—either the atrium or the ventricle may be enlarged

*Lung*

Enlarged lung—all lobes are usually enlarged
Small lung—all lobes are usually small; lung may appear immature

*Uterus/testes*

Undescended testes—testes are located anterior to the bladder instead of lateral; may be
   bilateral or unilateral
Agenesis of testes—one or both testes may be missing
Agenesis of uterus—one or both horns of the uterus may be missing

*Kidney*

Hydronephrosis—fluid-filled kidney, often grossly enlarged; may be accompanied by a
   hydroureter (enlarged, fluid-filled ureter)
Fused—kidneys fused, appearing as one misshapen kidney with two ureters
Agenesis—one or both kidneys missing
Misshapen—small, enlarged (usually internally), or odd-shaped kidneys

---

fetuses at C-section. Visceral anomalities are determined by examination of fetuses after fixation using either the dissection method of Staples or by the hand-sectioning method of Wilson. Common visceral findings are listed in Table 21.8. Fetal skeletons are examined after first fixing the fetus and then staining the bone with Alizarin red. Numerous skeletal variations occur in controls and may not have an adverse effect on the fetus (Table 21.9). Their frequency of occurrence may, however, be dose related and should be evaluated.

Almost all chemically induced malformations have been observed in control animals, and most malformations are known to be produced by more than one cause. Thus it is obvious that great care is necessary in the interpretation of teratology studies. For an agent to be classified as a development toxicant or teratogen, it must produce adverse effects on the conceptus at exposure levels that do not induce toxicity in the mother. Signs of maternal toxicity include reduction in weight gain, changes in eating patterns, hypo or hyperactivity, neurotoxic signs, and organ weight changes. Adverse effects on development under these conditions may be secondary to stress on the maternal system. Findings in the fetus, at dose levels that produce maternal toxicity, can not be easily separated from the maternal toxicity. Compounds can be deliberately administered at maternally toxic dose levels to determine the threshold for adverse effects on the offspring. In such cases conclusions can be qualified to indicate that adverse effects on the offspring were found at maternally toxic dose levels and may not be indicative of selective or unique developmental toxicity.

*Effect of Chemicals in Late Pregnancy and Lactation (Perinatal and Postnatal effects).* These tests are usually carried out on rats, and 20 pregnant females per

**Table 21.9  Skeletal Abnormalities Commonly Seen in Teratogenicity Tests**

*Digits*

Polydactyly—presence of extra digits, in mouse six or more, instead of five
Syndactyly—fusion of two or more digits
Oligodactyly—absence of one or more digits
Brachydactyly—smallness of one or more digits

*Ribs*

Wavy—ribs may be any aberrant shape
Extra—may have extra ribs on either side
Fused—may be fused anywhere along the length of the rib
Branched—single base and branched

*Tail*

Short—short tail, usually lack of vertebrae
Missing—absence of tail
Corkscrew—corkscrew-shaped tail

dosage group are treated during the final third of gestation and through lactation to weaning (day 15 of pregnancy through day 21 postpartum) (Figure 21.4). The duration of gestation, parturition problems, and the number and size of pups in the naturally delivered litter are observed, as is the growth performance or the offspring. Variations of this test are the inclusion of groups treated only to parturition and only postpartum in order to separate prenatal and postnatal effects. Cross-fostering of pups to untreated dams may also be used to the same end. Behavioral testing of the pups has been suggested, and this and other physiological testing is to be recommended.

### 21.5.4  Special Tests

This general heading is used to include brief assessments of tests that are not always required but that may be required in certain cases or that have been suggested as useful adjuncts to current testing protocols. Many are in areas of toxicology that are developing rapidly; as a result no consensus has yet evolved as to the best tests or sequence of tests, only an understanding that such evaluations may shed light on previously undefined aspects of chemical toxicity.

***Neurotoxicity.*** The nervous system is complex, both structurally and functionally, and toxicants can affect one or more units of this system in selective fashion. It is necessary therefore to devise tests, or sequences of tests, that measure not only changes in overall function but that also indicate which basic unit is affected and how the toxicant interacts with its target. This is complicated by the fact that the nervous system has a considerable functional reserve, and specific observable damage may not affect overall function until it becomes even more extensive. Types of damage to the nervous system are classified in various ways but include neuronal toxicity, axonopathy, toxic interruption of impulse transmission, myelinopathy, and synaptic alterations in transmitter release or receptor function. Signs of neuropathy are frequently revealed by the acute, subchronic, chronic, and other tests that are required by regulatory agencies.

Neurotoxicity is of great significance in toxicology, however, and tests have been devised to supplement those routinely required. These include acute and subchronic neurotoxicity studies as well as developmental neurotoxicity studies.

*Behavioral and Pharmacological Tests.*  Behavioral and pharmacological tests involve the observation of clinical signs and behavior. These include signs of changes in awareness, mood, motor activity, central nervous system excitation, posture, motor incoordination, muscle tone, reflexes, and autonomic functions. If these tests so indicate, more specialized tests can be carried out that evaluate spontaneous motor activity, conditioned avoidance responses, operant conditioning, as well as tests for motor incoordination such as the inclined plane or rotarod tests.

Tests for specific classes of chemicals include the measurement of transmitter stimulated adenyl cyclase and Na/K-ATPase for chemicals that affect receptor function or cholinesterase inhibition for organophosphates or carbamates. Electrophysiological techniques may detect chemicals such as DDT or pyrethroids, which affect impulse transmission.

*Acute and Subchronic Neurotoxicity Tests.*  These studies are designed to test a wide range of neurotoxicity effects including CNS stimulation or depression, reflex perturbation, peripheral nerve damage, cognitive effects on learning and memory, motor activity effects and neuropathology. The tests are conducted using rats or sometimes mice.

In the acute neurotoxicity study, approximately 10 to 15 animals per sex per dose group are administered a single gavage (bolus) dose of the test material. There are usually three dose groups and a control. Behavioral assessments are made on the day of dosing, and at 1 and 2 weeks post dose. The assessments include tests on motor activity, a functional observation battery (FOB), and neuropathology (at termination). The FOB screens for sensorimotor, neuromuscular, autonomic and general physiological effects of a test compound. Table 21.10 depicts component tests of the FOB. These functional tests have the advantage over biochemical measures that they permit repeated evaluation of individual animals over time to determine the onset, progression, duration, and reversibility of neurotoxic effects. Motor activity is also measured over time and can be evaluated by a variety of devices. One such device that has been frequently used is the figure 8 maze, which consists of a series of interconnected alleys (in the shape of the numeral 8) converging on a central open area and covered with transparent acrylic plastic. Motor activity is detected by photobeams, and an activity count is registered each time a photobeam is interrupted by the animal. Motor activity sessions are generally 60 minutes in length and each session is divided into 5- to 10-minute reporting intervals (epochs). "Habituation" is an end point evaluated in the motor activity test, and this is defined as a decrement in activity during the test session. Activity is expected to decrease toward the end of the test as the animal's exploratory activity normally lessens as the time in the maze increases. Neuropathological examinations are the same as those described below for the subchronic neurotoxicity test.

Before the acute neurotoxicity study is conducted, it is necessary to conduct a preliminary test to determine the time of peak effect after dosing of the test material. Preliminary tests may evaluate a selected group of end points in the FOB or other sensitive end points, if known, for a particular test material. Results of this preliminary test will determine the time when observations are performed on the day of dosing in the acute neurotoxicity study.

**Table 21.10   Example of Evaluations Made in a Functional Observation Battery**

| Home Cage and Open Field | Manipulative | Physiological |
|---|---|---|
| Arousal | Ease of removal | Body temperature |
| Gait | Ease of handling | Body weight |
| Posture | Touch response | |
| Vocalizations | Righting response | |
| Piloerection | Hindlimb foot splay | |
| Lacrimation | Forelimb grip strength | |
| Salivation | Hindlimb grip strength | |
| Urination/defecation | Finger-snap response | |
| Grooming behavior | Catalepsy | |
| Rearing | Palpebral closure | |
| Abnormal movements | Pupil function | |
| Tremors, convulsions | | |

In the subchronic neurotoxicity study, end points measured are similar to those measured in the acute neurotoxicity study. However, the duration of dosing is 90 days and exposure to the test material is usually via the diet. As for the acute neurotoxicity study, these studies consist of three test groups and a control group. The functional observation battery and motor activity tests are conducted at selected intervals such as weeks 5, 9, and 13, as well as pretest. At test termination, at least 6 animals per group are perfused via the heart with fixative to ensure optimal fixation of nervous tissues for histopathology examination. Nervous tissues examined include brain, spinal cord (various segments), and selected nerves such as the optic, sciatic, tibial, and sural nerves.

For all behavioral tests it is important that the person making the actual observations is unaware of the treatment group for each animal ("blind" to dose group assignment). In addition laboratories that conduct neurotoxicity studies for regulatory agencies must demonstrate that their methods are validated. Therefore these laboratories must conduct positive control studies using known neurotoxicants and provide this information to regulators as necessary. Also, since it is not feasible for one person to perform the observations on all animals on all test occasions, laboratories must maintain evidence of interobserver reliability (agreement) for individuals who are involved with performing the functional observation batteries.

*Developmental Neurotoxicity Tests.* A separate component of developmental toxicology focuses on potential behavioral or morphological modifications resulting from exposures to toxins during early development. These studies track the outcome of such exposures through the postnatal period and into early adulthood. In a developmental neurotoxicity study at least 20 pregnant female rats for each of three treatment groups plus a control are administered test material from gestation day (GD) 6 through weaning on lactation day 21. FOBs are conducted on the maternal animals at selected intervals such as GD 6, GD 17, lactation day 11 and lactation day 21. Evaluations include observations in the home cage, during handling, and outside the home cage in an open field. Body weights and food consumption are also monitored in the maternal animals. After birth the offspring are counted, weighed, and gender determined. On postnatal day (PND) 4, litters are culled to 8 pups per litter. Following culling, at least 10 pups/sex/group are assigned to one of the following tests: learning and memory, motor activity, or acoustic startle. Additional pups are assigned for neuropathology and brain weight evaluations on PND 11 and PND 70 (10/sex/group). FOBs are performed on the offspring at selected intervals such as PND 11, 17, 21, 35, and 60/70. Indicators of physical development such as preputial separation (male sexual maturation) and vaginal patency (female sexual maturation) are evaluated for all offspring as well as body weight and food consumption. Learning and memory can be evaluated with a variety of tests. Frequently a water maze is used where the rat learns to swim through a series of alleys to find a platform it can use to climb out of the water. The time it takes for the animal to swim through the maze to the platform (trial latency) and the number of mistakes made are some of the end points evaluated in this test. The trials are conducted over a series of days and the assay provides an index of the development of both working memory (with-in day performance) as well as reference memory (between-day performance). The startle test measures the animal's response to a burst of loud noise and also how quickly it becomes habituated to 10 pulses of startle-eliciting tones in 5 blocks. For the startle test special chambers lined with sound-attenuating and vibration-absorbing material are used. These chambers can measure the force exerted on a platform on which the animal stands during the test procedure. The startle test is conducted at 2 time points such as PND 22 and 60. When the offspring are approximately 70 days old, the test is terminated. Selected animals are perfused with fixative for neuropathology assessments. In addition to detailed microscopic evaluation of at least five different sections of the brain, simple morphometric analysis of the cerebrum, hippocampus, and cerebellum are conducted.

*Delayed Neuropathy (OPIDN).* The delayed neurotoxic potential of certain organophosphates such as tri-o-cresyl phosphate (TOCP) is usually evaluated by observation of clinical signs (paralysis of leg muscles in hens) or pathology (degeneration of the motor nerves in hens), but a biochemical test involving the ratio between the ability to inhibit cholinesterase and the ability to inhibit an enzyme that has been referred to as the neurotoxic esterase (NTE) has been suggested. The ability of chemicals to cause delayed neuropathy is generally correlated with their ability to inhibit this nonspecific esterase, found in various tissues, although the role, if any, of NTE in the sequence of events leading to nerve degeneration is not known. The preferred test organism is the mature hen because the clinical signs are similar to those in humans, and such symptoms cannot be readily elicited in the common laboratory rodents.

*Potentiation.* Potentiation and synergism represent interactions between toxicants that are potential sources of hazard because neither humans nor other species are usually exposed to one chemical at a time. The enormous number of possible combinations of chemicals makes routine screening for all such effects not only impractical but impossible.

One of the classic cases is the potentiation of the insecticide malathion by another insecticide, EPN, the LD50 of the mixture being dramatically lower than that of either compound alone. This potentiation can also be seen between malathion and certain contaminants that are formed during synthesis, such as isomalathion. For this reason quality control during manufacture is essential. This example of potentiation involves inhibition, by EPN or isomalathion, of the carboxylesterase responsible for the detoxication of malathion in mammals.

It is practical to test for potentiation only when there has been some preliminary indication that it might occur or when either or both compounds belong to chemical classes previously known to cause potentiation. Such a test can be conducted by comparing the LD50, or any other appropriate toxic end point, of a mixture of equitoxic doses of the chemicals in question with the same end point measured with the two chemicals administered alone.

In the case of synergism, in which one of the compounds is relatively nontoxic when given alone, the toxicity of the toxic compound can be measured when administered alone or after a relatively large dose of the nontoxic compound.

***Toxicokinetics and Metabolism.*** Routine toxicity testing without regard to the mechanisms involved is likely to be wasteful of time and of human, animal, and financial resources. A knowledge of toxicokinetics and metabolism can give valuable insights and provide for testing that is both more efficient and more informative. Such knowledge provides the necessary background to make the most appropriate selection of test animal species and of dose levels, and the most appropriate method for extrapolating from animal studies to the assessment of human hazard. Moreover they may provide information on possible reactive intermediates as well as information on induction or inhibition of the enzymes of xenobiotic metabolism, the latter being critical to an assessment of possible interaction.

The nature of metabolic reactions and their variations between species is detailed in Chapters 7, 8, and 9 with some aspects of toxicokinetics in Chapter 6. The methods used for the measurement of toxicants and their metabolites are detailed in Chapter 25. The present section is concerned with the general principles, use, and need for metabolic and toxicokinetics studies in toxicity testing.

Toxicokinetics studies are designed to measure the amount and rate of the absorption, distribution, metabolism, and excretion of a xenobiotic. These data are used to construct predictive mathematical models so that the distribution and excretion of other doses can be simulated. Such studies are carried out using radiolabeled compounds to facilitate measurement and total recovery of the administered dose. This can be done entirely in vivo by measuring levels in blood, expired air, feces, and urine; these procedures can be done relatively noninvasively and continuously in the same animal. Tissue levels can be measured by sequential killing and analysis of organ levels. It is important to measure not only the compound administered but also its metabolites, because simple radioactivity counting does not differentiate among them.

The metabolic study, considered separately, consists of treatment of the animal with the radiolabeled compound followed by chemical analysis of all metabolites formed in vivo and excreted via the lungs, kidneys, or bile. Although reactive intermediates are unlikely to be isolated, the chemical structure of the end products may provide vital clues to the nature of the intermediates involved in their formation. The use of tissue homogenates, subcellular fractions, and purified enzymes may serve to clarify events occurring during metabolic sequences leading to the end products.

Information of importance in test animal selection is the similarity in toxicodynamics and metabolism to that of humans. Although all of the necessary information may not be available for humans, it can often be inferred with reference to metabolism and excretion of related compounds, but it is clearly ill advised to use an animal that differs from most others in the toxicokinetics or metabolism of the compound in question or that differs from humans in the nature of the end products. Dose selection is influenced by knowledge of whether a particular dose saturates a physiological process such as excretion or whether it is likely to accumulate in a particular tissue, because these factors are likely to become increasingly important the longer a chronic study continues.

***Behavior.*** Although the primary emphasis in toxicity testing has long been the estimation of morphologic changes, much recent interest has focused on more fundamental evaluations. One such aspect has been the evaluation of chemical effects on behavior.

The categories of methods used in behavioral toxicology fall into two principal classes, stimulus-oriented behavior, and internally generated behavior. The former includes two types of conditioned behavior: operant conditioning, in which animals are trained to perform a task in order to obtain a reward or to avoid a punishment, and classical conditioning, in which an animal learns to associate a conditioning stimulus with a reflex action. Stimulus-oriented behavior also involves unconditioned responses in which the animal's response to a particular stimulus is recorded.

Internally generated behavior includes observation of animal behavior in response to various experimental situations, and includes exploratory behavior, circadian activity, social behavior, and so on. The performance of animals treated with a particular chemical is compared with that of untreated controls as a measure of the effect of the chemical.

Many of the variables associated with other types of testing must also be controlled in behavioral tests: sex, age, species, environment, diet, and animal husbandry. Behavior may vary with all of these. Norton describes a series of four tests that may form an appropriate series, inasmuch as they represent four different types of behavior; the series should therefore reflect different types of nervous system activity. They are as follows:

1. *Passive avoidance.* This test involves the use of a shuttle box, in which animals can move between a light side and a dark side. After an acclimatization period, in which the animal can move freely between the two sides, it receives a mild electric shock while in the dark (preferred) side. During subsequent trials, the time spent in the "safe side" is recorded.

2. *Auditory startle.* This test involves the response (movement) to a sound stimulus either without, or preceded by, a light-flash stimulus.

3. *Residential maze.* Movements of animals in a residential maze are automatically recorded during both light and dark photoperiods.

4. *Walking patterns.* Gait is measured in walking animals, including such characteristics as the length and width of stride and the angles formed by the placement of the feet.

Problems associated with behavioral toxicology include the functional reserve and adaptability of the nervous system. Frequently behavior is maintained despite clearly observable injury. Other problems are the statistical ones associated with multiple tests, multiple measurements, and the inherently large variability in behavior.

The use of human subjects occupationally exposed to chemicals is often attempted, but such tests are complicated by the subjective nature of the end points (dizziness, etc.).

***Covalent Binding.*** Toxicity has been associated with covalent binding in a number of ways. Organ-specific toxicants administered in vivo bind covalently to macromolecules, usually at a higher level in the target tissues than in nontarget tissues. Examples include acetaminophen in the liver, carbon tetrachloride in the liver, *p*-aminophenol in the kidney, and ipomeanol in the lung. Similarly many carcinogens are known to give rise to DNA adducts. In general, covalent binding occurs as a result of metabolism of the toxicant to highly reactive intermediates, usually, but not always, by cytochrome P450. Because these intermediates are highly reactive electrophiles, they bind to many nucleophilic sites on DNA, RNA, or protein molecules, not just the site of toxic action. Thus measurement of covalent binding may be a measure of toxic potential rather than a specific measurement, related directly to a mechanism of action. The occurrence of covalent binding at the same time as toxicity is so common an occurrence, however, that a measurement of covalent binding of a chemical may be regarded as an excellent although perhaps not infallible indication of potential for toxicity. Although such tests are not routine, considerable interest has been shown in their development.

The measurement of DNA adducts is an indirect indication of genotoxic (carcinogenic) potential, and DNA adducts in the urine are an indication, obtained by a noninvasive technique, of recent exposure. Protein adducts give an integrated measure of exposure because they accumulate over the life span of the protein and, at the same time, indicate possible organ toxicity.

Tissue protein adducts are usually demonstrated in experimental animals following injection of radiolabeled chemicals and, after a period to time, the organs are removed and homogenized, and by rigorous extraction, all the noncovalently bound material is removed. Extraction methods include lipid solvents, acids and bases, concentrated urea solutions, and solubilization and precipitation of the proteins. They tend to underestimate the extent of covalent binding because even covalent bonds may be broken by the rigorous procedures used. Newer methods involving dialysis against detergents and separation of adducted proteins will probably prove more appropriate.

Blood proteins, such as hemoglobin, may be used in tests of human exposure because blood is readily and safely accessible. For example, the exposure of mice to ethylene oxide or dimethylnitrosamine was estimated by measuring alkylated residues in hemoglobin. The method was subsequently extended to people exposed occupationally to ethylene oxide by measuring N-3-(2-hydroxyethyl) histidine residues in hemoglobin. Similarly methyl cysteine residues in hemoglobin can be used as a measure of methylation.

DNA-RNA adducts can also be measured in various ways, including rigorous extraction, separation, and precipitation following administration of labeled compounds in vivo, or use of antibodies raised to chemically modified DNA or RNA.

Although many compounds of different chemical classes have been shown to bind covalently when activated by microsomal preparations in vitro (e.g., aflatoxin, ipomeanol, stilbene, vinyl chloride), these observations have not been developed into routine testing procedures. Such procedures could be useful in predicting toxic potential.

***Immunotoxicity.*** Immunotoxicology comprises two distinct types of toxic effects: the involvement of the immune system in mediating the toxic effect of a chemical and the toxic effects of chemicals on the immune system. The former is shown, for example, in tests for cutaneous sensitization, whereas the latter is shown in impairment of the ability to resist infection.

Tests for immunotoxicity are not required by all regulatory agencies, but it is an area of great interest, both in the fundamental mechanisms of immune function and in the design of tests to measure impairment of immune function. Both of these aspects are discussed in detail in Chapter 19.

## 21.6  IN VITRO AND OTHER SHORT-TERM TESTS

### 21.6.1  Introduction

The toxicity tests that follow are tests conducted largely in vitro with isolated cell systems. Some are short-term tests carried out in vivo or are combinations of in vivo and in vitro systems. The latter are included because of similarities in approach, mechanism, or intent. In general, these tests measure effects on the genome or cell transformation; their importance lies in the relationship between such effects and the mechanism of chemical carcinogenesis. Mutagenicity of cells in the germ line is itself an expression of toxicity, however, and the mutant genes can be inherited and expressed in the next or subsequent generations.

The theory that the initiating step of chemical carcinogenesis is a somatic mutation is well recognized, and considerable evidence shows that mutagenic potential is correlated with carcinogenic potential. Thus the intent of much of this type of testing is to provide early warning of carcinogenic potential without the delay involved in conducting lifetime chronic feeding studies in experimental animals. Despite the numerous tests that have been devised, regulatory agencies have not yet seen it fit to substitute any of them, or any combination of them, for chronic feeding studies. Instead, they have been added as additional testing requirements. One function of such tests should be to identify those compounds with the greatest potential for toxicity and enable the amount of chronic testing to be reduced to more manageable proportions.

### 21.6.2  Prokaryote Mutagenicity

***Ames Test.*** The Ames test, developed by Bruce Ames and his coworkers at the University of California, Berkeley, depends on the ability of mutagenic chemicals to bring about reverse mutations in *Salmonella typhimurium* strains that have defects in the histidine biosynthesis pathway. These strains will not grow in the absence of histidine but can be caused to mutate back to the wild type, which can synthesize histidine and hence can grow in its absence. The postmitochondrial supernatant (S-9 fraction), obtained from homogenates of livers of rats previously treated with PCBs in order to induce certain cytochrome P450 isoforms, is also included in order to provide the activating enzymes involved in the production of the potent electrophiles often involved in the toxicity of chemicals to animals.

Bacterial tester strains have been developed that can test for either base-pair (e.g., strain TA-1531) or frameshift (e.g., strains TA-1537, TA-1538) mutations. Other, more sensitive strains such as TA-98 and TA-100 are also used, although they may be less specific with regard to the type of mutation caused.

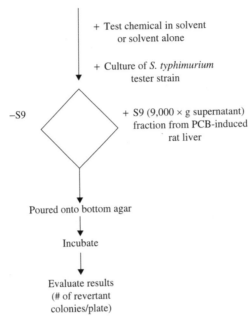

Molten soft agar at 45°C

+ Test chemical in solvent
or solvent alone

+ Culture of *S. typhimurium*
tester strain

−S9

+ S9 (9,000 × g supernatant)
fraction from PCB-induced
rat liver

Poured onto bottom agar

Incubate

Evaluate results
(# of revertant
colonies/plate)

**Figure 21.5**   Protocol for the Ames test for mutagenicity.

In brief, the test is carried out (Figure 21.5) by mixing a suspension of bacterial cells with molten top agar. This also contains cofactors, S-9 fraction, and the material to be tested. The mixture is poured onto Petri plates containing hardened minimal agar. The number of bacteria that revert and acquire the wild-type ability to grow in the absence of histidine can be estimated by counting the colonies that develop on incubation. To provide a valid test, a number of concentrations are tested, and positive controls with known mutagens are included along with negative controls that lack only the test compound. The entire test is replicated often enough to satisfy appropriate statistical tests for significance. Parallel tests without the S-9 fraction may help distinguish between chemicals with intrinsic mutagenic potential and those that require metabolic activation.

The question of correlation between mutagenicity and carcinogenicity is crucial in any consideration of the utility of this or similar tests. In general, this appears to be high, although a small proportion of both false positives and false negatives occurs. For example, certain base analogues and inorganics such as manganese are not carcinogens but are mutagens in the Ames test, whereas diethylstilbestrol (DES) is a carcinogen but not a bacterial mutagen (see Chapter 12 for additional detail).

***Related Tests.***   Related tests include tests based on reverse mutations, as in the Ames test, as well as tests based on forward mutations. Examples include:

1. *Reverse mutations in Escherichia coli.* This test is similar to the Ames test and depends on reversion of tryptophane mutants, which cannot synthesize this amino acid, to the wild type, which can. The S-9 fraction from the liver of induced rats

can also be used as an activating system in this test. Other *E. coli* reverse mutation tests utilize nicotinic acid and arginine mutants.

2. *Forward mutations in S. typhimurium.* One such assay, dependent on the appearance of a mutation conferring resistance to 8-azaguanine in a histidine revertant strain, has been developed and is said to be as sensitive as the reverse-mutation tests

3. *Forward mutations in E. coli.* These mutations depend on mutation of galactose nonfermenting *E. coli* to galactose fermenting *E. coli* or the change from 5-methyltryptophane to 5-methyltryptophane resistance.

4. *DNA repair.* Polymerase-deficient, and thus DNA repair-deficient, *E. coli* has provided the basis for a test that depends on the fact that the growth of a deficient strain is inhibited more by a DNA-damaging agent than is that of a repair-competent strain. The recombinant assay using *Bacillus subtilis* is conducted in much the same way because recombinant deficient strains are more sensitive to DNA-damaging agents.

## 21.6.3 Eukaryote Mutagenicity

***Mammalian Cell Mutation.*** The development of cell culture techniques that permit both survival and replication have led to many advances in cell biology, including the use of certain of these cell lines for detection of mutagens. Although such cells, if derived from mammals, would seem ideal for testing for toxicity toward mammals, there are several problems. Primary cells, which generally resemble those of the tissue of origin, are difficult to culture and have poor cloning ability. Because of these difficulties, certain established cell lines are usually used. These cells, such as Chinese hamster ovary cells and mouse lymphoma cells, clone readily and do not become senescent with passage through many cell generations. Unfortunately, they have little metabolic activity toward xenobiotics and thus do not readily activate toxicants. Moreover they usually show chromosome changes, such as aneuploidy (i.e., more or fewer than the usual diploid number of chromosomes).

The characteristics usually involved in these assays are resistance to 8-azaguanine or 6-thioguanine (the hypoxanthine guanine phosphoriboxyl transferase or HGPRT locus), resistance to bromodeoxyuridine or triflurothymidine (the thymidine kinase or TK locus), or resistance to ouabain (the OU or Na/K-ATPase locus). HGPRT is responsible for incorporation of purines from the medium into the nucleic acid synthesis pathway. Its loss prevents uptake of normal purines and also of toxic purines such as 8-azaguanine, which would kill the cell. Thus mutation at this locus confers resistance to these toxic purine analogues. Similarly TK permits pyrimidine transport, and its loss prevents uptake of toxic pyrimidine analogues and confers resistance to them. In the absence of HGPRT or TK, the cells can grow by de novo synthesis of purines and pyrimidines. Ouabain kills cells by combining with the Na/K-ATPase. Mutation at the OU locus alters the ouabain-binding site in a way that prevents inhibition and thus confers resistance.

A typical test system is the analysis of the TK locus in mouse lymphoma cells for mutations that confer resistance to bromodeoxyuracil. The tests are conducted with and without the S-9 fraction from induced rat liver because the lymphoma cells have little activating ability. Both positive and negative controls are included, and the parameter measured is the number of cells formed that are capable of forming colonies in the presence of bromodeoxyuridine.

***Drosophila Sex-Linked Recessive Lethal Test.*** The advantages of *Drosophila* (fruit fly) tests are that they involve an intact eukaryotic organism with all of its interrelated organ systems and activation mechanisms but, at the same time, are fast, relatively easy to perform, and do not involve mammals as test animals. The most obvious disadvantages are that the hormonal and immune systems of insects are significantly different from those of mammals and that the nature, specificity, and inducibility of the cytochrome P450s are not as well understood in insects as they are in mammals.

In a typical test, males that are 2 days postpuparium and that were raised from eggs laid within a short time period (usually 24 hours) are treated with the test compound in water to which sucrose has been added to increase palatability. Males from a strain carrying a gene for yellow body on the X chromosome are used. Preliminary tests determine that the number of offspring of the survivors of the treatment doses (usually 0.25 LD50 and 0.5 LD50) are adequate for future crosses. Appropriate controls, including a solvent control (with emulsifier if one was necessary to prepare the test solution), and a positive control, such as ethyl methane sulfonate, are routinely included with each test. Individual crosses of each surviving treated male with a series of three females are made on a 0- to 2-, 3- to 5-, and 6- to 8-day schedule. The progeny of each female is reared separately, and the males and females of the $F_1$ generation are mated in brother-sister matings. If there are no males with yellow bodies in a particular set of progeny, it should be assumed that a lethal mutation was present on the treated X chromosomes. A comparison of the $F_2$ progeny derived from females inseminated by males at different times after treatment allows a distinction to be made between effects on spermatozoa, spermatids, and spermatocytes.

In the Basc (Muller-5) test shown in Figure 21.6, the strain used for the females in the $F_1$ cross is a multiple-marked strain that carries a dominant gene for bar eyes and recessive genes for apricot eyes and a reduction of bristles on the thorax (scute gene). (Basc is an acronym for bar, apricot, and scute.)

***Related Tests.*** Many tests related to the two types of eukaryote-mutation tests are discussed earlier in this section, and many of them are simply variations of the tests described. Two distinct classes are worthy of mention: the first uses yeasts as the test organisms, and the second is the spot test for mutations in mice.

One group of yeast tests includes tests for gene mutations and strains that can be used to detect forward mutations in genes that code for enzymes in the purine biosynthetic pathway; other strains can be used to detect reversions. Yeasts can also be used to test for recombinant events such as reciprocal mitotic recombination (mitotic crossing over) and nonreciprocal mitotic recombination. Saccharomyces cerevisiae is the preferred organism in almost all these tests. Although they possess cytochrome P450s capable of metabolizing xenobiotics, their specificity and sensitivity are limited as compared with those of mammals, and an S-9 fraction is often included, as in the Ames test, to enhance activation.

The gene mutation test systems in mice include the specific locus test, in which wild-type treated males are crossed with females carrying recessive mutations for visible phenotypic effects. The $F_1$ progeny have the same phenotype as the wild-type parent unless a mutation, corresponding to a recessive mutant marker, has occurred. Such tests

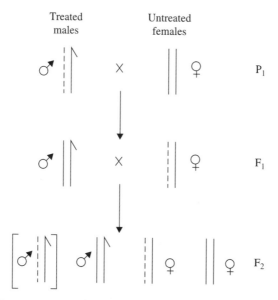

**Figure 21.6**  The Basc (Muller-5) mating scheme. Dashed lines represent the treated X chromosome of males. Brackets indicate males with yellow bodies, which would be absent if a lethal mutation occurred on the X chromosome of the treated male.

are accurate, and the spontaneous (background) mutation rate is very low, making them sound tests that are predictive for other mammals. Unfortunately, the large number of animals required has prevented extensive use. Similar tests involving the activity and electrophoretic mobility of various enzymes in the blood or other tissues in the $F_1$ progeny from treated males and untreated females have been developed. In the previously mentioned tests, as with many others, sequential mating of males with different females can provide information about the stage of sperm development at which the mutational event occurred.

### 21.6.4  DNA Damage and Repair

Many of the end points for tests described in this chapter, including gene mutation, chromosome damage, and oncogenicity, develop as a consequence of damage to or chemical modification of DNA. Most of these tests, however, also involve metabolic events that occur both prior to and subsequent to the modification of DNA. Some tests, however, use events at the DNA level as end points. One of these, the unscheduled synthesis of DNA in mammalian cells, is described in some detail; the others are summarized briefly.

***Unscheduled DNA Synthesis in Mammalian Cells.***  The principle of this test is that it measures the repair that follows DNA damage and is thus a reflection of the damage itself. It depends on the autoradiographic measurement of the incorporation of tritiated thymidine into the nuclei of cells previously treated with the test chemical.

The preferred cells are usually primary hepatocytes in cultures derived from adult male rats whose cells are dispersed and allowed to attach themselves to glass coverslips.

From this point on, the test is carried out on the attached cells. Both positive controls with agents known to stimulate unscheduled DNA synthesis, such as the carcinogen aflatoxin B1 or 2-acetylaminofluorene, and negative controls, which are processed through all procedures except exposure to the test compound, are performed routinely with every test. Cells are exposed by replacing the medium for a short time with one containing the test chemical. The dose levels are determined by a preliminary cell viability test (Trypan blue exclusion test) and consist of several concentrations that span the range from no apparent loss of viability to almost complete loss of viability. Following exposure, the medium is removed, and the cells are washed by several changes of fresh medium and finally placed in a medium containing tritiated thymidine. The cells are fixed and dried, and the coverslip with the cells attached is coated with photographic emulsion. After a suitable exposure period (usually several weeks), the emulsion is developed and the cells are stained with hemotoxylin and eosin. The number of grains in the nuclear region is corrected by subtracting nonnuclear grains, and the net grain count in the nuclear area is compared between treated and untreated cells.

This test has several advantages in that primary liver cells have considerable activation capacity and the test measures an event at the DNA level. It does not, however, distinguish between error-free repair and error-prone repair, the latter being itself a mutagenic process. Thus it cannot distinguish between events that might lead to toxic sequelae and those that do not. A modification of this test measures in vivo unscheduled DNA synthesis. In this modification animals are first treated in vivo, and primary hepatocytes are then prepared and treated as already described.

**Related Tests.** Tests for the measurement of binding of the test material to DNA have already been discussed under covalent binding (Section 21.5.4). Another method of assessing DNA damage is the estimation of DNA breakage following exposure to the test chemical; the DNA-strand length is estimated by using alkaline elution or sucrose density gradient centrifugation. This has been done with a number of cell lines and with freshly prepared hepatocytes, in the latter case following treatment either in vivo or in vitro. It may be regarded as promising but not yet fully validated. The polymerase-deficient *E. coli* tests as well as recombinant tests using yeasts are also related to DNA repair.

### 21.6.5 Chromosome Aberrations

Tests for chromosome aberrations involve the estimation of effects on extended regions of whole chromosomes rather than on single or small numbers of genes. Primarily they concern chromosome breaks and the exchange of material between chromosomes.

**Sister Chromatid Exchange.** Sister chromatid exchange (SCE) occurs between the sister chromatids that together make up a chromosome. It occurs at the same locus in each chromatid and is thus a symmetrical exchange of chromosome material. In this regard it is not strictly an aberration because the products do not differ in morphology from normal chromosome. SCE, however, is susceptible to chemical induction and appears to be correlated with the genotoxic potential of chemicals as well as with their oncogenic potential. The exchange is visualized by permitting the treated cells to pass through two DNA replication cycles in the presence of 5-bromo-2'-deoxyuridine, which

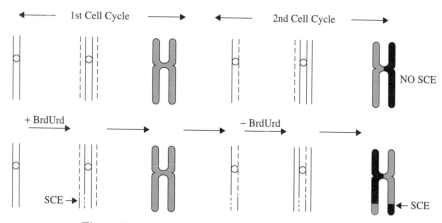

**Figure 21.7**   Visualization of sister chromatid exchange.

is incorporated in the replicated DNA. The cells are then stained with a fluorescent dye and irradiated with UV light, which permits differentiation between chromatids that contain bromodeoxyuridine and those that do not (Figure 21.7).

The test can be carried out on cultured cells or on cells from animals treated in vivo. In the former case the test chemical is usually evaluated in the presence and absence of the S-9 activation system from rat liver. Typically cells from a Chinese hamster ovary cell line are incubated in a liquid medium and exposed to several concentration of the test chemical, either with or without the S-9 fraction, for about 2 hours. Positive controls, such as ethyl methane sulfonate (a direct-acting compound) or dimethylnitrosamine (one that requires activation), as well as negative controls are also included. Test concentrations are based on cell toxicity levels determined by prior experiment and are selected in such a way that even at the highest dose excess growth does not occur. At the end of the treatment period the cells are washed, bromodeoxyuridine is added, and the cells are incubated for 24 hours or more. The cells are then fixed, stained with a fluorescent dye, and irradiated with UV light. Second division cells are scored under the microscope for SCEs (Figure 21.7).

The test can also be carried out on cells treated in vivo, and analyses have been made of SCEs in lymphocytes from cancer patients treated with chemotherapeutic drugs, smokers, and workers exposed occupationally; in several cases increased incidence of SCEs has been noted. This is a sensitive test for compounds that alkylate DNA, with few false positives. It may be useful for detecting promoters such as phorbol esters.

***Micronucleus Test.*** The micronucleus test is an in vivo test usually carried out in mice. The animals are treated in vivo, and the erythrocyte stem cells from the bone marrow are stained and examined for micronuclei. Micronuclei represent chromosome fragments or chromosomes left behind at anaphase. It is basically a test for compounds that cause chromosome breaks (clastogenic agents) and compounds that interfere with normal mitotic cell division, including compounds that affect spindle fiber function.

Male and female mice from an outbred strain are handled by the best animal husbandry techniques, as described for acute, subchronic, and chronic tests, and are treated either with the solvent, 0.5 LD50, or 0.1 LD50 of the test chemical. Animals are killed at several time intervals up to 2 days; the bone marrow is extracted, placed on microscope slides, dried, and stained. The presence of micronuclei is scored visually under the microscope.

***Dominant Lethal Test in Rodents.*** The dominant lethal test, which is performed using rats, mice, or hamsters, is an in vivo test to determine the germ-cell risk from a suspected mutagen. The test consists of treating males with the test compound for several days, followed by mating to different females each week for enough weeks to cover the period required for a complete spermatogenic cycle. Animals are maintained under optimal conditions of animal husbandry and are dosed, usually by a gavage, with several doses of less than 0.1 LD50. The females are killed after two weeks of gestation and dissected; corpora lutea and living and dead implantations are counted. The end points used to determine the occurrence of dominant lethal mutations in the treated males are the fertility index (ratio of pregnant females to mated females), preimplantation losses (the number of implantations relative to the number of corpora lutea), the number of females with dead implantations relative to the total number of pregnant females, and the number of dead implantations relative to the total number of implantations. Mutations in sperm that are dominant and lethal do not result in viable offspring.

***Related Tests.*** Many cells exposed to test chemicals can be scored for chromosome aberrations by staining procedures followed by visual examination with the aid of the microscope. These include Chinese hamster ovary cells in culture treated in a protocol very similar to that used in the test for SCEs, bone marrow cells from animals treated in vivo, or lymphocytes from animals treated in vivo. The types of aberrations evaluated include chromatid gaps, breaks, and deletions; chromosome gaps, breaks, and deletions; chromosome fragments; translocations; and ploidy.

Heritable translocations can be detected by direct examination of cells from male or female offspring in various stages of development or by crossing the treated animals to untreated animals and evaluating fertility, with males with reduced fertility being examined for translocations, and so on. Progeny from this or other tests, such as those for dominant lethals, can be permitted to survive and then examined for translocations and other abnormalities.

## 21.6.6 Mammalian Cell Transformation

Most cell transformation assays utilize fibroblast cultures derived from embryonic tissue. The original studies showed that cells from C3H mouse fibroblast cultures developed morphologic changes and changes in growth patterns when treated with carcinogens. Later similar studies were made with Syrian hamster embryo cells. The direct relationship of these changes to carcinogenesis was demonstrated by transplantation of the cells into a host animal and the subsequent development of tumors. The recent development of practical assay procedures involves two cell lines from mouse embryos, Balb/3T3 and C3H/10T1/2, in which transformation is easily recognized and scored. In a typical assay situation, cells, such as Balb/3T3 mouse fibroblasts, will multiply in culture until a monolayer is formed. At this point they cease dividing unless transformed. Chemicals that are transforming agents will, however, cause growth to occur in thicker layers above the monolayer. These clumps of transformed cells are known as foci. Despite many recommended controls the assay is only semiquantitative.

The doses are selected from the results of a preliminary experiment and range from a high dose that reduces colony formation (but not by >50%) to a low dose that has no measurable effect on colony formation. After exposure to the test chemical for 1 to 3 days, the cells are washed and incubation is continued for up to 4 weeks. At that time the monolayers are fixed, stained, and scored for transformed foci.

Transformation assays have several distinct advantages. Because transplanted foci give rise to tumors in congenic hosts (those from the same inbred strain from which the cells were derived) whereas untransformed cells do not, cell transformation is believed to be illustrative of the overall expression of carcinogenesis in mammalian tissues. The two cell types used most (Balb/3T3 and C3H/10T1/2) respond to promoters in the manner predicted by the multistage model for carcinogenesis in vivo and may eventually be useful in the development of assays for promotion. Unfortunately, a large number of false negative results are obtained because these cell lines do not show much activation capacity; it has not proved practical to combine them with the S-9 activation system. Furthermore the cells are aneuploidy and may be preneoplastic in the untreated state. Syrian hamster cells, which do have considerable activation capacity, have proved difficult to use in test procedures and are difficult to score.

### 21.6.7  General Considerations and Testing Sequences

Considering all of the tests for acute and chronic toxicity, long and short term, in vivo and in vitro, it is clearly impractical to apply a complete series of tests to all commercial chemicals and all their derivatives in food, water, and the environment. The challenge of toxicity testing is to identify the most effective set or sequence of tests necessary to describe the apparent and potential toxicity of a particular chemical or mixture of chemicals. The enormous emphasis on in vitro or short-term tests that has occurred since the mid-1970s had its roots in the need to find substitutes for lifetime feeding studies in experimental animals or, at the very least, to suggest a sequence of tests that would enable priorities to be set for which chemicals should be subjected to chronic tests. Such tests might also be used to eliminate the need for chronic testing for chemicals that either clearly possessed the potential for toxicity or clearly do not. Although there has been much success in test development, the challenge outlined here has not been met, primarily because of the failure of scientists and regulatory agencies, worldwide, to agree on test sequences or on the circumstances in which short-term tests may substitute for chronic tests. Thus not only are short-term tests often required, these tests are in addition to long-term tests. As an example, the US EPA requirements for the Federal Insecticide, Fungicide, and Rodenticide Act (FIFRA) include, besides a full battery of acute, subchronic, and chronic tests, tests to address the following three categories: gene mutations, structural chromosome aberrations, and other genotoxic tests as appropriate (e.g., DNA damage and repair and chromosome aberrations). It is important, however, that test sequences have been suggested and considered by regulatory agencies, but there must also be taken into account the fact that short-term tests do not provide all of the information needed from the longer term tests.

### 21.7  ECOLOGICAL EFFECTS

Tests for ecological effects include those designed to address the potential of chemicals to affect ecosystems and the population dynamics in the environment. The tests are conducted to estimate effects on field populations of vertebrates, invertebrates, and plants. The use of environmental risk assessment tests is discussed in detail in Chapter 28.

### 21.7.1  Laboratory Tests

There are two types of laboratory tests: toxicity determinations on wildlife and aquatic organisms and the use of model ecosystems to measure bioaccumulation and transport of toxicants and their degradation products.

Among the tests included in the first category are the avian oral LD50, the avian dietary LC50, wild mammal toxicity, and avian reproduction. The avian tests are usually carried out on bobwhite quail or mallard ducks, whereas the wild mammals may be species such as the pine mouse, *Paramyscus*. The tests are similar to those described under acute and chronic testing procedures but suffer from some drawbacks; the standards of animal husbandry used with rats and mice are probably unattainable with birds or wild mammals, even through bobwhite quail and mallards are easily reared in captivity. The genetics of the birds and mammals used are much more variable than are those of the traditional laboratory rodent strains.

Similar tests can be carried out with aquatic organisms (e.g., the LC50 for freshwater fish such as rainbow trout and bluegills), the LC50 for estuarine and marine organisms, the LC50 for invertebrates such as *Daphnia*, and the effect of chemicals on the early stages of fish and various invertebrates.

Model systems, first developed by ecologists to study basic ecological processes, have been adapted to toxicological testing. In toxicology these models were first used to determine the movement and concentration of pesticides. Typically the model has a water phase containing vertebrates and invertebrates, and a terrestrial phase containing at least one plant species and one herbivore species. First, the $^{14}$C-labeled pesticide or other environmental contaminant is applied to the leaves of the terrestrial plant sorghum *(Sorgum halpense)*, and then salt marsh caterpillars *(Estigmene acrea)* are placed on the plants. The larvae eat the plants and contaminate the water with feces and their dead bodies. The aquatic food chain is simulated with plankton (diatoms, rotifers, etc.), water fleas *(Daphnia)*, mosquito larvae *(Culex pipiens)*, and fish *(Gambusia affinis)*. From an analysis of the plants, animals, and substrates for the $^{14}$C-labeled compound and its degradation products, the biologic magnification or rate of degradation can be calculated.

More complex models involving several compartments, simulated rain, simulated soil drainage, simulated tidal flow, and so on, have been constructed and their properties investigated, but none have been brought to the stage of use in routine testing. Similarly aquatic models using static, re-circulating, and continuous flow have also been used, as have entirely terrestrial models: again, none have been developed for routine testing.

### 21.7.2  Simulated Field Tests

Simulated field tests may be quite simple, consisting of feeding treated prey to predators and studying the toxic effects on the predator, enabling some predictions concerning effects to nontarget organisms. In general, however, the term is used for greenhouse, small plot, small artificial pond, or small natural pond tests. These serve to test biologic accumulation and degradation under conditions somewhat more natural than in model ecosystems and the test chemicals are exposed to environmental as well as biologic degradation. Population effects may be noted, but these methods are more useful for soil invertebrates, plants, and aquatic organisms because other organisms are not easily contained in small plots.

### 21.7.3  Field Tests

In field-test situations, test chemicals are applied to large areas under natural conditions. The areas are at least several acres and may be either natural or part of some agroecosystem. Because the area is large and in the open, radiolabeled compounds cannot be used, it is not possible to obtain a balance between material applied and material recovered.

The effects are followed over a long period of time and two types of control may be used: first, a comparison with a similar area that is untreated; and second, and a comparison with the same area before treatment. In the first case it is difficult, if not impossible, to duplicate exactly a large natural area, and in the second, changes can occur that are unrelated to the test material.

In either case, studies of populations are the most important focus of this type of testing, although the disappearance of the test material, its accumulation in various life forms, and the appearance, accumulation, and disappearance of its degradation products are also important. The population of soil organisms, terrestrial organisms, and aquatic organisms as well as plants all must be surveyed and characterized, both qualitatively and quantitatively. After application of the test material the populations can be followed through two or more annual cycles to determine both acute and long-term population effects.

### 21.8  RISK ANALYSIS

The preceding tests for various kinds of toxicity can be used to measure adverse effects of many different chemical compounds in different species, organ, tissues, cells, or even populations, and under many different conditions. This information can be used to predict possible toxicity of related compounds from QSAR or of the same chemical under different conditions (e.g., mutagenicity as a predictor of carcinogenicity). It is considerably more difficult to use this information to predict possible risk to other species, such as humans, because little experimental data on this species is available. Some methods are available to predict risk to humans and to provide the risk factor in the risk-benefit assessment that provides the basis for regulatory action, however. Human health risk assessment is discussed in detail in Chapter 24. The benefit factor is largely economic in nature, and the final regulatory action is not, in the narrow sense, a scientific one. It also involves political and legal aspects and, in toto, represents society's evaluation of the amount of risk that can be tolerated in any particular case.

### 21.9  THE FUTURE OF TOXICITY TESTING

Because of the public awareness of the potentially harmful effects of chemicals, it is clear that toxicity testing will continue to be an important activity and that it will be required by regulatory agencies before the use of a particular chemical is permitted either in commercial processes or for use by the public. Because of the proliferation of testing procedures, the number of experimental species and other test systems available, as well as the high dose rates usually used, it is clear that eventually some expression of some type of toxicity will be obtained for most exogenous chemicals. Thus the identification of toxic effects with the intent of banning any chemical causing such effects is no longer a productive mode of attack. The aim of toxicity testing should be to identify those compounds that present an unacceptable potential for risk to humans or to the environment and thus ought to be banned but, at the same time, provide an accurate assessment of the risk to humans and the environment of less toxic compounds so that their use may be regulated.

Subjecting all chemicals to all possible tests is logistically impossible, and the future of toxicity testing must lie in the development of techniques that will narrow the testing process so that highly toxic and relatively nontoxic compounds can be identified early and either banned or permitted unrestricted use without undue waste of time, funds, and human resources. These vital commodities could then be concentrated on compounds whose fate and effects are less predictable.

Such progress will come from further development and validation of the newer testing procedures and the development of techniques to select, for any given chemical, the most suitable testing methods. Perhaps of most importance is the development of integrated test sequences that permit decisions to be made at each step, thereby either abbreviating the sequence or making the next step more effective and efficient.

## SUGGESTED READING

Balls, M., R. J. Riddell, and A. N. Worden. *Animals and Alternatives in Toxicity Testing.* London: Academic Press, 1983.

Clark, B., and D. A. Smith. Pharmacokinetics and toxicity testing. *CRC Crit. Rev. Toxicol.* **12**: 343, 1984.

Couch, J. A., and W. J. Hargis Jr. Aquatic animals in toxicity testing. *J. Am. Coll. Toxicol.* **3**: 331, 1984.

Dean, J. H., M. I. Luster, M. J. Murray, and L. D. Laver. Approaches and methodology for examining the immunological effects of xenobiotics. *Immunotoxicol.* **7**: 205, 1983

de Serres F. J., Ashby J. (eds): *Evaluation of Short Term Tests for Carcinogens.* New York: Elsevier, 1981.

Ecobichon, D. J.. *The Basis of Toxicity Testing.* Boca Raton, FL: CRC Press, 1992.

Enslein, K., and P. N. Craig. Carcinogenesis: A predictive structure-activity model. *J. Toxicol. Environ. Health* **10**: 521, 1982.

Gorrod, J. W., ed. *Testing for Toxicity.* London: Taylor and Francis, 1981. Relevant chapters include:

> Chapter 3. Brown, V. K. H.: Acute toxicity testing—a critique.
> Chapter 4. Roe, F. J. C.: Testing in vivo for general chronic toxicity and carcinogenicity.
> Chapter 7. Gorrod, J. W.: Covalent binding as an indication of drug toxicity.
> Chapter 15. Dewar, A. J.: Neurotoxicity testing—with particular reference to biochemical methods.
> Chapter 18. Cobb, L. M.: Pulmonary toxicity.

Chapter 20. Parish, W. E.: Immunological tests to predict toxicological hazards to man.

Chapter 21. Venitt, S.: Microbial tests in carcinogenesis studies.

Chapter 22. Styles, J. A.: Other short-term tests in carcinogenesis studies.

Hayes, A. W., ed. *Principles and Methods of Toxicology*, 2nd ed. New York: Raven Press. This volume contains the following chapters of particular relevance to this chapter:

Chapter 6. Chan, P. K., and A. W. Hayes: Principles and methods for acute and eye irritancy.

Chapter 7. Mosberg, A. T., and A. W. Hayes: Subchronic toxicity testing.

Chapter 8. Stevens, K. R., and M. A. Gallo: Practical considerations in the conduct of chronic toxicity studies.

Chapter 9. Roberts, J. F., W. W. Piegorsch, and R. L. Schueler: Methods in testing for carcinogenicity.

Chapter 10. Zenick, H., and E. D. Clegg: Assessment of male reproductive toxicology: a risk assessment approach.

Chapter 11. Manson, J. M., and Y. J. Kang: Test methods for assessing female reproductive and developmental toxicology.

Chapter 12. Kennedy, G. L., Jr: Inhalation toxicology.

Chapter 13. Patrick, E., and H. I. Maiback: Dermatotoxicology.

Chapter 14. Brusick, D.: Genetic toxicology.

Chapter 17. Burger, G. T., and L. C. Miller: Animal care and facilities.

Chapter 18. Norton, S.: Methods for behavioral toxicology.

Chapter 26. Dean, J. H., et al.: Immune system; evaluation of injury.

Chapter 30. Renwick, A. G.: Pharmacokinetics in toxicology.

Chapter 31. Hogan, M. D., and D. G. Hoel: Extrapolation to humans.

Moser, V. C., G. C. Becking, V. Cuomo, et al. The IPCS collaborative study on neurobehavioral screening methods: IV. Control data. *Neurotoxicol.* **18**: 947–967, 1997.

Weiss, B., and D. Cory-Slechta. Assessment of behavioral toxicity. In *Principles and Methods of Toxicology*, 3rd ed. A. Wallace Hayes, ed. New York: Raven Press, 1994.

# CHAPTER 22

# Forensic and Clinical Toxicology

STACY BRANCH

## 22.1 INTRODUCTION

Forensic toxicology refers to the use of toxicology for the purposes of the law. It is considered a hybrid of analytical chemistry and fundamental toxicology. The efforts or activities conducted to effectuate this purpose include but are not limited to the following:

- Urine testing to detect drug use
- Regulatory toxicology
- Occupational disease
- Identification of causative agents causing death or injury in humans and animals
- Courtroom testimony and consultation concerning toxicoses

Analytical toxicology in a clinical setting plays a role similar to that in forensic toxicology. Therefore, this chapter will be divided into forensic and clinical toxicology.

## 22.2 FOUNDATIONS OF FORENSIC TOXICOLOGY

Until the 1700s convictions associated with homicidal poisoning were based only on circumstantial evidence rather than the identification of the actual toxicant within the victim. In 1781, Joseph Plenic stated that the detection and identification of the poison in the organs of the deceased was the only true sign of poisoning. Years later, 1813, Mathieiv Orfila (considered the father of toxicology) published the first complete work on the subject of poisons and legal medicine. By 1836, James M. Marsh developed a test for the presence of arsenic in tissue. Then, in 1839, Orfila successfully used Marsh's test to identify arsenic extracted from human tissues. Fifty years later, Ernst Wilhelm Heinrich Gutzeit developed a method (Gutzeit test) to quantitate arsenic in tissues. In this process arsenic compounds are reduced by hydrogen produced when zinc and sulfuric acid react. The hydrogen then reduces the arsenic compounds to arsine, which is exposed to paper that has been treated with mercuric chloride solution. This produces a color range from yellow to brown depending on the arsenic concentration. By 1918, the Medical Examiner's Office and Toxicology Laboratory was established in New York. The chief forensic toxicologist was Alexander O. Gettler who is considered the father of American toxicology.

## 22.3  COURTROOM TESTIMONY

Reports provided by forensic toxicology personnel and expert consultants may ultimately be introduced as evidence in a court of law. These reporting individuals may be asked to interpret and substantiate their findings and any associated opinions. It is therefore necessary that the forensic toxicologist be thoroughly knowledgeable or familiar with legal practices and be professionally comfortable in a courtroom environment.

*Expert Witness.* An expert witness is one who possessed the knowledge or experience in subject matters beyond the range of ordinary or common knowledge or observation. The court considers the following when qualifying an expert witness:

- Education
- Work experience
- Job training
- Academic appointments
- Publications
- Acceptance of witness by other courts
- Professional board certifications and memberships

## 22.4  INVESTIGATION OF TOXICITY-RELATED DEATH/INJURY

The basic phases in conducting an investigation of a suspected toxicant-induced/related death can be viewed as follows:

Collection of information and specimens
Toxicological analysis
Data interpretation

The primary questions to be answered are when conducting an investigation include:

- What was the route of administration?
- What was administered dose?
- Is concentration enough to have caused death or injury or altered the victim's behavior enough to cause death or injury?

*Collection of Information and Specimens.* As much information as possible concerning the facts of the case must be collected. Due to the often limited amount of physical material available for analysis, it is essential to obtain as much historical information as possible. In addition to any witness accounts of events, one must accurately record information such as the age, sex, and weight of the victim, his medical history, identification of any medications or other drugs/substances taken before death, and the time interval that elapsed between the intake of these substances and death.

When collecting the specimens, many different body fluids and organs should be collected since xenobiotics have different affinities for body tissues and therefore multiple extractions (for specific analyses) may be needed. Specimens should be collected before applying processes that may destroy evidence, that is, before embalming. The process of embalming, for example, may destroy or dilute the xenobiotic and may yield a false positive result, for example, for the presence of ethanol (which is a constituent of embalming fluid). It is possible to obtain useful specimens from burned or burial remains. The tissue often collected under these circumstances include bone marrow, skeletal muscle, vitreous humor, hair, and maggots. For example, from hair samples, it is possible to detect the presence of antibiotics, antipsychotics, and drugs of abuse. However, the information is primarily qualitative in nature. From maggots, barbiturates, barbiturates, benzodiazepines, phenothiazines, morphine, and malathion can be detected.

It is often necessary to add preservatives to specimens to protect against postmortem changes. For example, the addition of sodium fluoride to a tissue specimen can prevent the production of bacterial ethanol (which can potentially yield a false positive result for the presence of ingested ethanol).

### 22.4.1 Documentation Practices

Labeling and all handling documentation must exist from the beginning of data/specimen collection to analysis. Figure 22.1 is a sample of a typical toxicology worksheet. The name of the victim (if known) is recorded along with the name of the medical examiner. The condition of the body when found is described and the date of death is recorded. Additionally the date of request of toxicological analyses is also stated on the report. Each collected specimen is identified as well as the tests to be performed. In the results section it is necessary that the analyst for each test signs the form identifying the actual results for the tissue tested and the date the results were obtained.

### 22.4.2 Considerations for Forensic Toxicological Analysis

The decision concerning which analytical methods to employ depend greatly on a number of factors. One can imagine that a given method may need more sample volume or weight than another method. Therefore the amount of specimen available is a critical determinant of methods to chose for proper toxicological analysis. It is necessary to know the nature of the toxicant to test. In a particular case, is it relevant to detect the parent compound, its metabolites, or all of these? Furthermore toxicant biotransformation must be taken into account when doing the analyses and making interpretations. A low concentration of a toxic parent compound may reflect biotransformation as opposed to a low level of exposure. Conversely, a low-level presence of a nontoxic parent compound may be associated with a sufficient concentration of a biotransformation product that was high enough to cause the insult. Furthermore both forms (parent and metabolites) may have contributed to the adverse outcome.

**TOXICOLOGY WORK SHEET**

**Name** - Johnston, Gato T.                    **Date of Death** - 8/08/2002

**Medical Examiner** - Bahiyah Sushaunna Ngozi, M.D.   **Date of Request** - 8/09/02

**Reference number** - 38470-38973-2          **Decomposed Corpse**: ~~Yes~~ No

**Brief History** - Subject was found with a syringe, white power at scene, a small transparent glass bottle, and 0.5 liters of alcohol.

| SPECIMENS | #OF SAMPLES | TEST |
|---|---|---|
| Blood (Heart) | 1 | Alcohol ___X___ |
| Blood (Femoral) | 1 | Acidic Drug _____ |
| Blood (Other: cephalic) | 1 | Basic Drug _____ |
| Urine | 1 | Other: analyze unknown white powder |
| Bile | | |
| Vitreous | 1 | |
| Kidney | | |
| Brain | | |
| Other: (      ) | | |

**RESULTS:**

| ANALYSIS | DATE | TEST |
|---|---|---|
| Estacia Sullivan | 8/9/02 | Heart Bld Alc 44.12 mmol/L |
| Francisco Ruiz | 8/10/02 | Femoral Bld Alc 33.22 mmol/L |
| Alejandra Santiago | 8/10/02 | White Powder Heroine - pos. GC/MS |
| Estacia Sullivan | 8/11/02 | Femoral Bld Heroine 12.08 μmol/L; GC/MC |

**Figure 22.1**  Toxicology work sheet.

### 22.4.3  Drug Concentrations and Distribution

As a rule, the highest concentrations of a poison are found at the site of administration. A large quantity of drug in the GI tract and liver indicates oral ingestion. The gastrointestinal (GI) tract may contain large amounts of unabsorbed toxicant. Cases that involve the oral administration of toxicants indicate analysis of GI contents. However, the presence of toxic material in the GI tract does not provide sufficient evidence that the agent is the cause of death. Absorption and transport of the toxicant to the site of action must be demonstrated. Blood and tissue analysis is necessary and would still be paramount.

Higher concentrations of drug or toxicant in the lungs compared to other tissues may indicate inhalation, while compounds located in tissue surrounding an injection site indicates a fresh intramuscular or intravenous injection. Detection of drug combustion breakdown products within fluids/tissues reveals that smoking was the route of drug administration. For example, the primary pyrolysis product of "crack" cocaine is anhydroecgonine methylester. A high concentration of this compound and the parent cocaine indicates smoking as the route of the cocaine administration. Urine analysis is also of great value since the kidney is the major organ of excretion for most toxicants. The liver is usually the first internal organ to be analyzed. After GI tract absorption, xenobiotics are transported to the liver. This is a major center of compound biotransformation. Finally blood specimens must be collected with care and thought. When collecting blood, it is advantageous to collect both heart and peripheral blood specimens. Postmortem blood drug concentrations are site-dependent. This site dependency is referred to as "anatomical site concentration differences" or "postmortem redistribution."

## 22.5   LABORATORY ANALYSES

The nonspecific initial tests in a series are valuable for determining the presence or absence of a particular class of compounds. Colorimetric tests to detect the presence of phenothiazines would give initial information about a drug class present. This would be followed by more specific tests to identify the actual compound as well as provide quantitative data. Another example of a type of initial test would be an immunoassay that determines the presence of barbiturates. Confirmatory tests are mandatory to identify the particular drug within the class detected.

### 22.5.1   Colorimetric Screening Tests

These tests require little sample preparation and are usually performed directly on the specimen. This is a rapid procedure but requires confirmation.

### 22.5.2   Thermal Desorption

In addition to the analysis of arson crime scene evidence, thermal desorption has been used for the analysis of residual volatile agents in street drugs and the analysis of stains on forensic evidence. Samples are heated to volatilize water and organic compounds. The organic analytes may then be separated by gas chromatography (Figure 22.2).

### 22.5.3   Thin-Layer Chromatography (TLC)

An extract of a specimen is spotted on a TLC plate, the plate is placed in a mobile phase. The solvent travels up plate via capillary action and the compounds separate depending on compound solubility. Detection is by observing color changes or by using UV light to observe bands. The $R_f$ value is calculated (the distance traveled by the compound divided by distance traveled by solvent). This value along with color reactions are used as qualitative results.

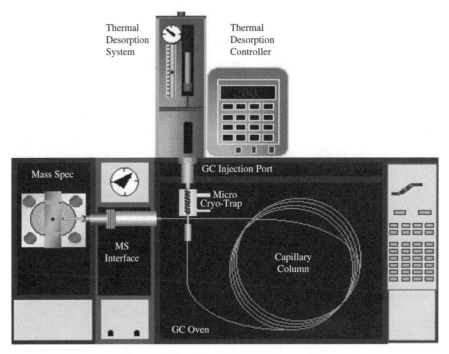

**Figure 22.2**   Thermal desorption system.

## 22.5.4   Gas Chromotography (GC)

Involves a sample being vaporized and injected into the head space of the chromato-graphic column. The sample is transported through the column by the flow of an inert gas (mobile phase). The column itself contains a liquid stationary phase which is adsorbed onto the surface of an inert solid. Retention time with detection techniques (spectrophotometer, mass spectrometry, fluorescence) identifies the compound.

## 22.5.5   High-Performance Liquid Chromatography (HPLC)

The mobile phase is a solvent that is pumped at high pressure through a packed col-umn. As described for GC, retention time with various detection techniques identifies the compound.

## 22.5.6   Enzymatic Immunoassay

An enzyme-linked drug derivative is added to the specimen to be tested. This competes with the drug in question for antibody. The more drug that binds to antibody, the less is bound to enzyme-linked drug. Enzyme activity is proportional to the amount of the drug that was already in the specimen.

## 22.6   ANALYTICAL SCHEMES FOR TOXICANT DETECTION

The circumstances surrounding the case will usually determine the types of toxicologi-cal tests that are required. There are different screens specific for the type of substance

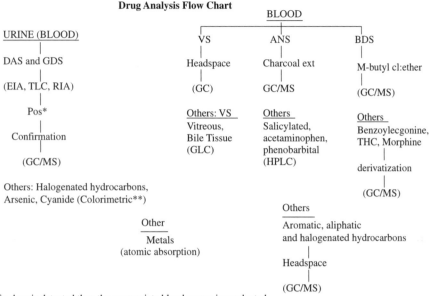

**Drug Analysis Flow Chart**

*If a drug is detected then the appropriate blood screen is conducted
**If a colorimetric test is positive, an appropriate confirmation is performed.

**Figure 22.3**  Drug analysis flowchart.

to be assayed. A given laboratory will follow an algorithm to handle the analysis. The Volatile Screen (VS) is frequently used for the detection of ethanol. A Drugs of Abuse Screen (DAS) is commonly used for amphetamines, cocaine, marijuana, and so on. When the cause of death is unclear, a General Drug Screen (GDS) is employed. Acidic/Neutral Screen (ANS) is primarily used to detect barbiturates, muscle relaxants, and so on. Basic Drug Screens (BDS) are more specific for the detection of drugs such as cocaine and antidepressants.

It is recommended that the presence of a drug or toxicant be verified in more than one specimen. However, if only one specimen is available, replicate analyses on different occasions should be performed with adequate concurrent positive and negative controls. However, it should be recognized that the compound in question may not necessarily be present in all specimen types. Figure 22.3 shows a typical drug analysis flowchart. The algorithmic approach is employed to adequately identify the drug (or at least drug class) that may be present in a tested specimen.

## 22.7  CLINICAL TOXICOLOGY

As was pointed out at the beginning of the chapter, the analytical toxicology approaches used in forensic toxicology play important roles in a clinical setting. The methods and instrumentation used in a clinical toxicology laboratory are similar to those used in forensic toxicology laboratories. The described approaches have the following benefits in clinical toxicology:

- Aids the diagnosis and treatment of toxicoses
- Allows the monitoring of treatment effectiveness

- Identification of the nature of exposure
- Quantification of toxicant

## 22.7.1 History Taking

Taking thorough general history aids in the effective treatment of an intoxicated patient. Often it may not be possible to communicate directly with the patient due to a lack of consciousness, altered mental state leading to inaccurate information, or deliberate submission of misleading information. The type of information that is essential and helpful include:

- Identifying what was taken and when, how much, and by what route
- Presence of preexisting conditions or allergies
- Whether the patient is currently using any medications or substances of any kind
- Whether the patient is pregnant

Historical information can include information obtained from family, friends, law enforcement and medical personnel, and any observers.

## 22.7.2 Basic Operating Rules in the Treatment of Toxicosis

The following concepts are central to approaching a toxicosis patient:

- Ensure airway so that breathing and circulation are adequate
- Remove unabsorbed material
- Limit the further absorption of toxicant
- Hasten toxicant elimination

Reducing further exposure to a toxicant is crucial and may include removal of the patient from a toxic environment and the application of decontamination procedures. Immediate decontamination reduces absorption of toxic compounds and represents a primary essential aspect of the treatment regimen.

*External/Skin Decontamination.* This entails the complete removal of clothing and gentle washing of the victim with copious amounts of lukewarm water. Mild soaps are often useful and may increase effectiveness of the removal of the offending substance.

*Internal Decontamination.* The most recommended methods of internal decontamination Include gastric lavage, whole bowel irrigation and administration of activated charcoal.

*Lavage.* This is utilized if a patient has ingested a life-threatening toxicant. It is recommended for use within a few short hours after ingestion and involves the use of a nasogastric or orogastric tube to flush the gastrointestinal tract. A large bore tube is inserted into the stomach and the contents removed with sequential administration and aspiration of small quantities of warm water or saline. This technique is contraindicated in cases of ingestion of corrosives (i.e., acids, bases) and hydrocarbons (i.e., fuels, essential oils).

*Whole-Bowel Irrigation.* This involves the infusion by nasogastric or orogastric tube of a lavage solution consisting of an isosmotic electrolyte (polyethylene glycol electrolyte solution is currently recommended). This procedure is indicated after ingestion of metals (e.g., iron, lithium), controlled release medications (terms to describe formulations that do not release the active compound immediately after oral ingestion), ingestion of a large amount of an anticholinergic drug (e.g., tricyclics, carbamazepine), and after ingestion of large numbers of tablets. A polyethylene glycol electrolyte solution is administer per hour per os (P.O.) or via a nasogastric tube. Antiemetics may be required to control vomiting. This is continued until the rectal effluent is clear (approximately 3 to 6 hours). The goal is to completely irrigate the gastrointestinal tract to prevent or decrease toxicant absorption. The use of an isosmotic compound such as polyethylene glycol results in minimal electrolyte loss and fluid changes.

*Activated Charcoal.* This is considered the most useful agent for the prevention of absorption of toxicants. Repeated administration (multiple dose activated charcoal) can impair the enteroenteric-enterohepatic circulation of drugs by binding to drugs that undergo significant enterohepatic or enteroenteric recycling, including carbamazepine, digoxin, phenobarbitone, theophylline, and verapamil. The prescribed amount of activated charcoal is administered every hour P.O or via nasogastric tube.

*Emesis and Catharsis.* Emesis is not recommended as a treatment measure for the toxicosis patient. The danger of aspiration of the gastric contents is great (leading to asphyxiation or aspiration pneumonia). Also a concern is the damage to esophageal and related tissue by ingested corrosive substances. Sorbitol is a commonly used cathartic. Often used in charcoal formulations, it increases the gut motility to improve excretion of poison-charcoal complexes. It is not recommended in poisonings by compounds that cause profuse diarrhea (e.g., organophosphates, carbamates, and arsenic). The use of a cathartic alone has no value in the management of the poisoned patient. Its use is even controversial as a treatment in combination with activated charcoal. Its use is contraindicated in hypotensive patients, when dehydration or electrolyte balance is present, when corrosive substances have been ingested and in cases of abdominal trauma or surgery, and intestinal perforation or obstruction.

### 22.7.3    Approaches to Selected Toxicoses

A number of antidotes are effective by altering the distribution or metabolism of a toxicant. Reducing the distribution of toxic substances to their sites of action can be achieved by a variety of methods including blocking access of specific poisons to their receptors with compounds that can compete with these receptors and by using chelating agents to bind the toxicants (e.g., dimercaprol for arsenic). Biotransformation of a toxic compound into a less toxic form can be achieved by certain agents. For example, thiosulphate is used to increase the conversion of cyanide into thiocyanate.

*Ethylene Glycol Toxicosis.* Ethylene glycol is commonly used as an antifreeze. It is metabolized by alcohol dehydrogenase to mixed aldehydes, carboxylic acids, and oxalic acid. Toxicosis results in nephrotoxicosis (kidney damage). Administration of fomepizole (4-methylpyrazole) or ethanol is effective due to their competition with ethylene glycol for the alcohol dehydrogenase enzyme. Although ethanol has a long-term history of clinical experience and is less costly to acquire, fomepizole treatment is associated with less adverse effects, predictable pharmacokinetics and has a validated efficacy. Hemodialysis is indicated in patients with severe kidney failure.

*Ethanol Toxicosis.* Absorption of ethanol by the GI tract absorption is rapid. Peak levels can be reached 30 to 60 minutes after ingestion. As a rule of thumb, 1 ml of absolute ethanol per kilogram weight results in a level of 100 mg/100 ml (0.1%) in 1 hour. Supportive treatment is directed toward the control of acidosis and hypoglycemia.

*Organophosphates and Carbamates.* The adverse effects of organophosphorous and carbamate pesticides are mediated through the inhibition of the cholinesterase enzymes. One form, acetylcholinesterase, is located at neurosynaptic junctions while butyryl cholinesterase is primarily located in the plasma and pancreas. Organophosphate pesticides inhibit cholinesterase by forming covalent bonds via phosphorylation. Enzymatic regeneration half-lives are long, taking days to months. Organophosphates affect both red blood cell and plasma cholinesterase activity. Carbamates primarily affect only the plasma derivative. Carbamate insecticides inhibit cholinesterase activity in reversible manner. Since carbamates interact with cholinesterase by weak, ionic bonding, the cholinesterases can regenerates itself more readily in matter of minutes to hours.

Since organophosphate toxicosis results in respiratory failure, the treatment approach for must include the maintenance of a patent airway. Artificial respiration may also need to be employed. The first pharmacological approach is the administration of atropine. Atropine competes with acetylcholine for its receptor site, thus reducing the effects of the neurotransmitter. *N*-methylpyridinium 2-aldoxime (2-PAM) is used in with atropine therapy as an effective means to restore the covalently bound enzyme to a normal state. It reacts with the phosphorylated cholinesterase enzyme removing the phosphate group. As previously mentioned, carbamates interact with cholinesterase by weak, ionic bonding; thus 2-PAM is of no use to combat toxicosis caused by these compounds. However, atropine is effective to prevent the effects on respiration.

*Arsenic Toxicosis.* Urine arsenic is the best indicator of current or recent exposure. Atomic absorption spectrophotometry is preferred as the detection method. Hair or fingernail sampling may also be helpful. Use of blood is useful if analyzed soon after exposure or in cases of continuous chronic exposure. After acute exposure, chelation therapy is instituted utilizing either (1) Dimercaprol BAL (British Anti-Lewisite) and analogues:

DMSA (dimercaptosuccinic acid)
DMPS (dimercaptopropane succinate)

or (2) d-penicillamine. Supportive/symptomatic therapy is also necessary. A higher protein diet and the alleviation of dehydration due to diarrhea and vomiting are beneficial.

*Chronic Exposure (Arsenic).* Primarily symptomatic treatment is chosen. Chelation therapy is practiced, but its usefulness in cases of chronic exposure is still questionable.

Supportive treatment is an essential component of the management of the intoxicated patient. Monitoring and assessment of all organ systems in conjunction with the use of appropriate pharmaceutical agents/antidotes increases therapeutic success. The nature of this care will depend on the toxicant in question and the patient's condition upon presentation.

## SUGGESTED READING

Ballantyne, B., T. Marrs, and P. Turner, eds. *General and Applied Toxicology*, college ed., New York: Macmillan, 1995.

R. E. Ferner, ed. *Forensic Pharmacology: Medicines, Mayhem, and Malpractice.* New York: Oxford University Press, 1996.

A. Furst, ed. *The Toxicologist as Expert Witness.* Washington, DC: Taylor and Francis, 1997.

Hardman, J. G., L. E. Limbird, P. B. Molinoff, R. W. Ruddon, and A. Goodman Gilman, eds. *Goodman and Gilman's The Pharmacological Basis of Therapeutics*, 10th ed., New York: McGraw-Hill, 2001.

Klaassen, C. D., *Cassarett and Doull's Toxicology: The Basic Science of Poisons*, 6th ed. New York: McGraw-Hill, 2001.

Trestrail, III, J. H. ed. *Criminal Poisoning: Investigational Guide For Law Enforcement, Toxicologists, Forensic Scientists, and Attorneys.* Totowa, NJ: Humana Press, 2000.

# CHAPTER 23

# Prevention of Toxicity

ERNEST HODGSON

## 23.1 INTRODUCTION

It is obvious, but often forgotten, that toxicity is always a consequence of exposure and that no matter what the results of hazard assessment, without exposure there cannot be a toxic effect. However, if both hazard and exposure are verified, and the risk appears to be significant, there are a range of possible actions available to reduce that risk. These actions range from outright banning, from both production and use, of the chemical in question, through measures to restrict exposure, to measures to restrict effect. Exposure can be restricted by prevention of manufacture, control of use patterns, control of application techniques, environmental manipulation, and education. Effects can be restricted by prophylactic and therapeutic methods and by education. Many of these approaches are controlled in whole, or in part, by legislation. All of them, taken together, comprise the subject matter of this chapter.

Laws and regulations provide the framework for organized efforts to prevent toxicity, and sanctions are necessary to prevent those without social conscience from deliberately exposing their fellows to risks from toxic hazards. That is not enough, however, without a population educated to toxic hazards and their prevention, the laws could never be administered properly. Moreover, in many circumstances, and particularly in the home, wisdom dictates courses of action not necessarily prescribed by law. The key to toxicity prevention lies in information and education with legislation, regulation, and penalties as final safeguards. In all probability, the better the general population is educated and informed, the less likely are laws to be necessary.

## 23.2 LEGISLATION AND REGULATION

In the best sense, legislation provides an enabling act describing the areas to be covered under the particular law and the general manner in which they are to be regulated, and designating an executive agency to write and enforce specific regulations within the intent of the legislative body. For example, the Toxic Substances Control Act (TSCA) was passed by Congress to regulate the introduction of chemicals into commerce, to determine their hazards to the human population and the environment, and to regulate or ban those deemed hazardous. The task of writing and enforcing specific regulations was assigned to the Environmental Protection Agency (EPA).

Legislative attempts to write specific regulations into laws usually fail. The resultant laws lack flexibility and, because they are written by lawyers rather than toxicologists, seldom address the problems in a scientifically rigorous manner.

It should be borne in mind that legislation is a synthesis of science, politics, and public and private pressure. It represents a society's best estimate, at that moment, of the risks it is prepared to take and those it wishes to avoid, as well as the price it is prepared to pay. Such decisions properly include more than science. The task of the toxicologist is to see that the science that is included is accurate and is interpreted logically.

This section is based primarily on regulations in the United States, not because these are the best but because, in toto, they are the most comprehensive. In many respects they are a complex mixture of overlapping laws and jurisdictions, providing unnecessary work for the legal profession. At the same time few, if any, toxic hazards in the home, workplace, or environment are not addressed.

### 23.2.1  Federal Government

Following is a summary of the most important federal statues concerned in whole or in part with the regulation of toxic substances.

*Clean Air Act.*  The Clean Air Act is administered by the EPA. Although the principal enforcement provisions are the responsibility of local governments, overall administrative responsibility rests with EPA. This act requires criteria documents for air pollutants and sets both national air quality standards and standards for sources that create air pollutants, such as motor vehicles, power plants, and so on. Important actions already taken under this law include standards for the now complete phased-out elimination of lead in gasoline, and the setting of sulfuric acid air emission guidelines for existing industrial plants.

*Clean Water Act.*  The Clean Water Act, which amends the Federal Water Pollution Control Act, is also administered b the EPA and provides for funding of municipal sewage treatment plants. However, with respect to toxicity prevention, it is more important that the act regulates emissions from municipal and industrial sources. It has as its goal the elimination of discharges of pollutants and the protection of rivers so that they are "swimmable and fishable" and applies to "waters of the United States" subsequently defined to include all waters that reach navigable waters, wetland, and intermittent streams. Some of the more important actions taken under this statute include setting standards for emissions of inorganics from smelter operations and publishing priority lists of toxic pollutants. This act allows the federal government to recover cleanup and other costs as damages from the polluting agency, company, or individual.

*Safe Drinking Water Act.*  (1974, 1986, 1996). Specifically applied to water supplied for humans consumption, this act requires the EPA to set maximum levels for contaminants in water delivered to users of public water systems. Two criteria are established for a particular contaminant: the *maximum containment level goal* (MCGL) and the *maximum contaminant level* (MCL). The former, the MCLG, is the level at which no known or anticipated adverse effects on the health of persons occur and within an allowed adequate margin of safety. The latter, the MCL, is the maximum permissible level of a contaminant in water that is delivered to any user of a public water system. MCLs are expected to be as close to the MCLG as is feasible.

*Comprehensive Environmental Response, Compensation, and Liability Act (CERCLA).* This is an attempt to deal with the many waste sites that exist across the nation. It covers remedial action, including the establishment of a National Priorities List to identify those sites that should have a high priority for remediation. This act authorizes the cleanup of hazardous waste sites, including those containing pesticides, that threaten human health or the environment. If they can be identified, the US EPA is authorized to recover cleanup costs from those parties responsible for the contamination. CERCLA provides a fund to pay for the cleanup of contaminated sites when no other parties are able to conduct the cleanup. The Superfund Amendments and Reauthorization Act (SARA) (1986) is an amendment to CERCLA that enables the US EPA to identify and cleanup inactive hazardous waste sites and to recover reimbursement of cleanup costs. One section of CERCLA authorizes the EPA to act whenever there is a release or substantial threat of release of a hazardous substance or "any pollutant or contaminant that may present an imminent or substantial danger to the public health or welfare" into the environment.

*Consumer Products Safety Act and Consumer Products Safety Commission Improvements Act.* Administered by a Consumer Products Safety Commission, the Consumer Products Safety Act is designed to protect the public against risk of injury from consumer products and to set safety standards for such products.

*Controlled Substances Act.* The Controlled Substances Act not only strengthens law enforcement in the field of drug abuse but also provides for research into the prevention and treatment of drug abuse.

*Federal Food, Drug, and Cosmetic Act.* The Federal Food, Drug, and Cosmetic Act is administered by the Food and Drug Administration (FDA). It establishes limits for food additives and cosmetic components, sets criteria for drug safety for both human and animal use, and requires the manufacturer to prove both safety and efficacy. The FDA is authorized to define the required toxicity testing for each product. This act contains the Delaney clause, which states that food additives that cause cancer in humans or animals at any level shall not be considered safe and are therefore prohibited from such use. This clause has recently been modified to permit the agency to use more flexible risk-benefit based guidelines. Under the Food Quality Protection Act of 1966 (see below) the Delaney clause is no longer applied to pesticide residues in food. This law also empowers the FDA to establish and modify the generally recognized as safe (GRAS) list and to establish good laboratory practice (GLP) rules.

*Occupational Safety and Health Act.* Administered by the Occupational Safety and Health Administration (OSHA), the Occupation Safety and Health Act concerns health and safety in the workplace, OSHA sets standards for worker exposure to specific chemicals, for air concentration values, and for monitoring procedures. Construction and environmental controls also come under this act. This act provides for research, information, education, and training in occupational safety and health.

By establishing the National Institute for Occupational Safety and Health (NIOSH), the act provided for appropriate studies to be conducted so that regulatory decisions could be based on the best available information.

*National Environmental Policy Act.* The National Environmental Policy Act is an umbrella act covering all US government agencies, requiring them to prepare environmental impact statements for all federal actions affecting the quality of the human environment. Environmental impact statements must include not only an assessment of the effect of the proposed action on the environment, but also alternatives to the proposed action, the relationship between local short-term use and enhancements of long-term productivity, and a statement of irreversible commitment of resources. This act also created the Council on Environmental Quality, which acts in an advisory capacity to the president on matters affecting or promoting environmental quality.

*Resource Conservation and Recovery Act.* Also administered by the EPA, the Resource Conservation and Recovery Act (RCRA) is the most important act governing the disposal of hazardous wastes including pesticide formulations, containers, and rinsates; it promulgates standards for identification of hazardous wastes, their transportation, and their disposal. Included in the latter are siting and construction criteria for landfills and other disposal facilities as well as the regulation of owners and operators of such facilities. The three principal areas covered are hazardous wastes, nonhazardous solid wastes, and underground storage tanks. Farmers and commercial pesticide applicators are subject to penalties if they fail to store or dispose of pesticides and pesticide containers properly. The Agency is responsible for enforcement.

*Toxic Substances Control Act.* Administered by the EPA, the TSCA is mammoth, covering almost all chemicals manufactured in the United States for industrial and other purposes, excluding certain compounds covered under other laws such as FIFRA. The EPA may control or stop production of compounds deemed hazardous. Producers must give notice or intent to manufacture new chemicals or increase significantly the production of existing chemicals. They may be required to conduct toxicity and other tests. This law is as yet incompletely applied due to the enormous number of existing chemicals that must be evaluated. Once fully applied, it will be the most important statute affecting toxicology.

*Statutes Affecting the Manufacture and Use of Agricultural Chemicals.* Because of the intense interest and concern over the use of agricultural chemicals, especially pesticides, and their possible effects on human health, these statutes are perhaps the most overregulated group of commercial xenobiotics in use today. A number of laws deal almost exclusively with this use class while several others also deal with them, to a greater or lesser extent. The first law directed specifically toward pesticides in the United States was the Insecticide Act of 1910. This act was passed to ensure that the percentages of ingredients were as stated and that the product was efficacious.

Surprisingly, it was 37 years before a law was written to replace the 1910 Act. This replacement was the Federal Insecticide, Fungicide, and Rodenticide Act (FIFRA). First passed in 1947 and amended many times since, this act is now administered by the EPA. FIFRA regulates all pesticides and other agricultural chemicals, such as plant growth regulators, used in the United States. Establishing the requirement "that the burden of proof of a product's acceptability rested with the manufacturer," it includes the authority to establish registration requirements, with appropriate chemical and toxicological tests prescribed by the agency. This act also permits the agency to specify labels, to restrict application to certified applicators, and to deny, rescind, or modify registration. Under this act the EPA also establishes tolerances for residues on raw agricultural products. FIFRA was amended in 1988 requiring a re-evaluation of all pesticides manufactured prior to 1984. The purposes of the 1988 amendment were to remove hazardous pesticides and to require additional testing, primarily toxicity tests that were not available when these early compounds were registered. Section 19 of the 1988 FIFRA amendments greatly expanded the Agency's authority to regulate pesticide storage, transport and disposal of pesticides, containers, and rinsates of containers.

The Food Quality Protection Act (FQPA) of 1996 is an amendment to FIFRA and provides a new standard for evaluating pesticides applied to food crops, in that there be "reasonable certainty of no harm" from residues found on food. US EPA is required to perform an aggregate risk assessment that combines dietary risk from a specific pesticide with those from resides in drinking water and from residential exposure. As a result of this law, the US EPA is required to reevaluate all existing food tolerance residue levels based on a number of criteria. One of these is to determine the cumulative (combined) risk of exposure to classes of pesticides having the same mechanism of toxicity, with special emphasis on infants and children. In some instances this has required adding an additional safety factor to the tolerance of between 3 and 10 for certain compounds to ensure the safety of children. This factor is in addition to the safety factor of 100 covering differences due to species and individual variation. Thus, if typical residue levels on a food crop are 1.0 ppm, then a tolerance of 0.01 ppm could be established, and if the additional factor of 10 were added, the tolerance could be set at 0.001 ppm. Currently organophosphate insecticides and several other classes are undergoing this reassessment process.

The Act established the Tolerance Reassessment Advisory Committee (TRAC), composed of individuals with a variety of backgrounds and interests to consult and make recommendations to both the EPA and USDA. When this committee went out of existence in 1999, the EPA and USDA established a new advisory committee, the Committee to Advise on Reassessment and Transition (CARAT) to provide strategic advice on issues raised by this Act.

An Endocrine Disruptor Screening and Testing Advisory Committee (EDSTAC) was established to develop a comprehensive screening and testing program for pesticides and other compounds to determine potential estrogenic effects on both humans and on wildlife. FQPA is one of the most significant amendments ever made to FIFRA and continues to generate considerable controversy as it is put into effect.

The Worker Protection Standard for Agricultural Pesticides (1994) was written to protect workers from pesticide exposures. Responsibility lies with the employer and involves two types of employees: agricultural workers (e.g., harvesters) and pesticide handlers (e.g., mixers). It requires that these people be provided safety training and access to labels, and that medical treatment be made available prior to and 30 days after the REI has expired. The types of protection offered include notification prior to applying pesticides, exclusion during applications and during an REI, and monitoring the worker's PPE. In addition the employer is required to provide a decontamination site equipped with water and a clean change of clothes.

The Act was written to cover pesticide use on farms, forests, nurseries, and greenhouses. It does not include applications to pastures, golf courses, parks, livestock, right-of-way, or home gardens, nor does it cover treatments for mosquito abatement and rodent control.

Many other legislative acts impact in whole or in part on pesticide use. They include the Endangered Species Act of 1973, an act written to protect endangered wildlife; it regulates pesticide use around wildlife sanctuaries. Pesticides might injure or kill endangered species if allowed to drift onto habitat, or runoff into streams, lakes, or wetlands might be found to significantly degrade endangered wildlife habitat. Also included are the Clean Water Act, the Safe Drinking Water Act, RCRA, CERCLA, and SARA, all discussed above.

*Other Statutes with Relevance to the Prevention of Toxicity.*  It should be noted that some of these statutes have been superseded by others, either in whole or in part.

- Comprehensive Employment and Training Act
- Dangerous Cargo Act
- Federal Coal Mine Safety and Health Amendment Act
- Federal Caustic Poison Act
- Federal Railroad Safety Authorization Act
- Hazardous Materials Transport Act
- Lead-Based Paint Poison Prevention Act
- Marine Protection Research and Sanctuaries Act
- Poison Prevention Packaging Act
- Ports and Waterways Safety Act

### 23.2.2   State Governments

Within the United States, states are free to adopt legislation with toxicological significance, although their jurisdiction does not extend beyond their geographic boundaries. In other cases the states may enforce federal statutes under certain circumstances. For example, if state regulations concerning hazardous waste disposal is neither less comprehensive nor less rigorous than the federal statute, enforcements is delegated to the states. Similarly certain aspects of FIFRA are enforced by individual states. In some cases (California is notable in this respect) states have passed laws considerably more comprehensive and more rigorous than the corresponding federal stature.

### 23.2.3   Legislation and Regulation in Other Countries

It would serve little purpose to enumerate all the laws affecting toxicology, toxicity testing, and the prevention of toxicity that have been promulgated in all countries that have such laws. Legislation in this area has been adopted in most countries of western Europe and in Japan. Although the laws in use in the United States are a complex mixture of overlapping statutes and enforcement agencies, they are probably the most comprehensive set of such laws in existence. Most other industrialized countries have legislation in the same areas, although the emphasis varies widely from one country to another. Many underdeveloped countries, due to the lack of both trained workers and financial resources, are unable to write and enforce their own code of regulations, and instead, many adopt the regulatory decisions of either the United States or some other industrialized nation. For example, they will permit the use, in their own territory, of pesticides registered under FIRA by the US EPA and will prohibit the use of pesticides not so registered.

## 23.3 PREVENTION IN DIFFERENT ENVIRONMENTS

Humans spend their time in many environments. Homes vary with climate, family income, and personal choice. The workplace varies from pristine mountains to industrial jungles, and the outdoor environment from which recreation, food, and water are derived varies through the same extremes. Each of these environments has its own specific complex of hazards, and thus requires its own set of rules and recommendations if these hazards are to be avoided.

### 23.3.1 Home

Approximately 50% of all accidental poisoning fatalities in the United States involve preschool children. Thus prevention of toxicity is particularly important in homes with young children.

Prescription drugs should always be kept in the original container (in the United States and some other countries, these are now required to have safety closures). They should be taken only by the person for whom they were prescribed, and excess drugs should be discarded safely when the illness is resolved. When children are present, prescription drugs should be kept in a locked cabinet because few cabinets are inaccessible to a determined child. Although nonprescription drugs are usually less hazardous, they are frequently flavored in an attractive way. Thus it is prudent to follow the same rules as for prescription drugs.

Household Chemicals such as lye, polishes, and kerosene should be kept in locked storage if possible; if not, they should be kept in as secure a place as possible, out of the reach of children. Such chemicals should never be stored in anything but the original containers. Certainly they should never be stored in beverage bottles, kitchen containers, and so on. Unnecessary materials should be disposed of safely in appropriate disposal sites.

Certain Household Operations such as interior painting, and so on, should be done only with adequate ventilation. Insecticide treatment should be done precisely in accordance with instructions on the label.

Increasing fuel costs have caused several changes in lifestyle, and some of these changes carry potential toxic hazards. They include more burning of wood and coal and the construction of heavily insulated houses with a concomitant reduction in ventilation. In the latter circumstances, improperly burning furnaces can generate high levels of CO and aromatic hydrocarbons, whereas even those burning properly may still generate oxides of nitrogen ($NO_x$) at levels high enough to cause respiratory tract irritation in sensitive individuals. These effects can be avoided by ensuring that all heating equipment (e.g., furnaces, wood stoves, heaters) is properly ventilated, maintained, and checked regularly. In addition some ventilation of the building itself should always be provided. Less ventilation is needed when the temperature is either excessively high or excessively low, and more is needed when the temperature is in the midrange, but under no circumstances should the homeowner strive for a completely sealed house.

### 23.3.2 Workplace

Exposure levels of hazardous chemicals in the air of work environments are mandated by OSHA as exposure limit values. The studies necessary to establish these limits are carried out by NIOSH. However, the more complete list of the better-known threshold limit values (TLVs) is established by the American Conference of Governmental Industrial Hygienists. Although TLVs are not binding in law, they are an excellent guide to the employer. In fact they are often adopted by OSHA as exposure limit values. The concentrations thus expressed are the weighted average concentrations normally considered safe for an exposure of 8 h/day, 5 days/week. Absolute upper limits (excursion values) may also be included. Some exposure limits are shown in Table 23.1.

Concentrations at or lower than those normal or working exposures are usually maintained by environmental engineering controls. Operations that generate large amounts of dusts or vapors are conducted in enclosed spaces that are vented separately or under hoods. Other spaces are ventilated adequately, and temperature and humidity controls are installed where necessary.

**Table 23.1   Some Selected Threshold Limit Values (1991)**

| Chemical | TLV-TWA[a] (ppm) | TLV-STEL[b] (ppm) | TLV-C[c] (ppm) |
|---|---|---|---|
| Acetaldehyde | 100 | 150 | — |
| Boron trifluoride | — | — | 1 |
| o-Dichlorobenzene | — | — | 50 |
| p-Dichlorobenzene | 75 | 110 | — |
| N-Ethylmorpholine | 5 | 20 | — |
| Fluorine | 1 | 2 | — |
| Phosgene | 0.1 | — | — |
| Trichloroethylene | 50 | 200 | — |

[a]TLV-TWA, threshold limit value: time-weighted average, concentration for a normal 8-hour workday and 40-hour workweek to which nearly all workers may be repeatedly exposed without adverse effect.

[b]TLV-STEL, threshold limit value: short-term exposure limit, concentration. This time-weighted 15-minute average exposure should not be exceeded at any time during a workday even if the TLV-TWA is within limits. Intended as supplement to TLV-TWA.

[c]TLV-C, threshold limit value–ceiling, concentration that should not be exceeded at any time.

Other precautions must be taken to prevent accidental or occasional increases in concentrations. Materials should be transported in "safe" containers, spilled material removed rapidly, and floor and wall materials selected to prevent contamination and allow easy cleaning.

Additional methods for the prevention of toxicity in the workplace include the use of personal safety equipment—protective clothing, gloves, and goggles are the most important. In particularly hazardous operations, closed-circuit air masks, gas masks, and so on, may also be necessary.

Pre-employment instruction and pre-employment physical examinations are of critical importance in most work situations involving hazardous chemicals. The former should make clear the hazards involved, the need to avoid exposure under normal working conditions, and the mechanisms by which exposure is limited. Furthermore employees should understand how and when to contain spills and how and when to evacuate the area around the spill. Locations and use of emergency equipment, showers, eye washes, and so on, should also be given, and the most important procedures should be posted in the work area.

### 23.3.3  Pollution of Air, Water, and Land

The toxicological significance of pollution of the environment may be work related, as in the case of agricultural workers, or related to the outside environment encountered in daily life. In the case of agricultural workers, numerous precautions are necessary for the prevention of toxicity. For example:

- Pesticides and other agricultural chemicals should be kept only in the original container, carrying the labels prescribed by EPA under FIFRA.
- Empty containers and excess chemicals should be disposed of properly in safe hazardous waste disposal sites, incinerated when possible, or, in some cases, decontaminated.
- Workers should not re-enter treated areas until the safe re-entry period has elapsed.
- Certain workers such as applicators and those preparing tank mixes should wear appropriate protection clothing, gloves, face masks, and so forth. The development of closed systems for mixing pesticides should help protect mixers and loaders of pesticides from exposure.
- Spraying operations should be carried out in such a way as to minimize drift, contamination of water, and so on.

Pesticides have caused a number of fatalities in the past. The current practice in some countries of restricting the most hazardous chemicals for use only by certified operators should greatly minimize pesticide poisoning in these locations.

Individuals can do little to protect themselves from poisoning by chemicals that pollute the air and water except to insist that discharge of toxicants into the environment be minimized. The exposure levels are low compared with those in acute toxicity cases, and the effects may be indirect, as in the increase in preexisting respiratory irritation during smog. Thus these effects can be determined only at the epidemiologic level. Because many persons are not affected or may not be affected for years, it is often argued that environmental contamination is not very important. However, a small percentage increase may represent a large number of people when the whole population is considered. Furthermore chronic toxicity is not often reversible. Because in most industrialized countries laws already exist to control emission problems, if such problems exist in these countries, they are usually problems of enforcement.

One of the most critical areas for the prevention of toxicity caused by environmental contamination is that of disposal of hazardous wastes. It is now apparent that past practices in many industrialized countries have created large numbers of waste sites in which the waste is often unidentified, improperly stored, and leaching into the environment. The task of rectifying these past errors is an enormous one just now being addressed.

The ideal situation for current and future practices is to reduce chemical waste to an irreducible minimum and then to place the remainder in secure storage. Waste reduction can be accomplished in many ways.

- Refine plant processes so that less waste is produced.
- Recycle waste into useful products.
- Concentrate wastes.
- Incinerate. The technology is available to incinerate essentially all waste to inorganic slag. Unfortunately, the technology is sophisticated and expensive. Inadequate incineration is itself a hazard because of the risk of generating dioxins and other toxicants and releasing them into the environment. Less complex and more easily maintained incinerators will be essential if this technology is to play a prominent role in waste reduction.

Safe storage for the remaining waste may be in dump sites or in above-ground storage. In either case such storage ideally should be properly sited, constructed, maintained, and monitored.

Because of the nature of commerce, probably none of these measures will be successful unless the laws, penalties, and incentives are manipulated in such a way as to make safe disposal more attractive economically than unsafe disposal.

## 23.4  EDUCATION

Because chemicals, many of them hazardous, are an inevitable part of life in industrialized countries, education is probably the most important method for the prevention of toxicity. Unfortunately, it is also one of the most neglected. In a typical public debate concerning a possible chemical hazard, the principle protagonists tend to fall into two extreme groups. The "everything is OK" protagonists and the "ban it completely" protagonists. The media seldom seem to educate the public, usually serving only to add fuel to the flames.

The educational role of the toxicologist should be the voice of reason, presenting a balanced view of risks and benefits, and outlining alternatives whenever possible. The simple lesson that science deals not in certainty, but rather degrees of certitude, must be learned by all involved.

In terms of ongoing educational programs, there should be opportunities at all levels: elementary schools, high schools, university, adult education, and media education. Several approaches can be used to educate the general public in ideal situations:

- *Elementary school.* Teach the rudiments of first aid and environmental concerns, proper disposal, and so on.
- *High school.* Teach concepts of toxicology (dose response, etc.) and environmental toxicology (bioaccumulation, etc.), as these concepts can be introduced into general science courses.
- *University.* In addition to toxicology degrees, general courses for nontoxicology and/or nonscience majors should stress a balanced approach, with both responsible use and toxicity prevention as desirable end points. General Toxicology should be a required course in all chemically related academic programs such as chemistry and chemical engineering.
- *Media.* Encourage a balanced approach to toxicity problems. Toxicologists should be available to media representatives and, where appropriate, should be involved directly.

## SUGGESTED READING

American Conference of Governmental Industrial Hygienists. *TLVs—Threshold Limit Values for Chemical Substances and Physical Agents in the Work Environment with Intended Changes.* Cincinnati (published annually).

American Conference of Governmental Industrial Hygienists. *Documentation of the Threshold Limit Values for Substances in Workroom Air.* Cincinnati (published annually).

Doull, J. Recommended limits for exposure to chemicals. In *Casarett and Doull's Toxicology: The Basic Science of Poisons*, 6th ed., C. D. Klaassen, ed. New York: McGraw-Hill, 2001, pp. 1115–1176.

Dreisbach, R. H. *Handbook of Poisoning: Prevention, Diagnosis, and Treatment*, 11th ed. Los Altos, CA: Lange Medical, 1983.

Ellenhorn, M. J., and D. G. Barceloux. *Medical Toxicology: Diagnosis and Treatment of Human Poisoning.* New York: Elsevier Science, 1998.

Environmental Regulation: An International View. A series of papers on: I. Britain, T. W. Hall; II. European Economic Community, S. P. Johnson; III. The United States, J. B. Ritch, Jr; IV. An Industry View, R. C. Tineknell. *Chem. Soc. Rev.* **5**: 431–771, 1976.

Fan, A. M., and L. W. Chang, eds. *Toxicology and Risk Assessment: Principles, Methods and Applications.* New York: Dekker, 1996.

Merrill, R. A. Regulatory toxicology. In *Casarett and Doull's Toxicology: The Basic Science of Poisons*, 6th ed., C. D. Klaassen, ed. New York: McGraw-Hill, 2001, pp. 1141–1153.

Sandmeyer, E. E. Regulatory toxicology. In *A Guide to General Toxicology.* F. Homberger, J. A. Hayes, E. W. Pelikan, eds. Basel: Karger, 1983, ch. 20.

# CHAPTER 24

# Human Health Risk Assessment

RONALD E. BAYNES

## 24.1 INTRODUCTION

We often perform toxicological research to better understand the mechanism and associated health risk following exposure to hazardous agents. Risk assessment is a systematic scientific characterization of potential *adverse health effects* following exposure to these hazardous agents. Risk assessment activities are designed to *identify, describe*, and *measure qualities and quantities* from these toxicological studies, which are often conducted with homogeneous animal models at doses and exposure duration not encountered in a more heterogeneous human population. Herein lie the challenge of risk assessment. The use of default assumptions because of some level of uncertainty in our extrapolations across species, doses, routes, and interindividual variability, the risk assessment process is often perceived as lacking scientific rigor. This chapter will cover traditional practices as well as new and novel approaches that utilize more of the available scientific data to identify and reduce uncertainty in the process. The advent of powerful computers and sophisticated software programs has allowed the development of quantitative models that better describe the dose-response relationship, refine biologically relevant dose estimates in the risk assessment process, and encourage departure from traditional default approaches (Conolly et al., 1999). Although the focus of this chapter is on current and novel risk assessment methods that are scientifically based, it is critical that the reader be aware of the differences between risk assessment and risk management, which are summarized in Table 24.1.

Results from the risk assessment are used to inform *risk management*. The risk manager uses the risk information in conjunction with factors such as the social importance of the risk, the social acceptability of the risk, the economic impacts of risk reduction, engineering, and legislative mandates when deciding on and implementing risk management approaches.

The risk assessment may be perceived as the source of a risk management decision, when in fact, social concerns, international issues, trade, public perception, or other non-risk considerations may be taken into consideration. Finally there is one activity known as *risk communication* that involves making the risk assessment and risk

*A Textbook of Modern Toxicology, Third Edition*, edited by Ernest Hodgson
ISBN 0-471-26508-X Copyright © 2004 John Wiley & Sons, Inc.

**Table 24.1   Comparison of Risk Assessment and Risk Management Activities**

| Risk Assessment | Risk Management |
| --- | --- |
| Nature of effects | Social importance of risk |
| Potency of agent | Acceptable risk |
| Exposure | Reduce/not reduce risk |
| Population at risk | Stringency of reduction |
| Average risk | Economics |
| High-end risk | Priority of concern |
| Sensitive groups | Legislative mandates |
| Uncertainties of science | Legal issues |
| Uncertainties of analysis | Risk perception |
| *Identify* | *Evaluate* |
| *Describe* | *Decide* |
| *Measure* | *Implement* |

management information comprehensible to lawyers, politicians, judges, business and labor, environmentalists, and community groups.

## 24.2   RISK ASSESSMENT METHODS

According to the National Research Council of the National Academy of Science, risk assessment consists of four broad but *interrelated* components: hazard identification, dose-response assessment, exposure assessment, and risk characterization, as depicted in Figure 24.1. The reader should, however, be aware that these risk assessment activities can provide research needs that improve the accuracy of estimating the "risk" or probability of an adverse outcome.

### 24.2.1   Hazard Identification

In this first component of risk assessment, the question of causality in a qualitative sense in addressed; that is, the degree to which evidence suggests that an agent elicits

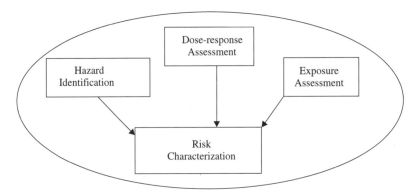

**Figure 24.1**   Risk assessment paradigm as per NAS and US EPA.

a given effect in an exposed population. Among many factors the quality of the studies and the severity of the health effects should be evaluated at this stage. The following are evaluated: (1) validity of the toxicity data, (2) weight-of-evidence summary of the relationship between the substance and toxic effects, and (3) estimates of the generalizability of data to exposed populations. Where there are limited in vivo toxicity data, *structural activity relationships* (SARs) and *short-term assays* may be indicative of a chemical hazard. Key molecular structures such as *n*-nitroso or aromatic amine groups and azo dye structures can be used for prioritizing chemical agents for further testing. SARs are useful in assessing relative toxicity of chemically related compounds, but there are several limitations. For example, toxicity equivalent factors (TEFs) based on induction of Ah receptor by dioxins demonstrated that SARs may not always be predictive. In vitro short-term inexpensive test such as bacterial mutation assays can help *identify* carcinogens, and there are other short-term tests that can help identify chemicals that potentially can be associated with neurotoxicity, developmental effects, or immunotoxicity. Many of these in vitro studies can provide some insight into mechanism(s) of action, but there may be some *false positives* and *false negatives*. Animal studies are usually route-specific and relevant to human exposure, and animal testing usually involves two species, both sexes, 50 animals/dose group, and near-lifetime exposures. Doses are usually 90, 50, and 10 to 25% of the maximum tolerated dose (MTD). In carcinogenicity studies, the aim is to observe significant increases in number of tumors, induction of rare tumors, and earlier induction of observed tumors. However, rodent bioassays may not be predictive of human carcinogenicity because of mechanistic differences For example, renal tumors in male rats is associated with $\alpha_{2\mu}$-globulin-chemical binding and accumulation leading to neoplasia; however, $\alpha_{2\mu}$-globulin is not found in humans, mice, or monkeys. There are differences in susceptibility to aflatoxin-induced tumors between rats and mice that can be explained by genetic differences in expression of cytochrome P450 and GST isoenzymes. Whereas humans may be as sensitive as rats to $AFB_1$-induced liver tumors, mice may not be predictive of $AFB_1$-induced tumors in humans. Epidemiological data from human epidemiological studies are the most convincing of an association between chemical exposure and disease, and therefore can very useful for hazard identification. Exposures are not often well defined and retrospective, and confounding factors such as genetic variations in a population and human lifestyle differences (e.g., smoking) present a further challenge. The three major types of epidemiological studies available are (1) *cross-sectional studies*, which involve sampling without regard to exposure or disease status, and these studies identify risk factors (exposure) and disease but not useful for establishing cause-effect relationships; (2) *cohort studies*, which involve sampling on the basis of exposure status, and they target individuals exposed and unexposed to chemical agent and monitored for development of disease, and these are *prospective studies*; (3) *case-control studies*, which involve sampling on the basis of disease status. These are retrospective studies, where diseased individuals are matched with disease-free individuals.

### 24.2.2  Exposure Assessment

This process is an integral part of the risk assessment process. However this will be introduced only briefly in this chapter, and the reader is encouraged to consult Chapter 28 in this text as well as numerous other texts that describe the process in more depth. In brief, exposure assessment attempts to identify potential or completed exposure pathways resulting in contact between the agent and at-risk populations. It also includes demographic analysis of at-risk populations describing properties and characteristics of the population that potentiate or mitigate concern and description of the magnitude, duration, and frequency of exposure. The reader should be aware that exposure may be aggregate (single event added across all media) and/or cumulative (multiple compounds that share a similar mechanism of toxicity). Various techniques such as biomonitoring, model development, and computations can be used to arrive at an estimate of chemical dose taken up by humans, that is, chemical exposure. For example, the lifetime average daily dose (LADD) is a calculation for individuals exposed at levels near the middle of the exposure distribution:

$$LADD = \frac{(\text{Conc. in media}) \times (\text{Contact rate}) \times (\text{Contact fraction}) \times (\text{Exposure duration})}{(\text{Body weight}) \times (\text{Lifetime})}.$$

Biological monitoring of blood and air samples represent new ways of reducing uncertainty in these extrapolations. For occupational exposures there are occupational exposure limits (OELs) that are guidelines or recommendations aimed at protecting the worker over their entire working lifetime (40 years) for 8 h/day, 5 days/week work schedule. Most OELs are presented as a time-weighted average concentration for an 8-hour day for a 40-hour work week. There are threshold limit values (TLVs) that refer to airborne concentrations and conditions under which workers may be exposed daily but do not develop adverse health effects. The short-term exposure limit (STEL) are recommended when exposures are of short duration to high concentrations known to cause acute toxicity.

### 24.2.3 Dose Response and Risk Characterization

Dose response is a quantitative risk assessment process, and primarily involves characterizing the relationship between chemical potency and incidence of adverse health effect. Approaches to characterizing dose-response relationships include effect levels such as LD50, LC50, ED50, no observed adverse effect levels (NOAELs), margins of safety, therapeutic index. The dose-response relationship provides an estimation of the relationship between the dose of a chemical agent and incidence of effects in a population. Intuitively, a steep dose-response curve may be indicative of a homogeneous population response, while less steep or almost flat slope may be indicative of greater distribution in response. In extrapolating from relatively high levels of exposure in experimental exposures (usually animals) to significantly lower levels that are characteristic of the ambient environment for humans, it is important to note the shape of the dose-response function below the experimentally observable range and therefore the range of inference. The shape of the slope may be linear or curvilinear and, it should be noted that the focus of risk assessment is generally on these lower regions of the dose-response curve (Figure 24.2).

There is a class of curvilinear dose-response relationships in toxicological and epidemiological studies that may be described as *U-shaped* or *J-shaped curves*. Other terms such as biphasic, and more recently *hormesis*, have been used to refer to paradoxical effects of low-level toxicants. In brief, these dose-response curves reflect an apparent improvement or reversal in the effect of an otherwise toxic agent. These

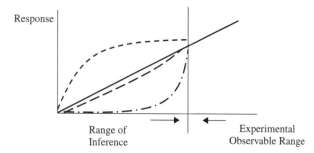

**Figure 24.2** Dose-response curve, with emphasis on the shape of the dose-response function below the experimentally observable range and therefore the range of inference where people are realistically exposed.

U-shaped effects can be explained in terms of homeostatic adjustments or overcorrections in the operation of feedback mechanisms. Examples of studies with data fitting a U-shaped curve include the hormetic effect of organic lead on body growth in rats (Cragg and Rees, 1984) and peripheral nerve conduction velocity in children at low doses (Ewert et al., 1986). Similar relationships have been observed with alcohol and nicotine in humans. It has been proposed that because thresholds are inherent in U-shaped dose-response curves, the linear no-threshold extrapolation method is not an appropriate approach for regulating hormetic agents. The current risk assessment paradigm used by US EPA and other federal agencies does not conflict with the concept of hormesis, but it has been proposed that the risk assessor's analyzes make an active consideration of the data and the application of that data in the low dose portion of the dose-response curve for hormetic agents.

## 24.3 NONCANCER RISK ASSESSMENT

The noncancer risk assessment process assumes a *threshold*. For many noncarcinogenic effects, protective mechanisms are believed to exist that must be overcome before an adverse effect is manifested. At the cellular level for some toxicant, a range of exposures exists from zero to some finite value that can be tolerated by the organism with essentially no chance of expression of adverse effects. The aim here in risk assessment is to identify the upper bound of this tolerance range (i.e., the maximum subthreshold level). This approach involves obtaining the no observed adverse effect level. NOAEL is the highest dose level that *does not produce a significant* elevated increase in an adverse response. Significance refers to biological and statistical criteria and is dependent on dose levels tested, number of animals, background incidence in the unexposed control groups. Sometimes there is insufficient data to arrive at a NOAEL, and a LOAEL (lowest observed adverse effect level) is derived. The NOAEL is the key datum obtained from the study of the dose-response relationship. The NOAEL is used to calculate reference doses (RfD) for chronic oral exposures and reference concentrations (RfC) for chronic inhalation exposures as per EPA. Other agencies, such as the ATSDR and WHO, use the NOAEL to calculate *minimum risk levels* (MRLs) and *acceptable daily intakes* (ADI). The US EPA describes the RfD as an estimate, with uncertainty spanning an order of magnitude, of a daily exposure to the human population, including sensitive subgroups, that is likely to be without appreciable deleterious effects during a lifetime. In deriving reference doses, ADIs, or MRLs, the NOAEL is divided by uncertainty factors (UF) as per EPA (EPA, 1989) and ATSDR (ATSDR, 1993) and by modifying factors (MF) as per EPA:

$$RfD = \frac{NOAEL}{(UF * MF)}, \qquad US\ EPA;$$

$$MRL = \frac{NOAEL}{UF}, \qquad ATSDR.$$

The calculated RfD or RfC is based on the selected critical study and selected critical end point. The risk assessor may obtain numerous studies where the toxicant may have more than one toxic end point, and thus there may be many NOAELs to choose from the literature. In some instances poor data quality may be used to exclude those end points from consideration. Also at issue is the determining what is considered an adverse effect, and this has been summarized with a few examples in Table 24.2. In sum, the MRL or RfD is based on the less serious effects and no serious effects. The following are example effects not used in obtaining a NOAEL: decrease in body weight less than 10%, enzyme induction with no pathologic changes, changes in organ weight with no pathologic changes, increased mortality over controls that is not significant ($p > 0.05$), and hyperplasia or hypertrophy with or with out changes in organ weights.

### 24.3.1  Default Uncertainty and Modifying Factors

Most extrapolations from animal experimental data in the risk assessments require the utilization of uncertainty factors. This is because we are not certain how to extrapolate across species, with species for the most sensitive population, and across duration. To account for variations in the general population and to protect sensitive subpopulations, an uncertainty factor of 10 is used by EPA and ATSDR. The value of 10 is derived from a threefold factor for differences in toxicokinetics and for threefold factor for toxicodynamics. To extrapolate from animals to humans and account for interspecies variability between humans and other mammals, an uncertainty factor of 10 is used by EPA and ATSDR, and as with intraspecies extrapolations, this 10-fold factor is assumed to be associated with in toxicodynamics and toxicokinetics. An uncertainty

**Table 24.2  Comparison of Less Serious Effects and Serious Effects**

| Less Serious | Serious |
| --- | --- |
| Reversible cellular changes | Death |
| Necrosis, metaplasia, or atrophy | Cancer |
| | Clinically significant organ impairment |
| Delayed ossifciation | Visceral or skeletal abnormalities |
| Alteration in offspring weight | Cleft palate, fused ribs |
| Altered T-cell activity | Necrosis inn immunologic components |
| Auditory disorders | Visual disorders |
| 50% Reduction in offspring | Abnormal sperm |

factor of 10 is used when a NOAEL derived from a subchronic study instead of a chronic study is used as the basis for a calculation of a chronic RfD (EPA only). Note that ATSDR does not perform this extrapolation but derive chronic and subchronic MRLs. An uncertainty factor of 10 is used in deriving an RfD or MRL from a LOAEL when a NOAEL is not available. It should be noted that there are no reference doses for dermal exposure, however when there is insufficient dermal absorption data, the EPA uses a default factor of 10% to estimate bioavailability for dermal absorption. A modifying factor ranging from 1 to 10 is included by EPA only to reflect a qualitative professional assessment of additional uncertainties in the critical study and in the entire data base for the chemical not explicitly addressed by preceding uncertainty factors.

Refinements of the RfC have utilized mechanistic data to modify the interspecies uncertainty factor of 10 (Jarabek, 1995). The reader should appreciate that with the inhalation route of exposure, dosimetric adjustments are necessary and can affect the extrapolations of toxicity data of inhaled agents for human health risk assessment. The EPA has included dosimetry modeling in RfC calculations, and the resulting dosimetric adjustment factor (DAF) used in determining the RfC is dependent on physiochemical properties of the inhaled toxicant as well as type of dosimetry model ranging from rudimentary to optimal model structures. In essence, the use of the DAF can reduce the default uncertainty factor for interspecies extrapolation from 10 to 3.16.

The 1996 Food Quality Protection Act (FQPA) now requires that an additional safety factor of 10 be used in the risk assessment of pesticides to ensure the safety of infants and children, unless the EPA can show that an adequate margin of safety is assured with out it (Scheuplein, 2000). The rational behind this additional safety factor is that infants and children have different dietary consumption patterns than adults and infants, and children are more susceptible to toxicants than adults. We do know from pharmacokinetics studies with various human pharmaceuticals that drug elimination is slower in infants up to 6 months of age than in adults, and therefore the potential exists for greater tissue concentrations and vulnerability for neonatal and postnatal effects. Based on these observations, the US EPA supports a default safety factor greater or less than 10, which may be used on the basis of reliable data. However, there are few scientific data from humans or animals that permit comparisons of sensitivities of children and adults, but there are some examples, such as lead, where children are the more sensitive population. It some cases qualitative differences in age-related susceptibility are small beyond 6 months of age, and quantitative differences in toxicity between children and adults can sometimes be less than a factor of 2 or 3.

Much of the research efforts in risk assessment are therefore aimed at reducing the need to use these default uncertainty factors, although the risk assessor is limited by data quality of the chemical of interest. With sufficient data and the advent of sophisticated and validated physiologically based pharmacokinetic models and biologically based dose-response models (Conolly and Butterworth, 1995), these default values can be replaced with science-based factors. In some instances there may be sufficient data to be able to obtain distributions rather than point estimates.

### 24.3.2 Derivation of Developmental Toxicant RfD

Developmental toxicity includes any detrimental effect produced by exposures during embryonic development, and the effect may be temporary or overt physical malformation. Adverse effects include death, structural abnormalities, altered growth, and functional deficiencies. Maternal toxicity is also considered. The evidence is assessed and assigned a weight-of-evidence designation as follows: category A, category B, category C, and category D. The scheme takes into account the ratio of minimum maternotoxic dose to minimum teratogenic dose, the incidence of malformations and thus the shape of the dose-response curve or dose relatedness of the each malformation, and types of malformations at low doses. A range of uncertainty factors are also utilized according to designated category as follows: category A $= 1-400$, category B $= 1-300$, category C $= 1-250$, and category D $= 1-100$. Developmental RfDs are based a short duration of exposure and therefore cannot be applied to lifetime exposure.

### 24.3.3 Determination of RfD and RfC of Naphthalene with the NOAEL Approach

The inhalation RfC for naphthalene was 0.003 mg/m$^3$, and this RfC was derived from a chronic (2-year) NTP inhalation study in mice using exposures of 0, 10, or 30 ppm (NTP, 1992). Groups of mice were exposed for 5 days a week and 6 hours a day. This study identified a LOAEL of 10 ppm. A dose-related incidence of chronic inflammation of the epithelium of the nasal passages and lungs was observed. This LOAEL concentration was normalized by adjusting for the 6-hour-per-day and 5-day-per-week exposure pattern. A LOAEL of 9.3 mg/m3 was obtained was derived by converting 10 ppm first to mg/m$^3$ and then duration-adjusted levels for 6 h/day and 5 days/week for 103 weeks. An UF of 3000 was used, where 10 was for the interspecies (mice to humans) extrapolations, 10 for intraspecies variation in humans, 10 for using a LOAEL instead of a NOAEL, and 3 for database deficiencies.

The oral RfD for naphthalene was 0.02 mg/kg/day, and a study by Battelle (1980) was used to calculate the RfD. Decreased body weight was the most sensitive end point in groups of Fischer 344 rats given 0, 25, 50, 100, 200, or 400 mg/kg for 5 days/week for 13 weeks. These doses were also duration-adjusted to 0, 17.9, 35.7, 71.4, 142.9, and 285.7 mg/kg/day, respectively. The NOAEL for $a > 10\%$ decrease in body weight in this study was 71 mg/kg/day. The UF of 3000 was based on 10 for rats to humans extrapolation, 10 for human variation, 10 to extrapolate from subchronic to chronic, and 3 for database deficiencies including lack of chronic oral exposure studies.

### 24.3.4 Benchmark Dose Approach

There are several problems associated with using the NOAEL approach to estimate RfDs and RfCs. The first obvious constraint is that the NOAEL must by definition be one of the experimental doses tested. Once this dose is identified, the rest of the dose-response curve is ignored. In some experimental designs where there is no identifiable NOAEL but LOAEL, the dose-response curve is again ignored, and the NOAEL is derived by application of uncertainty factors as described earlier. This NOAEL approach does not account for the variability in the estimate of the dose response, and furthermore experiments that test fewer animals result in larger NOAELs and thus larger RfDs and RfCs.

An alternative approach known as the benchmark dose (BMD) approach has been developed and implemented by risk assessors as an alternative to the NOAEL approach to estimate RfDs and RfCs. This approach is not constrained by experimental design as the NOAEL approach, and it incorporates information on the sample size and shape of the dose-response curve. In fact this approach can be used for both threshold and nonthreshold adverse effects as well as continuous and quantal data sets. This requires use of Benchmark Dose Software where the dose-response is modeled and the lower confidence bound for a dose at a specified response level (benchmark response) is calculated. The benchmark response is usually specified as a 1–10% response; that is, it corresponds to a dose associated with a low level of risk such as 1–10%.

Figure 24.3 shows how an effective dose that corresponds to a specific change of effect/response (e.g., 10%) over background and a 95% lower confidence bound on the dose is calculated. The latter is often referred to as the BMDL or LBMD, as opposed to the BMD, which does not have this confidence limited associated with it.

Because the benchmark represents a statistical lower limit, larger experiments will tend, on average, to give larger benchmarks, thus rewarding good experimentation. This is not the case with NOAELs, as there is an inverse relationship between NOAEL and size of experiments. For example, poorer experiments possessing less sensitivity for detecting statistically significant increases in risk inappropriately result in higher NOAELs and RfDs, which may have an unknown unacceptable level of risk. In essence, the NOAEL is very sensitive to sample size, and there can also be high variability between experiments. With the benchmark dose approach, all the doses and slopes of the curve influence the calculations, variability of the data is considered, and the BMD is less variable between experiments. In the BMD approach quantitative toxicological data such as continuous data (organ weights serum levels, etc.) and quantal or incidence data (pathology findings, genetic anomalies, etc.) are fitted to numerous dose-response models described in the literature. The resulting benchmark dose that, for example, corresponds to a tumor risk of 10% generally can be estimated with adequate precision and not particularly dependent on the dose-response model used to fit the data. Note that dose intervals are not required for BMD estimation. This will be greatly appreciated in the cancer risk assessment section of this chapter.

### 24.3.5 Determination of BMD and BMDL for ETU

The BMD method has been quite extensively in assessing quantal data, and very often this has involved analysis of data from developmental and reproductive toxicity

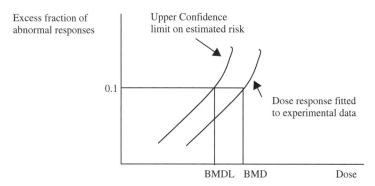

**Figure 24.3** Benchmark dose determination from dose response relationship with the BMDL corresponding to the lower end of a one-sided 95% confidence interval for the BMD.

studies. In this study example (Crump, 1984), rats were exposed to ethylenethiourea (ETU) at 0, 5, 10, 20, 40, and 80 mg/kg doses, and the number affected with fetal anomalies per number of rats were 0/167, 0/132, 1/138, 14/81, 142/178, and 24/24, respectively. The benchmark dose computation can involve utilization of any given dose-response probability model, but in this example the quantal Weibull model was used and the specified effect was set at 0.01 (1%) with confidence level of 0.95. The BMD was determined to be 8.9 mg/kg, and the BMDL was 6.9 mg/kg. This value is close to the NOAEL, which is 5 mg/kg, but it does demonstrate that the NOAEL approximates a lower confidence limit on the BMD corresponding to an excess risk of about 1% for proportions of fetal anomalies. In fact an empirical analysis of some 486 developmental toxicity studies has demonstrated that the NOAEL can result in an excess risk of 5% for proportions of dead or malformed fetuses per litter. The reader should at this stage recognize that the BMD approach can also be used in cancer risk assessment as we are often times working with quantal data that are ideally suited for BMD modeling.

### 24.3.6 Quantifying Risk for Noncarcinogenic Effects: Hazard Quotient

The measure used to describe the potential for noncarcinogenic toxicity to occur is not expressed as the probability. Probabilistic approach is used in cancer RA. For noncancer RA, the potential for noncarcinogenic effects is evaluated by comparing an exposure level (E) over a specified time period with a reference dose (RfD). This ratio is called a hazard quotient:

$$\text{Hazard quotient} = \frac{E}{RfD}.$$

In general, the greater the value of E/RfD exceeds unity, the greater is the level of concern. Note that this is a ratio and not to be interpreted as a statistical probability.

### 24.3.7 Chemical Mixtures

Human populations are more likely to be exposed simultaneously or sequentially to a mixture of chemicals rather than to one single chemical. Standard default approaches to mixture risk assessment consider doses and responses of the mixture components to be additive. However, it should also be recognized that components in the mixture can also result in synergistic, antagonistic, or no toxicological effect following exposure to a chemical mixture. Therefore mixture toxicity cannot always be predicted even if we know the mechanisms of all toxic components in a defined mixture. Furthermore tissue dosimetry can be complicated by interactions at the route of entry (e.g., GIT, skin surface) and clearance mechanisms in the body. In essence, there are considerable uncertainties involved in trying to extrapolate effects following exposure to chemical mixtures. Several PBPK models have been used to quantitate these effects and also provide some information useful for risk assessment of chemical mixtures (Krishnan et al., 1994; Haddad et al. 2001).

The 1996 FQPA has also mandated that the EPA should also consider implementing cumulative risk assessments for pesticides. Cumulative risk assessments usually involve integration of the hazard and cumulative exposure analysis, and it primarily involves cumulative nonoccupational exposure by multiple routes or pathways to two or more pesticides or chemicals sharing a common mechanism of toxicity.

Calculation procedures differ for carcinogenic and noncarcinogenic effects, but both sets of procedures *assume dose additivity* in the absence of information on mixtures:

$$\text{Cancer risk equation for mixtures}: \quad \text{Risk}_T = \Sigma \text{Risk}_I,$$

$$\text{Noncancer hazard index} = \frac{E_1}{RfD_1} + \frac{E_2}{RfD_2} + \cdots + \frac{E_i}{RfD_i}.$$

This hazard index (HI) approach as well as other indexes (e.g., relative potency factors) are applied for mixture components that induce the same toxic effect by identical mechanism of action. In cases where there are different mechanisms, separate HI values can be calculated for each end point of concern. As the equation above indicates, the HI is easy to calculate, as there is simply scaling of individual component exposure concentrations by a measure of relative potency such as the RfD or RfC, and adding scaled concentrations to get an indicator of risk from exposure to the mixture of concern. However, as noted above, this additivity approach does not take into account tissue dosimetry and pharmacokinetic interactions. Recent published risk assessments have utilized mixture PBPK models to account for multiple pharmacokinetic interactions among mixture constituents. These interaction-based PBPK models can quantify change in tissue dose metrics of chemicals during exposure to mixtures and thus improve the mechanistic basis of mixture risk assessment. Finally the reader should be aware that this HI is different from the a term known as the margin of safety (MOS), which is the ratio of the critical or chronic NOAEL for a specific toxicological end point to an estimate of human exposure. MOS values greater than 100 are generally considered protective if the NOAEL is derived from animal data.

## 24.4  CANCER RISK ASSESSMENT

For cancer risk assessment an assumption is held that a threshold for an adverse effect does not exist with most individual chemicals. It is assumed that a small number of molecular events can evoke changes in a single cell that can lead to uncontrolled cellular proliferation and eventually to a clinical state of disease. This mechanism is referred to as "nonthreshold" because there is believed to be essentially no level of exposure to such a chemical that does not pose a finite probability, however small, of generating a carcinogenic response. That is, no dose is though to be risk free. Therefore, in evaluations of cancer risks, an effect threshold cannot be estimated. For carcinogenic effects, the US EPA uses a two-part evaluation: (1) the substance is first assigned a weight-of-evidence classification and then (2) a slope factor is calculated.

*1. Assigning a weight-of-evidence.* The aim here is to determine the likelihood that the agent is a human carcinogen. The *evidence* is characterized separately for human studies and animal studies as *sufficient, limited, inadequate, no data*, or *evidence of no effect*. Based on this characterization and on the extent to which the chemical has been shown to be a carcinogen in animals or humans or both, the chemical is given a provisional *weight-of-evidence* classification. The US EPA classification system (EPA,

**Table 24.3    Weight of Evidence Designation Based on EPA (1986) Guidelines**

| Group | Description |
|-------|-------------|
| A | Human carcinogen |
| B1 or B2 | Probable human carcinogen |
| C | Possible human carcinogen |
| D | Not classifiable as to human carcinogenicity |
| E | Evidence of noncarcinogenicity for humans |

*Note*: B1 indicates that limited human data are available; B2 indicates sufficient evidence in animals and inadequate or no evidence in humans.

1986) shown in Table 24.3 has been revised in the EPA (1996) proposed guidance and more recent draft guidance (EPA, 1999).

This system was also adapted from the approach taken by the International Agency for Research on Cancer (IARC). This alphanumeric classification system has been replaced with a narrative and the following descriptor categories: *known/likely, cannot be determined,* or *not likely.* These EPA (1996) guidelines indicate that not only are tumor findings an important consideration, but also structure-activity relationships, modes of action of carcinogenic agents at cellular or subcellular level and toxicokinetic and metabolic processes. These revised guidelines also indicate that the weighing of evidence should address the conditions under which the agent may be expressed. For example, an agent may "likely" be carcinogenic via inhalation exposure but "not likely" via oral exposure. The narrative will summarize much of this information as well as the mode of action information.

*2. Quantifying risk for carcinogenic effects.* In the second part of the evaluation, the EPA (1986) guidelines required that quantitative risk be based on the evaluation that the chemical is a known or probable human carcinogen, a toxicity value that defined quantitatively the relationship between dose and response (slope factor) is calculated. Slope factors have been calculated for chemicals in classes A, B1, and B2. Sometimes a value is derived for those in class C on a case-by-case basis. The slope factor is a plausible upper-bound estimate of the probability of a response per unit intake of chemical over a lifetime. Slope factors have been accompanied by the weight-of-evidence classification to indicate the strength of evidence that the chemical is a human carcinogen.

Development of a slope factor entails applying a model to the available data set and using the model to extrapolate from high doses to lower exposure levels expected for human contact. There are a number of low-dose extrapolation models that can be divided into distribution models (e.g., log-probit, Weibull) and mechanistic models (e.g., one-hit, multi-hit, and *linearized multistage*). EPA 1986 guidelines for carcinogen risk assessment are currently being revised, and it is very likely that the new guidelines will encourage the use of biologically based models for cancer risk assessment. The previous guidelines (EPA, 1986) recommended that the linearized multistage model, which is a mechanistic model, be employed in as the default model in most cases. Most of the other models are less conservative. The proposed biologically based models attempt to incorporate as much mechanistic information as possible to arrive at an estimate of slope factors. In essence, after the data are fit to the selected model, the upper 95th percent confidence limit of the slope of the resulting dose response curve is calculated. *This represents the probability of a response per unit intake over a lifetime*, or that there is a 5% chance that the probability of a response could be greater than the estimated value on the basis of experimental data and model used. In some cases, the slope factors based on human dose-response data are based on "best" estimate instead of upper 95th percent confidence limit. The toxicity values for carcinogenic effects can be expressed in several ways.

The slope factor is expressed as $q_1{}^*$:

$$\text{Slope factor} = \text{Risk per unit dose}$$

$$= \text{Risk per mg/kg-day}.$$

The slope factor can therefore be used to calculate the upper bound estimate on risk (R)

$$\text{Risk} = q_1{}^*[\text{risk} \times (\text{mg/kg/day})^{-1}] \times \text{exposure (mg/kg/day)}.$$

Here risk is a unitless probability (e.g., $2 \times 10^{-5}$) of an individual developing cancer and exposure is really chronic daily intake averaged over 70 years: mg/kg/day. This can be determined if we can determine the slope factor and human exposure at the waste site or occupational site. The EPA usually sets a goal of limiting lifetime cancer risks in the range of $10^{-6}$ to $10^{-4}$ for chemical exposures, while the FDA typically aims for risks below $10^{-6}$ for general population exposure. It is therefore quite likely for very high exposures for the accepted EPA range of risk to be exceeded. The EPA range is considered protective of the general and sensitive human population. It should be noted that these orders of magnitude are substantially greater than those used in estimating RfD and RfCs in noncancer risk assessment.

Because relatively low intakes (compared to those experienced by test animals) are most likely from environmental exposure at Superfund hazardous waste sites, it generally can be assumed that the dose-response relationship will be linear on the low-dose portion of the multistage model dose-response curve. The equation above can apply to these linear low-dose situations. This linear equation is valid only at low risk levels (i.e., below the estimated risk of 0.01). For risk above 0.01 the one-hit equation should be used:

$$\text{Risk} = 1 - \exp(-\text{exposure} \times \text{slope factor}).$$

As indicated above, biologically based extrapolation models are the preferred approach for quantifying risk to carcinogens, although it is possible that all the necessary data will not be available for many chemicals. The EPA (1986) guidelines have been modified to include the response data on effects of the agent on carcinogenic processes in addition to data on tumor incidence. Precursor effects and tumor incidence data may be combined to extend the dose response curve below the tumor data; that is, below the range of observation. Thus a biologically based or case-specific dose-response model is developed when there is sufficient data, or a standard default procedure is used when there is insufficient data to adequately curve-fit the data. In brief, the dose-response assessment is considered in two parts or steps, range of observation and range of extrapolation, and the overriding preferred approach is to use the biologically based or case-specific model for both of these ranges. In the first step of this process, the lower 95% confidence limit on a dose associated with an estimated 10% increase in tumor or nontumor response ($LED_{10}$) is identified. When human real world exposures are outside the range of the observed or experimental data, this serves as the point of departure or marks the beginning for the extrapolating to these low environmental exposure levels. Note that these procedures are very similar to the benchmark procedure for quantitating risk to noncarcinogenic chemicals. In the second step, the biologically based or case-specific model is preferred for use in extrapolations to lower dose levels provided that there are sufficient data. If the latter is not the case, then default approaches consistent with agent chemical mode of action are implemented with the assumption of linearity or nonlinearity of the dose-response relationship. The linear default approach is a departure from the 1986 guidelines, which used the linearized multistage (LMS) procedure, but is based on mode of action or alternatively if there is insufficient data to support a nonlinear mode of action. In brief, it involves drawing a straight line from the point of departure ($LED_{10}$) to the origin (i.e., zero). When there is no evidence of linearity or there is a nonlinear mode of action, the default approach is the margin of exposure (MOE) analysis. The MOE approach computes the ratio between the $LED_{10}$ and the environmental exposure, and the analysis begins from the point of departure that is adjusted for toxicokinetic differences between species to give a human equivalent dose.

Finally it should be noted that prior to the FQPA in 1996, the Delaney clause prohibited the establishment of tolerances or maximum allowable levels for food additives if it has been shown to induce cancer in human or animal. This is an important change in regulations because pesticide residues were considered as food additives. Because of the FQPA, pesticide residues are no longer regarded as food additives, and there is no prohibition against setting tolerances for carcinogens.

## 24.5 PBPK MODELING

Physiologically based pharmacokinetic (PBPK) modeling has been used in risk assessment to make more scientifically based extrapolations, and at the same time to help explore and reduce inherent uncertainties. Historically pharmacokinetics has relied on empirical models, and in many instances this process offers little insight into mechanisms of absorption, distribution, and clearance of hazardous agents and does not facilitate translation from animal experiments to human exposures. For example, dose scaling using by body weight or size may often time overestimate or underestimate toxicant levels at the target tissue. PBPK models can help predict tissue concentrations in different species under various conditions based on *independent* anatomical, physiological, and biochemical parameters. In these analyzes physiological parameters such as organ volumes, tissue-blood partition coefficients, and blood flow to specific tissue compartments described by the model, are calculated or obtained from the literature and integrated into the model. Monte Carlo analysis, a form of uncertainty analysis, can now be performed, and this allows for the propagation of uncertainty through a model that results in estimation of the variance of model output. This can be achieved by randomly sampling model parameters from defined distributions; some parameters such as cardiac output, metabolic, and log $P$ parameters, may have a lognormal distribution, while other parameters may be normal or uniform. In essence, the Monte Carlo analysis when coupled with PBPK characterizes the distribution of potential risk in a population by using a *range* of potential values for each input parameter (not single values) as well as an estimate of how these values are distributed (Clewell and Andersen, 1996). By these approaches, uncertainty is identifiable and quantifiable, and can reduce inappropriate levels of concern in reporting the risk of chemical exposure. These mathematical modeling approaches also help identify areas of potential scientific research that could improve the human health assessment.

In recent years there have been significant efforts at harmonizating noncancer and cancer risk assessments (Barton et al., 1998; Clewell et al., 2002), and in this respect PKPD modeling can be a very useful tool in the risk assessment process. For example, recall that noncancer risk assessment addresses variability in a population by dividing the NOAEL by 10, whereas the cancer risk assessment does not address this quantitatively. PBPK modeling coupled with Monte Carlo analysis is one approach as described in the previous paragraph that will help address this level of uncertainty in the risk assessment. In conclusion, it should be noted that PBPK modeling has been utilized with very few toxicants. It is hoped that risk assessment policy will encourage the use of this tool as well as other appropriate models to integrate mechanistic information and the pharmacokinetics (dosimetry), and pharmacodynamics (dose response) of toxicants. Improved quantitative risk assessments will ultimately provide scientifically sound information that will influence the risk management decision process.

## SUGGESTED READING

ATSDR. *Guidance for the Preparation of Toxicological Profiles.* ATSDR, US Public Health Service. 1993.

Battelle. Subchronic toxicity study: Naphthalene (C52904), Fischer 344 rats. Report to US Department of Health and Human Services, National Toxicology Program, Research Triangle Park, NC, by Battelle Columbus Laboratories, Columbus, OH, 1980.

Borton, H. A., M. E. Anderson, and H. J. Clewell. Harmonization: Developing consistent guidelines for applying mode of action and dosimetry information for applying mode of action and dosimetry information to cancer and noncancer risk assessment. *Hum. Ecol. Risk Assess.* **4**: 75–115, 1998.

Clewell, H. J., and M. E. Andersen. Use of physiologically based pharmacokinetic modeling to investigate individual versus population risk. *Toxicol.* **111**(1–3): 315–329, 1996.

Clewell, H. J., M. E. Andersen, and H. A. Barton. A consistent approach for the application of pharmacokinetic modeling in cancer and noncancer risk assessment. *Envion. Health. Perspect.* **110**: 85–93, 2002.

Conolly, R. B., and B. E. Butterworth. Biologically based dose response model for hepatic toxicity: a mechanistically based replacment for traditional estimates of noncancer risk. *Toxicol. Lett.* **82–83**: 901–906, 1995.

Conolly, R. B., B. D. Beck, and J. I. Goodman. Stimulating research to improve the scientific basis of risk assessment. *Toxicol. Sci.* **49**: 1–4, 1999.

Crump, K. S. A new method for determining allowable daily intakes. *Fundam. Appl. Toxicol.* **4**: 854–871, 1984.

EPA. Guidelines for Carcinogen Risk Assessment. *Fed. Reg.* **51**: 33992–34003, 1986.

EPA. Proposed Guidelines for Carcinogen Risk Assessment. *Fed. Reg.* **61**: 17960–18011, 1996.

EPA. Guidelines for Carcinogen Risk Assessment. Risk Assessment Forum, US Environmental Protection Agency, NCEA-F-0644, July 1999, Review Draft.

EPA. *Risk Assessment Guidance for Superfund*, Vol. 1. *Human Health Evaluation Manual* (Part A). US EPA. 1989.

Haddad, S., M. Beliveau, R. Tardif, and K. Krishnan. A PBPK modeling-based approach to account for interactions in the health risk assessment of chemical mixtures. *Toxicol. Sci.* **63**: 125–131, 2001.

Jarabek, A. M., The application of dosimetry models to identify key processes and parameters for default dose-response assessment approaches. *Toxicol. Lett.* **79**(1–3): 171–184, 1995.

Krishnan, K., H. J. Clewell, and M. E. Andersen Physiologically based pharmacokinetic analyzes of simple mixtures. *Environ. Health Perspec.* **102** (supp. 9): 151–155, 1994.

NTP. National Toxicology Program. Technical report series No. 410. Toxicology and carcinogenesis studies of naphthalene (CAS No. 91-20-3) in B6C3F1 mice (inhalation studies). Research Triangle Park, NC. US Department of Health and Human Services, Public Health Services, National Institutes of Health. NIH Publication No. 92–3141, 1992.

Page, N. P.,  D. V Singh,  W. Farland,  J. I. Goodman,  R. B. Conolly,  M. E Andersen, H. J. Clewell, C. B. Frederick, H. Yamasaki, G. Lucier. Implementation of EPA Revised Cancer Assessment Guidelines: Incorporation of Mechanistic and Pharmacokinetic Data. *Fundam. Appl. Toxicol.* **37**: 16–36, 1997.

Scheuplein, R. J. Pesticides and infant risk: Is there a need for an additional safety margin? *Regul. Toxicol. Pharmacol.* **31**(3): 267–279, 2000.

# PART VII

# ENVIRONMENTAL TOXICOLOGY

# CHAPTER 25

# Analytical Methods in Toxicology

ROSS B. LEIDY

## 25.1 INTRODUCTION

Some 200,000 chemicals are synthesized annually worldwide, and the toxicity of most of them is unknown. Few of these chemicals reach the stage of further development and use, but those that do usually find their way into the environment. Some are persistent and remain adsorbed to soil particles or soil organic matter, some find their way into water through soil movement or aerial deposition, others are metabolized by microorganisms into compounds of greater toxicity that move up the food chain. Over time, their accumulation in higher life forms could result in debilitating alterations in metabolism, leading to illness. It might be years before such illness could be attributed to specific compounds because of the difficulty involved in identifying and quantitating them. The concern over the role of persistent organochlorines in the food chain and their possible role as human xenoestrogens is an example. The identification and quantitation of chemicals in both the environment and in living beings relies on the development of analytical techniques and instruments.

Advances in analytical techniques continue to multiply in all fields of toxicology, and as mentioned, many of these focus on the environmental area. Whether looking for new techniques to sample water or for an automated instrument to determine quantities of sulfur-containing compounds in air, such devices are available. In many instances, developments in environmental analyses are adaptable to experimental work related to drug toxicity, or in forensic medicine, to determine the cause of poisoning.

Although new techniques and instruments continue to enter the commercial market, the basic analytical process has not changed: define the research goal(s), develop a sampling scheme to obtain representative samples, isolate the compound(s) of interest, remove potential interfering components, and quantitate and evaluate the data in relation to the initial hypothesis. Based on the data generated, many options are available. For example, was the sampling scheme complete? Would further refinement of the analytical procedure be required? Should other sample types be analyzed? Thus it is obvious that within these general categories particular methods vary considerably depending on the chemical characteristics of the toxicant (Table 25.1).

*A Textbook of Modern Toxicology, Third Edition,* edited by Ernest Hodgson
ISBN 0-471-26508-X Copyright © 2004 John Wiley & Sons, Inc.

**Table 25.1  Typical Protocols for Analysis of Toxicants**

| | Toxicant | | |
|---|---|---|---|
| Step | Arsenic | TCDD | Chlorpyrifos |
| Sampling | Grind solid sample homogenize tissue to homogeneity; subsample | Grind solid sample or homogenize tissue to homogeneity; subsample | Grind solid sample or homogenize tissue to homogeneity; subsample Soxhlet extract with hexane:acetone (1:1) |
| Extraction and cleanup | Dry ash; redissolve residue; generate arsine and absorb into solution | Extract with ethanol and KOH; remove saponified lipids; column chromatography on $H_2SO_4$/silica gel followed by basic alumina and then by $AgNO_3$/silica gel followed by basic alumina; reverse-phase HPLC | Remove co-extractives on Florisil using ether: petroleum ether |
| Analysis | AA spectroscopy | GC/MS | GC/NPD or FPD |

*Source*: Modified from R. J. Everson and F. W. Oehme, *Analytical Toxicology Manual*, New York: KS American College of Veterinary Toxicologists, 1981.
*Note*: TLC, thin-layer chromatography; HPLC, high-performance liquid chromatography; GLC or GC, gas-liquid chromatography; AA, atomic adsorption; NPD, nitrogen phosphorus detector; FPD, flame photometric detector; GC/MS, gas chromatography/mass spectrometry.

This chapter is concerned with the sampling, isolation, separation, and measurement of toxicants, including bioassay methods. Bioassay does not measure toxic effects; rather, it is the quantitation of the relative effect of a substance on a test organism as compared with the effect of a standard preparation of a basic toxicant. Although bioassay has many drawbacks, particularly lack of specificity, it can provide a rapid analysis of the relative potency of toxicants in environmental samples.

## 25.2  CHEMICAL AND PHYSICAL METHODS

### 25.2.1  Sampling

Even with the most sophisticated analytical equipment available, the resulting data are only as representative as the samples from which the results are derived. This is particularly true for environmental samples. In sampling, care must be taken to ensure that the result meets the objectives of the study. Often special attention to sampling procedures is necessary. Sampling accomplishes a number of objectives, depending on the type of area being studied. In environmental areas (e.g., wilderness regions, lakes, rivers) sampling can provide data not only on the concentration of pollutants but also on the extent of contamination. In urban areas, sampling can provide information on the types of pollutants, to which one is exposed, by dermal contact, by inhalation, or by ingestion over a given period of time.

In industrial areas, hazardous conditions can be detected and sources of pollution can be identified. Sampling is used in the process of designing pollution controls and can provide a chronicle of the changes in operational conditions as controls are implemented. Another important application of sampling in industrial areas in the United States is the documentation of compliance with existing Occupational Safety and Health Administration (OSHA) and US Environmental Protection Agency (US EPA) regulations. The many methods available for sampling the environment can be divided into categories of air, soil, water, and tissue sampling. The fourth category is of particular interest in experimental and forensic studies.

*Air.* Most pollutants entering the atmosphere come from fuel combustion, industrial processes, and solid waste disposal. Additional miscellaneous sources, such as nuclear explosions, forest fires, dusts, volcanoes, natural gaseous emissions, agricultural burning, and pesticide drift, contribute to the level of atmospheric pollution. To affect terrestrial animals and plants, particulate pollutants must be in a size range that allows them to enter the body and remain there; that is, they must be in an aerosol (defined as an airborne suspension of liquid droplets) or on solid particles small enough to possess a low settling velocity. Suspensions can be classified as liquids including fogs (small particles) and mists (large particles) produced from atomization, condensation, or entrapment of liquids by gases; and solids including dusts, fumes, and smoke produced by crushing, metal vaporization, and combustion of organic materials, respectively.

At rest, an adult human inhales 6 to 8 L of air each minute ($1 \text{ L} = 0.001 \text{ m}^3$) and, during an 8-hour workday, can inhale from 5 to 20 $m^3$ depending on the level of physical activity. The optimum size range for aerosol particles to get into the lungs and remain there is 0.5 to 5.0 $\mu$m. As instrumentation used to collect atmospheric dust have become more precise, particulate matter (PM) in the size range of 2.5 to 10 $\mu$m have come under increasing scrutiny, because many potential toxicants are adsorbed to their surfaces. These particles are inhaled and will remain in the lungs and allow the compounds to pass into the bloodstream.

Thus air samplers have been miniaturized and adsorbents have been developed to collect either particulate matter in the size range most detrimental to humans or to "trap" organic toxicants from air. An air sampler generally consists of an inlet to direct air through a filter to entrap particles that might be of interest (e.g., dust); through the adsorbent, which collects organic vapors, a flowmeter and valve to calibrate airflow, and a pump to pull air through the system. Personnel samplers are run by battery power and can be attached to an individual's clothing, thus allowing continual monitoring while performing assigned tasks in the work environment. This allows the estimation of individual exposure.

Many air samplers use various types of filters to collect solid particulate matter, such as asbestos, which is collected on glass fiber filters with pores 20 $\mu$m or less in diameter. Membrane filters with pores 0.01 to 10 $\mu$m in diameter are used to collect dusts and silica. Liquid-containing collectors, called impingers, are used to trap mineral dusts and pesticides. Mineral dusts are collected in large impingers that have flow rates of 10 to 50 L of air per minute, and insecticides can be collected in smaller "midget" impingers that handle flows of 2 to 4.5 L of air per minute. Depending on the pollutant being sought, the entrapping liquid might be distilled water, alcohol, ethylene glycol, hexylene glycol (2-methyl, 2,4-pentane diol) or some other solvent. Because of the ease of handling and the rapid desorption of compounds, polyurethane foam (PUF) has become a popular trapping medium for pesticides and is rapidly replacing the use of midget impingers. A large volume air sampler has been developed by the US EPA for detection of pesticides and polychlorinated biphenyls (PCBs). Air flows at rates of around 225.0 L/min are drawn through a PUF pad, and the insecticides and PCBs are trapped in the foam. Small glass tubes approximately $7.0 \times 0.5$ cm in diameter containing activated charcoal are used to entrap organic vapors in air.

A number of specialty companies have and are continuing to develop adsorbents to collect organic molecules from air samples. Industrial chemicals resulting, from syntheses or used in production processes, pesticides and emissions from exhaust towers are monitored routinely with commercially available adsorbents. Personnel monitoring can be accomplished without a pump using a system composed of a porous membrane through which air diffuses and compounds of interest are collected by an adsorbent.

Minute quantities of gaseous pollutants (e.g., $CO_2$, $HNO_3$), are monitored with direct reading instruments, using infrared spectroscopy, and have been in use for a number of years. These instruments passively monitor large areas and rely on extensive statistical evaluations to remove substances like water vapor, which can mask the small quantities of these pollutants. Research into the millimeter/submillimeter area of spectroscopy coupled with Russian technologies is leading to the development of a direct reading instrument that will quantitate any atmospheric gas or a mixture of gases containing a dipole moment within 10 seconds, regardless of the presence or quantity of water vapor in the atmosphere. Such devices are expected to be commercially available within the next five years.

**Soil.** When environmental pollutants are deposited on land areas, their subsequent behavior is complicated by a series of simultaneous interactions with organic and inorganic components, existing liquid-gas phases, microscopic organisms, and other soil constituents. Depending on the chemical composition and physical structure, pollutants might remain in one location for varying periods of time, be absorbed into plant tissue, or move through the soil profile from random molecular motion. Movement is also affected by mass flow as a result of external forces such as the pollutant being dissolved in or suspended in water or adsorbed onto both inorganic and organic soil components. Thus sampling for pollutants in soils is complex and statistical approaches must be taken to ensure representative samples.

To obtain such samples, the chemical and physical characteristics of the site(s) must be considered, as well as possible reactions between the compound(s) of interest and soil components and the degree of variability (i.e., variation in soil profiles) within the sampling site. With these data, the site(s) can then be divided into homogeneous areas and the required number of samples can be collected. The required number of samples depends on the functions of variance and degree of accuracy. Once the correct procedure has been determined, sampling can proceed.

Many types of soil samplers are available, but coring devices are preferable because this collection method allows determination of a pollutant's vertical distribution. These devices can be either stainless steel tubes, varying in both diameter from 2.5 to 7.6 cm and length from 60 to 100 cm (hand operated). Large, mechanically operated boring tubes, 200 cm in length are also used. It is possible to sample to uniform depths with these devices, and one can subdivide the cores into specific depths (e.g., 0–7.6 cm, 7.6–15.2 cm, etc.) to determine movement. Another type of coring device is a wheel to which are attached tubes so that large numbers of small subsamples can be collected, thus allowing a more uniform sampling over a given area. Soils from specific depths can be collected using a large diameter cylinder (ca. 25 cm) that incorporates a blade to slice a core of soil after placing the sampler at the desired depth.

*Water.* Many factors must be considered to obtain representative samples of water. The most important are the pollutant and the point at which it entered the aquatic environment. Pollutants can be contributed by agricultural, industrial, municipal, or other sources, such as spills from wrecks or train derailments. The prevailing wind direction and speed, the velocity of stream or river flow, temperature, thermal and salinity stratification, and sediment content are other important factors.

Two questions, where to monitor or sample and how to obtain representative samples are both important. Surface water samples often are collected by automatic sampling devices controlled by a variety of sensors. The simplest method of collecting water is the "grab" technique, whereby a container is lowered into the water, rinsed, filled, and capped. Specialized samplers frequently are used to obtain water at greater depths.

With the implementation in the United States of the Clean Water Act of 1977, continuous monitoring is required to obtain data for management decisions. A number of continuous monitoring wells are in operation throughout the United States. Sampling from potable wells can be accomplished by collecting from an existing tap, either in the home or from an outside fixture. However, multistep processes are required to collect samples from wells used to monitor pollutants. Standing water must be removed after measuring the water table elevation. If wells are used to monitor suspected pollutants, two criteria are used to determine the amount of water removed prior to sampling: conductivity and pH. Removal of a specific number of well volumes by bailers or pumps is done until both pH and conductivity are constant. A triple-rinsed bottle is then used to collect the sample.

Because large numbers of samples can be generated by such devices, collectors containing membranes with small pores (e.g., 45.0 μm) to entrap metal-containing pollutants, cartridges containing ion-exchange resins, or long-chain hydrocarbons (e.g., $C_{18}$) bonded to silica to adsorb organic pollutants. These devices often are used to diminish the number and bulk of the samples by allowing several liters of water to pass through and leave only the pollutants entrapped in a small cylinder or container. In addition disk technologies use a filter containing a Teflon matrix in which $C_{18}$ hydrocarbon chains are embedded to concentrate pollutants as water is passed through the membrane. Polar solvents (e.g., methanol) are used to elute them from the disk.

Once samples have been collected, they should be frozen immediately in solid $CO_2$ (dry ice) and returned to the laboratory. If they are not analyzed at that time, they should be frozen at temperatures of $-20°C$ or lower. Sufficient head space must be left in the container to prevent breakage.

*Tissues.* When environmental areas are suspected of being contaminated, surveys of plants and animals are conducted. Many of the surveys, conducted during hunting and fishing seasons by federal and state laboratories, determine the number of animals killed and often, organs and other tissues are removed for analysis of suspected contaminants. Sampling is conducted randomly throughout an area, and the analyses can help determine the concentration, extent of contamination within a given species and areas of contamination.

Many environmental pollutants are known to concentrate in bone, certain organs, or specific tissues (e.g., adipose). These organs are removed from recently killed animals for analysis. In many instances, the organs are not pooled with others from the same species but are analyzed separately as single sub-samples to determine the extent of possible contamination in the area sampled.

When plant material is gathered for analysis, it is either divided into roots, stems, leaves, and flowers and/or fruit or the whole plant is analyzed as a single entity. Pooling of samples from a site can also provide a single sample for analysis. The choice depends on the characteristics of the suspected contaminant.

### 25.2.2 Experimental Studies

Experimental studies, particularly those involving the metabolism or mode of action of toxic compounds in animals (or, less often, plants), can be conducted either in vivo or in vitro. Because organisms or enzyme preparations are treated with known compounds, the question of random sampling techniques does not arise as it does with environmental samples. Enough replication is needed for statistical verification of significance, and it should always be borne in mind that repeated determinations carried out on aliquots of the same preparation do not represent replication of the experiment; at best, they test the reproducibility of the analytical method.

In environmental studies, the analyst is concerned with stable compounds or stable products; in metabolic studies, the question of reactive (therefore unstable) products and intermediates is of critical concern. Thus the reaction must be stopped, and the sample must be processed using techniques that minimize degradation. This is facilitated by the fact that the substrate is known, and the range of possible products can be determined by a variety of methods.

The initial sampling step is to stop the reaction, usually by a protein precipitant. Although traditional compounds such as trichloroacetic acid are effective protein precipitants, they are usually undesirable. The use of a single water-miscible organic solvent such as ethanol or acetone are milder, whereas a mixture of solvents (e.g., chloroform/methanol) not only denatures the protein but also effects a preliminary separation into water-soluble and organic-soluble products. Rapid freezing is a mild method of stopping reactions, but low temperature during the subsequent handling is necessary.

In toxicokinetic studies involving sequential animal sacrifice and tissue examination, it is critical to obtain uncontaminated organ samples. Apart from contamination by blood, suitable samples can be obtained by careful dissection and rinsing of the organs in ice-cold buffer, saline, or other appropriate solution. Blood samples themselves are obtained by cardiac puncture, and blood contamination of organ samples is minimized by careful bleeding of the animal at the time of sacrifice or, if necessary, by perfusion of the organ in question.

### 25.2.3 Forensic Studies

Because forensic toxicology deals primarily with sudden or unexpected death, the range of potential toxicants is extremely large. The analyst does not usually begin examination of the samples until all preliminary studies are complete, including autopsy and microscopic examination of all tissues. Thus the analyst is usually able to begin with some working hypothesis of the possible range of toxicants involved.

Because further sampling usually involves exhumation and is therefore unlikely or, in the case of cremation, impossible, adequate sampling and sample preservation is essential. For example, various body fluids must be collected in a proper way: blood by cardiac puncture, never from the body cavity; urine from the urinary bladder; bile collected intact as part of the ligated gallbladder; and so on. Adequate sample size is important. Blood can be analyzed for carbon monoxide, ethanol, and other alcohols, barbiturates, tranquilizers, and other drugs; at least 100 mL should be collected. Urine is useful for analysis of both endogenous and exogenous chemicals and the entire content of the bladder is retained. The liver frequently contains high levels of toxicants and/or their metabolites, and it and the kidney are the most important solid tissues for forensic analysis; 100 to 200 g of the former and the equivalent of one kidney usually are retained. DNA analysis has made tremendous strides through the use of polymerase chain reaction (PCR) that allows old samples (e.g., exhumation and sampling of bone marrow) to be analyzed and compared to living relatives; thus these data provide valuable information to law authorities and others.

An unusual requirement with important legal ramifications is that of possession. An unbroken chain of identifiable possession (i.e., chain of custody) must be maintained. All transfers are marked on the samples as to time and date of collection, arrival at the laboratory, and all transfers must be signed by both parties. The security and handling of samples during time of possession must be verifiable as a matter of law.

### 25.2.4  Sample Preparation

*Extraction.* In most cases the analysis of a pollutant or other toxicant depends on its physical removal from the sample medium. In order to ensure that the sample used is homogeneous, it is chopped, ground or blended to a uniform consistency and then subsampled. This subsample is extracted, which involves bringing a suitable solvent into intimate contact with the sample, generally in a ratio of 5 to 25 volumes of solvent to 1 volume of sample. One or more of four different procedures can be used, depending on the chemical and physical characteristics of the toxicant and the sample matrix. Other extraction methods such as boiling, grinding, or distilling the sample with appropriate solvents are used less frequently.

*Blending.* The use of an electric or air-driven blender is currently the most common method of extraction of biologic materials. The weighed sample is placed in a container, solvent is added, and the tissue is homogenized by motor-driven blades. Blending for 5 to 15 minutes followed by a repeat blending will extract most environmental toxicants. A homogenate in an organic solvent can be filtered through anhydrous sodium sulfate to remove water that might cause problems in the quantitation phase of the analysis. The use of sonication is a popular method for extracting tissue samples, particularly when the binding of toxicants to subcellular fractions is of interest. Sonicator probes rupture cells rapidly, thus allowing the solvent to come into intimate contact with all cell components. Differential centrifugation can then be used to isolate fractions of interest. Large wattage (e.g., 450 watt) sonifiers are used to extract compounds from environmental samples, and several US EPA methods list sonication as a valid method of extraction.

***Shaking.*** Pollutants are generally extracted from water samples, and in some cases soil samples, by shaking with an appropriate solvent or solvent combination. Mechanical shakers are used to handle several water or soil samples at once. These devices allow the analyst to conduct long-term extractions (e.g., 24 h) if required. Two or more shakings normally are required for complete removal (i.e., >98%) of the toxicant from the sample matrix.

***Washing.*** Washing with water-detergent combinations or with solvents can be used to remove surface contamination from environmental samples such as fruits or plants or from a worker's hands, if dermal exposure from industrial chemicals or pesticides is suspected.

***Continuous Extraction.*** The procedure, called Soxhlet extraction, is performed on solid samples (e.g., soil) and involves the use of an organic solvent or combination of solvents. The sample is weighed into a cup (thimble) of specialized porous material such as cellulose or fiberglass and placed in the apparatus. This consists of a boiling flask, in which the solvent is placed: an extractor, which holds the thimble, and a water-jacketed condenser. When heated to boiling, the solvent vaporizes, is condensed, and fills the extractor, thus bathing the sample and extracting the toxicant. A siphoning action drains the solvent back into the boiling flask, and the cycle begins again. Depending on the nature of the toxicant and sample matrix, the extraction can be completed in as little as 2 hours but may take as long as 3 to 4 days. Automated instruments have been introduced that perform the same operation in a shorter period of time (e.g., 30 min) and use much less solvent (e.g., 15–30 mL compared to 250 mL). They are expensive compared to the older method but are cost effective.

***Supercritical Fluid Extraction.*** Conditions can be generated that allow materials to behave differently from their native state. For example, boiling points are defined as that temperature at which a liquid changes to a gas. If the liquid is contained and pressure exerted, the boiling point changes. For a particular liquid, a combination of pressure and temperature will be reached, called the critical point, at which the material is neither a liquid nor a gas. Above this point exists a region, called the supercritical region, at which increases in both pressure and temperature will have no effect on the material (i.e., it will neither condense nor boil). This so-called supercritical fluid will exhibit properties of both a liquid and a gas. The supercritical fluid penetrates materials as if it were a gas and has solvent properties like a liquid.

Of all the materials available for use as a supercritical fluid, $CO_2$ has become the material of choice because of its chemical properties. Instruments have been developed to utilize the principles described to effect extractions of compounds from a variety of sample matrices including asphalt, plant material, and soils (Figure 25.1). The supercritical fluid is pumped through the sample, through a filter or column to a trap where the fluid vaporizes and solvent is added to transfer the analyses to a vial for analysis. More recent instruments combine the supercritical fluid extraction system with a variety of columns and detectors to acquire data from complex samples.

### 25.2.5   Separation and Identification

During extraction processes, many undesirable compounds are also released from the sample matrix; these must be removed to obtain quantitative results from certain

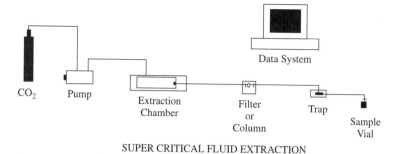

SUPER CRITICAL FLUID EXTRACTION

**Figure 25.1**  Supercritical fluid extraction.

instruments. These components include plant and animal pigments, lipids, organic material from soil and water, and inorganic compounds. If not removed, the impurities decrease the sensitivity of the detectors and columns in the analytical instrument, mask peaks, or produce extraneous peaks on chromatograms. Although some more recently developed instruments automatically remove these substances and concentrate the samples to small volumes for quantitative analysis, they are expensive. Thus most laboratories rely on other methods. These include adsorption chromatography, thin-layer chromatography (TLC), and solvent partitioning. Generally, adsorption chromatography is the method of choice to remove co-extractives from the compound in question.

Because most techniques use large volumes of solvent, the solvent must be removed to obtain a working volume (e.g., 5–10 mL) that is easy to manipulate by the analyst. This is accomplished by distillation, evaporation under a stream of air or an inert gas such as nitrogen, or evaporation under reduced pressure. Once the working volume is reached, extracts can be further purified by one or more procedures. In addition to the use of adsorbents, many organic toxicants will distribute between two immiscible solvents (e.g., chloroform and water or hexane and acetonitrile). When shaken in a separatory funnel and then allowed to equilibrate into two original solvent layers, some of the toxicant will have transferred from the original extracting solvent into the other layer. With repeated additions (e.g., 4 to 5 volumes), mixing, and removal, most or all of the compound of interest will have been transferred, leaving many interfering compounds in the original solvent. Regardless of the separation method or combination of methods used, the toxicant will be in a large volume of solvent in relation to its amount that is removed as described. Final volumes used to identify and quantitate compounds generally range from 250 μL to 10.0 mL.

Recent advances in circuit miniaturization and column technology, the development of microprocessors and new concepts in instrument design have allowed sensitive measurement at the parts per billion and parts per trillion levels for many toxicants. This increased sensitivity has focused public attention on the extent of environmental pollution, because many toxic materials present in minute quantities could not be detected until technological advances reached the present state of the art. At present, most pollutants are identified and quantified by chromatography, spectroscopy, and bioassays.

Once the toxicant has been extracted and separated from extraneous materials, the actual identification procedure can begin, although it should be remembered that the purification procedures are themselves often used in identification (e.g., peak position in gas-liquid chromatography [GLC] and high-performance liquid chromatography [HPLC]). Thus no definite line can be drawn between the two procedures.

***Chromatography.*** All chromatographic processes, such as TLC, GLC, HPLC, or capillary electrophoresis (CE), use a mobile and immobile phase to effect a separation of components. In TLC, the immobile phase is a thin layer of adsorbent placed on glass, resistant plastic, or fiberglass, and the mobile phase is the solvent. The mobile phase can be a liquid or gas, whereas the immobile phase can be a liquid or solid. Chromatographic separations are based on the interactions of these phases or surfaces. All chromatographic procedures use the differential distribution or partitioning of one or more components between the phases, based on the absorption, adsorption, ion-exchange, or size exclusion properties of one of the phases.

***Paper Chromatography.*** When the introduction of paper chromatography to common laboratory use occurred in the mid-1930s, it revolutionized experimental biochemistry and toxicology. This technique is still used in laboratories that lack the expensive instruments necessary for GLC or HPLC. The stationary phase is represented by the aqueous constituent of the solvent system, which is adsorbed onto the paper; the moving phase is the organic constituents. Separation is effected by partition between the two phases as the solvent system moves over the paper. Although many variations exist, including reverse-phase paper chromatography in which the paper is treated with a hydrophobic material, ion-exchange cellulose paper, and so on, all have been superseded by equivalent systems involving thin layers of adsorbents bonded to an inert backing.

***Thin-Layer Chromatography.*** Many toxicants and their metabolites can be separated from interfering substances with TLC. In this form of chromatography, the adsorbent is spread as a thin layer (250–2000 μm) on glass, resistant plastic or fiberglass backings. When the extract is placed near the bottom of the plate and the plate is placed in a tank containing a solvent system, the solvent migrates up the plate, and the toxicant and other constituent move with the solvent; differential rates of movement result in separation. The compounds can be scraped from the plate and eluted from the adsorbent with suitable solvents. Recent developments in TLC adsorbents allow toxicants and other materials to be quantitated at the nanogram ($10^{-9}$ g) and picogram ($10^{-12}$ g) levels.

***Column: Adsorption, Hydrophobic, Ion Exchange.*** A large number of adsorbents are available to the analyst. The adsorbent can be activated charcoal, aluminum oxide, Florisil, silica, silicic acid, or mixed adsorbents. The characteristics of the toxicant determine the choice of adsorbent. When choosing an adsorbent, select conditions that either bind the co-extractives to it, allowing the compound of interest to elute, and vice versa. The efficiency of separation depends on the flow rate of solvent through the column (cartridge) and the capacity of the adsorbent to handle the extract placed on it. This amount depends on the type and quantity of adsorbent, the capacity factor ($k'$) and concentration of sample components, and the type and strength of the solvents used to elute the compound of interest. Many environmental samples contain a sufficient amount of interfering materials so that the analyst must prepare a column using a glass chromatography tube into which the adsorbent is added. In the most common sequence the column is packed in an organic solvent of low polarity, the sample is added in the same solvent, and the column is then developed with a sequence of solvents or solvent mixtures of increasing polarity. Such a sequence might include (in order of increasing polarity) hexane, benzene, chloroform, acetone, and methanol. Once removed, the eluate containing the toxicant is reduced to a small volume for quantitation.

However, cartridge technologies are improving to allow similar concentrations of sample to be added that result in a less expensive and more rapid analysis. A number of miniaturized columns have been introduced since the early 1980s. Most contain 0.5 to 2.0 g of the adsorbent in a plastic tube with fitted ends. The columns can be attached to standard Luer Lock syringes. Other companies have designed vacuum manifolds that hold the collecting device. The column is placed on the apparatus, a vacuum is applied, and the solvent is drawn through the column. Some advantages of these systems include preweighed amounts of adsorbent for uniformity, easy disposal of the co-extractives remaining in the cartridge, no breakage and decreased cost of the analysis because less solvent and adsorbent are used. Other forms of column chromatography can be used. They include ion-exchange chromatography, permeation chromatography, and affinity chromatography. Ion-exchange chromatography depends on the attraction between charged molecules and opposite charges on the ion exchanger, usually a resin. Compounds so bound are eluted by changes in pH and, because the net charge depends on the relationship between pH of the solution and the isoelectric point of the compounds, compounds of different isoelectric point can be eluted sequentially. Both ionic and anionic exchangers are available. Permeation chromatography utilizes the molecular sieve properties of porous materials. Molecules large enough to be excluded from the pores of the porous material will move through the column faster than will smaller molecules not excluded, thus separating them. Cross-linked dextrans such as Sephadex or agarose (Sepharose) are commonly used materials. Affinity chromatography is a potent tool for biologically active macromolecules but is seldom used for purifying small molecules, such as most toxicants. It depends on the affinity of an enzyme for a substrate (or substrate analogue) that has been incorporated into a column matrix or the affinity of a receptor for a ligand.

### Gas-Liquid Chromatography (GLC).

GLC is used most commonly for the separation and quantitation of organic toxicants. This system consists of an injector port, oven, detector, amplifier (electrometer), and supporting electronics (Figure 25.2). Current modern gas chromatographs use a capillary column to effect separation of complex mixtures of organic molecules and has replaced, to a large extent, the "packed" column. Instead of coating an inert support, the stationary phase is coated onto the inside of the column. The mobile phase is an inert gas (called the carrier gas), usually helium or nitrogen that passes through the column.

When a sample is injected, the injector port is at a temperature sufficient to vaporize the sample components. Based on the solubility and volatility of these components with respect to the stationary phase, the components separate and are swept through the column by the carrier gas to a detector, which responds to the concentration of each component. The detector might not respond to all components. The electronic signal produced as the component passes through the detector is amplified by the electrometer, and the resulting signal is sent to a recorder, computer, or electronic data-collecting device for quantitation.

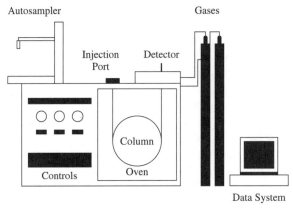

**Figure 25.2**   Gas-liquid chromatograph.

*Column Technology.* Increased sensitivity and component resolution have resulted from advances in solid-state electronics and column and detector technologies. In the field of column technology, the capillary column has revolutionized toxicant detection in complex samples. This column generally is made of fused silica 5 to 60 m in length with a very narrow inner diameter (0.23–0.75 mm) to which a thin layer (e.g., 1.0 μm) of polymer is bonded. The polymer acts as the immobile or stationary phase. The carrier gas flows through the column at flow rates of 1 to 2 ml/min.

Two types of capillary columns are used: the support-coated, open tubular (SCOT) column and the wall-coated, open tubular (WCOT) column. The SCOT column has a very fine layer of diatomaceous earth coated with liquid phase, that is deposited on the inside wall. The WCOT column is pretreated and then coated with a thin film of liquid phase. Of the two columns, the SCOT is claimed to be more universally applicable because of large sample capacity, simplicity in connecting it to the chromatograph, and lower cost. However, for difficult separations or highly complex mixtures, the WCOT is more efficient and is used to a much greater extent. Many older chromatographs are not designed to accommodate capillary columns, and because of these design restrictions, manufacturers offer the wide-bore capillary column along with the fittings and valving required to adapt the columns to older instruments. These columns also can be used on current instruments. With inner diameters of 0.55 to 0.75 mm, flow rates of 5.0 to 10.0 ml/min of carrier gas can be used to affect separations of components approaching that of the narrow-bore columns. Water samples chromatographed on capillary columns routinely separate 400 to 500 compounds, as compared with 90 to 120 resolved compounds from the packed column.

*Detector Technology.* The second advance in GLC is detector technology. Five detectors are used widely in toxicant detection: the flame ionization (FID), flame photometric (FPD), electron capture (ECD), conductivity, and nitrogen-phosphorous detectors. Other detectors have application to toxicant analysis and include the Hall conductivity detector and the photoionization detector.

The FID operates on the principle of ion formation from compounds being burned in a hydrogen flame as they elute from a column. The concentrations of ions formed are several orders of magnitude greater than those formed in the uncontaminated flame. The ions cause a current to flow between two electrodes held at a constant potential, thus sending a signal to the electrometer.

The FPD is a specific detector in that it detects either phosphorous- or sulfur-containing compounds. When atoms of a given element are burned in a hydrogen-rich flame, the excitation energy supplied to these atoms produces a unique emission spectrum. The intensity of the wavelengths of light emitted by these atoms is directly proportional to the number of atoms excited. Larger concentrations cause a greater number of atoms to reach the excitation energy level, thus increasing the intensity of the emission spectrum. The change in intensity is detected by a photomultiplier, amplified by the electrometer, and recorded. Filters that allow only the emission wavelength of phosphorous (526 nm) or sulfur (394 nm) are inserted between the flame and the photomultiplier to give this detector its specificity.

The ECD is used to detect halogen-containing compounds, although it will produce a response to any electronegative compound. When a negative DC voltage is applied to a radioactive source (e.g., $^{63}Ni$, $^{3}H$), low-energy $\beta$ particles are emitted, producing secondary electrons by ionizing the carrier gas as it passes through the detector. The secondary electron stream flows from the source (cathode) to a collector (anode), where the amount of current generated (called a standing current) is amplified and recorded. As electronegative compounds pass from the column into the detector, electrons are removed or "captured," and the standing current is reduced. The reduction is related to both the concentration and electronegativity of the compound passing through, and this produces a response that is recorded. The sensitivity of ECD is greater than that of any other detectors currently available.

Early electrolytic conductivity detectors operated on the principle of component combustion, which produced simple molecular species that readily ionized, thus altering the conductivity of deionized water. The changes were monitored by a dc bridge circuit and recorded. By varying the conditions, the detector could be made selective for different types of compounds (e.g., chlorine containing, nitrogen containing).

The alkali flame detector can also be made selective. Enhanced response to compounds containing arsenic, boron, halogen, nitrogen and phosphorous results when the collector (cathode) of an FID is coated with different alkali metal salts such as KBr, KCl, $Na_2SO_4$. As with conductivity detectors, by varying gas flow rates, types of salt, and electrode configuration, enhanced responses are obtained. The nitrogen-phosphorous alkali detector is used widely for analysis of herbicides. Alkali salts are embedded in a silica gel matrix and are heated electrically. The detector allows routine use of chlorinated solvents and derivatizing reagents that can be detrimental to other detectors.

The Hall electrolytic conductivity uses advanced designs in the conductivity cell, furnace, and an ac conductivity bridge to detect chlorine, nitrogen, and sulfur-containing compounds at sensitivities of 0.01 ng. It operates on the conductivity principle described previously. Another detector, the photoionization detector, uses an ultraviolet (UV) light source to ionize molecules by absorption of a photon of UV light. The ion formed has an energy greater than the ionization potential of the parent compound, and the formed ions are collected by an electrode. The current, which is proportional to concentration, is amplified and recorded. The detector can measure a number of organic and inorganic compounds in air, biologic fluids, and water. A number of instrument manufacturers have introduced portable GLCs that can be transported for use on field sites.

***High-Performance Liquid Chromatography (HPLC).*** HPLC has become very popular in the field of analytical chemistry for the following reasons: it can be run at ambient temperatures; it is nondestructive to the compounds of interest, which can be collected intact; in many instances, derivatization is not necessary for response; and columns can be loaded with large quantities of the material for detection of low levels.

The instrument consists of a solvent reservoir, gradient-forming device. high-pressure pumping device, injector, column, and detector (Figure 25.3). The principle of operation is very similar to that of GLC except that the mobile phase is a liquid instead of a gas. The composition of the mobile phase and its flow rate effect separations. The columns being developed for HPLC are too numerous to discuss in detail. Most use finely divided packing (3–10 μm in diameter), some have bonded phases and others are packed with alumina or silica. The columns normally are 15 to 25 cm in length, with small diameters. (ca. 4.6 mm number diameter). A high-pressure pump is required to fi the solvent through this type of column. The major detectors presently used for HPLC are UV or fluorescent spectrophotometers or differential refractometers.

***Capillary Electrophoresis (CE).*** A relatively new analytical technique, CE, is receiving considerable attention in the field of toxicology. Its uses appear endless, and methods have been developed to analyze a diversity of compounds, including DNA adducts, drugs, small aromatic compounds, and pesticides. Commercial instruments are available that are composed of an autosampler, high-voltage power supply, two buffer reservoirs, the capillary (approximately 70 cm × 75 μm in diameter) and a detector (Figure 25.4). The versatility of the process lies in the ability to separate compounds of interest by a number of modes, including affinity, charge/mass ratios, chiral compounds, hydrophobicity, and size. The theory of operation is simple. Because the capillary is composed of silica, silanol groups are exposed in the internal surface, which can become ionized as the pH of the eluting buffer is increased. The ionization attracts cations to the silica surface, and when current is applied, these cations migrate toward the cathode, which causes a fluid migration through the capillary. This flow can be adjusted by changing the dielectric strength of the buffer, altering the pH, adjusting the voltage, or changing the viscosity.

Under these conditions both anions and cations are separated in a single separation, with cations eluting first. Neutral molecules (e.g., pesticides) can be separated by adding a detergent (e.g., sodium dodecyl sulfate) to the buffer, forming micelles into which

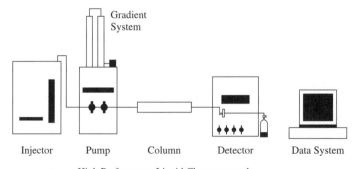

High Performance Liquid Chromatograph

**Figure 25.3** High-performance liquid chromatograph.

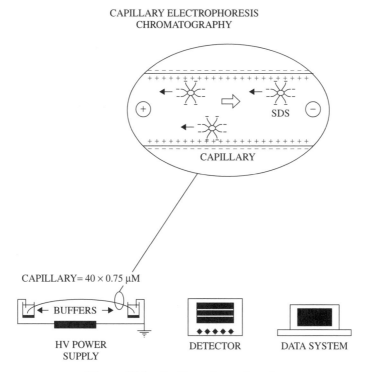

CAPILLARY ELECTROPHORESIS
CHROMATOGRAPHY

**Figure 25.4** Capillary electrophoresis.

neutral molecules will partition based on their hydrophobicity. Because the micelles are attracted to the anode, they move toward the cathode at a slower rate than does the remainder of fluid in the capillary, thus allowing separation. This process is called micellar electrokinetic capillary chromatography (MECK) (Figure 25.4). Many of these analyses can be carried out in 5 to 10 minutes with sensitivities in the low parts per billion (ppb) range. A UV detector is usually used, but greatly sensitivities can be obtained using fluorescent laser detectors.

## 25.2.6  Spectroscopy

In certain experiments involving radiation, observed results cannot be explained on the basis of the wave theory of radiation. It must be assumed that radiation comes in discrete units, called quanta. Each quantum of energy has a definite frequency, $v$, and the quantum energy can be calculated by the equation $E = hv$, where $h$ *is* Planck's constant $(6.6 \times 10^{-27}$ erg-s). Matter absorbs radiation one quantum at a time, and the energy of radiation absorbed becomes greater as either the frequency of radiation increases or the wavelength decreases. Therefore radiation of shorter wavelength causes more drastic changes in a molecule than does that of longer wavelength. Spectroscopy is concerned with the changes in atoms and molecules when electromagnetic radiation is absorbed or emitted. Instruments have been designed to detect these changes, and these instruments are important to the field of toxicant analysis. Discussions of atomic absorption (AA) spectroscopy, mass spectroscopy (MS), infrared (IR), and UV spectroscopy follow. A summary of spectroscopic techniques is given in Table 25.2.

**Table 25.2    Characteristics of Spectroscopic Techniques**

*Visible and UV spectrometry*

Principle:  Energy transitions of bonding and nonbonding outerelectrons of molecules, usually delocalized electrons.

Use:    Routine qualitative and quantitative biochemical analysis including many colorimetric assays. Enzyme assays, kinetic studies, and difference spectra.

*Spectrofluorimetry*

Principle:  Absorbed radiation emitted at longer wavelengths.

Use:    Routine quantitative analysis, enzyme analysis and kinetics. More sensitive at lower concentrations than visible and UV absorption.

*Infrared and Raman spectroscopy*

Principle:  Atomic vibrations involving a change in dipole moment and a change in polarizability, respectively.

Use:    Qualitative analysis and fingerprinting of purified molecules of intermediate size.

*Flame spectrophotometry (emission and absorption)*

Principle:  Energy transitions of outer electrons of atoms after volatilization in a flame.

Use:    Qualitative and quantitative analysis of metals; emission techniques; routine determination of alkali metals; absorption technique extends range of metals that may be determined and the sensitivity.

*Electron spin resonance*

Principle:  Detection of magnetic moment associated with unpaired electrons.

Use:    Research on metalloproteins, particularly enzymes and changes in the environment of free radicals introduced into biological structures (e.g., membranes).

*Nuclear magnetic resonance*

Principle:  Detection of magnetic moment associated with an odd number of protons in an atomic nucleus.

Use:    Determination of structure of organic molecules of molecular weight < 20,000 daltons.

*Mass spectrometry*

Principle:  Determination of the abundance of positively ionized molecules and fragments.

Use:    Qualitative analysis of small quantities of material ($10^{-6}$–$10^{-9}$g), particularly in conjunction with gas-liquid chromatography, HPLC and ICP.

*Source*: Modified from B. W. Williams and K. Wilson, *Principles and Techniques of Practical Biochemistry*, London: Edward Arnold, 1975.

***Atomic Absorption Spectroscopy.*** One of the more sensitive instruments used to detect metal-containing toxicants is the AA spectrophotometer. Samples are vaporized either by aspiration into an acetylene flame or by carbon rod atomization in a graphite cup or tube (flameless AA). The atomic vapor formed contains free atoms of an element in their ground state, and when illuminated by a light source that radiates light of a frequency characteristic of that element, the atom absorbs a photon of wavelength corresponding to its AA spectrum, thus exciting it. The amount of absorption is a function of concentration. The flameless instruments are much more sensitive than conventional flame AA. For example, arsenic can be detected at levels of 0.1 ng/mL and selenium at 0.2 mg/mL, which represent sensitivity three orders of magnitude greater than that of conventional flame AA.

***Induced Coupled Plasma Spectrometry (ICP).*** An even more sensitive instrument has been developed to detect and quantitate, simultaneously, all inorganic species contained with a sample matrix. One such system is the ICP-OES (optical emission spectrometer) (Figure 25.5). The ICP-OES takes an aliquot of sample that has been acid digested and mixes it with a gas (e.g., argon) forming a plasma (i.e., an ionized gas) that is channeled into a nebulizer. Energy is applied to excite the atoms that are converted by the optics of the instrument into individual wavelengths. The

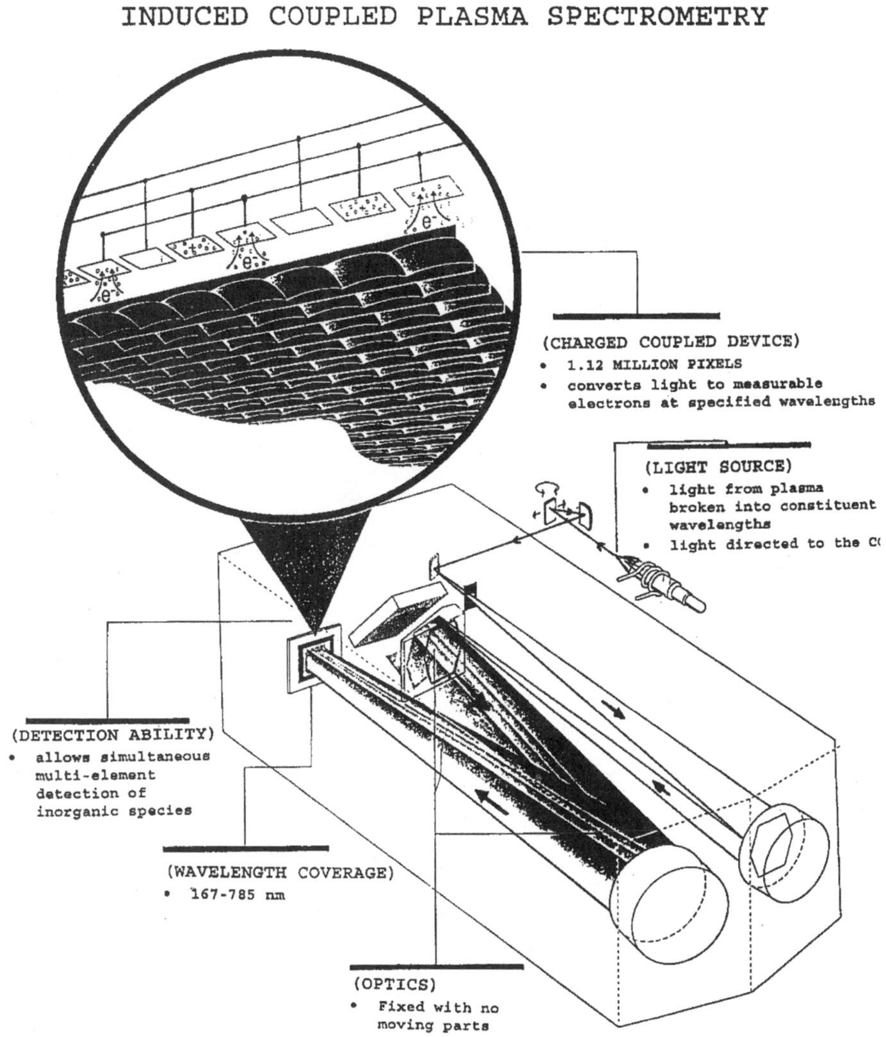

## INDUCED COUPLED PLASMA SPECTROMETRY

(CHARGED COUPLED DEVICE)
• 1.12 MILLION PIXELS
• converts light to measurable electrons at specified wavelengths

(LIGHT SOURCE)
• light from plasma broken into constituent wavelengths
• light directed to the C(

(DETECTION ABILITY)
• allows simultaneous multi-element detection of inorganic species

(WAVELENGTH COVERAGE)
• 167-785 nm

(OPTICS)
• Fixed with no moving parts

**Figure 25.5**  Induced coupled plasma spectrometry.

spectra are captured by a charged coupled device (CCD) that converts the light to measurable electrons at specific wavelengths. Wavelength coverage ranges from 175 to 785 nm. In addition, other instruments couple the ICP to a mass spectrometer (ICP-MS) to collect information on the analyte being sought within the sample matrix. These instruments utilize high throughput of samples and are used in both research and industrial settings.

***Mass Spectroscopy (MS).*** The mass spectrometer is an outstanding instrument for the identification of compounds (Figure 25.6). In toxicant analysis, MS is widely used as a highly sensitive detection method for GLC and is increasingly used with HPLC, CE, and ICP because these instruments can be interfaced to the mass spectrometer. Chromatographic techniques (e.g., GLC, CE, HPLC) are used to separate individual components as previously described. A portion of the column effluent passes into the mass spectrometer, where it is bombarded by an electron beam. Electrons or negative groups are removed by this process, and the ions produced are accelerated. After acceleration they pass through a magnetic field, where the ion species are separated by the different curvatures of their paths under gravity. The resulting pattern is characteristic of the molecule under study. Two detectors are used primarily in pollutant analysis: the quadripole and the ion trap. Both produce reliable and reproducible data, and if routine maintenance is performed, both are reliable. Computer libraries of mass spectral data continue to expand, and data are generated rapidly with current software. Instrument costs have gone down, and tabletop instruments can be purchased for $70,000 although

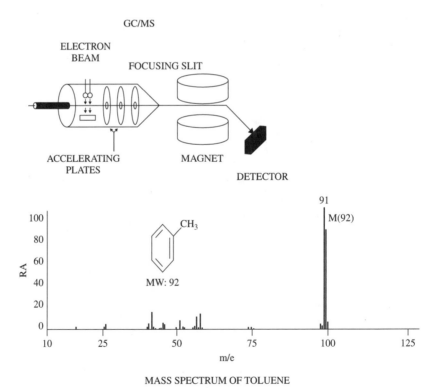

**Figure 25.6** Mass spectroscopy.

research-grade instruments can cost several hundred thousand dollars. By interfacing the detector with a computer system, data reduction, analysis, and quantitation are performed automatically.

***Bimolecular Interaction Analysis–Mass Spectrometry (BIA-MS).*** An exciting new field that is utilizing mass spectrometry as a tool in biological and toxicological research to investigate protein interactions is that of proteomics. This rapidly expanding science explores proteins within the cellular environment, their various forms, interacting partners (e.g., cofactors), and those processes that affect their regulation and processing. The BIA-MS can determine such things as the kinetics of protein interactions, selectively retrieve and concentrate specific proteins from biological media, quantitate target proteins, identify protein : ligand interactions, and recognize protein variants (e.g., point mutations). BIA-MS uses two technologies, surface plasmon resonance (SPR) sensing and matrix assisted laser desorption ionization time-of-flight mass spectrometry (MALDI-TOF MS) (Figure 25.7). Cells are fragmented and come in contact with a gold-plated glass slide, called a chip. The chip has highly defined sites containing a number of immobilized ligands to which the proteins of interest bind and are quantitated by SPR that monitors the interaction and quantifies the amount of protein localized at precise locations on the surface of the chip. The chip is then subjected to MALDI-TOF MS, which yields the masses of retained analytes and other bound biomolecules.

***Infrared Spectrophotometry (IR).*** Atoms are in constant motion within molecules, and associated with these motions are molecular energy levels that correspond to the energies of quanta of IR radiation. These motions can be resolved into rotation of the whole molecule in space and into motions corresponding to the vibration of atoms with

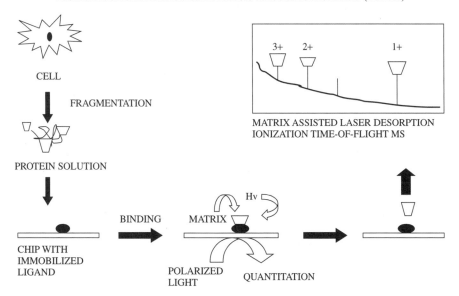

**Figure 25.7** Bimolecular interaction analysis–mass spectrometry.

respect to one another by bending or stretching covalent bonds. The vibrational motions are very useful in identifying complex molecules, because functional groups (e.g., OH, C, O, SH) within the molecule have characteristic absorption bands. The principle functional groups can be determined and used to identify compounds in cases in which chemical evidence permits relatively few possible structures. Standard IR spectrophotometers cover the spectral range from 2.5 to 15.4 Nm (wave number equivalent to $4000-650$ cm$^{-1}$) and use a source of radiation that passes through the sample and reference cells into a monochromator (a device to isolate spectral regions). The radiation is then collected, amplified, and recorded. Current instruments use microprocessors, allowing a number of refinements that have increased the versatility of IR instruments so that more precise qualitative and quantitative data can be obtained.

*Ultraviolet/Visible–Spectrophotometry (UV/VIS).* Transitions occur between electronic levels of molecules producing absorptions and emissions in the visible (VIS) and UV portions of the electromagnetic spectrum. Many inorganic and organic molecules show maximum absorption at specific wavelengths in the UV/VIS range, and these can be used to identify and quantitate compounds. Instruments designed to measure absorbance in the UV/VIS portions of the spectrum (190–700 nm) have been used in many specific purposes, such as detectors in HPLC and CE. These detectors use small flow cells having short path lengths (approximately 10 mm) and hold small volumes (e.g., 10.0 μL) through which light at a specific wavelength passes. Basic spectrophotometers have the same components as the IR instruments described previously, including a source (usually a deuterium lamp) monochromator, beam splitter, sampler and reference cells, and detector.

*Nuclear Magnetic Resonance (NMR).* Nuclear magnetic resonance (NMR) detects atoms that have nuclei and possess a magnetic moment. These are usually atoms containing nuclei with an odd number of protons (charges). Such nuclei can exist in two states: a low-energy state with the nuclear spin aligned parallel to the magnetic field and a high-energy state with the spin perpendicular to the field. Basically the instrument measures the absorption or radiowave necessary to change the nuclei from a low- to a high-energy state as the magnetic field is varied. It is used most commonly for hydrogen atoms, although $^{13}$C and $^{31}$P are also suitable. Because the field seen by a proton varies with its molecular environment, such molecular arrangements as $CH_3$, $CH_2$, and CH give different signs, providing much information about the structure of the molecule in question.

### 25.2.7 Other Analytical Methods

The instruments discussed earlier are the primary ones used in toxicant analysis, but an enormous number of analytical techniques are used in the field. Many of the instruments are expensive (e.g., Raman spectrometers, X-ray emission spectrometers) and few laboratories possess them. Many other instruments are available, however, such as the specific-ion electrode, which is both sensitive and portable. Specific-ion electrodes have many other advantages in that sample color, suspended matter, turbidity, and viscosity do not interfere with analysis; therefore many of the sample preparation steps are not required. Some of the species that can be detected at ppb levels are ammonia, carbon dioxide, chloride, cyanide, fluoride, lead, potassium, sulfide, and urea. Analytical pH meters or meters designed specifically for this application are used to calculate concentrations.

Finally an increasing number of portable and direct reading instruments are now available to detect and quantitate environmental pollutants. Most of these measure airborne particulates and dissolved molecules and operate on such diverse principles as aerosol photometry, chemiluminescence, combustion, and polarography. Elemental analyzers have been developed for carbon, nitrogen, and sulfur using IR, chemiluminescence, and fluorescence, respectively. Analyses can be completed in about 1 minute if the samples are gases, liquids, or small solids, and within 10 minutes if solid samples are larger. These devices are microprocessor controlled, contain built-in printers, and are used to analyze materials including gasolines, pesticides, protein solutions, and wastewater.

## SUGGESTED READING

Hayes, A. W., ed. *Principles and Methods of Toxicology*. London: Taylor and Francis, 2000.

Morel, F. M. M., and J. G. Heirig. *Principles and Applications of Aquatic Chemistry*. New York: Wiley-Interscience, 1993.

Nelson, A. H. *Organic Chemicals in the Aquatic Environment*. Boca Raton, FL: CRC Press, 1994.

Pickering, W. F. *Pollution Evaluation: The Quantitative Aspects*. New York: Dekker 1977.

Rand, G. M., and S. N. Petrocelli. *Fundamentals of Aquatic Toxicology*. Washington, DC: Hemisphere, 1985.

Rouessae, F., A. Rouessae, and M. B. Waldron. *Chemical Analysis: Modern Instrumentation, Methods and Techniques*. New York: Wiley-Interscience, 2000.

Thomas, J. J., P. B. Bond, and I. Sunshine, eds. *Guidelines for Analytical Toxicology Programs*, vols. 1 and 2. Cleveland: CRC Press, 1977.

Wagner, R. E. ed. *Guide to Environmental Analytical Methods*, 4th ed. Scheenectady, NY: Genium Publishing, 1998.

Ware, G. W., ed. *Reviews of Environmental Contamination and Toxicology*. New York: Springer-Verlag. (This excellent series of review articles is approaching Volume 180 and covers all aspects of toxicology.)

Williams, P. L., R. C. Jones and S. M. Roberts, eds. *The Principles of Toxicology: Environmental and Industrial Applications*, 2nd ed. New York: Wiley-Interscience, 2000.

Zweig, G. ed. *Analytical Methods for Pesticides and Plant Growth Regulators*. New York: Academic Press. (This was a multi-volume series appearing between 1973 and 1989 that contains analytical methods for the analysis of food and food additives, fungicides, herbicides, nematicides, pheromones, rodenticides, and soil fumigants.)

### Web Sites

Instrument manufacturers all have detailed Web sites containing considerable information, not only on their equipment but on theory of operations, methods to maximize sensitivity, etc.

The following are some government Web sites that can be searched for analytical methods:

http://www.epa.gov/pesticides/ (A number of links to US EPA analytical methods)

http://www.nal.usda.gov/

http://npic.orst.edu/

# CHAPTER 26

# Basics of Environmental Toxicology

GERALD A. LEBLANC

## 26.1 INTRODUCTION

Industrial and agricultural endeavors are intimately associated with the extensive use of a wide array of chemicals. Historically chemical wastes generated through industrial processes were disposed of through flagrant release into the environment. Gasses quickly dispersed into the atmosphere; liquids were diluted into receiving waters and efficiently transported away from the site of generation. Similarly pesticides and other agricultural chemicals revolutionized farm and forest productivity. Potential adverse effects of the application of such chemicals to the environment were viewed as insignificant relative to the benefits bestowed by such practices. Then in 1962, a science writer for the US Fish and Wildlife Service, Rachel Carson, published a book that began by describing a world devoid of birds and from which the title *The Silent Spring* was inspired. In her book Ms. Carson graphically described incidents of massive fish and bird kills resulting from insecticide use in areas ranging from private residences to national forests. Further she inferred that such pollutant effects on wildlife may be heralding similar incipient effects on human health.

The resulting awakening of the general public to the hazards of chemicals in the environment spurred several landmark activities related to environmental protection, including Earth Day, organization of the US Environmental Protection Agency, and the enactment of several pieces of legislation aimed at regulating and limiting the release of chemicals into the environment. Appropriate regulation of the release of chemicals into the environment without applying unnecessarily stringent limitation on industry and agriculture requires a comprehensive understanding of the toxicological properties and consequences of release of the chemicals into the environment. It was from this need that modern environmental toxicology evolved.

Environmental toxicology is defined as the study of the fate and effects of chemicals in the environment. Although this definition would encompass toxic chemicals naturally found in the environment (i.e., animal venom, microbial and plant toxins), environmental toxicology is typically associated with the study of environmental chemicals of anthropogenic origin. Environmental toxicology can be divided into two subcategories:

*A Textbook of Modern Toxicology, Third Edition,* edited by Ernest Hodgson
ISBN 0-471-26508-X  Copyright © 2004 John Wiley & Sons, Inc.

environmental health toxicology and ecotoxicology. Environmental health toxicology is the study of the adverse effects of environmental chemicals on human health, while ecotoxicology focuses upon the effects of environmental contaminants upon ecosystems and constituents thereof (fish, wildlife, etc.). Assessing the toxic effects of chemicals on humans involves the use of standard animal models (i.e., mouse and rat) as well as epidemiological evaluations of exposed human populations (i.e., farmers and factory workers). In contrast, ecotoxicology involves the study of the adverse effects of toxicants on myriad of organisms that compose ecosystems ranging from microorganisms to top predators. Further, comprehensive insight into the effects of chemicals in the environment requires assessments ancillary to toxicology such as the fate of the chemical in the environment (Chapter 27), and toxicant interactions with abiotic (nonliving) components of ecosystems. Comprehensive assessments of the adverse effects of environmental chemicals thus utilize expertise from many scientific disciplines. The ultimate goal of these assessments is elucidating the adverse effects of chemicals that are present in the environment (retrospective hazard assessment) and predicting any adverse effects of chemicals before they are discharged into the environment (prospective hazard assessment). The ecological hazard assessment process is discussed in Chapter 28.

Historically chemicals that have posed major environmental hazards tend to share three insidious characteristics: environmental persistence, the propensity to accumulate in living things, and high toxicity.

## 26.2 ENVIRONMENTAL PERSISTENCE

Many abiotic and biotic processes exist in nature that function in concert to eliminate (i.e., degrade) toxic chemicals. Accordingly many chemicals released into the environment pose minimal hazard simply because of their limited life span in the environment. Chemicals that have historically posed environmental hazard (i.e., DDT, PCBs, TCDD) resist degradative processes and accordingly persist in the environment for extremely long periods of time (Table 26.1). Continued disposal of persistent chemicals into the environment can result in their accumulation to environmental levels sufficient to pose toxicity. Such chemicals can continue to pose hazard long after their disposal into the environment has ceased. For example, significant contamination of Lake Ontario by the pesticide mirex occurred from the 1950s through the 1970s. Mass balance studies performed 20 years later revealed that 80% of the mirex deposited into the lake

**Table 26.1   Environmental Half-lives of Some Chemical Contaminants**

| Contaminant | Half-life | Media |
|---|---|---|
| DDT | 10 Years | Soil |
| TCDD | 9 Years | Soil |
| Atrazine | 25 Months | Water |
| Benzoperylene (PAH) | 14 Months | Soil |
| Phenanthrene (PAH) | 138 Days | Soil |
| Carbofuran | 45 Days | Water |

persisted. One decade following the contamination of Lake Apopka, Florida, with pesticides including DDT and diclofol, populations of alligators continued to experience severe reproductive impairment. Both biotic and abiotic processes contribute to the degradation of chemicals.

### 26.2.1  Abiotic Degradation

A plethora of environmental forces compromise the structural integrity of chemicals in the environment. Many prominent abiotic degradative processes occur due to the influences of light (photolysis) and water (hydrolysis).

*Photolysis.* Light, primarily in the ultraviolet range, has the potential to break chemical bonds and thus can contribute significantly to the degradation of some chemicals. Photolysis is most likely to occur in the atmosphere or surface waters where light intensity is greatest. Photolysis is dependent upon both the intensity of the light and the capacity of the pollutant molecules to absorb the light. Unsaturated aromatic compounds such as the polycyclic aromatic hydrocarbons tend to be highly susceptible to photolysis due to their high capacity to absorb light energy. Light energy can also facilitate the oxygenation of environmental contaminants via hydrolytic or oxidative processes. The photooxidation of the organophosphorus pesticide parathion is depicted in Figure 26.1.

*Hydrolysis.* Water, often in combination with light energy or heat, can break chemical bonds. Hydrolytic reactions commonly result in the insertion of an oxygen atom into the molecule with the commensurate loss of some component of the molecule. Ester bonds, such as those found in organophosphate pesticides (i.e., parathion; Figure 26.1), are highly susceptible to hydrolysis which dramatically lowers the environmental half-lives of these chemicals. Hydrolytic rates of chemicals are influenced by the temperature and pH of the aqueous media. Rates of hydrolysis increase with increasing temperature and with extremes in pH.

### 26.2.2  Biotic Degradation

While many environmental contaminants are susceptible to abiotic degradative processes, such processes often occur at extremely slow rates. Environmental degradation of chemical contaminants can occur at greatly accelerated rates through the action of microorganisms. Microorganisms (primarily bacteria and fungi) degrade chemicals in an effort to derive energy from these sources. These biotic degradative processes are enzyme mediated and typically occur at rates that far exceed abiotic degradation. Biotic degradative processes can lead to complete mineralization of chemicals to water, carbon dioxide, and basic inorganic constituents. Biotic degradation includes those processes associated with abiotic degradation (i.e., hydrolysis, oxidation) and processes such as the removal of chlorine atoms (dehalogenation), the scission of ringed structures (ring cleavage), and the removal of carbon chains (dealkylation). The process by which microorganisms are used to facilitate the removal of environmental contaminants is called bioremediation.

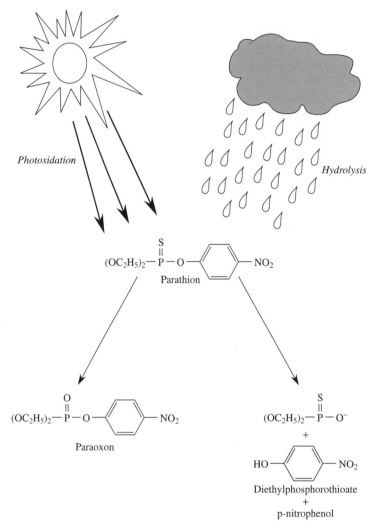

**Figure 26.1** The effect of sunlight (photooxidation) and precipitation (hydrolysis) on the degradation of parathion.

### 26.2.3 Nondegradative Elimination Processes

Many processes are operative in the environment that contribute to the regional elimination of a contaminant by altering its distribution. Contaminants with sufficiently high vapor pressure can evaporate from contaminated terrestrial or aquatic compartments and be transferred through the atmosphere to new locations. Such processes of global distillation are considered largely responsible for the worldwide distribution of relatively volatile organochlorine pesticides such as lindane and hexachlorobenzene. Entrainment by wind and upper atmospheric currents of contaminant particles or dust onto which the contaminants are sorbed also contribute to contaminant redistribution. Sorption of contaminant to suspended solids in an aquatic environment with commensurate sedimentation can result with the removal of contaminants from the water column and its redistribution into bottom sediments. Sediment sorption of contaminants greatly reduces bioavailability, since the propensity of a lipophilic chemical to partition from sediments to organisms is significantly less than its propensity to partition from water to organism. More highly water soluble contaminants can be removed and redistributed through runoff and soil percolation. For example, the herbicide atrazine is one of the most abundantly used pesticides in the United States. It is used to control broadleaf and weed grasses in both agriculture and landscaping. Atrazine is ubiquitous in surface waters due to its extensive use. A study of midwestern states revealed that atrazine was detectable in 92% of the reservoirs assayed. In addition atrazine has the propensity to migrate into groundwater because of its relatively high water solubility and low predilection to sorb to soil particles. Indeed, field studies have shown that surface application of atrazine typically results in the contamination of the aquifer below the application site. A more detailed account of the fate of chemicals in the environment is presented in Chapter 27.

## 26.3 BIOACCUMULATION

Environmental persistence alone does not render a chemical problematic in the environment. If the chemical cannot enter the body of organisms, then it would pose no threat of toxicity (see Chapter 6). Once absorbed, the chemical must accumulate in the body to sufficient levels to elicit toxicity. Bioaccumulation is defined as the process by which organisms accumulate chemicals both directly from the abiotic environment (i.e., water, air, soil) and from dietary sources (trophic transfer). Environmental chemicals are largely taken up by organisms by passive diffusion. Primary sites of uptake include membranes of the lungs, gills, and gastrointestinal tract. While integument (skin) and associated structures (scales, feathers, fur, etc.) provide a protective barrier against many environmental insults, significant dermal uptake of some chemicals can occur. Because the chemicals must traverse the lipid bilayer of membranes to enter the body, bioaccumulation potential of chemicals is positively correlated with lipid solubility (lipophilicity).

The aquatic environment is the major site at which lipophilic chemicals traverse the barrier between the abiotic environment and the biota. This is because (1) lakes, rivers, and oceans serve as sinks for these chemicals, and (2) aquatic organisms pass tremendous quantities of water across their respiratory membranes (i.e., gills) allowing for the efficient extraction of the chemicals from the water. Aquatic organisms can bioaccumulate lipophilic chemicals and attain body concentrations that are several orders of magnitude greater that the concentration of the chemical found in the environment (Table 26.2). The degree to which aquatic organisms accumulate xenobiotics from the environment is largely dependent on the lipid content of the organism, since body lipids serve as the primary site of retention of the chemicals (Figure 26.2).

Chemicals can also be transferred along food chains from prey organism to predator (trophic transfer). For highly lipophilic chemicals, this transfer can result in increasing concentrations of the chemical with each progressive link in the food chain (biomagnification). As depicted in Figure 26.3, a chemical that bioaccumulates by a factor of 2 regardless of whether the source of the contaminant is the water or food would have the potential to magnify at each trophic level leading to high levels in the birds of

**Table 26.2  Bioaccumulation of Some Environmental Contaminants by Fish**

| Chemical | Bioaccumulation Factor[a] |
|---|---|
| DDT | 127,000 |
| TCDD | 39,000 |
| Endrin | 6,800 |
| Pentachlorobenzene | 5,000 |
| Lepthophos | 750 |
| Trichlorobenzene | 183 |

*Source*: Data derived from G. A. LeBlanc, 1994, *Environ. Sci. Technol.* **28**: 154–160.
[a]Bioaccumulation factor is defined as the ratio of the chemical concentration in the fish and in the water at steady-state equilibrium.

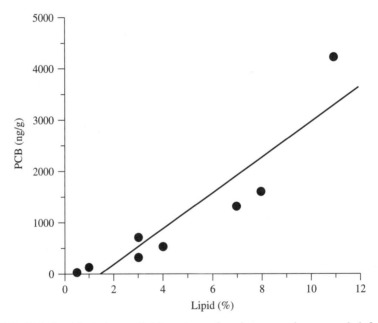

**Figure 26.2**  Relationship between lipid content of various organisms sampled from Lake Ontario and whole body PCB concentration (Data derived from B. G. Oliver and A. J. Niimi, *Environ. Sci. Technol.* **22**: 388–397, 1988.)

prey relative to that found in the abiotic environment. It should be noted that bioaccumulation is typically much greater from water than from food, and it is unlikely that an organism would accumulate a chemical to the same degree from both sources. The food-chain transfer of DDT was responsible for the decline in many bird-eating raptor populations that contributed to the decision to ban the use of this pesticide in the United States.

Bioaccumulation can lead to a delayed onset of toxicity, since the toxicant may be initially sequestered in lipid deposits but is mobilized to target sites of toxicity

BIOACCUMULATION OF ENVIRONMENTAL CHEMICALS

**Figure 26.3** Bioaccumulation of a chemical along a generic food chain. In this simplistic paradigm, the amount of the chemical in the water is assigned an arbitrary concentration of 1, and it is assumed that the chemical will bioaccumulate either from the water to the fish or from one trophic level to another by a factor of 2. Circled numbers represent the concentration of chemical in the respective compartment. Numbers associated with arrows represent the concentration of chemical transferred from one compartment to another.

when these lipid stores are utilized. For example, lipid stores are often mobilized in preparation for reproduction. The loss of the lipid can result in the release of lipophilic toxicants rendering them available for toxic action. Such effects can result in mortality of adult organisms as they approach reproductive maturity. Lipophilic chemicals also can be transferred to offspring in lipids associated with the yolk of oviparous organisms or the milk of mammals, resulting in toxicity to offspring that was not evident in the parental organisms.

## 26.3.1  Factors That Influence Bioaccumulation

The propensity for an environmental contaminant to bioaccumulate is influenced by several factors. The first consideration is environmental persistence. The degree to

**Table 26.3   Measured and Predicted Bioaccumulation Factors in Fish of Chemicals That Differ in Susceptibility to Biotransformation**

| Chemical | Susceptibility to Biotransformation | Bioaccumulation Factor | |
|---|---|---|---|
| | | Predicted | Measured |
| Chlordane | Low | 47,900 | 38,000 |
| PCB | Low | 36,300 | 42,600 |
| Mirex | Low | 21,900 | 18,200 |
| Pentachloro-phenol | High | 4,900 | 780 |
| Tris(2,3-dibromo-propyl)phosphate | High | 4,570 | 3 |

*Source*: Predicted bioaccumulation factors were based upon their relative lipophilicity as described by, D. Mackay, *Environ. Sci. Technol.* 1982, **16**: 274–278.

which a chemical bioaccumulates is dictated by the concentration present in the environment. Contaminants that are readily eliminated from the environment will generally not be available to bioaccumulate. An exception would be instances where the contaminant is continuously introduced into the environment (i.e., receiving water of an effluent discharge).

As discussed above, lipophilicity is a major determinant of the bioaccumulation potential of a chemical. However, lipophilic chemicals also have greater propensity to sorb to sediments, thus rendering them less available to bioaccumulate. For example, sorption of benzo[a]pyrene to humic acids reduced its propensity to bioaccumulate in sunfish by a factor of three. Fish from oligotrophic lakes, having low suspended solid levels, have been shown to accumulate more DDT than fish from eutrophic lakes that have high suspended solid contents.

Once absorbed by the organism, the fate of the contaminant will influence its bioaccumulation. Chemicals that are readily biotransformed (Chapter 7) are rendered more water soluble and less lipid soluble. The biotransformed chemical is thus less likely to be sequestered in lipid compartments and more likely to be eliminated from the body. As depicted in Table 26.3, chemicals that are susceptible to biotransformation, bioaccumulate much less than would be predicted based on lipophilicity. Conjugation of xenobiotics to glutathione and glucuronic acid (Chapter 7) can target the xenobiotic for biliary elimination through active transport processes thus greatly increasing the rate of elimination (Chapter 10). Differences in chemical elimination rates contribute to species differences in bioaccumulation.

## 26.4   TOXICITY

### 26.4.1   Acute Toxicity

Acute toxicity is defined as toxicity elicited as a result of short-term exposure to a toxicant. Incidences of acute toxicity in the environment are commonly associated with accident (i.e., derailment of a train resulting in leakage of a chemical into a river) or imprudent use of the chemical (i.e., aerial drift of a pesticide to nontarget areas). Discharge limits placed upon industrial and municipal wastes, when adhered to, have been generally successful in protecting against acute toxicity to organisms in waste-receiving areas. As discussed in Chapter 11, the acute toxicity of a chemical is commonly quantified as the LC50 or LD50. These measures do not provide any insight

**Table 26.4    Ranking Scheme for Assessing the Acute Toxicity of Chemicals to Fish and Wildlife**

| Fish LC50 (mg/L) | Avian/Mammalian LD50 (mg/kg) | Toxicity Rank | Example Contaminant |
|---|---|---|---|
| >100 | >5000 | Relatively nontoxic | Barium |
| 10–100 | 500–5000 | Moderately toxic | Cadmium |
| 1–10 | 50–500 | Very toxic | 1,4-Dichlorobenzene |
| <1 | <50 | Extremely toxic | Aldrin |

into the environmentally acceptable levels of contaminants (a concentration that kills 50% of the exposed organisms is hardly tolerable). However, LC50 and LD50 values do provide statistically sound, reproducible measures of the relative acute toxicity of chemicals. LC50 and LD50 ranges for aquatic and terrestrial wildlife, respectively, and their interpretation are presented in Table 26.4.

Acute toxicity of environmental chemicals is determined experimentally with select species that serve as representatives of particular levels of trophic organization within an ecosystem (i.e., mammal, bird, fish, invertebrate, vascular plant, algae). For example, the US Environmental Protection Agency requires acute toxicity tests with representatives of at least eight different species of freshwater and marine organisms (16 tests) that include fish, invertebrates, and plants when establishing water quality criteria for a chemical. Attempts are often made to rank classes of organisms by toxicant sensitivity; however, no organism is consistently more or less susceptible to the acute toxicity of chemicals. Further the use of standard species in toxicity assessment presumes that these species are "representative" of the sensitivity of other members of that level of ecological organization. Such presumptions are often incorrect.

### 26.4.2  Mechanisms of Acute Toxicity

Environmental chemicals can elicit acute toxicity by many mechanisms. Provided below are example mechanisms that are particularly relevant to the types of chemicals that are more commonly responsible for acute toxicity in the environment at the present time.

***Cholinesterase Inhibition.*** The inhibition of cholinesterase activity is characteristic of acute toxicity associated with organophosphate and carbamate pesticides (see Chapter 11 for more detail on cholinesterase inhibition). Forty to 80% inhibition of brain cholinesterase activity is typically reported in lethally poisoned fish. Acute toxicity resulting from cholinesterase inhibition is relatively common among incidents of acute poisoning of fish and birds due to the high volume usage of organophosphates and carbamates in applications such as lawn care, agriculture, and golf course maintenance. Cholinesterase inhibition in fish may occur following heavy rains in aquatic habitats adjacent to areas treated with the pesticides and subject to runoff from these areas. Acute toxicity to birds commonly occurs in birds that feed in areas following application of the pesticides.

***Narcosis.*** A common means by which industrial chemicals elicit acute toxicity, particularly to aquatic organisms, is through narcosis. Narcosis occurs when a chemical accumulates in cellular membranes interfering with the normal function of the membranes. Typical responses to the narcosis are decreased activity, reduced reaction to external stimuli, and increased pigmentation (in fish). The effects are reversible, and nonmoribund organisms typically return to normal activity once the chemical is removed from the organism's environment. Prolonged narcosis can result in death. Approximately 60% of industrial chemicals that enter the aquatic environment elicit acute toxicity through narcosis. Chemicals that elicit toxicity via narcosis typically do not elicit toxicity at specific target sites and are sufficiently lipophilic to accumulate in the lipid phase or the lipid-aqueous interface of membranes to sufficient levels to disrupt membrane function. Chemicals that induce narcosis include alcohols, ketones, benzenes, ethers, and aldehydes.

***Physical Effects.*** Perhaps most graphic among recent incidents of environmental acute toxicity is the physical effects of petroleum following oil spills. Slicks of oil on the surface of contaminated waters results in the coating of animals, such as birds and marine mammals, that frequent the air-water interface. Such a spill of unprecedented magnitude and consequence in the United States occurred on March 24, 1989, when the hull of the Exxon Valdez was ruptured on Bligh Reef in Prince William Sound, Alaska. Nearly 11 million gallons of crude oil spilled onto the nearshore waters killing more wildlife than any prior oil spill in history. Thousands of sea birds and mammals succumbed to the acute effects of the oil.

Hypothermia is considered a major cause of death of oiled marine birds and mammals. These organisms insulate themselves from the frigid waters by maintaining a layer of air among the spaces within their coat of fur or feathers. The oil penetrates the fur/feather barrier and purges the insulating air. As a result the animals rapidly succumb to hypothermia. In addition to hypothermia, these animals can also experience oil toxicosis. Inhalation of oil, as well as ingestion through feeding and preening, can result in the accumulation of hydrocarbons to toxic levels. Toxicity to sea otters has been correlated to degree of oiling and is characterized by pulmonary emphysema (bubbles of air within the connective tissue of the lungs), gastric hemorrhages, and liver damage.

### 26.4.3 Chronic Toxicity

Chronic toxicity is defined as toxicity elicited as a result of long-term exposure to a toxicant. Sublethal end points are generally associated with chronic toxicity. These include reproductive, immune, endocrine, and developmental dysfunction. However, chronic exposure also can result in direct mortality not observed during acute exposure. For example, chronic exposure of highly lipophilic chemicals can result in the eventual bioaccumulation of the chemical to concentrations that are lethal to the organisms. Or as discussed previously, mobilization of lipophilic toxicants from lipid compartments during reproduction may result in lethality. It is important to recognize that, while theoretically, all chemicals elicit acute toxicity at a sufficiently high dose, all chemicals are not chronically toxic. Chronic toxicity is measured by end points such as the highest level of the chemical that does not elicit toxicity during continuous, prolonged exposure (no observed effect level, NOEL), the lowest level of the chemical that elicits

**Table 26.5   Acute and Chronic Toxicity of Pesticides Measured from Laboratory Exposures of Fish Species**

| Pesticide | LC50 (μg/L) | Acute Toxicity | Chronic Value (μg/L) | ACR | Chronic Toxicity |
|---|---|---|---|---|---|
| Endosulfan | 166 | Extremely toxic | 4.3 | 39 | Yes |
| Chlordecone | 10 | Extremely toxic | 0.3 | 33 | Yes |
| Malathion | 3,000 | Very toxic | 340 | 8.8 | No |
| Carbaryl | 15,000 | Moderately toxic | 378 | 40 | Yes |

toxicity during continuous, prolonged exposure (lowest observed effect level, LOEL), or the chronic value (CV) which is the geometric mean of the NOEL and the LOEL. Chronic toxicity of a chemical is often judged by the acute : chronic ratio (ACR), which is calculated by dividing the acute LC50 value by the CV. Chemicals that have an ACR of less than 10 typically have low to no chronic toxicity associated with them (Table 26.5).

The following must always be considered when assessing the chronic toxicity of a chemical: (1) Simple numerical interpretations of chronic toxicity based on ACRs serve only as gross indicators of the potential chronic toxicity of the chemical. Laboratory exposures designed to establish chronic values most often focus upon a few general endpoints such as survival, growth, and reproductive capacity. Examination of more subtle end points of chronic toxicity may reveal significantly different chronic values. (2) Laboratory exposures are conducted with a few test species that are amenable to laboratory manipulation. The establishment of chronic and ACR values with these species should not be considered absolute. Toxicants may elicit chronic toxicity in some species and not in others. (3) Interactions among abiotic and biotic components of the environment may contribute to the chronic toxicity of chemicals, while such interactions may not occur in laboratory assessments of direct chemical toxicity. These considerations are exemplified in the following incidence of chronic toxicity of chemicals in the environment.

### 26.4.4   Species-Specific Chronic Toxicity

*Tributyltin-Induced Imposex in Neogastropods.* Scientists noted in the early 1970s that dogwhelks inhabiting the coast of England exhibited a hermaphroditic-like condition whereby females possessed a penis in addition to normal female genitalia. While hermaphroditism is a reproductive strategy utilized by some molluscan species, dogwhelks are dioecious. This pseudohermaphroditic condition, called imposex, has since been documented worldwide in over 140 species of neogastropods. Imposex has been implicated in reduced fecundity of neogastropod populations, population declines, and local extinction of affected populations.

The observation that imposex occurred primarily in marinas suggested causality with some contaminant originating from such facilities. Field experiments demonstrated that neogastropods transferred from pristine sites to marinas often developed imposex. Laboratory studies eventually implicated tributyltin, a biocide used in marine paints, as the cause of imposex. Tributyltin is toxic to most marine species evaluated in the

**Table 26.6    Toxicity of Tributyltin to Aquatic Organisms**

| Species | Acute Toxicity (LC50, $\mu$g/L) | Chronic Toxicity (LOEL, $\mu$g/L) | Imposex ($\mu$g/L) |
|---|---|---|---|
| Daphnid | 1.7 | — | — |
| Polychaete worm | — | 0.10 | — |
| Copepod | 1.0 | 0.023 | — |
| Oyster | 1.3 | 0.25 | — |
| Dogwhelk | — | — | $\leq$0.0010 |

laboratory at low parts-per-billion concentrations (Table 26.6). However, exposure of neogastropods to low parts-per-trillion concentrations can cause imposex (Table 26.6). Thus neogastropods are uniquely sensitive to the toxicity of tributyltin, with effects produced that were not evident in standard laboratory toxicity characterizations.

***Atrazine-Induced Hermaphroditism in Frogs.***  The herbicide atrazine historically has been considered environmentally safe for use since the material has proved to be only slightly to moderately toxic in standard fish and wildlife toxicity evaluations. Measured atrazine levels in surface waters rarely exceed 0.04 mg/L. The acute and chronic toxicities of atrazine to aquatic organisms are typically in excess of 1 mg/L. Thus ample safety margins appear to exist for this compound. Recent studies with frogs have revealed that exposure to 0.0001 mg/L atrazine through the period of larval development caused the frogs to develop both a testis and an ovary. The toxicological significance of this chemical-induced hermaphroditic condition is not known. However, environmentally relevant levels of the herbicide appear to have the potential to adversely impact reproductive success of these organisms.

### 26.4.5  Abiotic and Biotic Interactions

***Chlorofluorocarbons–Ozone–UV-B Radiation–Amphibian Interactions.***  The atmospheric release of chlorofluorocarbons has been implicated in the depletion of the earth's stratospheric ozone layer which serves as a filter against harmful ultraviolet radiation. Temporal increases in UV-B radiation have been documented and pose increasing risks of a variety of maladies to both plant and animal life.

Commensurate with the increase in UV-B radiation levels at the earth's surface has been the decline in many amphibian populations. Multiple causes may be responsible for these declines including loss of habitat, pollutants, and increased incidence of disease; however, recent studies suggest that increases in UV-B radiation may be a major contributor to the decline in some populations. Field surveys in the Cascade Mountains, Oregon, revealed a high incidence of mortality among embryos of the Cascades frog and western toad. Incubation of eggs, collected from the environment, in the laboratory along with the pond water in which the eggs were collected resulted in low mortality, suggesting that contaminants in the water were not directly responsible for the mortality. Furthermore placement of UV-B filters over the embryos, incubated under ambient environmental conditions, significantly increased viability of the embryos.

Several amphibian species were examined for photolyase activity. This enzyme is responsible for the repair of DNA damage caused by UV-B radiation. A more than

80-fold difference in photolyase activity was observed among the species examined. Photolyase activity was appreciably lower in species known to be experiencing population decline as compared to species showing stable population levels. Recent studies have also suggested that ambient UV-B radiation levels can enhance the susceptibility of amphibian embryos to mortality originating from fungal infection.

These observations suggest that chlorofluorocarbons may be contributing to the decline in amphibian populations. However, this toxicological effect is the result of abiotic interactions (i.e., chlorofluorocarbons depleting atmospheric ozone levels, which increase UV-B radiation penetration resulting in embryo mortality) (Figure 26.4). In addition abiotic (UV-B) and biotic (fungus) interactions may also be contributing to the toxicity. Such effects would not be predicted from direct laboratory assessments of the toxicity of chlorofluorocarbons to amphibians and highlight the necessity to consider possible indirect toxicity associated with environmental contaminants.

***Masculinization of Fish due to Microbial Interactions with Kraft Pulpmill Effluent.*** Field surveys of mosquito fish populations in the state of Florida revealed populations containing females that exhibited male traits such as male-type mating behavior and the modification of the anal fin to resemble the sperm-transmitting gonopodium of males. Masculinized females were found to occur downstream of kraft pulpmill effluents suggesting that components of the effluent were responsible for the masculinizing effect. Direct toxicity assays performed with the effluent did not produce such effects. However, the inclusion of microorganisms along with the effluent resulted in masculinization. Further studies revealed that phytosterols present in the kraft pulpmill effluent can be converted to androgenic C19 steroids by microorganisms and these steroids are capable of masculinizing female fish (Figure 26.5).

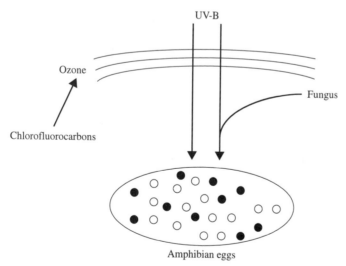

**Figure 26.4** Abiotic and biotic interactions leading to the indirect toxicity of chlorofluorocarbons to amphibians. Atmospheric release of chlorofluorocarbons causes the depletion of the stratospheric ozone layer (abiotic-abiotic interaction). Depleted ozone allows for increased penetration of UV-B radiation (abiotic-abiotic interaction). UV-B radiation alone and in combination with fungus (abiotic-biotic interaction) causes increased mortality of amphibian embryos.

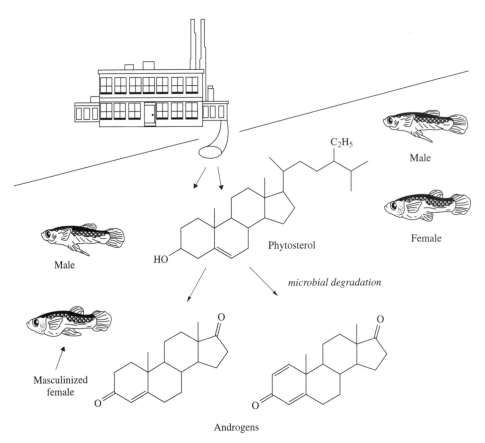

**Figure 26.5**  Indirect toxicity of kraft pulpmill effluent to mosquito fish. Phytosterols in the mill effluent are converted to C19 steroidal androgens through the action of microorganisms in the environment. These androgens masculinize both anatomy and behavior of female mosquito fish. An arrow identifies the modified anal fin on the masculinized female.

Thus abiotic (phytosterols) : biotic (microorganisms) interactions in the environment must occur before this occult toxicity associated with the kraft pulpmill effluent is unveiled.

***Environmental Contaminants and Disease among Marine Mammals.***  Massive mortality have occurred over the past 20 years among populations of harbor seals, bottlenose dolphins, and other marine mammals worldwide. In many instances this mortality has been attributed to disease. For example, nearly 18,000 harbor seals died in the North, Irish, and Baltic seas in the late 1980s due to phocine distemper virus. Incidences of the disease outbreak were highest in areas containing high levels of pollutants, and seals that succumbed to the disease were found to have high tissue levels of polychlorinated biphenyls (PCBs). PCBs and other organochlorine chemicals such as DDT, hexachlorobenzene, and dieldrin have been shown to immunosuppress laboratory animals, and accumulation of these chemicals by the seals may have increased their susceptibility to the virus. This hypothesis was tested by feeding fish, caught either from a relatively pristine area or from a polluted coastal area, to seals for 93 weeks then assessing the integrity of the immune system in the seals. Seals fed the contaminated fish did indeed have impaired immune responses lending credence to the hypothesis that organochlorine contaminants in the marine environment are rendering some species immunodeficient. Mortality occurs, not as a direct result of chemical toxicity, but due to increased susceptibility to pathogens.

## 26.5 CONCLUSION

Environmental toxicologists have learned a great deal about the effects of chemicals in the environment and the characteristics of chemicals that are responsible for the hazards they pose. Much of the information gained has been due to retrospective analyzes of the environmental consequences of the deposition of chemicals into the environment. Such analyzes have resulted in curtailing the release of demonstrated hazardous chemicals into the environment and provide benchmark information upon which the regulation of chemicals proposed for release into the environment can be based. The recognition that environmentally hazardous chemicals commonly share characteristics of persistence, potential to bioaccumulate, and high toxicity has resulted in development and use of chemicals that lack one or more of these characteristics yet fulfill societal needs previously served by hazardous chemicals. For example, recognition that persistence and propensity to bioaccumulate were largely responsible for the environmental hazards posed by many organochlorine pesticides led to the development and use of alternative classes of pesticides such as organophosphates, carbamates, and pyrethroids. While these chemicals all possess the toxicity necessary to function as pesticides, their lack of persistence and reduced propensity to bioaccumulate makes them more suitable for use in the environment.

Such advances in our understanding of the fate and effects of chemicals in the environmental does not imply that the role of environmental toxicologists in the twenty-first century will diminish. A dearth of information persist in areas vital to continued protection of natural resources against chemical insult. These include understanding (1) the unique susceptibilities of key species to the toxicity of different classes of chemicals, (2) the interactions of chemical contaminants with abiotic components of the environment that lead to increased toxicity, (3) the toxicological consequences of exposure to complex chemical mixtures, and (4) the consequences of toxicant effects on individuals with respect to ecosystem viability. Additionally continued research is needed to develop molecular and cellular biomarkers of toxicant exposure and effect that could be used to predict dire consequences to ecosystem before such effects are manifested at higher levels of biological organization. The role of the environmental toxicologist undoubtedly will increase in prospective activities aimed at reducing the risk associated with chemical contaminants in the environments before problems arise, and hopefully will decrease with respect to assessing damage caused by such environmental contaminants.

## SUGGESTED READING

### Persistence

Burns, L. A., and G. L. Baughman. Fate modeling. In *Fundamentals of Aquatic Toxicology*, G. M. Rand, and S. R. Petrocelli, eds. New York: Hemisphere, 1985, pp. 558–586.

Larson, R. A., and E. J. Weber. *Reaction Mechanisms in Environmental Organic Chemistry.* Boca Raton, FL: Lewis Publishers, 1994.


## Bioaccumulation

Banerjee, S., and G. A. Baughman. Bioconcentration factors and lipid solubility. *Environ. Sci. Technol.* **25**: 536–539, 1991.

Barron, M. G. Bioconcentration. *Environ. Sci. Technol.* **24**: 1612–1618, 1990.

Barron, M. G. Bioaccumulation and bioconcentration in aquatic organisms. In *Handbook of Ecotoxicology*, D. J. Hoffman, B. A. Rattner, G. A. Burton Jr., and J. Cairns Jr., eds. Boca Raton, FL: Lewis Publishers, 1995, pp. 652–666.

LeBlanc, G. A. Trophic-level differences in the bioconcentration of chemicals: Implications in assessing environmental biomagnification. *Environ. Sci. Technol.* **28**: 154–160, 1995.


## Acute Toxicity

Kelso, D. D., and M. Kendziorek. Alaska's response to the Exxon Valdez oil spill. *Environ. Sci. Technol.* **25**: 183–190, 1991.

Parrish, P. R. Acute toxicity tests. In *Fundamentals of Aquatic Toxicology*, G. M. Rand, and S. R. Petrocelli, eds. New York: Hemisphere, 1985, pp. 31–57.

Stansley, W. Field results using cholinesterase reactivation techniques to diagnose acute anti-cholinesterase poisoning in birds and fish. *Arch. Environ. Contam. Toxicol.* **25**: 315–321, 1993.

van Wezel, A. P., and A. Opperhuizen. Narcosis due to environmental pollutants in aquatic organisms: residue-based toxicity, mechanisms, and membrane burdens. *Crit. Rev. Toxicol.* **25**: 255–279, 1995.

Wilson, V. S., and G. A. LeBlanc. Petroleum pollution. *Rev. Toxicol.* **3**: 1–36, 1999.


## Chronic Toxicity

Adams, W. J. Aquatic toxicology testing methods. In *Handbook of Ecotoxicology*, D. J. Hoffman, B. A. Rattner, G. A. Burton Jr., and J. Cairns Jr., eds. Boca Raton, FL: Lewis Publishers, 1995, pp. 25–46.

Blaustein, A. R., P. D. Hoffman, D. G. Hokit, J. M. Kiesecker, S. C. Walls, J. B. Hays. UV repair and resistance to solar UV-B in amphibian eggs: A link to population declines? *Proc. Nat. Acad. Sciences. USA* **91**: 1791–1795, 1994.

Colborn, T., and C. Clement, eds. *Chemically-Induced Alterations in Sexual and Functional Development: The Wildlife/Human Connection.* Princeton, NJ: Princeton Scientific, 1992.

LeBlanc, G. A., and L. J. Bain. Chronic toxicity of environmental contaminants: Sentinels and biomarkers. *Environ. Health Perspect.* **105**(suppl. 1): 65–80, 1997.

Hayes, T. B., A. Collins, M. Lee, M. Mendoza, N. Noriega, A. A. Stuart, and A. Vonk. Hermaphroditic, demasculinized frogs after exposure to the herbicide atrazine at low ecologically relevant doses. *Proc. Nat. Acad. Sciences.* **99**: 5476–5480, 2002.

# CHAPTER 27

# Transport and Fate of Toxicants in the Environment

DAMIAN SHEA

## 27.1 INTRODUCTION

More than 100,000 chemicals are released into the global environment every year through their normal production, use, and disposal. To understand and predict the potential risk that this environmental contamination poses to humans and wildlife, we must couple our knowledge on the toxicity of a chemical to our knowledge on how chemicals enter into and behave in the environment. The simple box model shown in Figure 27.1 illustrates the relationship between a toxicant source, its fate in the environment, its effective exposure or dose, and resulting biological effects. A *prospective* or *predictive* assessment of a chemical hazard would begin by characterizing the source of contamination, modeling the chemical's fate to predict exposure, and using exposure/dose-response functions to predict effects (moving from left to right in Figure 27.1). A common application would be to assess the potential effects of a new waste discharge. A *retrospective* assessment would proceed in the opposite direction starting with some observed effect and reconstructing events to find a probable cause. Assuming that we have reliable dose/exposure-response functions, the key to successful use of this simple relationship is to develop a qualitative description and quantitative model of the sources and fate of toxicants in the environment.

Toxicants are released into the environment in many ways, and they can travel along many pathways during their lifetime. A toxicant present in the environment at a given point in time and space can experience three possible outcomes: it can be *stationary* and add to the toxicant inventory and exposure at that location, it can be *transported* to another location, or it can be *transformed* into another chemical species. Environmental contamination and exposure resulting from the use of a chemical is modified by the transport and transformation of the chemical in the environment. Dilution and degradation can attenuate the source emission, while processes that focus and accumulate the chemical can magnify the source emission. The actual fate of a chemical depends on the chemical's use pattern and physical-chemical properties, combined with the characteristics of the environment to which it is released.

*A Textbook of Modern Toxicology, Third Edition,* edited by Ernest Hodgson
ISBN 0-471-26508-X Copyright © 2004 John Wiley & Sons, Inc.

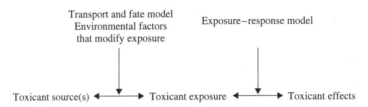

**Figure 27.1** Environmental fate model. Such models are used to help determine how the environment modifies exposure resulting from various sources of toxicants.

Conceptually and mathematically, the transport and fate of a toxicant in the environment is very similar to that in a living organism. Toxicants can enter an organism or environmental system by many routes (e.g., dermal, oral, and inhalation versus smoke stack, discharge pipe, or surface runoff). Toxicants are redistributed from their point of entry by fluid dynamics (blood flow vs. water or air movement) and intermedia transport processes such as partitioning (blood-lipid partitioning vs. water-soil partitioning) and complexation (protein binding vs. binding to natural organic matter). Toxicants are transformed in both humans and the environment to other chemicals by reactions such as hydrolysis, oxidation, and reduction. Many enzymatic processes that detoxify and activate chemicals in humans are very similar to microbial biotransformation pathways in the environment.

In fact, physiologically based pharmacokinetic models are similar to environmental fate models. In both cases we divide a complicated system into simpler compartments, estimate the rate of transfer between the compartments, and estimate the rate of transformation within each compartment. The obvious difference is that environmental systems are inherently much more complex because they have more routes of entry, more compartments, more variables (each with a greater range of values), and a lack of control over these variables for systematic study. The discussion that follows is a general overview of the transport and transformation of toxicants in the environment in the context of developing qualitative and quantitative models of these processes.

## 27.2 SOURCES OF TOXICANTS TO THE ENVIRONMENT

Environmental sources of toxicants can be categorized as either *point sources* or *nonpoint sources* (Figure 27.2). Point sources are discrete discharges of chemicals that are usually identifiable and measurable, such as industrial or municipal effluent outfalls, chemical or petroleum spills and dumps, smokestacks and other stationary atmospheric discharges. Nonpoint sources are more diffuse inputs over large areas with no identifiable single point of entry such as agrochemical (pesticide and fertilizer) runoff, mobile sources emissions (automobiles), atmospheric deposition, desorption or leaching from very large areas (contaminated sediments or mine tailings), and groundwater inflow. Nonpoint sources often include multiple smaller point sources, such as septic tanks or automobiles, that are impractical to consider on an individual basis. Thus the identification and characterization of a source is relative to the environmental compartment or system being considered. For example, there may be dozens of important toxicant sources to a river, each must be considered when assessing the hazards of toxicants to aquatic life in the river or to humans who might drink the water or consume the

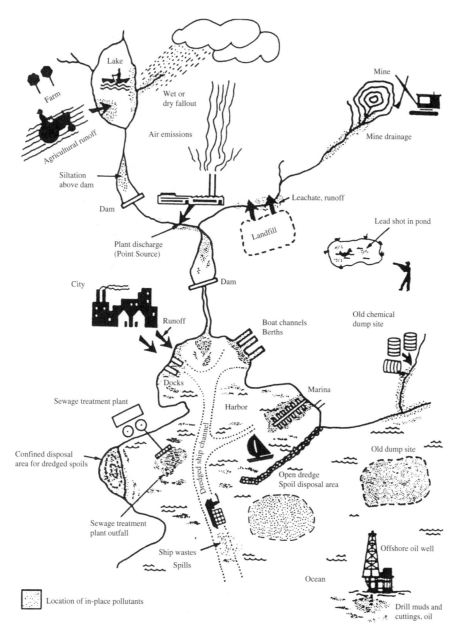

**Figure 27.2**  Entry of toxicants into environment through many point and nonpoint sources.

fish and shellfish. However, these toxicant sources can be well mixed in the river, resulting in a rather homogeneous and large point source to a downstream lake or estuary (Figure 27.2).

The rate (units of g/h) at which a toxicant is emitted by a source (*mass emission rate*) can be estimated from the product of the toxicant concentration in the medium ($g/m^3$) and the flow rate of the medium ($m^3/h$). This would appear to be relatively simple for point sources, particularly ones that are routinely monitored to meet environmental regulations. However, the measurement of trace concentrations of chemicals in complex effluent matrices is not a trivial task (see Chapter 25). Often the analytical methods prescribed by environmental agencies for monitoring are not sensitive or selective enough to measure important toxicants or their reactive metabolites. Estimating the mass emission rates for nonpoint sources is usually very difficult. For example, the atmospheric deposition of toxicants to a body of water can be highly dependent on both space and time, and high annual loads can result from continuous deposition of trace concentrations that are difficult to measure. The loading of pesticides from an agricultural field to an adjacent body of water also varies with time and space as shown in Figure 27.3 for the herbicide atrazine. Rainfall following the application of atrazine results in drainage ditch loadings more than 100-fold higher than just two weeks following the rain. A much smaller, but longer lasting, increase in atrazine loading occurs at the edge of the field following the rain. Again, we see the need to define the spatial scale of concern when identifying and characterizing a source. If one is concerned with the fate of atrazine within a field, the source is defined by the application rate. If one is concerned with the fate and exposure of atrazine in an adjacent body of water, the source may be defined as the drainage ditch and/or as runoff from the edge of field. In the latter case one either needs to take appropriate measurements in the field or model the transport of atrazine from the field.

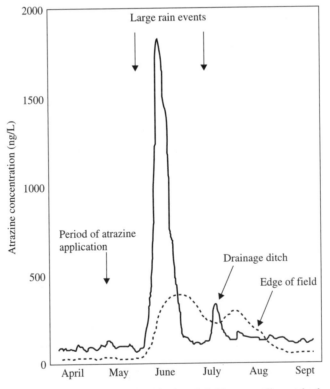

**Figure 27.3**  Loading of atrazine from an agricultural field to an adjacent body of water. The loading is highly dependent on rainfall and the presence of drainage ditches that collect the chemical and focus its movement in the environment.

## 27.3  TRANSPORT PROCESSES

Following the release of a toxicant into an environmental compartment, transport processes will determine its spatial and temporal distribution in the environment. The transport medium (or fluid) is usually either air or water, while the toxicant may be in dissolved, gaseous, condensed, or particulate phases. We can categorize physical transport as either *advection* or *diffusion*.

### 27.3.1  Advection

Advection is the passive movement of a chemical in bulk transport media either within the same medium (intraphase or homogeneous transport) or between different media (interphase or heterogeneous transport). Examples of homogeneous advection include transport of a chemical in air on a windy day or a chemical dissolved in water moving in a flowing stream, in surface runoff (nonpoint source), or in a discharge effluent (point source). Examples of heterogeneous advection include the deposition of a toxicant sorbed to a suspended particle that settles to bottom sediments, atmospheric deposition to soil or water, and even ingestion of contaminated particles or food by an organism (i.e., bioaccumulation). Advection takes place independently from the presence of a chemical; the chemical is simply going along for the ride. Advection is not influenced by diffusion and can transport a chemical either in the same or opposite direction as diffusion. Thus advection is often called *nondiffusive transport*.

***Homogeneous Advection.*** The homogeneous advective transport rate ($N$, g/h) is simply described in mathematical terms by the product of the chemical concentration in the advecting medium ($C$, g/m$^3$) and the flow rate of the medium ($G$, m$^3$/h):

$$N = GC.$$

For example, if the flow of water out of a lake is 1000 m$^3$/h and the concentration of the toxicant is 1 $\mu$g/m$^3$, then the toxicant is being advected from the lake at a rate of 1000 $\mu$g/h (or 1 mg/h). The emission rates for many toxicant sources can be calculated in the same way.

As with source emissions, advection of air and water can vary substantially with time and space within a given environmental compartment. Advection in a stream reach might be several orders of magnitude higher during a large rain event compared to a prolonged dry period, while at one point in time, advection within a stagnant pool might be several orders of magnitude lower than a connected stream. Thus, as with source characterization, we must match our estimates of advective transport to the spatial and temporal scales of interest. Again, a good example is the movement of atrazine from an agricultural field (Figure 27.3). Peak flow advective rates that follow the rain might be appropriate for assessing acute toxicity during peak flow periods but not for estimating exposure at other times of the year. Conversely, an annual mean advective rate would underestimate exposure during peak flow but would be more appropriate for assessing chronic toxicity.

In surface waters advective currents often dominate the transport of toxicants, and they can be estimated from hydrodynamic models or current measurements. In many cases advective flow can be approximated by the volume of water exchanged per unit time by assuming conservation of mass and measuring flow into or out of the system. This works only for well mixed systems that have no or only small volumes of stagnant water. In water bodies that experience density stratification (i.e., thermocline) separate advective models or residence times can be used for each water layer. In air, advection also dominates the transport of chemicals, with air currents being driven by pressure gradients. The direction and magnitude of air velocities are recorded continuously in many areas, and daily, seasonal, or annual means can be used to estimate advective air flow.

Advective air and water currents are much smaller in soil systems but still influence the movement of chemicals that reside in soil. Advection of water in the saturated zone is usually solved numerically from hydrodynamic models. Advection of air and water in the unsaturated zone is complicated by the heterogeneity of these soil systems. Models are usually developed for specific soil property classes, and measurements of these soil properties are made at a specific site to determine which soil-model layers to link together.

***Heterogeneous Advection.***   Heterogeneous advective transport involves a secondary phase within the bulk advective phase, such as when a particle in air or water acts as a carrier of a chemical. In many cases we can treat heterogeneous advection the same as homogeneous advection if we know the flow rate of the secondary phase and the concentration of chemical in the secondary phase. In the lake example above, if the volume fraction of suspended particles in the lake water is $10^{-5}$, the flow rate of suspended particles is 0.01 m$^3$/h, and the concentration of the toxicant in the solid particles is 100 mg/m$^3$, then the advective flow of the toxicant on suspended particles will be 1 mg/h or the same as the homogeneous advection via water. Although the flow rate of particles is much lower than that of water, the concentration of the toxicant is much higher in the suspended particles than dissolved in the water. This is typical of a hydrophobic toxicant such as DDT or benzo[a]pyrene. In soil and sedimentary systems, colloidal particles (often macromolecular detritus) can play a very important role in heterogeneous advective transport because they have greater mobility than larger particles, and they often have greater capacity to sorb many toxicants because of their higher organic carbon content and higher surface area/mass ratio. In highly contaminated sites, organic co-solvents can be present in the water (usually groundwater) and act as a high-capacity and high-efficiency carrier of toxicants through heterogeneous advection in the water.

Unfortunately, the dynamics of heterogeneous transport are rarely simple, particularly over shorter scales of time and space. In addition to advection of particles with flowing water, aqueous phase heterogeneous transport includes particle settling, resuspension, burial in bottom sediments, and mixing of bottom sediments. Particle settling can be an important mechanism for transporting hydrophobic toxicants from the water to the bottom sediments. Modeling this process can be as simple as using an overall mass transfer coefficient or can include rigorous modeling of particles with different size, density, and organic carbon content. Estimates of particle settling are usually obtained through the use of laboratory settling chambers, in situ sediment traps, or by calculation using Stoke's law. Resuspension of bottom sediments occurs when sufficient energy is transferred to the sediment bed from advecting water, internal waves, boats, dredging, fishing, and the movement of sediment dwelling organisms (i.e., bioturbation). Resuspension rates are difficult to measure and often are highly variable in both time and space. Much as the annual runoff of pesticides from an agricultural field may be dominated by a few rain events, annual resuspension rates can be dominated by a major storm, and in smaller areas by a single boat or school of bottom fish. Resuspension rates can be estimated from sediment traps deployed just above the sediment surface or from the difference between particle settling and permanent burial or sedimentation. Sedimentation is the net result of particle settling and resuspension and can be measured using radionuclide dating methods (e.g., $^{210}$Pb). Sediment dating itself becomes difficult when there is significant mixing of the surface sediments (e.g., through bioturbation). Thus the heterogeneous transport of toxicants on aqueous particles can be rather complicated, though many aquatic systems have been modeled reasonably well.

Heterogeneous advective transport in air occurs primarily through the absorption of chemicals into falling water droplets (wet deposition) or the sorption of chemicals into solid particles that fall to earth's surface (dry deposition). Under certain conditions both processes can be treated as simple first-order advective transport using a flow rate and concentration in the advecting medium. For example, wet deposition is usually characterized by a washout coefficient that is proportional to rainfall intensity.

### 27.3.2 Diffusion

Diffusion is the transport of a chemical by random motion due to a state of disequilibrium. For example, diffusion causes the movement of a chemical within a phase (e.g., water) from a location of relatively high concentration to a place of lower concentration until the chemical is homogeneously distributed throughout the phase. Likewise diffusive transport will drive a chemical between media (e.g., water and air) until their equilibrium concentrations are reached and thus the chemical potentials or fugacities are equal in each phase.

***Diffusion within a Phase.*** Diffusional transport within a phase can result from random (thermal) motion of the chemical (molecular diffusion), the random turbulent mixing of the transport medium (turbulent diffusion), or a combination of both. Turbulent diffusion usually dominates the diffusive (but not necessarily the advective) chemical transport in air and water due to the turbulent motions or eddies that are common in nature. In porous media (sediment and soil) the water velocities are typically too low to create eddies, but random mixing still occurs as water tortuously flows around particles. This mechanical diffusion is often called dispersion by hydrologists and dispersion on larger scales, such as when groundwater detours around large areas of less permeable soil, is called macrodispersion. Note, however, that the term dispersion often is used by meteorologists and engineers to describe any turbulent diffusion.

Although different physical mechanisms can cause diffusive mixing, they all cause a net transport of a chemical from areas of higher concentration to areas of lower concentration. All diffusive processes are also referred to as *Fickian* transport because they all can be described mathematically by Fick's first law, which states that the flow (or flux) of a chemical ($N$, g/h) is proportional to its concentration gradient ($dC/dx$):

$$N = -DA\left(\frac{dC}{dx}\right),$$

where $D$ is the diffusivity or mass transfer coefficient ($m^2/h$), $A$ is the area through which the chemical is passing ($m^2$), $C$ is the concentration of the diffusing chemical ($g/m^3$), and $x$ is the distance being considered (m). The negative sign is simply the convention that the direction of diffusion is from high to low concentration (diffusion is positive when $dC/dx$ is negative). Note that many scientists and texts define diffusion as an area specific process with units of $g/m^2h$ and thus the area term ($A$) is not included in the diffusion equation. This is simply an alternative designation that describes transport as a flux density ($g/m^2h$) rather than as a flow ($g/h$). In either case the diffusion equation can be integrated numerically and even expressed in three dimensions using vector notation. However, for most environmental situations we usually have no accurate estimate of $D$ or $dx$, so we combine the two into a one-dimensional mass transfer coefficient ($k_M$) with units of velocity (m/h). The chemical flux is then the product of this velocity, area, and concentration:

$$N = -k_M A C.$$

Mass transfer coefficients can be estimated from laboratory, mesocosm, and field studies and are widely used in environmental fate models. Mass transfer coefficients can be derived separately for molecular diffusion, turbulent diffusion, and dispersion in porous media, and all three terms can be added to the chemical flux equation. This is usually not necessary because one term often dominates the transport in specific environmental regions. Consider the vertical diffusion of methane gas generated by methanogenic bacteria in deep sediments. Molecular diffusion dominates in the highly compacted and low porosity deeper sediments. Dispersion becomes important as methane approaches the more porous surface sediments. Following methane gas ebulation from the sediment porewater, turbulent diffusion will dominate transport in a well-mixed water column (i.e., not a stagnant pool or beneath a thermocline where molecular diffusion will dominate). At the water surface, eddies tend to be damped and molecular diffusion may again dominate transport. Under stagnant atmospheric conditions (i.e., a temperature inversion) molecular diffusion will continue to dominate but will yield to more rapid mixing when typical turbulent conditions are reached. The magnitude and variability of the transport rate generally increase as the methane moves vertically through the environment, except when very stagnant conditions are encountered in the water or air. Modeling the transport of a chemical in air is particularly difficult because of the high spatial and temporal variability of air movement. Note also that advective processes in water or air usually transport chemicals at a faster rate than either molecular or turbulent diffusion.

***Diffusion between Phases.*** The transport of a chemical between phases is sometimes treated as a third category of transport processes or even as a transformation reaction. Interphase or intermedia transport is not a transformation reaction because the chemical is moving only between phases; it is not reacting with anything or changing its chemical structure. Instead, intermedia transport is simply driven by diffusion between two phases. When a chemical reaches an interface such as air–water, particle–water, or (biological) membrane–water, two diffusive regions are created at either side of the interface. The classical description of this process is the Whitman two-film or two-resistance mass transfer theory, where chemicals pass through two stagnant boundary layers by molecular diffusion, while the two bulk phases are assumed to be homogeneously mixed. This allows us to use a first-order function of the concentration gradient in the two phases, where the mass transfer coefficient will depend only on the molecular diffusivity of the chemical in each phase and the thickness of the boundary layers. This is fairly straightforward for transfer at the air–water interface (and often at the membrane–water interface), but not for the particle–water or particle–air interfaces.

Diffusive transport between phases can be described mathematically as the product of the departure from equilibrium and a kinetic term:

$$N = kA(C_1 - C_2K_{12}),$$

where $N$ is the transport rate (g/h), $k$ is a transport rate coefficient (m/h), $A$ is the interfacial area (m$^2$), $C_1$ and $C_2$ are the concentrations in the two phases, and $K_{12}$ is the equilibrium partition coefficient. At equilibrium $K_{12}$ equals $C_1/C_2$, the term describing the departure from equilibrium ($C_1 - C_2K_{12}$) becomes zero, and thus the net rate of transfer also is zero. The partition coefficients are readily obtained from thermodynamic data and equilibrium partitioning experiments. The transport rate coefficients are usually estimated from the transport rate equation itself by measuring intermedia transport rates ($N$) under controlled laboratory conditions (temperature, wind and water velocities) at known values of $A$, $C_1$, $C_2$, and $K_{12}$. These measurements must then be extrapolated to the field, sometimes with great uncertainty. This uncertainty, along with the knowledge that many interfacial regions have reached or are near equilibrium, has led many to simply assume that equilibrium exists at the interface. Thus the net transport rate is zero and the phase distribution of a chemical is simply described by its equilibrium partition coefficient.

## 27.4 EQUILIBRIUM PARTITIONING

When a small amount of a chemical is added to two immiscible phases and then shaken, the phases will eventually separate and the chemical will partition between the two phases according to its solubility in each phase. The concentration ratio at equilibrium is the partition coefficient:

$$\frac{C_1}{C_2} = K_{12}.$$

In the laboratory, we usually determine $K_{12}$ from the slope of $C_1$ versus $C_2$ over a range of concentrations. Partition coefficients can be measured for essentially any two-phase system: air–water, octanol–water, lipid–water, particle–water, and so on. In situ partition coefficients also can be measured where site-specific environmental conditions might influence the equilibrium phase distribution.

### 27.4.1 Air–Water Partitioning

Air–water partition coefficients ($K_{\text{air–water}}$) are essentially Henry's law constants ($H$):

$$K_{\text{air–water}} = H = \frac{C_{\text{air}}}{C_{\text{water}}},$$

where $H$ can be expressed in dimensionless form (same units for air and water) or in units of pressure divided by concentration (e.g., $Pa$ m$^3$/mol). The latter is usually written as

$$H = \frac{P_{\text{air}}}{C_{\text{water}}},$$

where $P_{\text{air}}$ is the partial vapor pressure of the chemical. When $H$ is not measured directly, it can be estimated from the ratio of the chemical's vapor pressure and aqueous solubility, although one must be careful about using vapor pressures and solubilities that apply to the same temperature and phase. Chemicals with high Henry's law constants (e.g., alkanes and many chlorinated solvents) have a tendency to escape from water to air and typically have high vapor pressures, low aqueous solubilities, and low boiling points. Chemicals with low Henry's law constants (e.g., alcohols, chlorinated phenols, larger polycyclic aromatic hydrocarbons, lindane, atrazine) tend to have high water solubility and/or very low vapor pressure. Note that some chemicals that are considered to be "nonvolatile," such as DDT, are often assumed to have low Henry's law constants. However, DDT also has a very low water solubility yielding a rather high Henry's law constant. Thus DDT readily partitions into the atmosphere as is now apparent from the global distribution of DDT.

### 27.4.2 Octanol–Water Partitioning

For many decades chemists have been measuring the octanol–water partition coefficient ($K_{\text{OW}}$) as a descriptor of hydrophobicity or how much a chemical "hates" to be in water. It is now one of the most important and frequently used physicochemical properties in toxicology and environmental chemistry. In fact toxicologists often simply use the symbol $P$, for partition coefficient, as if no other partition coefficient is important. Strong correlations exist between $K_{\text{OW}}$ and many biochemical and toxicological properties. Octanol has a similar carbon:oxygen ratio as lipids, and the $K_{\text{OW}}$ correlates particularly well with lipid–water partition coefficients. This has led many to use $K_{\text{OW}}$ as a measure of lipophilicity or how much a chemical "loves" lipids. This is really not the case because most chemicals have an equal affinity for octanol and other lipids (within about a factor of ten), but their affinity for water varies by many orders of magnitude. Thus it is largely aqueous solubility which determines $K_{\text{OW}}$ not octanol or lipid solubility. We generally express $K_{\text{OW}}$ as log $K_{\text{OW}}$ because $K_{\text{OW}}$ values range from less than one (alcohols) to over one billion (larger alkanes and alkyl benzenes).

### 27.4.3 Lipid–Water Partitioning

In most cases we can assume that the equilibrium distribution and partitioning of organic chemicals in both mammalian and nonmammalian systems is a function of lipid content in the animal and that the lipid–water partition coefficient ($K_{\text{LW}}$) is equal to $K_{\text{OW}}$. Instances where this is not the case include specific binding sites (e.g., kepone in the liver) and nonequilibrium conditions caused by slow elimination rates of higher level organisms or structured lipid phases that sterically hinder accumulation of very hydrophobic chemicals. For aquatic organisms in constant contact with water, the bioconcentration factor or fish-water partition coefficient ($K_{\text{FW}}$) is simply:

$$K_{\text{FW}} = f_{\text{lipid}} K_{\text{OW}},$$

where $f_{lipid}$ is the mass fraction of lipid in the fish (g lipid/g fish). Several studies have shown this relationship works well for many fish and shellfish species and an aggregate plot of $K_{FW}$ versus $K_{OW}$ for many different fish species yields a slope of 0.048, which is about the average lipid concentration of fish (5%). Again, nonequilibrium conditions will cause deviation from this equation. Such deviations are found at both the top and bottom of the aquatic food chain. Phytoplankton can have higher apparent lipid–water partition coefficients because their large surface area : volume ratios increase the relative importance of surface sorption. Top predators such as marine mammals also have high apparent lipid-water partition coefficients because of very slow elimination rates. Thus the deviations occur not because "there is something wrong with the equation" but because the underlying assumption of equilibrium is not appropriate in these cases.

### 27.4.4 Particle–Water Partitioning

It has been known for several decades that many chemicals preferentially associate with soil and sediment particles rather than the aqueous phase. The particle–water partition coefficient ($K_P$) describing this phenomenon is

$$K_P = \frac{C_S}{C_W},$$

where $C_S$ is the concentration of chemical in the soil or sediment (mg/kg dry weight) and $C_W$ is the concentration in water (mg/L). In this form, $K_P$ has units of L/kg or reciprocal density. Dimensionless partition coefficients are sometimes used where $K_P$ is multiplied by the particle density (in kg/L). It has also been observed, first by pesticide chemists in soil systems and later by environmental engineers and chemists in sewage effluent and sediment systems, that nonionic organic chemicals were primarily associated with the organic carbon phase(s) of particles. A plot of $K_P$ versus the mass fraction of organic carbon in the soil ($f_{OC}$, g/g) is linear with a near-zero intercept yielding the simple relationship

$$K_P = f_{OC} K_{OC},$$

where $K_{OC}$ is the organic carbon–water partition coefficient (L/kg). Studies with many chemicals and many sediment/soil systems have demonstrated the utility of this equation when the fraction of organic carbon is about 0.5% or greater. At lower organic carbon fractions, interaction with the mineral phase becomes relatively more important (though highly variable) resulting in a small positive intercept of $K_P$ versus $f_{OC}$. The strongest interaction between organic chemicals and mineral phases appears to be with dry clays. Thus $K_P$ will likely change substantially as a function of water content in low organic carbon, clay soils.

Measurements of $K_{OC}$ have been taken directly from partitioning experiments in sediment–and soil–water systems over a range of environmental conditions in both the laboratory and the field. Not surprisingly, the $K_{OC}$ values for many organic chemicals are highly correlated with their $K_{OW}$ values. Plots of the two partition coefficients for hundreds of chemicals with widely ranging $K_{OW}$ values yield slopes from about 0.3 to 1, depending on the classes of compounds and the particular methods included. Most fate modelers continue to use a slope of 0.41, which was reported by the first definitive study on the subject in the early 1980s. Thus we now have a means of estimating the partitioning of a chemical between a particle and water by using the $K_{OW}$ and $f_{OC}$:

$$K_P = f_{OC} K_{OC} = f_{OC} \, 0.41 K_{OW}.$$

This relationship is commonly used in environmental fate models to predict aqueous concentrations from sediment measurements by substituting the equilibrium expression for $K_P$ and rearranging to solve for $C_W$:

$$K_P = \frac{C_S}{C_W} = f_{OC} \, 0.41 K_{OW},$$

$$C_W = \frac{C_S}{f_{OC} 0.41 K_{OW}}.$$

This last equation forms the basis for the EPA's sediment quality guidelines that are used to assess the potential toxicity of contaminated sediments. The idea is to simply measure $C_S$ and $f_{OC}$, look up $K_{OW}$ in a table, compute the predicted $C_W$, and compare this result to established water quality criteria for the protection of aquatic life or human life (e.g., carcinogenicity risk factors). The use of this simple equilibrium partitioning expression for this purpose is currently the subject of much debate among scientists as well as policy makers.

## 27.5  TRANSFORMATION PROCESSES

The potential environmental hazard associated with the use of a chemical is directly related to it's persistence in the environment (see Chapter 26), which in turn depends on the rates of chemical transformation reactions. Transformation reactions can be divided into two classes: reversible reactions that involve continuous exchange among chemical states (ionization, complexation) and irreversible reactions that permanently transform a parent chemical into a daughter or reaction product (photolysis, hydrolysis, and many redox reactions). Reversible reactions are usually abiotic, although biological processes can still exert great influence over them (e.g., via production of complexing agents or a change in pH). Irreversible reactions can be abiotic or mediated directly by biota, particularly bacteria.

### 27.5.1  Reversible Reactions

*Ionization.* Ionization refers to the dissociation of a neutral chemical into charged species. The most common form of neutral toxicant dissociation is acid-base equilibria. The hypothetical monoprotic acid, HA, will dissociate in water to form the conjugate acid-base pair ($H^+$, $A^-$) usually written as

$$HA + H_2O = H_3O^+ + A^-.$$

The equilibrium constant for this reaction, the acidity constant ($K_a$), is defined by the law of mass action and is given by

$$K_a = \frac{[H_3O^+][A^-]}{[HA]}.$$

For convenience we often express equilibrium constants as the negative logarithm, or pK value. Thus the relative proportion of the neutral and charged species, will be a function of the pKa and solution pH. When the pH is equal to pKa, equal concentrations of the neutral and ionized forms will be present. When pH is less than the pKa, the neutral species will be predominant; when pH is greater than pKa, the ionized species will be in excess. The exact equilibrium distribution can be calculated from the equilibrium expression above and the law of mass conservation.

The fate of a chemical is often a function of the relative abundance of a particular chemical species as well as the total concentration. For example, the neutral chemical might partition into biological lipids or organic carbon in soil to a greater extent than the ionized form. Many acidic toxicants (pentachlorophenol) exhibit higher toxicities to aquatic organisms at lower pH where the neutral species predominates. However, specific ionic interactions will take place only with the ionized species. A classic example of how pH influences the fate and effects of a toxicant is with hydrogen cyanide. The pKa of HCN is about 9 and the toxicity of $CN^-$ is much higher than that of HCN for many aquatic organisms. Thus the discharge of a basic (high pH) industrial effluent containing cyanide would pose a greater hazard to fish than a lower pH effluent (everything else being equal). The effluent could be treated to reduce the pH well below the pKa according to the reaction:

$$CN^- + H^+ = HCN_{(aq)},$$

thus reducing the hazard to the fish. However, HCN has a rather high Henry's law constant and will partition into the atmosphere:

$$HCN_{(aq)} = HCN_{(air)}.$$

This may be fine for the fish, but birds in the area and humans working at the industrial plant will now have a much greater exposure to HCN. Thus both the fate and toxicity of a chemical can be influenced by simple ionization reactions.

***Precipitation and Dissolution.*** A special case of ionization is the dissolution of a neutral solid phase into soluble species. For example, the binary solid metal sulfide, CuS, dissolves in water according to

$$CuS(s) + H^+ = Cu^{2+} + HS^-.$$

The equilibrium constant for this reaction, the solubility product ($K_{sp}$), is given by

$$K_{sp} = \frac{[Cu^{2+}][HS^-]}{[H^+]}.$$

The solubility product for CuS is very low ($K_{sp} = 10^{-19}$ as written) so that the presence of sulfide in water acts to immobilize Cu (and many other metals) and reduce effective exposure. The formation of metal sulfides is important in anaerobic soil and sediment, stagnant ponds and basins, and many industrial and domestic sewage treatment plants and discharges. Co-precipitation of metals also can be a very important removal process in natural waters. In aerobic systems, the precipitation of hydrous oxides of manganese and iron often incorporate other metals as impurities. In anaerobic systems, the precipitation of iron sulfides can include other metals as well. These co-precipitates are usually not thermodynamically stable, but their conversion to stable mineral phases often takes place on geological time scales.

***Complexation and Chemical Speciation.*** Natural systems contain many chemicals that undergo ionic or covalent interactions with toxicants to change toxicant speciation and chemical speciation can have a profound effect on both fate and toxicity. Again, in the case of copper, inorganic ions ($Cl^-$, $OH^-$) and organic detritus (humic acids, peptides) will react with dissolved $Cu^{2+}$ to form various metal-ligand complexes. Molecular diffusivities of complexed copper will be lower than uncomplexed (hydrated) copper and will generally decrease with the size and number of ligands. The toxicity of free, uncomplexed $Cu^{2+}$ to many aquatic organisms is much higher than $Cu^{2+}$ that is complexed to chelating agents such as EDTA or glutathione. Many transition metal toxicants, such as Cu, Pb, Cd, and Hg, have high binding constants with compounds that contain amine, sulfhydryl, and carboxylic acid groups. These groups are quite common in natural organic matter. Even inorganic complexes of $OH^-$ and $Cl^-$ reduce $Cu^{2+}$ toxicity. In systems where a mineral phase is controlling $Cu^{2+}$ solubility, the addition of these complexing agents will shift the solubility equilibrium according to LeChatelier's principle as shown here for CuS and $OH^-$, $Cl^-$, and GSH (glutathione):

$$CuS(s) + H^+ = Cu^{2+} + HS^-, \qquad K_{sp},$$

$$Cu^{2+} + OH^- = CuOH^+, \qquad K_{CuOH},$$

$$Cu^{2+} + Cl^- = CuCl^+, \qquad K_{CuCl},$$

$$Cu^{2+} + GSH = CuGS^+ + H^+, \qquad K_{CuGS}.$$

Each successive complexation reaction "leaches" $Cu^{2+}$ from the solid mineral phase, thereby increasing the total copper in the water but not affecting the concentration of (or exposure to) $Cu^{2+}$. These equilibria can be combined into one reaction:

$$4CuS(s) + 3H^+ + OH^- + Cl^- + GSH = Cu^{2+} + 4HS^-$$
$$+ CuOH^+ + CuCl^+ + CuGS^+,$$

and the overall equilibrium constant derived as shown:

$$K_{overall} = (4)K_{sp} \times K_{CuOH} \times K_{CuCl} \times K_{CuGS}$$
$$= [Cu^{2+}] [CuOH^+] [CuCl^+] [CuGS^+] [HS^-]^4 / [H^+]^3 [OH^-] [Cl^-] [GS^-].$$

A series of simultaneous equations can be derived for these reactions to compute the concentration of individual copper species, and the total concentration of copper, $[Cu]_T$, would be given by

$$[Cu]_T = [Cu^{2+}] + [CuOH^+] + [CuCl^+] + [CuGS^+].$$

Thus the total copper added to a toxicity test or measured as the exposure (e.g., by atomic absorption spectroscopy) may be much greater than that which is available to an organism to induce toxicological effects.

Literally hundreds of complex equilibria like this can be combined to model what happens to metals in aqueous systems. Numerous speciation models exist for this application that include all of the necessary equilibrium constants. Several of these models include surface complexation reactions that take place at the particle–water interface. Unlike the partitioning of hydrophobic organic contaminants into organic carbon, metals actually form ionic and covalent bonds with surface ligands such as sulfhydryl groups on metal sulfides and oxide groups on the hydrous oxides of manganese and iron. Metals also can be biotransformed to more toxic species (e.g., conversion of elemental mercury to methyl-mercury by anaerobic bacteria), less toxic species (oxidation of tributyl tin to elemental tin), or temporarily immobilized (e.g., via microbial reduction of sulfate to sulfide, which then precipitates as an insoluble metal sulfide mineral).

### 27.5.2  Irreversible Reactions

The reversible transformation reactions discussed above alter the fate and toxicity of chemicals, but they do not irreversibly change the structure or properties of the chemical. An acid can be neutralized to its conjugate base, and vice versa. Copper can precipitate as a metal sulfide, dissolve and from a complex with numerous ligands, and later re-precipitate as a metal sulfide. Irreversible transformation reactions alter the structure and properties of a chemical forever.

*Hydrolysis.* Hydrolysis is the cleavage of organic molecules by reaction with water with a net displacement of a leaving group (X) with $OH^-$:

$$RX + H_2O = ROH + HX.$$

Hydrolysis is part of the larger class of chemical reactions called nucleophilic displacement reactions in which a nucleophile (electron-rich species with an unshared pair of electrons) attacks an electrophile (electron deficient), cleaving one covalent bond to form a new one. Hydrolysis is usually associated with surface waters but also takes place in the atmosphere (fogs and clouds), groundwater, at the particle–water interface of soils and sediments, and in living organisms.

Hydrolysis can proceed through numerous mechanisms via attack by $H_2O$ (neutral hydrolysis) or by acid ($H^+$) or base ($OH^-$) catalysis. Acid and base catalyzed reactions proceed via alternative mechanisms that require less energy than neutral hydrolysis. The combined hydrolysis rate term is a sum of these three constituent reactions and is given by

$$\frac{d[RX]}{dt} = k_{obs}[RX] = k_a[H^+][RX] + k_n[RX] + k_b[OH^-][RX],$$

where [RX] is the concentration of the hydrolyzable chemical, $k_{obs}$ is the macroscopic observed hydrolysis rate constant, and $k_a$, $k_n$, and $k_b$ are the rate constants for the acid-catalyzed, neutral, and base-catalyzed hydrolysis. If we assume that the hydrolysis can be approximated by first-order kinetics with respect to RX (which is usually true), the rate term is reduced to

$$k_{obs} = k_a[H^+] + k_n + k_b[OH^-].$$

Neutral hydrolysis is dependent only on water which is present in excess, so $k_n$ is a simple pseudo-first-order rate constant (with units $t^{-1}$). The acid- and base-catalyzed hydrolysis depend on the molar quantities of $[H^+]$ and $[OH^-]$, respectively, so $k_a$ and $k_b$ have units of $M^{-1}t^{-1}$.

The observed or apparent hydrolysis half-life at a fixed pH is then given by

$$t_{1/2} = \frac{\ln 2}{k_{obs}}.$$

Compilations of hydrolysis half-lives at pH and temperature ranges encountered in nature can be found in many sources. Reported hydrolysis half-lives for organic compounds at pH 7 and 298 K range at least 13 orders of magnitude. Many esters hydrolyze within hours or days, whereas some organic chemicals will never hydrolyze. For halogenated methanes, which are common groundwater contaminants, half-lives range from about 1 year for $CH_3Cl$ to about 7000 years for $CCl_4$. The half-lives of halomethanes follows the strength of the carbon-halogen bond with half-lives decreasing in the order F > Cl > Br. Small structural changes can dramatically alter hydrolysis rates. An example is the difference between tetrachloroethane ($Cl_2HC–CHCl_2$) and tetrachloroethene ($Cl_2C{=}CCl_2$) which have hydrolysis half-lives of about 0.5 year and $10^9$ years, respectively. In this case the hydrolysis rate is affected by the C–Cl bond strength and the steric bulk at the site of nucleophilic substitution.

The apparent rate of hydrolysis and the relative abundance of reaction products is often a function of pH because alternative reaction pathways are preferred at different pH. In the case of halogenated hydrocarbons, base-catalyzed hydrolysis will result in elimination reactions while neutral hydrolysis will take place via nucleophilic displacement reactions. An example of the pH dependence of hydrolysis is illustrated by the base-catalyzed hydrolysis of the structurally similar insecticides DDT and methoxychlor. Under a common range of natural pH (5 to 8) the hydrolysis rate of methoxychlor is invariant while the hydrolysis of DDT is about 15-fold faster at pH 8 compared to pH 5. Only at higher pH (>8) does the hydrolysis rate of methoxychlor increase. In addition the major product of DDT hydrolysis throughout this pH range is the same (DDE), while the methoxychlor hydrolysis product shifts from the alcohol at pH 5–8 (nucleophilic substitution) to the dehydrochlorinated DMDE at pH > 8 (elimination). This illustrates the necessity to conduct detailed mechanistic experiments as a function of pH for hydrolytic reactions.

***Photolysis.*** Photolysis of a chemical can proceed either by direct absorption of light (direct photolysis) or by reaction with another chemical species that has been produced or excited by light (indirect photolysis). In either case photochemical transformations such as bond cleavage, isomerization, intramolecular rearrangement, and various intermolecular reactions can result. Photolysis can take place wherever sufficient light energy exists, including the atmosphere (in the gas phase and in aerosols and fog/cloud droplets), surface waters (in the dissolved phase or at the particle–water interface), and in the terrestrial environment (on plant and soil/mineral surfaces).

Photolysis dominates the fate of many chemicals in the atmosphere because of the high solar irradiance. Near the earth's surface, chromophores such as nitrogen oxides, carbonyls, and aromatic hydrocarbons play a large role in contaminant fate in urban areas. In the stratosphere, light is absorbed by ozone, oxygen, organohalogens, and hydrocarbons with global environmental implications. The rate of photolysis in surface waters depends on light intensity at the air–water interface, the transmittance through this interface, and the attenuation through the water column. Open ocean waters ("blue water") can transmit blue light to depths of 150 m while highly eutrophic or turbid waters might absorb all light within 1 cm of the surface.

***Oxidation-Reduction Reactions.*** Although many redox reactions are reversible, they are included here because many of the redox reactions that influence the fate of toxicants are irreversible on the temporal and spatial scales that are important to toxicity.

Oxidation is simply defined as a loss of electrons. Oxidizing agents are electrophiles and thus gain electrons upon reaction. Oxidations can result in the increase in the oxidation state of the chemical as in the oxidation of metals or oxidation can incorporate oxygen into the molecule. Typical organic chemical oxidative reactions include dealkylation, epoxidation, aromatic ring cleavage, and hydroxylation. The term autooxidation, or weathering, is commonly used to describe the general oxidative degradation of a chemical (or chemical mixture, e.g., petroleum) upon exposure to air. Chemicals can react abiotically in both water and air with oxygen, ozone, peroxides, free radicals, and singlet oxygen. The last two are common intermediate reactants in indirect photolysis. Mineral surfaces are known to catalyze many oxidative reactions. Clays and oxides of silicon, aluminum, iron, and manganese can provide surface active sites that increase rates of oxidation. There are a variety of complex mechanisms associated with this catalysis, so it is difficult to predict the catalytic activity of soils and sediment in nature.

Reduction of a chemical species takes place when an electron donor (reductant) transfers electrons to an electron acceptor (oxidant). Organic chemicals typically act as the oxidant, while abiotic reductants include sulfide minerals, reduced metals or sulfur compounds, and natural organic matter. There are also extracellular biochemical reducing agents such as porphyrins, corrinoids, and metal-containing coenzymes. Most of these reducing agents are present only in anaerobic environments where anaerobic bacteria are themselves busy reducing chemicals. Thus it is usually very difficult to distinguish biotic and abiotic reductive processes in nature. Well-controlled, sterile laboratory studies are required to measure abiotic rates of reduction. These studies indicate that many abiotic reductive transformations could be important in the environment, including dehalogenation, dealkylation, and the reduction of quinones, nitrosamines, azoaromatics, nitroaromatics, and sulfoxides. Functional groups that are resistant to reduction include aldehydes, ketones, carboxylic acids (and esters), amides, alkenes, and aromatic hydrocarbons.

***Biotransformations.*** As we have seen throughout much of this textbook, vertebrates have developed the capacity to transform many toxicants into other chemicals, sometimes detoxifying the chemical and sometimes activating it. The same or similar biochemical processes that hydrolyze, oxidize, and reduce toxicants in vertebrates also take place in many lower organisms. In particular, bacteria, protozoans, and fungi provide a significant capacity to biotransform toxicants in the environment. Although many vertebrates can metabolize toxicants faster than these lower forms of life, the aggregate capacity of vertebrates to biotransform toxicants (based on total biomass and exposure) is insignificant to the overall fate of a toxicant in the environment. In this section we use the term *biotransformation* to include all forms of direct biological transformation reactions.

Biotransformations follow a complex series of chemical reactions that are enzymatically mediated and are usually irreversible reactions that are energetically favorable, resulting in a decrease in the Gibbs free energy of the system. Thus the potential for biotransformation of a chemical depends on the reduction in free energy that results from reacting the chemical with other chemicals in its environment (e.g., oxygen). As with inorganic catalysts, microbes simply use enzymes to lower the activation energy of the reaction and increase the rate of the transformation. Each successive chemical reaction further degrades the chemical, eventually mineralizing it to inorganic compounds ($CO_2$, $H_2O$, inorganic salts) and continuing the carbon and hydrologic cycles on earth.

Usually microbial growth is stimulated because the microbes capture the energy released from the biotransformation reaction. As the microbial population expands, overall biotransformation rates increase, even though the rate for each individual microbe may be constant or even decrease. This complicates the modeling and prediction of biotransformation rates in nature. When the toxicant concentrations (and potential energy) are small relative to other substrates or when the microbes cannot efficiently capture the energy from the biotransformation, microbial growth is not stimulated but biotransformation often still proceeds inadvertently through cometabolism.

Biotransformation can be modeled using simple Michaelis-Menten enzyme kinetics, Monod microbial growth kinetics, or more complex numerical models that incorporate various environmental parameters, and even the formation of microbial mats or slime, which affects diffusion of the chemical and nutrients to the microbial population. Microbial ecology involves a complex web of interaction among numerous environmental processes and parameters. The viability of microbial populations and the rates of biotransformation depend on many factors such as genetic adaptation, moisture, nutrients, oxygen, pH, and temperature. Although a single factor may limit biotransformation rates at a particular time and location, we cannot generalize about what limits biotransformation rates in the environment. Biotransformation rates often increase with temperature (according to the Arrhenius law) within the optimum range that supports the microbes, but many exceptions exist for certain organisms and chemicals. The availability of oxygen and various nutrients (C, N, P, Fe, Si) often limits microbial growth, but the limiting nutrient often changes with space (e.g., downriver) and time (seasonally and even diurnally).

Long-term exposure of microbial populations to certain toxicants often is necessary for adaptation of enzymatic systems capable of degrading those toxicants. This was the case with the Exxon Valdez oil spill in Alaska in 1989. Natural microbial populations in Prince William Sound, Alaska, had developed enzyme systems that oxidize petroleum hydrocarbons because of long-term exposure to natural oil seeps and to hydrocarbons that leached from the pine forests in the area. Growth of these natural microbial populations was nutrient limited during the summer. Thus the application of nutrient formulations to the rocky beaches of Prince William Sound stimulated microbial growth and helped to degrade the spilled oil.

In terrestrial systems with high nutrient and oxygen content, low moisture and high organic carbon can control biotransformation by limiting microbial growth and the availability of the toxicant to the microbes. For example, biotransformation rates of certain pesticides have been shown to vary two orders of magnitude in two separate agricultural fields that were both well aerated and nutrient rich, but spanned the common range of moisture and organic carbon content.

## 27.6  ENVIRONMENTAL FATE MODELS

The discussion above provides a brief qualitative introduction to the transport and fate of chemicals in the environment. The goal of most fate chemists and engineers is to translate this qualitative picture into a conceptual model and ultimately into a quantitative description that can be used to predict or reconstruct the fate of a chemical in the environment (Figure 27.1). This quantitative description usually takes the form of a mass balance model. The idea is to compartmentalize the environment into defined units (control volumes) and to write a mathematical expression for the mass balance within the compartment. As with pharmacokinetic models, transfer between compartments can be included as the complexity of the model increases. There is a great deal of subjectivity to assembling a mass balance model. However, each decision to include or exclude a process or compartment is based on one or more assumptions—most of which can be tested at some level. Over time the applicability of various assumptions for particular chemicals and environmental conditions become known and model standardization becomes possible.

The construction of a mass balance model follows the general outline of this chapter. First, one defines the spatial and temporal scales to be considered and establishes the environmental compartments or control volumes. Second, the source emissions are identified and quantified. Third, the mathematical expressions for advective and diffusive transport processes are written. And last, chemical transformation processes are quantified. This model-building process is illustrated in Figure 27.4. In this example we simply equate the change in chemical inventory (total mass in the system) with the difference between chemical inputs and outputs to the system. The inputs could include numerous point and nonpoint sources or could be a single estimate of total chemical load to the system. The outputs include all of the loss mechanisms: transport

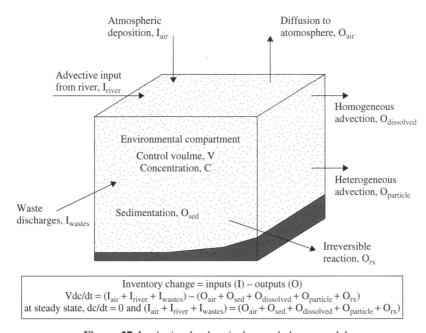

Inventory change = inputs (I) − outputs (O)

$$V dc/dt = (I_{air} + I_{river} + I_{wastes}) - (O_{air} + O_{sed} + O_{dissolved} + O_{particle} + O_{rx})$$

at steady state, $dc/dt = 0$ and $(I_{air} + I_{river} + I_{wastes}) = (O_{air} + O_{sed} + O_{dissolved} + O_{particle} + O_{rx})$

**Figure 27.4**  A simple chemical mass balance model.

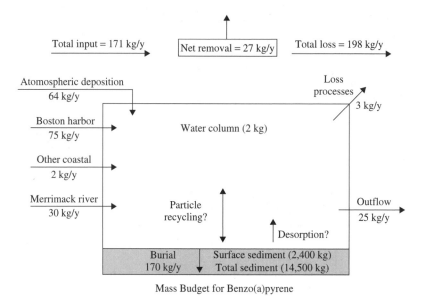

Mass Budget for Benzo(a)pyrene

**Figure 27.5**    Information provided by a chemical mass balance model. The annual mass budget of benzo[*a*]pyrene in Massachusetts Bay is shown.

out of the compartment and irreversible transformation reactions. If steady state can be assumed (i.e., the chemical's concentration in the compartment is not changing over the time scale of the model), the inventory change is zero and we are left with a simple mass balance equation to solve. Unsteady-state conditions would require a numerical solution to the differential equations.

There are many tricks and shortcuts to this process. For example, rather than compiling all of the transformation rate equations (or conducting the actual kinetic experiments yourself), there are many sources of typical chemical half-lives based on pseudo-first-order rate expressions. It is usually prudent to begin with these "best estimates" of half-lives in air, water, soil, and sediment and perform a sensitivity analysis with the model to determine which processes are most important. One can return to the most important processes to assess whether more detailed rate expressions are necessary. An illustration of this mass balance approach is given in Figure 27.5 for benzo[*a*]pyrene. This approach allows a first-order evaluation of how chemicals enter the environment, what happens to them in the environment, and what the exposure concentrations will be in various environmental media. Thus the chemical mass balance provides information relevant to toxicant exposure to both humans and wildlife.

## SUGGESTED READING

Hemond, H. F., and E. J. Fechner. *Chemical Fate and Transport in the Environment*. New York: Academic Press, 1994.

Mackay, D., W. Y. Shiu, and K. C. Ma. *Physical-Chemical Properties and Environmental Fate and Degradation Handbook*. CRCnetBASE 2000 CR-ROM. Boca Raton, FL: CRC Press, 2000.

Mackay, D. *Multimedia Environmental Models: The Fugacity Approach*, 2nd ed. Boca Raton, FL: Lewis Publishers, 2001.

Rand, G. M., ed. *Fundamentals of Aquatic Toxicology: Part II Environmental Fate*. Washington, DC: Taylor and Francis, 1995.

Schnoor, J. L. *Environmental Modeling: Fate and Transport of Pollutants in Water, Air, and Soil*. New York: Wiley, 1996.

Schwarzenbach, R. P., P. M. Gschwend, and D. M. Imboden. *Environmental Organic Chemistry*, 2nd ed. New York: Wiley, 2002.

# CHAPTER 28

# Environmental Risk Assessment

DAMIAN SHEA

## 28.1 INTRODUCTION

Risk assessment is the process of assigning magnitudes and probabilities to adverse effects associated with an event. The development of risk assessment methodology has focused on accidental events (e.g., an airplane crash) and specific environmental stresses to humans (exposure of humans to chemicals), and thus most risk assessment is characterized by discrete events or stresses affecting well-defined endpoints (e.g., incidence of human death or cancer). This *single stress–single end point* relationship allows the use of relatively simple statistical and mechanistic models to estimate risk and is widely used in human health risk assessment. However, this simple paradigm has only partial applicability to ecological risk assessment because of the inherent complexity of ecological systems and the exposure to numerous physical, chemical, and biological stresses that have both direct and indirect effects on a diversity of ecological components, processes, and endpoints. Thus, although the roots of ecological risk assessment can be found in human health risk assessment, the methodology for ecological risk assessment is not well developed and the estimated risks are highly uncertain. Despite these limitations, resource managers and regulators are looking to ecological risk assessment to provide a scientific basis for prioritizing problems that pose the greatest ecological risk and to focus research efforts in areas that will yield the greatest reduction in uncertainty.

To this end the US Environmental Protection Agency has issued guidelines for planning and conducting ecological risk assessments. Because of the complexity and uncertainty associated with ecological risk assessment the EPA guidelines provide only a loose framework for organizing and analyzing data, information, assumptions, and uncertainties to evaluate the likelihood of adverse ecological effects. However, the guidelines represent a broad consensus of the present scientific knowledge and experience on ecological risk assessment. This chapter presents a brief overview of the ecological risk assessment process as presently described by the EPA.

Ecological risk assessment can be defined as:

> The process that evaluates the likelihood that adverse ecological effects may occur or are occurring as a result of exposure to one or more stressors.

*A Textbook of Modern Toxicology, Third Edition,* edited by Ernest Hodgson
ISBN 0-471-26508-X  Copyright © 2004 John Wiley & Sons, Inc.

Estimating the *likelihood* can range from qualitative judgments to quantitative probabilities, though quantitative risk estimates still are rare in ecological risk assessment. The *adverse ecological effects* are changes that are considered undesirable because they alter valued structural or functional characteristics of ecological systems and usually include the type, intensity, and scale of the effect as well as the potential for recovery. The statement that effects *may occur or are occurring* refers to the dual *prospective* and *retrospective* nature of ecological risk assessment. The inclusion of *one or more stressors* is a recognition that ecological risk assessments may address single or multiple chemical, physical, and/or biological stressors. Because risk assessments are conducted to provide input to management decisions, most risk assessments focus on stressors generated or influenced by anthropogenic activity. However, natural phenomena also will induce stress that results in adverse ecological effects and cannot be ignored.

The overall ecological risk assessment process is shown in Figure 28.1 and includes three primary phases: (1) problem formulation, (2) analysis, and (3) risk characterization. Problem formulation includes the development of a conceptual model

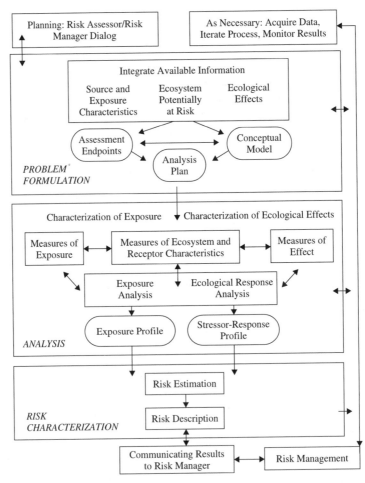

**Figure 28.1**   The ecological risk assessment framework as set forth by the US Environmental Protection Agency.

of stressor-ecosystem interactions and the identification of risk assessment end points. The analysis phase involves evaluating exposure to stressors and the relationship between stressor characteristics and ecological effects. Risk characterization includes estimating risk through integration of exposure and stressor-response profiles, describing risk by establishing lines of evidence and determining ecological effects, and communicating this description to risk managers. While discussions between risk assessors and risk managers are emphasized both at risk assessment initiation (planning) and completion (communicating results), usually a clear distinction is drawn between risk assessment and risk management. Risk assessment focuses on scientifically evaluating the likelihood of adverse effects, and risk management involves the selection of a course of action in response to an identified risk that is based on many factors (e.g., social, legal, or economic) in addition to the risk assessment results. Monitoring and other data acquisition is often necessary during any phase of the risk assessment process and the entire process is typically iterative rather than linear. The evaluation of new data or information may require revisiting a part of the process or conducting a new assessment.

## 28.2  FORMULATING THE PROBLEM

Problem formulation is a process for generating and evaluating preliminary hypotheses about why ecological effects have occurred, or may occur, because of human activities. During problem formulation, management goals are evaluated to help establish objectives for the risk assessment, the ecological problem is defined, and the plan for analyzing data and characterizing risk is developed. The objective of this process is to develop (1) assessment end points that adequately reflect management goals and the ecosystem they represent and (2) conceptual models that describe critical relationships between a stressor and assessment end point or among several stressors and assessment end points. The assessment end points and the conceptual models are then integrated to develop a plan or proposal for risk analysis.

### 28.2.1  Selecting Assessment End Points

Assessment end points are *explicit expressions of the actual environmental value that is to be protected* and they link the risk assessment to management concerns. Assessment end points include both a valued or key ecological entity and an attribute of that entity that is important to protect and that is potentially at risk. The scientific basis for a risk assessment is enhanced when assessment end points are both ecologically relevant and susceptible to the stressors of concern. Assessment endpoints that also logically represent societal values and management goals will increase the likelihood that the risk assessment will be understood and used in management decisions.

***Ecological Relevance.*** Ecologically relevant end points reflect important attributes of the ecosystem and can be functionally related to other components of the ecosystem; they help sustain the structure, function, and biodiversity of an ecosystem. For example, ecologically relevant end points might contribute to the food base (e.g., primary production), provide habitat, promote regeneration of critical resources (e.g., nutrient cycling), or reflect the structure of the community, ecosystem, or landscape (e.g., species diversity). Ecological relevance becomes most useful when it is possible to identify the potential cascade of adverse effects that could result from a critical initiating effect such as a change in ecosystem function. The selection of assessment end points that address both specific organisms of concern and landscape-level ecosystem processes becomes increasingly important (and more difficult) in landscape-level risk assessments. In these cases it may be possible to select one or more species and an ecosystem process to represent larger functional community or ecosystem processes. Extrapolations like these must be explicitly described in the conceptual model (see Section 28.2.2).

***Susceptibility to Stressors.*** Ecological resources or entities are considered susceptible if they are sensitive to a human-induced stressor to which they are exposed. *Sensitivity* represents how readily an ecological entity responds to a particular stressor. Measures of sensitivity may include mortality or decreased growth or fecundity resulting from exposure to a toxicant, behavioral abnormalities such as avoidance of food-source areas or nesting sites because of the proximity of stressors such as noise or habitat alteration. Sensitivity is directly related to the mode of action of the stressors. For example, chemical sensitivity is influenced by individual physiology, genetics, and metabolism. Sensitivity also is influenced by individual and community life-history characteristics. For example, species with long life cycles and low reproductive rates will be more vulnerable to extinction from increases in mortality than those with short life cycles and high reproductive rates. Species with large home ranges may be more sensitive to habitat fragmentation compared to those species with smaller home ranges within a fragment. Sensitivity may be related to the life stage of an organism when exposed to a stressor. Young animals often are more sensitive to stressors than adults. In addition events like migration and molting often increase sensitivity because they require significant energy expenditure that make these organisms more vulnerable to stressors. Sensitivity also may be increased by the presence of other stressors or natural disturbances.

*Exposure* is the other key determinant in susceptibility. In ecological terms, exposure can mean co-occurrence, contact, or the absence of contact, depending on the stressor and assessment end point. The characteristics and conditions of exposure will influence how an ecological entity responds to a stressor and thus determine what ecological entities might be susceptible. Therefore one must consider information on the proximity of an ecological entity to the stressor along with the timing (e.g., frequency and duration relative to sensitive life stages) and intensity of exposure. Note that adverse effects may be observed even at very low stressor exposures if a necessary resource is limited during a critical life stage. For example, if fish are unable to find suitable nesting sites during their reproductive phase, risk is significant even when water quality is high and food sources are abundant.

Exposure may take place at one point in space and time, but effects may not arise until another place or time. Both life history characteristics and the circumstances of exposure influence susceptibility in this case. For example, exposure of a population to endocrine-modulating chemicals can affect the sex ratio of offspring, but the population impacts of this exposure may not become apparent until years later when the cohort of affected animals begins to reproduce. Delayed effects and multiple stressor exposures add complexity to evaluations of susceptibility. For example, although toxicity tests may determine receptor sensitivity to one stressor, the degree of susceptibility may depend on the co-occurrence of another stressor that significantly alters receptor response. Again, conceptual models need to reflect these additional factors.

***Defining Assessment End Points.*** Assessment end points provide a transition between management goals and the specific measures used in an assessment by helping identify measurable attributes to quantify and model. However, in contrast to management goals, no intrinsic value is assigned to the end point, so it does not contain words such as *protect* or *maintain* and it does not indicate a desirable direction for change. Two aspects are required to define an assessment end point. The first is the valued ecological entity such as a species, a functional group of species, an ecosystem function or characteristic, or a specific valued habitat. The second is the characteristic about the entity of concern that is important to protect and potentially at risk.

Expert judgment and an understanding of the characteristics and function of an ecosystem are important for translating general goals into usable assessment end points. End points that are too broad and vague (ecological health) cannot be linked to specific measurements. End points that are too narrowly defined (hatching success of bald eagles) may overlook important characteristics of the ecosystem and fail to include critical variables. Clearly defined assessment end points provide both direction and boundaries for the risk assessment.

Assessment end points directly influence the type, characteristics, and interpretation of data and information used for analysis and the scale and character of the assessment. For example, an assessment end point such as "fecundity of bivalves" defines local population characteristics and requires very different types of data and ecosystem characterization compared with "aquatic community structure and function." When concerns are on a local scale, the assessment end points should not focus on landscape concerns. But if ecosystem processes and landscape patterns are being considered, survival of a single species would provide inadequate representation of this larger scale.

The presence of multiple stressors also influences the selection of assessment end points. When it is possible to select one assessment end point that is sensitive to many of the identified stressors, yet responds in different ways to different stressors, it is possible to consider the combined effects of multiple stressors while still discriminating among effects. For example, if recruitment of a fish population is the assessment end point, it is important to recognize that recruitment may be adversely affected at several life stages, in different habitats, through different ways, by different stressors. The measures of effect, exposure, and ecosystem and receptor characteristics chosen to evaluate recruitment provide a basis for discriminating among different stressors, individual effects, and their combined effect.

Although many potential assessment end points may be identified, practical considerations often drive their selection. For example, assessment end points usually must reflect environmental values that are protected by law or that environmental managers and the general public recognize as a critical resource or an ecological function that would be significantly impaired if the resource were altered. Another example of a practical consideration is the extrapolation across scales of time, space, or level of biological organization. When the attributes of an assessment end point can be measured directly, extrapolation is unnecessary and this uncertainty is avoided. Assessment end points that cannot be linked with measurable attributes are not appropriate for a risk assessment. However, assessment end points that cannot be measured directly but can be represented by surrogate measures that are easily monitored and modeled can still provide a good foundation for the risk assessment.

## 28.2.2  Developing Conceptual Models

Conceptual models link anthropogenic activities with stressors and evaluate the relationships among exposure pathways, ecological effects, and ecological receptors. The models also may describe natural processes that influence these relationships. Conceptual models include a set of risk hypotheses that describe predicted relationships between stressor, exposure, and assessment end point response, along with the rationale for their selection. Risk hypotheses are hypotheses in the broad scientific sense; they do not necessarily involve statistical testing of null and alternative hypotheses or any particular analytical approach. Risk hypotheses may predict the effects of a stressor, or they may postulate what stressors may have caused observed ecological effects.

Diagrams can be used to illustrate the relationships described by the conceptual model and risk hypotheses. Conceptual model diagrams are useful tools for communicating important pathways and for identifying major sources of uncertainty. These diagrams and risk hypotheses can be used to identify the most important pathways and relationships to consider in the analysis phase. The hypotheses considered most likely to contribute to risk are identified for subsequent evaluation in the risk assessment.

The complexity of the conceptual model depends on the complexity of the problem, number of stressors and assessment end points being considered, nature of effects, and characteristics of the ecosystem. For single stressors and single assessment end points, conceptual models can be relatively simple relationships. In cases where conceptual models describe, besides the pathways of individual stressors and assessment end points, the interaction of multiple and diverse stressors and assessment end points, several submodels would be required to describe individual pathways. Other models may then be used to explore how these individual pathways interact. An example of a conceptual model for a watershed in shown in Figure 28.2.

## 28.2.3  Selecting Measures

The last step in the problem formulation phase is the development of an analysis plan or proposal that identifies measures to evaluate each risk hypothesis and that describes the assessment design, data needs, assumptions, extrapolations, and specific methods for conducting the analysis. There are three categories of measures that can be selected. *Measures of effect* (also called *measurement end points*) are measures used to evaluate the response of the assessment end point when exposed to a stressor. *Measures of exposure* are measures of how exposure may be occurring, including how a stressor moves through the environment and how it may co-occur with the assessment end point. *Measures of ecosystem and receptor characteristics* include ecosystem characteristics that influence the behavior and location of assessment end points, the distribution of a stressor, and life history characteristics of the assessment end point that may affect exposure or response to the stressor. These diverse measures increase in importance as the complexity of the assessment increases.

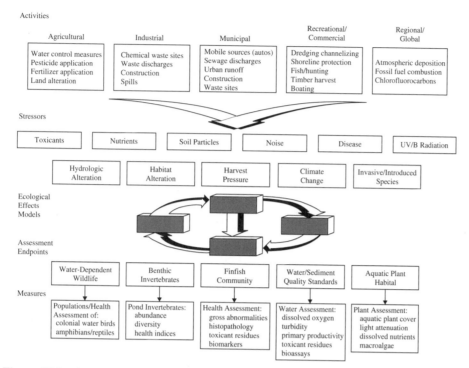

**Figure 28.2**   An example of a conceptual model for a watershed. Human activities, shown at the top of the diagram, result in various stressors that induce ecological effects. Assessment end points and related measures that are associated with these effects are shown at the bottom of the diagram.

An important consideration in the identification of these measures is their response sensitivity and ecosystem relevance. Response sensitivity is usually highest with measures at the lower levels of biological organization, but the ecosystem relevance is highest at the higher levels of biological organization. This dichotomy is illustrated in Figure 28.3. In general, the time required to illicit a response also increases with the level of biological organization. Note that toxicologists focus on measures at lower levels of biological organization, relying on an extrapolation of the toxicant effects on populations and communities that are initiated at the molecular/cellular level and, if this insult is not corrected for, or adapted to, then effects on physiological systems and individual organisms. For certain toxic modes of action (e.g., reproductive toxicity), this could result in effects at the population and community levels. In contrast, ecologists focus on measures at the population level or higher for obvious reasons of ecological relevance. A combination of measures often is necessary to provide reasonable sensitivity, ecosystem relevance, and causal relationships.

## 28.3   ANALYZING EXPOSURE AND EFFECTS INFORMATION

The second phase of ecological risk assessment, the analysis phase, includes two principal activities: characterization of exposure and characterization of ecological effects (Figure 28.1).

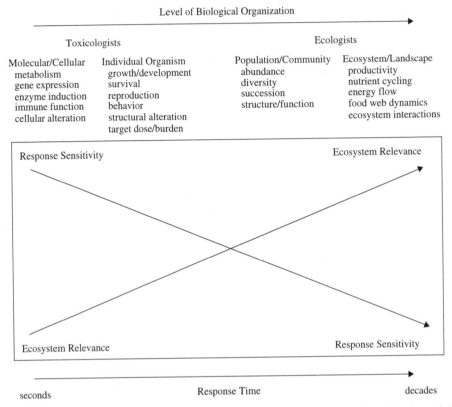

**Figure 28.3**   The response time and sensitivity of an ecological receptor is a function of the level of biological organization. Higher levels of organization have greater ecosystem relevance. However, as the level of biological organization increases, response time increases, sensitivity decreases, and causal relationships become more uncertain. Ecological risk assessments must balance the need for sensitive, timely, and well-established responses with ecological relevance.

## 28.3.1  Characterizing Exposure

In exposure characterization, credible and relevant data are analyzed to describe the source(s) of stressors, the distribution of stressors in the environment, and the contact or co-occurrence of stressors with ecological receptors. An exposure profile is developed that identifies receptors and exposure pathways, describes the intensity and spatial and temporal extent of exposure, describes the impact of variability and uncertainty on exposure estimates, and presents a conclusion about the likelihood that exposure will occur.

A source description identifies where the stressor originates, describes what stressors are generated, and considers other sources of the stressor. Exposure analysis may start with the source when it is known, but some analyses may begin with known exposures and attempt to link them to sources, while other analyses may start with known stressors and attempt to identify sources and quantify contact or co-occurrence. The source description includes what is known about the intensity, timing, and location of the stressor and whether other constituents emitted by the source influence transport, transformation, or bioavailability of the stressor of interest.

Many stressors have natural counterparts and/or multiple sources that must be considered. For example, many chemicals occur naturally (e.g., most metals), are generally widespread due to multiple sources (e.g., polycyclic aromatic hydrocarbons), or may have significant sources outside the boundaries of the current assessment (e.g., regional atmospheric deposition of PCBs). Many physical stressors also have natural counterparts such as sedimentation from construction activities versus natural erosion. In addition human activities may change the magnitude or frequency of natural disturbance cycles such as the frequency and severity of flooding. Source characterization can be particularly important for new biological stressors (e.g., invasive species), since many of the strategies for reducing risks focus on preventing entry in the first place. Once the source is identified, the likelihood of entry may be characterized qualitatively.

Because exposure occurs where receptors co-occur with or contact stressors in the environment, characterizing the spatial and temporal distribution of a stressor is a necessary precursor to estimating exposure. The stressor's spatial and temporal distribution in the environment is described by evaluating the pathways that stressors take from the source as well as the formation and subsequent distribution of secondary stressors. For chemical stressors, the evaluation of pathways usually follows the type of transport and fate modeling described in Chapter 27. Some physical stressors such as sedimentation also can be modeled, but other physical stressors require no modeling because they eliminate entire ecosystems or portions of them, such as when a wetland is filled, a resource is harvested, or an area is flooded.

The movement of biological stressors have been described as diffusion and/or jump-dispersal processes. Diffusion involves a gradual spread from the site of introduction and is a function primarily of reproductive rates and motility. Jump-dispersal involves erratic spreads over periods of time, usually by means of a vector. The gypsy moth and zebra mussel have spread this way; the gypsy moth via egg masses on vehicles and the zebra mussel via boat ballast water. Biological stressors can use both diffusion and jump-dispersal strategies, which makes it difficult to predict dispersal rates. An additional complication is that biological stressors are influenced by their own survival and reproduction.

The creation of secondary stressors can greatly alter risk. Secondary stressors can be formed through biotic or abiotic transformation processes and may be of greater or lesser concern than the primary stressor. Physical disturbances can generate secondary stressors, such as when the removal of riparian vegetation results in increased nutrients, sedimentation, and altered stream flow. For chemicals, the evaluation of secondary stressors usually focuses on metabolites or degradation products. In addition secondary stressors can be formed through ecosystem processes. For example, nutrient inputs into an estuary can decrease dissolved oxygen concentrations because they increase primary production and subsequent decomposition. A changeover from an aerobic to an anaerobic environment often is accompanied by the production of sulfide via sulfate-reducing bacteria. Sulfide can act as a secondary stressor to oxygen-dependent organisms, but it also can reduce exposure to metals through the precipitation of metal sulfides (see Chapter 27).

The distribution of stressors in the environment can be described using measurements, models, or a combination of the two. If stressors have already been released, direct measurements of environmental media or a combination of modeling and measurement is preferred. However, a modeling approach may be necessary if the assessment is intended to predict future scenarios or if measurements are not possible or practicable.

### 28.3.2   Characterizing Ecological Effects

In ecological effects characterization, relevant data are analyzed to evaluate stressor-response relationships and/or to provide evidence that exposure to a stressor causes an observed response. The characterization describes the effects that are elicited by a stressor, links these effects with the assessment endpoints, and evaluates how the effects change with varying stressor levels. The conclusions of the ecological effects characterization are summarized in a stressor-response profile.

***Analyzing Ecological Response.*** Ecological response analysis has three primary components: determining the relationship between stressor exposure and ecological effects, evaluating the plausibility that effects may occur or are occurring as a result of the exposure, and linking measurable ecological effects with the assessment end points.

Evaluating ecological risks requires an understanding of the relationships between stressor exposure and resulting ecological responses. The stressor-response relationships used in a particular assessment depend on the scope and nature of the ecological risk assessment as defined in problem formulation and reflected in the analysis plan. For example, a point estimate of an effect (e.g., an LC50) might be compared with point estimates from other stressors. The stressor-response function (e.g., shape of the curve) may be critical for determining the presence or absence of an effects threshold or for evaluating incremental risks, or stressor-response functions may be used as input for ecological effects models. If sufficient data are available, cumulative distribution functions can be constructed using multiple point estimates of effects. Process models that already incorporate empirically derived stressor-response functions also can be used. However, many stressor-response relationships are very complex, and ecological systems frequently show responses to stressors that involve abrupt shifts to new community or system types.

In simple cases the response will be one variable (e.g., mortality) and quantitative univariate analysis can be used. If the response of interest is composed of many individual variables (e.g., species abundances in an aquatic community), multivariate statistical techniques must be used. Multivariate techniques (e.g., factor and cluster analysis) have a long history of use in ecology but have not yet been extensively applied in risk assessment. Stressor-response relationships can be described using any of the dimensions of exposure (i.e., intensity, time, space). Intensity is probably the most familiar dimension and is often used for chemicals (e.g., dose, concentration). The duration of exposure also can be used for chemical stressor-response relationships; for example, median acute effects levels are always associated with a time parameter (e.g., 24 h, 48 h, 96 h). Both the time and spatial dimensions of exposure can be important for physical disturbances such as flooding. Single-point estimates and stressor-response curves can be generated for some biological stressors. For pathogens such as bacteria and fungi, inoculum levels may be related to the level of symptoms in a host or actual signs of the pathogen. For other biological stressors such as introduced species, developing simple stressor-response relationships may be inappropriate.

Causality is the relationship between cause (one or more stressors) and effect (assessment end point response to one or more stressors). Without a sound basis for linking cause and effect, uncertainty in the conclusions of an ecological risk assessment will be high. Developing causal relationships is especially important for risk assessments driven by observed adverse ecological effects such as fish kills or long-term declines in a population. Criteria need to be established for evaluating causality. For chemicals, ecotoxicologists have slightly modified Koch's postulates to provide evidence of causality:

1. The injury, dysfunction, or other putative effect of the toxicant must be regularly associated with exposure to the toxicant and any contributory causal factors.
2. Indicators of exposure to the toxicant must be found in the affected organisms.
3. The toxic effects must be seen when normal organisms or communities are exposed to the toxicant under controlled conditions, and any contributory factors should be manifested in the same way during controlled exposures.
4. The same indicators of exposure and effects must be identified in the controlled exposures as in the field.

While useful as an ideal, this approach may not be practical if resources for experimentation are not available or if an adverse effect may be occurring over such a wide spatial extent that experimentation and correlation may prove difficult or yield equivocal results. In most cases extrapolation will be necessary to evaluate causality. The scope of the risk assessment also influences extrapolation through the nature of the assessment end point. Preliminary assessments that evaluate risks to general trophic levels, such as fish and birds, may extrapolate among different genera or families to obtain a range of sensitivity to the stressor. On the other hand, assessments concerned with management strategies for a particular species may employ population models.

Whatever methods are employed to link assessment end points with measures of effect, it is important to apply the methods in a manner consistent with sound ecological and toxicological principles. For example, it is inappropriate to use structure-activity relationships to predict toxicity from chemical structure unless the chemical under consideration has a similar mode of toxic action to the reference chemicals. Similarly extrapolations from upland avian species to waterfowl may be more credible if factors such as differences in food preferences, physiology, and seasonal behavior (e.g., mating and migration habits) are considered.

Finally, many extrapolation methods are limited by the availability of suitable databases. Although these databases are generally largest for chemical stressors and aquatic species, even in these cases data do not exist for all taxa or effects. Chemical effects databases for mammals, amphibians, or reptiles are extremely limited, and there is even less information on most biological and physical stressors. Extrapolations and models are only as useful as the data on which they are based and should recognize the great uncertainties associated with extrapolations that lack an adequate empirical or process-based rationale.

***Developing a Stressor-Response Profile.*** The final activity of the ecological response analysis is developing a stressor-response profile to evaluate single species, populations, general trophic levels, communities, ecosystems, or landscapes—whatever is appropriate for the defined assessment end points. For example, if a single species is affected, effects should represent appropriate parameters such as effects on mortality, growth, and reproduction, while at the community level, effects may be summarized in terms of structure or function depending on the assessment end point. At the landscape level, there may be a suite of assessment end points, and each should be addressed separately. The stressor-response profile summarizes the nature and intensity of effect(s), the time scale for recovery (where appropriate), causal information linking the stressor with observed effects, and uncertainties associated with the analysis.

## 28.4   CHARACTERIZING RISK

Risk characterization is the final phase of an ecological risk assessment (Figure 28.1). During risk characterization, risks are estimated and interpreted and the strengths, limitations, assumptions, and major uncertainties are summarized. Risks are estimated by integrating exposure and stressor-response profiles using a wide range of techniques such as comparisons of point estimates or distributions of exposure and effects data, process models, or empirical approaches such as field observational data. Risks are described by evaluating the evidence supporting or refuting the risk estimate(s) and interpreting the adverse effects on the assessment end point. Criteria for evaluating adversity include the nature and intensity of effects, spatial and temporal scales, and the potential for recovery. Agreement among different lines of evidence of risk increases confidence in the conclusions of a risk assessment.

### 28.4.1   Estimating Risk

Risk estimation determines the likelihood of adverse effects to assessment end points by integrating exposure and effects data and evaluating any associated uncertainties. The process uses the exposure and stressor-response profiles. Risks can be estimated by one or more of the following approaches: (1) estimates based on best professional judgment and expressed as qualitative categories such as low, medium, or high; (2) estimates comparing single-point estimates of exposure and effects such as a simple ratio of exposure concentration to effects concentration (quotient method); (3) estimates incorporating the entire stressor-response relationship often as a nonlinear function of exposure; (4) estimates incorporating variability in exposure and effects estimates providing the capability to predict changes in the magnitude and likelihood of effects at different exposure scenarios; (5) estimates based on process models that rely partially or entirely on theoretical approximations of exposure and effects; and (6) estimates based on empirical approaches, including field observational data. An example of the first approach, using qualitative categorization, is shown in Figure 28.4.

### 28.4.2   Describing Risk

After risks have been estimated, available information must be integrated and interpreted to form conclusions about risks to the assessment endpoints. Risk descriptions include an evaluation of the lines of evidence supporting or refuting the risk estimate(s) and an interpretation of the adverse effects on the assessment end point. Confidence in the conclusions of a risk assessment may be increased by using several lines of evidence to interpret and compare risk estimates. These lines of evidence may be derived from different sources or by different techniques relevant to adverse effects on the assessment end points, such as quotient estimates, modeling results, field experiments,

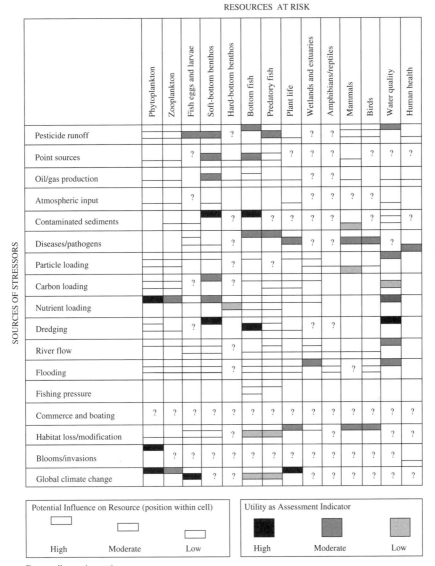

RESOURCES AT RISK

**Figure 28.4** An example of a qualitative categorization of ecological risk for a hypothetical matrix of stressors and resources at risk.

or field observations. Some of the factors to consider when evaluating separate lines of evidence are:

- Relevance of evidence to the assessment end points.
- Relevance of evidence to the conceptual model.
- Sufficiency and quality of data and experimental designs used in supporting studies.
- Strength of cause/effect relationships.
- Relative uncertainties of each line of evidence and their direction.

At this point in risk characterization, the changes expected in the assessment end points have been estimated and described. The next step is to interpret whether these changes are considered adverse and meaningful. Meaningful adverse changes are defined by ecological and/or social concerns, and thus usually depend on the best professional judgment of the risk assessor. Five criteria have been proposed by EPA for evaluating adverse changes in assessment end points:

1. Nature of effects
2. Intensity of effects
3. Spatial scale
4. Temporal scale
5. Potential for recovery

The extent to which the five criteria are evaluated depends on the scope and complexity of the ecological risk assessment. However, understanding the underlying assumptions and science policy judgments is important even in simple cases. For example, when exceedence of a previously established decision rule such as a benchmark stressor level or water quality criterion is used as evidence of adversity, the reasons why exceedences of the benchmark are considered adverse should be clearly understood.

To distinguish ecological changes that are adverse from those ecological events that are within the normal pattern of ecosystem variability or result in little or no meaningful alteration of biota, it is important to consider the nature and intensity of effects. For example, an assessment end point involving survival, growth, and reproduction of a species must consider whether predicted effects involve survival and reproduction or only growth. Or if survival of offspring are affected, the relative loss must be considered.

It is important to consider both the ecological and statistical contexts of an effect when evaluating intensity. For example, a statistically significant 1% decrease in fish growth may not be relevant to an assessment end point of fish population viability, and a 10% decline in reproduction may be worse for a population of slowly reproducing marine mammals than for rapidly reproducing planktonic algae.

Natural ecosystem variation can make it very difficult to observe (detect) stressor-related perturbations. For example, natural fluctuations in marine fish populations are often very large and cyclic events (e.g., fish migration) are very important in natural systems. Predicting the effects of anthropogenic stressors against this background of variation can be very difficult. Thus a lack of statistically significant effects in a field study does not automatically mean that adverse ecological effects are absent. Rather, factors such as statistical power to detect differences, natural variability, and other lines of evidence must be considered in reaching conclusions about risk.

Spatial and temporal scales also need to be considered in assessing the adversity of the effects. The spatial dimension encompasses both the extent and pattern of effect as well as the context of the effect within the landscape. Factors to consider include the absolute area affected, the extent of critical habitats affected compared with a larger area of interest, and the role or use of the affected area within the landscape. Adverse effects to assessment end points vary with the absolute area of the effect. A larger affected area may be (1) subject to a greater number of other stressors, increasing the complications from stressor interactions; (2) more likely to contain sensitive species or habitats; or (3) more susceptible to landscape-level changes because many ecosystems may be altered by the stressors.

Nevertheless, a smaller area of effect is not always associated with lower risk. The function of an area within the landscape may be more important than the absolute area. Destruction of small but unique areas, such as submerged vegetation at the land-water margin, may have important effects on local wildlife populations. Also, in river systems, both riffle and pool areas provide important microhabitats that maintain the structure and function of the total river ecosystem. Stressors acting on some of these microhabitats may present a significant risk to the entire system. Spatial factors also are important for many species because of the linkages between ecological landscapes and population dynamics. Linkages between one or more landscapes can provide refuge for affected populations, and species may require adequate corridors between habitat patches for successful migration.

The temporal scale for ecosystems can vary from seconds (photosynthesis, prokaryotic reproduction) to centuries (global climate change). Changes within a forest ecosystem can occur gradually over decades or centuries and may be affected by slowly changing external factors such as climate. The time scale of stressor-induced changes operates within the context of multiple natural time scales. In addition temporal responses for ecosystems may involve intrinsic time lags, so responses from a stressor may be delayed. Thus it is important to distinguish the long-term impacts of a stressor from the immediately visible effects. For example, visible changes resulting from eutrophication of aquatic systems (turbidity, excessive macrophyte growth, population decline) may not become evident for many years after initial increases in nutrient levels.

From the temporal scale of adverse effects we come to a consideration of recovery. Recovery is the rate and extent of return of a population or community to a condition that existed before the introduction of a stressor. Because ecosystems are dynamic and even under natural conditions are constantly changing in response to changes in the physical environment (weather, natural catastrophes, etc.) or other factors, it is unrealistic to expect that a system will remain static at some level or return to exactly the same state that it was before it was disturbed. Thus the attributes of a "recovered" system must be carefully defined. Examples might include productivity declines in an eutrophic system, re-establishment of a species at a particular density, species re-colonization of a damaged habitat, or the restoration of health of diseased organisms.

Recovery can be evaluated despite the difficulty in predicting events in ecological systems. For example, it is possible to distinguish changes that are usually reversible (e.g., recovery of a stream from sewage effluent discharge), frequently irreversible (e.g., establishment of introduced species), and always irreversible (e.g., species extinction). It is important to consider whether significant structural or functional changes have occurred in a system that might render changes irreversible. For example, physical alterations such as deforestation can change soil structure and seed sources such that forests cannot easily grow again.

Natural disturbance patterns can be very important when evaluating the likelihood of recovery from anthropogenic stressors. Ecosystems that have been subjected to repeated natural disturbances may be more vulnerable to anthropogenic stressors (e.g., overfishing). Alternatively, if an ecosystem has become adapted to a disturbance pattern, it may be affected when the disturbance is removed (fire-maintained grasslands). The lack of natural analogues makes it difficult to predict recovery from novel anthropogenic stressors such as exposure to synthetic chemicals.

The relative rate of recovery also can be estimated. For example, fish populations in a stream are likely to recover much faster from exposure to a degradable chemical than from habitat alterations resulting from stream channelization. It is critical to use knowledge of factors such as the temporal scales of organisms' life histories, the availability of adequate stock for recruitment, and the interspecific and trophic dynamics of the populations in evaluating the relative rates of recovery. A fisheries stock or forest might recover in several decades, a benthic infaunal community in years, and a planktonic community in weeks to months.

## 28.5 MANAGING RISK

When risk characterization is complete, a description of the risk assessment is communicated to the risk manager (Figure 28.1) to support a risk management decision. This communication usually is a report and might include:

- A description of risk assessor/risk manager planning results.
- A review of the conceptual model and the assessment end points.
- A discussion of the major data sources and analytical procedures used.
- A review of the stressor-response and exposure profiles.
- A description of risks to the assessment endpoints, including risk estimates and adversity evaluations.
- A summary of major areas of uncertainty and the approaches used to address them.
- A discussion of science policy judgments or default assumptions used to bridge information gaps, and the basis for these assumptions.

After the risk assessment is completed, risk managers may consider whether additional follow-up activities are required. Depending on the importance of the assessment, confidence level in the assessment results, and available resources, it may be advisable to conduct another iteration of the risk assessment in order to facilitate a final management decision. Ecological risk assessments are frequently designed in sequential tiers that proceed from simple, relatively inexpensive evaluations to more costly and complex assessments. Initial tiers are based on conservative assumptions, such as maximum exposure and ecological sensitivity. When an early tier cannot sufficiently define risk to support a management decision, a higher assessment tier that may require either additional data or applying more refined analysis techniques to available data may be needed. Higher tiers provide more ecologically realistic assessments while making less conservative assumptions about exposure and effects.

Another option is to proceed with a management decision based on the risk assessment and develop a monitoring plan to evaluate the results of the decision. For example, if the decision is to mitigate risks through exposure reduction, monitoring will help determine whether the desired reduction in exposure (and effects) is being achieved. Monitoring is also critical for determining the extent and nature of any ecological recovery that may be occurring.

Ecological risk assessment is important for environmental decision making because of the high cost of eliminating environmental risks associated with human activities and the necessity of making regulatory decisions in the face of uncertainty. Ecological risk assessment provides only a portion of the information required to make risk management decisions, but this information is critical to scientifically defensible risk management. Thus ecological risk assessments should provide input to a diverse set of environmental decision-making processes, such as the regulation of hazardous waste sites, industrial chemicals, and pesticides, and improve the management of watersheds affected by multiple nonchemical and chemical stressors.

## SUGGESTED READING

Bartell, S. M., R. H. Gardner, and R. V. O'Neill. *Ecological Risk Estimation*. Boca Raton, FL: Lewis Publishers, 1992.

Cardwell, R. D., B. R. Parkhurst, W. Warren-Hicks, and J. S. Volosin. Aquatic ecological risk. *Water Environ. Technol.* **5**: 47–51, 1993.

Harwell, M. A., W. Cooper, and R. Flaak. Prioritizing ecological and human welfare risks from environmental stresses. *Environ. Manag.* **16**: 451–464, 1992.

Kendall, R. J., T. E. Lacher, C. Bunck, B. Daniel, C. Driver, C. E. Grue, F. Leighton, W. Stansley, P. G. Watanabe, and M. Whitworth. An ecological risk assessment of lead shot exposure in non-waterfowl avian species: Upland game birds and raptors. *Environ. Toxicol. Chem.* **15**: 4–20, 1996.

National Research Council. A paradigm for ecological risk assessment. In *Issues in Risk Assessment*. Washington, DC: National Academy Press, 1993.

National Research Council. *Science and Judgment in Risk Assessment*. Washington, DC: National Academy Press, 1994.

National Research Council. *Understanding Risk: Informing Decisions in a Democratic Society*. Washington, DC: National Academy Press, 1996.

Ruckelshaus, W. D. Science, risk, and public policy. *Science* **221**: 1026–1028, 1983.

Solomon, K. R., D. B. Baker, R. P. Richards, K. R. Dixon, S. J. Klaine, T. W. La Point, R. J. Kendall, C. P. Weisskopf, J. M. Giddings, J. P. Geisy, L. W. Hall, W. M. Williams. Ecological risk assessment of atrazine in North American surface waters. *Environ. Toxicol. Chem.* **15**(1): 31–76, 1996.

Suter, G. W., II. Endpoints for regional ecological risk assessments. *Environ. Manag.* **14**: 19–23, 1990.

Suter, G. W., II. *Ecological Risk Assessment*. Boca Raton, FL: Lewis Publishers, 1993.

Suter, G. W., II. A critique of ecosystem health concepts and indexes. *Environ. Toxicol. Chem.* **12**: 1533–1539, 1993.

US Environmental Protection Agency. Summary report on issues in ecological risk assessment. Washington, DC: Risk Assessment Forum, USEPA, 1991. EPA/625/3-91/018.

US Environmental Protection Agency. Framework for ecological risk assessment. Washington, DC: Risk Assessment Forum, USEPA, 1992. EPA/630/R-92/001.

US Environmental Protection Agency. Peer review workshop report on a framework for ecological risk assessment. Washington, DC: Risk Assessment Forum, USEPA, 1992. EPA/625/3-91/022.

US Environmental Protection Agency. A review of ecological assessment case studies from a risk assessment perspective. Washington, DC: Risk Assessment Forum, USEPA, 1993. EPA/630/R-92/005.

US Environmental Protection Agency. Ecological risk assessment issue papers. Washington, DC: Risk Assessment Forum, USEPA, 1994. EPA/630/R-94/009.

US Environmental Protection Agency. Proposed guidelines for ecological risk assessment. Washington, DC: Risk Assessment Forum, USEPA, 1996. EPA/630/R-95/002B.

# PART VIII

# SUMMARY

# CHAPTER 29

# Future Considerations for Environmental and Human Health

ERNEST HODGSON

## 29.1 INTRODUCTION

Since the publication of the second edition of this textbook there has been rapid, and in some cases dramatic, progress not only in toxicology but in the sciences that contribute methods and insights to toxicology. However, it is still true that speculation concerning future developments in toxicology can be made only against an assessment of where the science has come from and its current status. Toxicology, despite its use of many state-of-the-art techniques and explorations of the most fundamental molecular mechanisms of toxic action, is, at its heart, an applied science serving the needs of society. Society is served in two principal ways: the protection of human health and the protection of the environment. In both of these aspects two avenues are explored: studies of chemicals in use and the development of new chemicals that are both safe and effective. These studies range from studies of the mechanisms of toxic action to in vivo toxicity testing, but the ultimate goal is a meaningful assessment of risk resulting from exposure to the chemicals in question.

The vast increase in public awareness of the potential of chemicals to cause harmful effects and the propensity of the print and electronic media to fan the flames of controversy in this area make certain the continued need for toxicologists. We need to ask what they will be doing during the next few decades compared to what they have been doing in the immediate past.

Through the 1950s and 1960s toxicology tended to be a largely descriptive science, relating the results of in vivo dosing to a variety of toxic end points, in many cases little more that the medial lethal dose (LD50) or median lethal concentration (LC50). However, ongoing studies of xenobiotic-metabolizing enzymes were attracting more attention and techniques for chemical analysis of toxicants were starting to undergo a remarkable metamorphosis. The 1970s were most remarkable for developments in metabolism and the beginnings of a boom in mode of toxic action studies, whereas the 1980s and 1990s saw the incorporation of the techniques of molecular biology into many aspects of toxicology, but perhaps to greatest effect in studies of the mechanisms of chemical carcinogenesis and the induction of xenobiotic-metabolizing enzymes.

*A Textbook of Modern Toxicology, Third Edition,* edited by Ernest Hodgson
ISBN 0-471-26508-X Copyright © 2004 John Wiley & Sons, Inc.

It should be emphasized that all of these activities proceed simultaneously, and that increased emphasis and interest in any particular area is often preceded by the development of new techniques—for example, the tremendous increase in specificity and sensitivity of chemical methods has proceeded simultaneously with the introduction of molecular biologic techniques into studies of mechanisms of toxic action.

The success of the project to describe the human genome along with progress in the definition of polymorphisms in human xenobiotic-metabolizing enzymes and other proteins will certainly lead to the ability to define populations and individuals at increased risk from a particular chemical insult. This ability will be extended and put on a more mechanistic basis by advances in the new disciplines of proteomics and metabonomics.

The future, both immediate and long term, will provide important information an all aspects of toxic action and the role of toxicology in public life will mature as the importance of toxicology is perceived by the population in general, first in developed countries and ultimately around the world. The fundamental role of the toxicologist, namely the acquisition and dissemination of information about all aspects of the deleterious effects of chemicals on living organisms, will not change; however, the manner in which it is carried out will almost certainly change. The next several decades will be exciting times for toxicologists, and those in training at this time have much to anticipate.

Change can be expected in almost every aspect of both the applied and the fundamental aspects of toxicology. Risk communication, risk assessment, hazard and exposure assessment, in vivo toxicity, development of selective chemicals, in vitro toxicology, and biochemical and molecular toxicology will all change, as will the integration of all of these areas into new paradigms of risk assessment and of the ways in which chemicals affect human health and the environment.

The importance of a new group of potential toxicants, genetically modified plants (GMPs) and their constituents, has emerged in the last decade. Potentially a boon to the human race, they have already generated considerable controversy. While these products of applied molecular biology appear to be relatively harmless, both to human health and to the environment, they will need to be monitored as they increase in number and complexity.

## 29.2  RISK MANAGEMENT

Public decisions concerning the use of chemicals will continue to be a blend of science, politics, and law, with the media spotlight continuing to shine on the most contentious aspects: the role of the trained toxicologist to serve as the source of scientifically sound information and as the voice of reason will be even more critical. As the chemist extends our ability to detect smaller and smaller amounts of toxicants in food, air, and water, the concept that science, including toxicology, does not deal in certainty but only in degrees of certitude must be made clear to all. Although this concept is easy for most scientists to grasp, it appears difficult, even arcane, to the general public and almost impossible to the average attorney or politician. Risk will have to be managed in the light of our new found ability to identify individuals and populations at increased and to accommodate new legislation such as the Food Quality Protection Act.

## 29.3   RISK ASSESSMENT

In the past, risk assessment consisted largely of computer-based models written to start from hazard assessment assays, such as chronic toxicity assays on rodents, encompass the necessary extrapolations between species and between high and low doses, and then produce a numerical assessment of the risk to human health. Although the hazard assessment tests and the toxic end points are different, an analogous situation exists in environmental risk assessment. A matter of considerable importance, now getting some belated attention, is the integration of human health and environmental risk assessments.

Although many of these risk assessment programs were statistically sophisticated, they frequently did not rise above the level of numbers crunching, and more often than not, different risk assessment programs, starting with the same experimental values, produced very different numerical assessments of risk to human health or to the environment. Although having risk assessment become more science based has been a stated goal of regulators for decades, its scientific basis has not been advanced significantly. The need to incorporate mechanistic data, including mode of action studies and physiologically based pharmacokinetics, has been realized to some extent. Apart from epidemiology and exposure analysis, human studies have not, despite the fact that many such studies can now be performed in noninvasive and ethical experiments.

The immediate future in risk assessment will focus on the difficult but necessary task of integrating experimental data from all levels into the risk assessment process. A continuing challenge to toxicologists engaged in hazard or risk assessment is that of risk from chemical mixtures. Neither human beings nor ecosystems are exposed to chemicals one at a time, yet logic dictates that the initial assessment of toxicity start with individual chemicals. The resolution of this problem will require considerable work at all levels, in vivo and in vitro, into the implications of chemical interactions for the expression to toxicity, particularly chronic toxicity.

## 29.4   HAZARD AND EXPOSURE ASSESSMENT

The enormous cost of multiple-species, multiple-dose, lifetime evaluations of chronic effects has already made the task of carrying out hazard assessments of all chemicals in commercial use impossible. At the same time, quantitative structure activity relationship (QSAR) studies are not yet predictive enough to indicate which chemicals should be so tested and which chemicals need not be tested. In exposure assessment, continued development of analytical methods will permit ever more sensitive and selective determinations of toxicants in food and the environment, as well as the effects of chemical mixtures and the potential for interactions that affect the ultimate expression of toxicity. Developments in QSARs, in short-term tests based on the expected mechanism of toxic action and simplification of chronic testing procedures, will all be necessary if the chemicals to which the public and the environment are exposed are to be assessed adequately for their potential to cause harm.

## 29.5   IN VIVO TOXICITY

Although developments continue in elucidating the mechanisms of chemical carcinogenicity, much remains to be done with regard to this and other chronic end points, particularly developmental and reproductive toxicity, chronic neurotoxicity, and immuno-toxicity. The further utilization of the methods of molecular biology will bring rapid advances in all of these areas. It will be a challenge to integrate all of this information into useful paradigms for responsible and meaningful risk assessments.

## 29.6   IN VITRO TOXICITY

In vitro studies of toxic mechanisms will depend heavily on developments in molecular biology, and great advances can be expected. Many of the ethical problems associated with carrying out studies on the effects of toxicants on humans will be circumvented at the in vitro level by the use of cloned and expressed human enzymes, receptors, and so on, although the integration of these data into intact organism models will still require experimental animals. High-throughput technology in genomics, proteomics, and metabonomics will greatly facilitate these studies.

## 29.7   BIOCHEMICAL AND MOLECULAR TOXICOLOGY

As indicated previously, contributions to all aspects of the mechanistic study of toxic action from the use of biochemical and molecular techniques can be expected. No doubt new techniques will be developed, answers will be found to many questions that did not yield to earlier techniques and new questions will be raised. The challenge, as always, will be to integrate the results form these studies—and reach new levels of sophistication—into useful and productive approaches to reduce chemical effects on human health and the environment.

## 29.8   DEVELOPMENT OF SELECTIVE TOXICANTS

Almost all aspects of contemporary human society depend on the use of numerous chemicals. Except in the unlikely event that society decides to return to a more sim-plistic and, in fact, more primitive, more unhealthy, and more demanding lifestyle, the challenge is in learning how to live with anthropomorphic chemicals, and not in learning how to live without them. In many aspects, such as the production of food and fiber and the maintenance of human health, the development of selective pesticides, drugs, and so on, is needed. New techniques in molecular biology, in particular, the availability of cloned and expressed human enzymes and receptors and new knowl-edge of human polymorphisms, will make this task easier, as will similar knowledge of target species, including microorganisms causing human disease, and insects and weeds affecting the production of food and fiber, and so on.

High-throughput techniques will not only speed up the search, but in this area, as in other aspects of toxicology, bioinformatics will be necessary, not only for correlating the data from many sources but also for reducing it for practical applications.

**acceptable daily intake (ADI)** Amount of exposure determined to be "safe"; usually derived from the lowest No-Effect Level in an experimental study, divided by a safety factor such as 100. Also known as the Reference Dose (RfD).

**acetylation** The addition of an acetyl group from acetyl coenzyme A to a xenobiotic or xenobiotic metabolite by the enzyme $N$-acetyltransferase. Polymorphisms in this enzyme can be important in the expression of toxicity in humans.

**acetylator phenotype** Variation in the expression of $N$-acetyltransferase isoforms in humans gives rise to two subpopulations—fast and slow acetylators. Slow acetylators are more susceptible to the toxic effects of toxicants that are detoxified by acetylation.

**acid deposition** Wet and dry air pollutants that lower the pH of deposition and subsequently the pH of the environment. Acid rain with a pH of 4 or lower refers to the wet components. Normal rain has a pH of about 5.6. Sulfuric acid from sulfur and nitric acid from nitrogen oxides are the major contributors. In lakes in which the buffering capacity is low, the pH becomes acidic enough to cause fish kills, and the lakes cannot support fish populations. A contributing factor is the fact that acidic conditions concurrently release toxic metals, such as aluminum, into the water.

**activation (bioactivation)** In toxicology, this term is used to describe metabolic reactions of a xenobiotic in which the product is more toxic than is the substrate. Such reactions are most commonly monooxygenations, the products of which are electrophiles that, if not detoxified by phase II (conjugation) reactions, may react with nucleophilic groups on cellular macromolecules such as proteins and DNA.

**active oxygen** Used to describe various short-lived highly reactive intermediates in the reduction of oxygen. Active oxygen species such as superoxide anion and hydroxyl radical are known or believed to be involved in several toxic actions. Superoxide anion is detoxified by superoxide dismutase.

**acute toxicity tests** The most common tests for acute toxicity are the LC50 and LD50 tests, which are designed to measure mortality in response to an acute toxic insult. Other tests for acute toxicity include dermal irritation tests, dermal sensitization tests, eye irritation tests, photoallergy tests, and phototoxicity tests. *See also* eye irritation tests; LC50; and LD50.

**acute toxicity** Refers to adverse effects on, or mortality of, organisms following soon after a brief exposure to a chemical agent. Either a single exposure or multiple exposures within a short time period may be involved, and an acute effect is generally regarded as an effect that occurs within the first few days after exposure, usually less than two weeks.

*A Textbook of Modern Toxicology, Third Edition,* edited by Ernest Hodgson
ISBN 0-471-26508-X Copyright © 2004 John Wiley & Sons, Inc.

**adaptation to toxicants** Refers to the ability of an organism to show insensitivity or decreased sensitivity to a chemical that normally causes deleterious effects. The terms resistance and tolerance are closely related and have been used in several different ways. However, a consensus is emerging to use the term *resistance* to mean that situation in which a change in the genetic constitution of a population in response to the stressor chemical enables a greater number of individuals to resist the toxic action than were able to resist it in the previous unexposed population. Thus an essential feature of resistance is selection and then inheritance by subsequent generations. In microorganisms, this frequently involves mutations and induction of enzymes by the toxicant; in higher organisms, it usually involves selection for genes already present in the population at low frequency. The term *tolerance* is then reserved for situations in which individual organisms acquire the ability to resist the effect of a toxicant, usually as a result of prior exposure.

**Ah locus** A gene(s) controlling the trait of responsiveness for induction of enzymes by aromatic hydrocarbons. In addition to aromatic hydrocarbons such as the polycyclics, the chlorinated dibenzo-*p*-dioxins, dibenzofurans, and biphenyls, as well as the brominated biphenyls, are involved. This trait, originally defined by studies of induction of hepatic aryl hydrocarbon hydroxylase activity following 3-methylcholanthrene treatment, is inherited by simple autosomal dominance in crosses and backcrosses between C57BL/6 (Ah-responsive) and DBA/2 (Ah-nonresponsive) mice.

**Ah receptor (AHR)** A protein coded for by a gene of the Ah locus. The initial location of the Ah receptor is believed to be in the cytosol and, after binding to a ligand such as TCDD, is transported to the nucleus. Binding of aromatic hydrocarbons to the Ah receptor of mice is a prerequisite for the induction of many xenobiotic metabolizing enzymes, as well as for two responses to TCDD; epidermal hyperplasia and thymic atrophy. Ah-responsive mice have a high-affinity receptor, whereas the Ah-nonresponsive mice have a low-affinity receptor.

**air pollution** In general, the principal air pollutants are carbon monoxide, oxides of nitrogen, oxides of sulfur, hydrocarbons, and particulates. The principal sources are transportation, industrial processes, electric power generation, and the heating of buildings. Hydrocarbons such as benzo(*a*)pyrene are produced by incomplete combustion and are associated primarily with the automobile. They are usually not present at levels high enough to cause direct toxic effects but are important in the formation of photochemical air pollution, formed as a result of interactions between oxides of nitrogen and hydrocarbons in the presence of ultraviolet light, giving rise to lung irritants such as peroxyacetyl nitrate, acrolein, and formaldehyde. Particulates are a heterogeneous group of particles, often seen as smoke, that are important as carriers of absorbed hydrocarbons and as irritants to the respiratory system.

**alkylating agents** These are chemicals that can add alkyl groups to DNA, a reaction that can result either in mispairing of bases or in chromosome breaks. The mechanism of the reaction involves the formation of a reactive carbonium ion that combines with electron-rich bases in DNA. Thus alkylating agents such as dimethylnitrosomine are frequently carcinogens and/or mutagens.

**Ames test** An in vitro test for mutagenicity utilizing mutant strains of the bacterium *Salmonella typhimurium* that is used as a preliminary screen of chemicals for assessing potential carcinogenicity. Several strains are available that cannot grow in the absence of histidine because of metabolic defects in histidine biosynthesis. Mutagens and presumed carcinogens can cause mutations that enable the strains to regain their ability to grow in a histidine deficient medium. The test can be formed in the presence of the S-9 fraction from rat liver to allow the metabolic activation of promutagens. There is a high correlation between bacterial mutagenicity and carcinogenicity of chemicals.

**antagonism** In toxicology, antagonism is usually defined as that situation in which the toxicity to two or more compounds administered together is less than that expected from consideration of their toxicities when administered alone. Although this definition includes lowered toxicity resulting from induction of detoxifying enzymes, antagonism is frequently considered separately because of the time that must elapse between treatment with the inducer and subsequent treatment with the toxicant. Antagonism not involving induction is often at a marginal level of detection and is consequently difficult to explain. Such antagonism may involve competition for receptor sites or nonenzymatic combination of one toxicant with another to reduce the toxic effect. Physiological antagonism, in which two agonists act on the same physiological system but produce opposite effects, may also occur.

**antibody** A large protein first expressed on the surface of the B cells of the immune system, followed by a series of events resulting in a clone of plasma cells that secrete the antibody into body fluids. Antibodies bind to the substance (generally a protein) that stimulated their production but may cross-react with related proteins. The natural function is to bind foreign substances such as microbes or microbial products, but because of their specificity, antibodies are used extensively in research and in diagnostic and therapeutic procedures.

**antidote** A compound administered in order to reverse the harmful effect(s) of a toxicant. They may be toxic mechanism specific, as in the case of 2-pyridine aldoxime (2-PAM) and organophosphate poisoning, or nonspecific, as in the case of syrup of ipecac, used to induce vomiting and, thereby, elimination of toxicants from the stomach.

**behavioral toxicity** Behavior may be defined as an organism's motor or glandular response to changes in its internal or external environment. Such changes may be simple or highly complex, innate or learned, but in any event represent one of the final integrated expressions of nervous system function. Behavioral toxicity is adverse or potentially adverse effects on such expression brought about by exogenous chemicals.

**binding, covalent** *See* covalent binding.

**bioaccumulation** The accumulation of a chemical either from the medium (usually water) directly or from consumption of food containing the chemical. Biomagnification is often used as a synonym for bioaccumulation, but it is more correctly used to describe an increase in concentration of a chemical as it passes from organisms at one tropic level to organisms at higher tropic levels.

**bioactivation** *See* activation.

**bioassay** This term is used in two distinct ways. The first and most appropriate is the use of a living organism to measure the amount of a toxicant present in a sample or the toxicity of a sample. This is done by comparing the toxic effect of the sample with that of a graded series of concentrations of a known standard. The second and less appropriate meaning is the use of animals to investigate the toxic effects of chemicals as in chronic toxicity tests.

**burden of proof** Responsibility for determining whether a substance is safe or hazardous; a range of approaches can be seen when comparing laws. For example, for

OSHA, regulators show a substance is hazardous before exposure is restricted, with the government conducting the tests; for FDA, manufacturers must show lack of hazard before marketing.

**biomagnification** *See* bioaccumulation.

**carcinogen** Any chemical or process involving chemicals that induces neoplasms that are not usually observed, the earlier induction of neoplasms than are commonly observed, and/or the induction of more neoplasms than are usually found.

**carcinogen, epigenetic** Cancer-causing agents that exert their carcinogenic effect by mechanisms other than genetic, such as by immunosuppression, hormonal imbalance, or cytotoxicity. They may act as cocarcinogens or promoters. Epigenetic carcinogenesis is not as well understood a phenomenon as is genotoxic carcinogenesis.

**carcinogen, genotoxic** Cancer-causing agents that exert their carcinogenic effect by a series of events that is initiated by an interaction with DNA, either directly or through an electrophilic metabolite.

**carcinogen, proximate** *See* carcinogen, ultimate.

**carcinogen, ultimate** Many, if not most, chemical carcinogens are not intrinsically carcinogenic but require metabolic activation to express their carcinogenic potential. The term *procarcinogen* describes the initial reactive compound, the term *proximate* carcinogen describes its more active products, and the term *ultimate* carcinogen describes the product that is actually responsible for carcinogenesis by its interaction with DNA.

**carcinogenesis** This is the process encompassing the conversion of normal cells to neoplastic cells and the further development of these neoplastic cells into a tumor. This process results from the action of specific chemicals, certain viruses, or radiation. Chemical carcinogens have been classified into those that are genotoxic and those that are epigenetic (i.e., not genotoxic).

**chronic toxicity** This term is used to describe adverse effects manifested after a long time period of uptake of small quantities of the toxicant in question. The dose is small enough that no acute effects are manifested, and the time period is frequently a significant part of the expected normal lifetime of the organism. The most serious manifestation of chronic toxicity is carcinogenesis, but other types of chronic toxicity are also known (e.g., reproductive effects, behavioral effects).

**chronic toxicity tests** Chronic tests are those conducted over a significant part of the lifespan of the test species or, in some cases, more than one generation. The most important tests are carcinogenicity tests, and the most common test species are rats and mice.

**cocarcinogenesis** Cocarcinogenesis is the enhancement of the conversion of normal cells to neoplastic cells. This process is manifested by enhancement of carcinogenesis when the agent is administered either before or together with a carcinogen. Cocarcinogenesis should be distinguished from promotion as, in the latter case, the promoter must be administered after the initiating carcinogen.

**comparative toxicology** The study of the variation in the expression of the toxicity of exogenous chemicals toward organisms of different taxonomic groups or of different genetic strains.

**compartment** In pharmaco(toxico)kinetics a compartment is a hypothetical volume of an animal system wherein a chemical acts homogeneously in transport and transformation. These compartments do not correspond to physiological or anatomic areas but are abstract mathematical entities useful for predicting drug concentrations. Transport into, out of, or between compartments is described by rate constants, which are used in models of the intact animal.

**conjugation reactions** *See* phase II reactions.

**covalent binding** This involves the covalent bond or "shared electron pair" bond. Each covalent bond consists of a pair of electrons shared between two atoms and occupying two stable orbitals, one of each atom. Although this is distinguished from the ionic bond or ionic valence, chemicals bonds may in fact show both covalent and ionic character. In toxicology the term covalent binding is used less precisely to refer to the binding of toxicants or their reactive metabolites to endogenous molecules (usually macromolecules) to produce stable adducts resistant to rigorous extraction procedures. A covalent bond between ligand and macromolecule is generally assumed. Many forms of chronic toxicity involve covalent binding of the toxicant to DNA or protein molecules within the cell.

**cross resistance, cross tolerance** These terms describe the situation where either resistance or tolerance to a particular toxicant (as defined under adaptation to toxicants) is induced by exposure to a different toxicant. This is commonly seen in resistance of insects to insecticides in which selection with one insecticide brings about a broad spectrum of resistance to insecticides of the same or different chemical classes. Such cross resistance is usually caused by the inheritance of a high level of nonspecific xenobiotic-metabolizing enzymes.

**cytotoxicity** Cellular injury or death brought about by chemicals external to the cell. Such chemicals may be soluble mediators produced by the immune system, or they may be chemicals (toxicants) to which the organism has been exposed.

**Delaney Amendment** *See* Food, Drug, and Cosmetics Act.

**detoxication** A metabolic reaction or sequence of reactions that reduces the potential for adverse effect of a xenobiotic. Such sequences normally involve an increase in water solubility that facilitates excretion and/or the reaction of a reactive product with an endogenous substrate (conjugation), thereby not only increasing water solubility but also reducing the possibility of interaction with cellular macromolecules. Not to be confused with detoxification. *See also* detoxification.

**detoxification** Treatment by which toxicants are removed from intoxicated patients or a course of treatment during which dependence on alcohol or other drugs of abuse is reduced or eliminated. Not to be confused with detoxication. *See also* detoxication.

**distribution** The movement of a toxicant from the portal of entry to the tissue and also the different concentrations reached in different locations. The toxicant's movement involves the study of transport mechanisms primarily in the blood, and both movement and concentration are subject to mathematical analysis in toxicokinetic studies.

**dosage** The amount of a toxicant drug or other chemical administered or taken expressed as some function of the organism, (e.g., mg/kg body weight/day).

**dose** The total amount of a toxicant, drug, or other chemical administered to or taken in by the organism.

**dose-response relationship** In toxicology the quantitative relationship between the amount of a toxicant administered or taken and the incidence or extent of the adverse effect.

**dose-response assessment** A step in the risk assessment process to characterize the relationship between the dose of a chemical administered to a population of test animals and the incidence of a given adverse effect. It involves mathematical modeling techniques to extrapolate from the high-dose effects observed in test animals to estimate the effects expected from exposure to the typically low doses that may be encountered by humans.

**Draize test** *See* eye irritation test.

**drugs of abuse** Although all drugs may have deleterious effects on humans, drugs of abuse either have no medicinal function or are taken at higher than therapeutic doses. Some drugs of abuse may affect only higher nervous functions (i.e., mood, reaction time, coordination), but many produce physical dependence and have serious physical effects, with fatal overdose being a common occurrence. The drugs of abuse include central nervous system (CNS) depressants such as ethanol, methaqualone (Quaalude), and secobarbital; CNS stimulants such as cocaine, methamphetamine (speed), caffeine, and nicotine; opioids such as heroin and morphine; hallucinogens such as lysergic acid diethylamide (LSD), phencyclidine (PCP), and tetrahydrocannabinol (THC), the most important active principle of marijuana.

**drugs, therapeutic** All therapeutic drugs can be toxic at some dose. The danger to the patient is dependent on the nature of the toxic response, the dose necessary to produce the toxic response, and the relationship between the therapeutic and the toxic dose. Drug toxicity is affected by all of those factors that affect the toxicity of xenobiotics, including (genetic) variation, diet, age, and the presence of other exogenous chemicals. The risk of toxic side effects from a particular drug must be weighed against the expected benefits; the use of a quite dangerous drug with only a narrow tolerance between the therapeutic and toxic doses might well be justified if it is the sole treatment for an otherwise fatal disease. For example, cytotoxic agents used in the treatment of cancer are known carcinogens.

**ecotoxicology** *See* environmental toxicology.

**electron transport system (ETS)** This term is often restricted to the mitochondrial system, although it applies equally well to other systems, including that of microsomes and chloroplasts. The mitochondrial ETS (also termed respiratory chain or cytochrome chain) consists of a series of cytochromes and other electron carriers arranged in the inner mitochondrial membrane. These components transfer the electrons from NADH or $FADH_2$ generated in metabolic oxidations, to oxygen, the final electron acceptor, through a series of alternate oxidations and reductions. The energy that these electrons lose during these transfers is used to pump $H^+$ from the matrix into the intermembrane space, creating an electrochemical proton gradient that drives oxidative phosphorylation. The energy is conserved as adenosine triphosphate (ATP).

**electron transport system (ETS) inhibitors** The three major respiratory enzyme complexes of the mitochondrial electron transport system can all be blocked by inhibitors. For example, rotenone inhibits the NADH dehydrogenase complex, antimycin A inhibits the b-c complex, and cyanide and carbon monoxide inhibits the cytochrome oxidase complex. Although oxidative phosphorylation inhibitors prevent phosphorylation while allowing electron transfers to proceed, ETS inhibitors prevent both electron transport and ATP production.

**electrophilic** Electrophiles are chemicals that are attracted to and react with electron-rich centers in other molecules in reactions known as electrophilic reactions. Many activation reactions produce electrophilic intermediates such as epoxides, which exert their toxic action by forming covalent bonds with nucleophilic centers in cellular macromolecules such as DNA or proteins.

**endoplasmic reticulum** The endoplasmic reticulum (ER) is an extensive branching and anastomosing double membrane distributed in the cytoplasm of eukaryotic cells. The ER is of two types: rough ER (RER) contains attached ribosomes on the cytosolic surface and smooth ER (SER), devoid of ribosomes are involved in protein biosynthesis; RER is abundant in cells specialized for protein synthesis. Many xenobiotic-metabolizing enzymes are integral components of both SER and RER, such as the cytochrome P450-dependent monooxygenase system and the flavin-containing monooxygenase, although the specific content is usually higher in SER. When tissue or cells are disrupted by homogenization, the ER is fragmented into many smaller (ca. 100 nm in diameter) closed vesicles called microsomes, which can be isolated by differential centrifugation.

**enterohepatic circulation** This term describes the excretion of a compound into the bile and its subsequent reabsorption from the small intestine and transport back to the liver, where it is available again for biliary excretion. The most important mechanism is conjugation in the liver, followed by excretion into the bile. In the small intestine the conjugation product is hydrolyzed, either nonenzymatically or by the microflora, and the compound is reabsorbed to become a substrate for conjugation and reexcretion into the bile.

**environmental toxicology** This is concerned with the movement of toxicants and their metabolites in the environment and in food chains and the effect of such toxicants on populations of organisms.

**epigenetic carcinogen** *See* carcinogen, epigenetic.

**exposure assessment** A component of risk assessment. The number of individuals likely to be exposed to a chemical in the environment or in the workplace is assessed, and the intensity, frequency, and duration of human exposure are estimated.

**eye irritation test (Draize test)** Eye irritation tests measure irritancy of compounds applied topically to the eye. These tests are variations of the Draize test, and the experimental animal used is the albino rabbit. The test consists of adding the material to be tested directly into the conjunctival sac of one eye of each of several albino rabbits, the other eye serving as the control. This test is probably the most controversial of all toxicity tests, being criticized primarily on the ground that it is inhumane. Moreover, because both concentrations and volumes used are high and show high variability, it has been suggested that these tests cannot be extrapolated to humans. However, because visual impairment is a critical toxic end point, tests for ocular toxicity are essential. Attempts to solve the dilemma have taken two forms: to find substitute in vitro tests and to modify the Draize test so that it becomes not only more humane but also more predictive for humans.

**Federal Insecticide, Fungicide, and Rodenticide Act (FIFRA)** This law is the basic US law under which pesticides and other agricultural chemicals distributed in interstate commerce are registered and regulated. First enacted in 1947, FIFRA placed the regulation of agrochemicals under the control of the US Department of Agriculture. In 1970 this responsibility was transferred to the newly created Environmental Protection Agency (EPA). Subsequently FIFRA has been revised extensively by the Federal Environmental Pesticide Control Act (FEPCA) of 1972 and by the FIFRA amendments of 1975, 1978, and 1980. Under FIFRA all new pesticide products used in the United States must be registered with the EPA. This requires the registrant to submit information on the composition, intended use, and efficacy of the product, along with a comprehensive database establishing that the material can be used without causing unreasonable adverse effects on humans or on the environment. The Food Quality Protection Act of 1996 is an amendment to FIFRA.

**fetal alcohol syndrome (FAS)** FAS refers to a pattern of defects in children born to alcoholic mothers. Three criteria for FAS are prenatal or postnatal growth retardation; characteristic facial anomalies such as microcephaly, small eye opening, and thinned upper lip; and central nervous system dysfunction, such as mental retardation and developmental delays.

**food additives** Chemicals may be added to food as preservatives (either antibacterial or antifungal compounds or antioxidants) to change the physical characteristics, for processing, or to change the taste or odor. Although most food additives are safe and without chronic toxicity, many were introduced when toxicity testing was relatively unsophisticated and some have been shown subsequently to be toxic. The most important inorganic additives are nitrate and nitrite. Well-known examples of food additives include the antioxidant butylatedhydroxyanisole (BHA), fungistatic agents such as methyl *p*-benzoic acid, the emulsifier propylene glycol, sweeteners such as saccharin and aspartame, and dyes such as tartrazine and Sunset Yellow.

**food contaminants (food pollutants)** Food contaminants, as opposed to food additives, are those compounds included inadvertently in foods that are raw, cooked, or processed. They include bacterial toxins such as the exotoxin of *Clostridium botulinum*, mycotoxins such as aflatoxins from *Aspergillus flavus*, plant alkaloids, animal toxins, pesticide residues, residues of animal food additives such as diethylstilbestrol (DES) and antibiotics, and a variety of industrial chemicals such as polychlorinated biphenyls (PCBs) and polybrominated biphenyls (PBBs).

**Food, Drug, and Cosmetics Act** The Federal Food, Drug, and Cosmetic Act is administered by the Food and Drug Administration (FDA). It establishes limits for food additives, sets criteria for drug safety for both human and animal use, and requires proof of both safety and efficacy. This act contains the Delaney Amendment, which states that food additives that cause cancer in humans or animals at any level shall not be considered safe and are, therefore, prohibited. The Delaney Amendment has been modified to permit more flexible use of mechanistic and cost-benefit data. This law also empowers the FDA to establish and modify the Generally Recognized as Safe (GRAS) list and to establish Good Laboratory Practice (GLP) rules. As a result of the Food Quality Protection (1996), the Delaney Amendment no longer applies to chemicals regulated under FIFRA.

**forensic toxicology** Forensic toxicology is concerned with the medicolegal aspects of the adverse effects of chemicals on humans and animals. Although primarily devoted to the identification of the cause and circumstances of death and the legal issues arising there from, forensic toxicologists also deal with sublethal poisoning cases.

**free radicals** Molecules that have unpaired electrons. Free radicals may be produced metabolically from xenobiotics and, because they are extremely reactive, may be involved in interactions with cellular macromolecules, giving rise to adverse effects. Examples include the trichloromethyl radical produced from carbon tetrachloride or the carbene radical produced by oxidation of the acetal carbon of methylenedioxyphenyl synergists.

**genotoxic carcinogen** *See* carcinogen, genotoxic.

**genotoxicity** Genotoxicity is an adverse effect on the genetic material (DNA) of living cells that, on the replication of the cells, is expressed as a mutagenic or a carcinogenic event. Genotoxicity results from a reaction with DNA that can be measured either biochemically or in short-term tests with end points that reflect DNA damage.

**Good Laboratory Practice (GLS)** In the United States this is a code of laboratory procedures laid down under federal law and to be followed by laboratories undertaking toxicity tests whose results will be used for regulatory or legal purposes.

**Generally Regarded as Safe (GRAS) list** *See* Food, Drug, and Cosmetics Act.

**hazard identification** Considered the first step in risk assessment, hazard identification involves the qualitative determination of whether exposure to a chemical causes an increased incidence of an adverse effect, such as cancer or birth defects, in a population of test animals and an evaluation of the relevance of this information to the potential for causing similar effects in humans.

**hepatotoxicity** Hepatotoxicants are those chemicals causing adverse effects on the liver. The liver may be particularly susceptible to chemical injury because of its anatomic relationship to the most important portal of entry, the gastrointestinal (GI) tract, and its high concentration of xenobiotic-metabolizing enzymes. Many of these enzymes, particularly cytochrome P450, metabolize xenobiotics to produce reactive intermediates that can react with endogenous macromolecules such as proteins and DNA to produce adverse effects.

**immunotoxicity** This term can be used in either of two ways. The first refers to toxic effects mediated by the immune system, such as dermal sensitivity reactions to compounds like 2,4-dinitrochlorobenzene. The second, and currently most acceptable definition, refers to toxic effects that impair the functioning of the immune system—for example, the ability of a toxicant to impair resistance to infection.

**in vitro tests** Literally these are tests conducted outside of the body of the organism. In toxicity testing they would include studies using isolated enzymes, subcellular organelles, or cultured cells. Although technically the term would not include tests involving intact eukaryotes (e.g., the Ames test), it frequently is used by toxicologists to include all short-term tests for mutagenicity that are normally used as indictors of potential carcinogenicity.

**in vivo tests** Tests carried out on the intact organism, although the evaluation of the toxic end point almost always requires pathological or biochemical examination of the test organism's tissues. They may be acute, subchronic, or chronic. The best known are the lifetime carcinogenesis tests carried out on rodents.

**induction** The process of causing a quantitative increase in an enzyme as a result of de novo protein synthesis following exposure to an inducing agent. This can occur either by a decrease in the degradation rate or an increase in the synthesis rate or both. Increasing the synthesis rate is the most common mechanism for induction by xenobiotics. Coordinate (pleiotypic) induction is the induction of multiple enzymes by a single inducing agent. For example, phenobarbital can induce isozymes of both cytochrome P450 and glutathione *S*-transferase.

**industrial toxicology** A specific area of environmental toxicology dealing with the work environment; it includes risk assessment, establishment of permissible levels of exposure, and worker protection.

**inhibition** In its most general sense, inhibition means a restraining or a holding back. In biochemistry and biochemical toxicology, inhibition is a reduction in the rate of an enzymatic reaction, and an inhibitor is any compound causing such a reduction. Inhibition of enzymes important in normal metabolism is a significant mechanism of toxic action of xenobiotics, whereas inhibition of xenobiotic-metabolizing enzymes can have important consequences in the ultimate toxicity of their substrates. Inhibition is sometimes used in toxicology in a more general and rather ill-defined way to refer to the reduction of an overall process of toxicity, as in the inhibition of carcinogenesis by a particular chemical.

**initiation** The initial step in the carcinogenic process involving the conversion of a normal cell to a neoplastic cell. Initiation is considered to be a rapid and essentially irreversible change involving the interaction of the ultimate carcinogen with DNA; this change primes the cell for subsequent neoplastic development via the promotion process.

**intoxication** In the general sense, this term refers primarily to inebriation with ethyl alcohol, secondarily to excitement or delirium cased by other means, including other chemicals. In the clinical sense it refers to poisoning or becoming poisoned. In toxicology it is sometimes used as a synonym for activation—that is, the production of a more toxic metabolite from a less toxic parent compound. This latter use of intoxication is ambiguous and should be abandoned in favor of the aforementioned general meanings, and activation or bioactivation used instead.

**isozymes (isoenzymes)** Isozymes are multiple forms of a given enzyme that occur within a single species or even a single cell and that catalyze the same general reaction but are coded for by different genes. Different isozymes may occur at different life stages and/or in different organs and tissues, or they may coexist within the same cell. The first well-characterized isozymes were those of lactic dehydrogenase. Several xenobiotic-metabolizing enzymes exist in multiple isozymes, including cytochrome P450 and glucuronosyltransferase.

**LC50 (median lethal concentration)** The concentration of a test chemical that, when a population of test organisms is exposed to it, is estimated to be fatal to 50% of the organisms under the stated conditions of the test. Normally used in lieu of the LD50 test in aquatic toxicology and inhalation toxicology.

**LD50 (median lethal dose)** The quantity of a chemical compound that, when applied directly to test organisms, is estimated to be fatal to 50% of those organisms under the stated conditions of the test. The LD50 value is the standard for comparison of acute toxicity between toxicants and between species. Because the results of LD50 determinations may vary widely, it is important that both biological and physical conditions be narrowly defined (e.g., strain, gender, and age of test organism; time and route of exposure; environmental conditions). The value may be determined graphically from a plot of log dose against mortality expressed in probability units (probits) or, more recently, by using one of several computer programs available.

**lethal synthesis** This term is used to describe the process by which a toxicant, similar in structure to an endogenous substrate, is incorporated into the same metabolic pathway as the endogenous substrate, ultimately being transformed into a toxic or lethal product. For example, fluoroacetate simulates acetate in intermediary metabolism, being transformed via the tricarboxylic acid cycle to fluorocitrate, which then inhibits aconitase, resulting in disruption of the TCA cycle and energy metabolism.

**lipophilic** The physical property of chemical compounds that causes them to be soluble in nonpolar solvents (e.g., chloroform and benzene) and, generally, relatively insoluble in polar solvents such as water. This property is important toxicologically because lipophilic compounds tend to enter the body easily and to be excretable only when they have been rendered less lipophilic by metabolic action.

**maximum tolerated dose (MTD)** The MTD has been defined for testing purposes by the US environmental Protection Agency as the highest dose that causes no more than a 10% weight decrement, as compared to the appropriate control groups, and does not produce mortality, clinical signs of toxicity, or pathologic lesions (other than those that may be related to a neoplastic response) that would be predicted to shorten the animals' natural life span. It is an important concept in chronic toxicity testing; however, the relevance of results produced by such large doses has become a matter of controversy.

**mechanism of action** *See* mode of action.

**membranes** Membranes of tissues, cells, and cell organelles are all basically similar in structure. They appear to be bimolecular lipid leaflets with proteins embedded in the matrix and also arranged on the outer polar surfaces. This basic plan is present despite many variations, and it is important in toxicological studies of uptake of toxicants by passive diffusion and active transport.

**microsomes** Microsomes are small closed vesicles (ca.110 nm in diameter) that represent membrane fragments formed from the endoplasmic reticulum when cells are disrupted by homogenization. Microsomes are separated from other cell organelles by differential centrifugation. The cell homogenate contains rough microsomes that are studded with ribosomes and are derived from rough endoplasmic reticulum, and smooth microsomes that are devoid of ribosomes and are derived from smooth endoplasmic reticulum. Microsomes are important preparations for studying the many processes carried out by the endoplasmic reticulum, such as protein biosynthesis and xenobiotic metabolism.

**mode of action (mode of toxic action)** Terms used to describe the mechanism(s) that enables a toxicant to exert its toxic effect. The term(s) may be narrowly used to describe only those events at the site of action (perhaps better referred to as mechanism of action), or more broadly, to describe the sequence of events from uptake from the environment, through metabolism, distribution, and so on, up to and including events at the site of action.

**monooxygenase (mixed-function oxidase)** An enzyme for which the co-substrates are an organic compound and molecular oxygen. In reactions catalyzed by these enzymes, one atom of a molecule of oxygen is incorporated into the substrate, whereas the other atom is reduced to water. Monooxygenases of importance in toxicology include cytochrome P450 and the flavin-containing monooxygenase, both of which initiate the metabolism of lipophilic xenobiotics by the introduction of a reactive polar group into the molecule. Such reactions may represent detoxication or may generate reactive intermediates of importance in toxic action. The term mixed-function oxidase is now considered obsolete and should not be used. The term multifunction oxidase was never widely adopted and also should not be used.

**mutagenicity** Mutations are heritable changes produced in the genetic information stored in the DNA of living cells. Chemicals capable of causing such changes are known as mutagens, and the process is known as mutagenesis.

**mycotoxins** Toxins produced by fungi. Many, such as aflatoxins, are particularly important in toxicology.

**nephrotoxicity** A pathologic state that can be induced by chemicals (nephrotoxicants) and in which the normal homeostatic functioning of the kidney is impaired. It is often associated with necrosis of the proximal tubule.

**neurotoxicity** This is a general term referring to all toxic effects on the nervous system, including toxic effects measured as behavioral abnormalities. Because the nervous system is complex, both structurally and functionally, and has considerable functional reserve, the study of neurotoxicity is a many-faceted branch of toxicology. It involves electrophysiology, receptor function, pathology, behavior, and other aspects.

**No Observed Effect Level (NOEL)** This is the highest dose level of a chemical that, in a given toxicity test, causes no observable effect in the test animals. The NOEL for a given chemical varies with the route and duration of exposure and the nature of the adverse effect (i.e., the indicator of toxicity). The NOEL for the most sensitive test species and the most sensitive indicator of toxicity is usually employed for regulatory purposes. Effects considered are usually adverse effects, and this value may be called the No Observed Adverse Effects Level (NOAEL).

**Occupational Safety and Health Administration (OSHA)** In the United States, OSHA is the government agency concerned with health and safety in the workplace. Through the administration of the Occupational Safety and Health Act, OSHA sets the standards for worker exposure to specific chemicals, for air concentration values, and for monitoring procedures. OSHA is also concerned with research (through the National Institute for Occupational Safety and Health, NIOSH), information, education, and training in occupational safety and health.

**oncogenes** Oncogenes are genes that, when activated in cells, can transform the cells from normal to neoplastic. Sometimes oncogenes are carried into normal cells by infecting viruses, particularly, RNA viruses or retroviruses. In some cases, however, the oncogene is already present in the normal human cell, and it needs only a mutation or other activating event to change it from a harmless and possible essential gene, called *proto-oncogene*, into a cancer-producing gene. More than 30 oncogenes have been identified in humans.

**oxidative phosphorylation** The conservation of chemical energy extracted from fuel oxidations by the phosphorylation of adenosine diphosphate (ADP) by inorganic phosphate to form adenosine triphosphate (ATP) is accomplished in several ways. The majority of ATP is formed by respiratory chain-linked oxidative phosphorylation associated with the electron transport system in the mitochondrial inner membrane. The oxidations are tightly coupled to phosphorylations through a chemiosmotic mechanism in which $H^+$ are pumped across the inner mitochondrial membrane. Uncouplers of oxidative phosphorylation serve as $H^+$ ionophores to dissipate the $H^+$ gradient, and thus uncouple the phosphorylations from the oxidations.

**oxidative stress** Damage to cells and cellular constituents and processes by reactive oxygen species generated in situ. Oxidative stress may be involved in such toxic interactions as DNA damage, lipid peroxidation, and pulmonary and cardiac toxicity. Because of the transitory nature of most reactive oxygen species, although oxidative stress is often invoked as a mechanism of toxicity, rigorous proof may be lacking.

**partition coefficient** This is a measure of the relative lipid solubility of a chemical and is determined by measuring the partitioning of the compound between an organic phase and an aqueous phase (e.g., octanol and water). The partition coefficient is important in studies of the uptake of toxicants because compounds with high coefficients (lipophilic compounds) are usually taken up more readily by organisms and tissues.

**pharmacokinetics** The study of the quantitative relationship between absorption, distribution, and excretion of chemicals and their metabolites. It involves derivation of rate constants for each of these processes and their integration into mathematical models that can predict the distribution of the chemical throughout the body compartments at any point in time after administration. Pharmacokinetics have been carried out most extensively in the case of clinical drugs. When applied specifically to toxicants, the term *toxicokinetics* is often used.

**phase I reactions** These reactions introduce a reactive polar group into lipophilic xenobiotics. In most cases this group becomes the site for conjugation during phase II reactions. Such reactions include microsomal monooxygenations, cytosolic and mitochondrial oxidations, cooxidations in the prostaglandin synthetase reaction, reductions, hydrolyses, and epoxide hydrolases. The products of phase I reactions may be potent electrophiles that can be conjugated and detoxified in phase II reactions or that may react with nucleophilic groups on cellular constituents, thereby causing toxicity.

**phase II reactions** Reactions involving the conjugation with endogenous substrates of phase I products and other xenobiotics that contain functional groups such as hydroxyl, amino, carboxyl, epoxide, or halogen. The endogenous metabolites include sugars, amino acids, glutathione, and sulfate. The conjugation products, with rare exceptions, are more polar, less toxic, and more readily excreted than are their parent compounds. There are two general types of conjugations: type I (e.g., glycoside and sulfate formation), in which an activated conjugating agent combines with substrate to yield the conjugated product, and type II (e.g., amino acid conjugation), in which the substrate is activated and then combines with an amino acid to yield a conjugated product.

**poison (toxicant)** A poison (toxicant) is any substance that causes a harmful effect when administered to a living organism. Due to a popular connotation that poisons are, by definition, fatal in their effects and that their administration is usually involved with attempted homicide or suicide, most toxicologists prefer the less prejudicial term toxicant. Poison is a quantitative concept. Almost any substance is harmful at some dose and, at the same time, is harmless at a very low dose. There is a range of possible effects, from subtle long-term chronic toxicity to immediate lethality.

**pollution** This is contamination of soil, water, food, or the atmosphere by the discharge or admixture of noxious materials. A pollutant is any chemical or substance contamination the environment and contributing to pollution.

**portals of entry** The sites at which xenobiotics enter the body. They include the skin, the gastrointestinal (GI) tract, and the respiratory system.

**potentiation** *See* synergism and potentiation.

**procarcinogen** *See* carcinogen, ultimate.

**promotion** The facilitation of the growth and development of neoplastic cells into a tumor. This process is manifested by enhancement of carcinogenesis when the agent is given after a carcinogen.

**pulmonary toxicity** This term refers to the effects of compounds that exert their toxic effects on the respiratory system, primarily the lungs.

**quantitative structure activity relationships (QSAR)** The relationship between the physical and/or chemical properties of chemicals and their ability to cause a particular effect, enter into particular reactions, and so on. The goal of QSAR studies in toxicology is to develop procedures whereby the toxicity of a compound can be predicted from its chemical structure by analogy with the properties of other toxicants of known structure and toxic properties.

**reactive intermediates (reactive metabolites)** Chemical compounds, produced during the metabolism of xenobiotics, that are more chemically reactive than is the parent compound. Although they are susceptible to detoxication by conjugation reactions, these metabolites, as a consequence of their increased reactivity, have a greater potential for adverse effects than does the parent compound. A well-known example is the metabolism of benzo(a)pyrene to its carcinogenic dihydrodiol epoxide derivative as a result of metabolism by cytochrome P450 and epoxide hydrolase. Reactive intermediates involved in toxic effects include epoxides, quinones, free radicals, reactive oxygen species, and a small number of unstable conjugation products.

**reference dose (RfD)** See acceptable daily intake (ADI).

**resistance** *See* adaptation to toxicants.

**Resource Conservation and Recovery Act (RCRA)** Administered by the EPA, the RCRA is the most important act governing the disposal of hazardous wastes in the United States; it promulgates standards for identification of hazardous wastes, their transportation, and their disposal. Included in the last are siting and construction criteria for landfills and other disposal facilities as well as the regulation of owners and operators of such facilities.

**risk assessment (risk analysis)** The process by which the potential adverse health effects of human exposure to chemicals are characterized; it includes the development of both qualitative and quantitative expression of risk. The process of risk assessment may be divided into four major components: hazard identification, dose-response assessment (high-dose to low-dose extrapolation), exposure assessment, and risk characterization.

**risk, toxicologic** The probability that some adverse effect will result from a given exposure to a chemical is known as the risk. It is the estimated frequency of occurrence of an event in a population and may be expressed in absolute terms (e.g., 1 in 1 million) or in terms of relative risk (i.e., the ratio of the risk in question to that in an equivalent unexposed population).

**safety factor (uncertainty factor)** A number by which the no observed effect level (NOEL) is divided to derive the reference dose (RfD), the reference concentration (RfC) or minimum risk level (MRL) of a chemical from experimental data. The safety factor is intended to account for the uncertainties inherent in estimating the potential effects of a chemical on humans from results obtained with test species. The safety factor allows for possible difference insensitivity between the test species and humans, as well as for variations in the sensitivity within the human population. The size of safety factor (e.g., 100–1000) varies with confidence in the database and the nature of the adverse effects. Small safety factors indicate a high degree of confidence in the data, an extensive database, and/or the availability of human data. Large safety factors are indicative of an inadequate and uncertain database and/or the severity of the unexpected toxic effect.

**selectivity (selective toxicity)** A characteristic of the relationship between toxic chemicals and living organisms whereby a particular chemical may be highly toxic to one species but relatively innocuous to another. The search for and study of selective toxicants is an important aspect of comparative toxicology because chemicals toxic to target species but innocuous to nontarget species are extremely valuable in agriculture and medicine. The mechanisms involved vary from differential penetration rates through different metabolic pathways to differences in receptor molecules at the site of toxic action.

**solvents** In toxicology this term usually refers to industrial solvents. These belong to many different chemical classes and a number of these are known to cause problems of toxicity to humans. They include aliphatic hydrocarbons (e.g., hexane), halogenated aliphatic hydrogens (e.g., methylene chloride), aliphatic alcohols (e.g., methanol), glycols and glycol ethers (e.g., propylene and propylene glycol), and aromatic hydrocarbons (e.g., toluene).

**subchronic toxicity** Toxicity due to chronic exposure to quantities of a toxicant that do not cause any evident acute toxicity for a time period that is extended but is not so long as to constitute a significant part of the lifespan or the species in question. In subchronic toxicity tests using mammals, a 30- to 90-day period is considered appropriate.

**synergism and potentiation** The terms synergism and potentiation have been variously used and defined but in any case involve a toxicity that is greater when two compounds are given simultaneously or sequentially than would be expected from a consideration of the toxicities of the compounds given alone. In an attempt to make the use of these terms uniform, it is a suggested that, insofar as toxic effects are concerned, they be used as defined as follows: both involve toxicity greater that would be expected from the toxicities of the compounds administered separately, but in the case of synergism one compound has little or no intrinsic toxicity administered alone, whereas in the case of potentiation both compounds have appreciable toxicity when administered alone.

**teratogenesis** This term refers to the production of defects in the reproduction process resulting in either reduced productivity due to fetal or embryonic mortality or the birth of offspring with physical, mental, behavioral, or developmental defects. Compounds causing such defects are known as teratogens.

**therapy** Poisoning therapy may be nonspecific or specific. Nonspecific therapy is treatment for poisoning that is not related to the mode of action of the particular toxicant. It is designed to prevent further uptake of the toxicant and to maintain vital signs. Specific therapy, however, is therapy related to the mode of action of the toxicant and not simply to the maintenance of vital signs by treatment of symptoms. Specific therapy may be based on activation and detoxication reactions, on mode of action, or on elimination of the toxicant. In some cases more than one antidote, with different modes of action, is available for the same toxicant.

**threshold dose** This is the dose of a toxicant below which no adverse effect occurs. The existence of such a threshold is based on the fundamental tenet of toxicology that, for any chemical, there exists a range of doses over which the severity of the observed effect is directly related to the dose, the threshold level representing the lower limit of this dose range. Although practical thresholds are considered to exist for most adverse effects, for regulatory purposes it is assumed that there is no threshold dose for carcinogens.

**threshold limit value (TLV)** The upper permissive limit of airborne concentrations of substances represents conditions under which it is believed that nearly all workers may be exposed repeatedly, day after day, without adverse effect. Threshold limits are based on the best available information from industrial experience, from experimental human and animal studies, and, when possible, from a combination of the three.

**threshold limit value–ceiling (TLV-C)** This is the concentration that should not be exceeded even momentarily. For some substances (e.g., irritant gases), only one TLV category, the TLV-C, may be relevant. For other substances, two or three TLV categories may need to be considered.

**threshold limit value–short-term exposure limit (TLV-STEL)** This is the maximal concentration to which workers can be exposed for a period up to 15 minutes continuously without suffering from (1) irritation, (2) chronic or irreversible tissue change, or (3) narcosis of sufficient degree to increase accident proneness.

**threshold limit value–time-weighted average (TLV-TWA)** This is the TWA concentration for a normal 8-hour workday or 40-hour workweek to which nearly all workers may be exposed repeatedly day after day, without adverse effect. Time-weighted averages allow certain permissible excursions above the limit, provided that they are compensated by equivalent excursions below the limit during the workday. In some instances the average concentration is calculated for a workweek rather than for a workday.

**tolerance** *See* adaptation to toxicants.

**Toxic Substances Control Act (TSCA)** Enacted in 1976, the TSCA provides the EPA with the authority to require testing and to regulate chemicals, both old and new, entering the environment. It was intended to supplement sections of the Clean Air Act, the Clean Water Act, and the Occupational Safety and Health Act that already provide for regulation of chemicals. Manufacturers are required to submit information to allow the EPA to identify and evaluate the potential hazards of a chemical prior to its introduction into commerce. The act also provides for the regulation of production, use, distribution and disposal of chemicals.

**toxicant** *See* poison.

**toxicokinetics** *See* pharmacokinetics.

**toxicology** Toxicology is defined as that branch of science that deals with poisons (toxicants) and their effects; a poison is defined as any substance that causes a harmful effect when administered, either by accident or design, to a living organism. There are difficulties in bringing a more precise definition to the meaning of poison and in the definition and measurement of toxic effect. The range of deleterious effects is wide and varies with species, gender, developmental stage, and so on, while the effects of toxicants are always dose dependent.

**toxin** This is a toxicant produced by a living organism. Toxin should never be used as a synonym for toxicant.

**transport** In toxicology this term refers to the mechanisms that bring about movement of toxicants and their metabolites from one site in the organism to another. *Transport* usually involves binding to either blood albumins or blood lipoproteins.

**ultimate carcinogen** *See* carcinogen, ultimate.

**venom** A venom is a toxin produced by an animal specifically for the poisoning of other species via a mechanism designed to deliver the toxin to its prey. Examples include the venom of bees and wasps, delivered by a sting, and the venom of snakes, delivered by fangs.

**water pollution** Water pollution is of concern in both industrialized and nonindustrialized nations. Chemical contamination is more common in industrialized nations, whereas microbial contamination is more important in nonindustrialized areas. Surface water contamination has been the primary cause for concern, but since the discovery of agricultural and industrial chemicals in groundwater, contamination of water from this source is also a problem. Water pollution may arise from runoff of agricultural chemicals, from sewage, or from specific industrial sources. Agricultural chemicals found in water include insecticides, herbicides, fungicides, and nematocides; fertilizers, although less of a toxic hazard, contribute to such environmental problems as eutrophication. Other chemicals of concern include low molecular-weight halogenated hydrocarbons such as chloroform, dichloroethane, and carbon tetrachloride; polychlorinated biphenyls (PCBs); chlorophenols; 2,3,7,8-tetrachlorodibenzo-*p*-dioxin (TCDD); phthalate ester plasticizers; detergents; and a number of toxic inorganics.

**xenobiotic** A general term used to describe any chemical interacting with an organism that does not occur in the normal metabolic pathways of that organism. The use of this term in lieu of "foreign compound," among others, has gained wide acceptance.

*A Textbook of Modern Toxicology, Third Edition,* edited by Ernest Hodgson
ISBN 0-471-26508-X  Copyright © 2004 John Wiley & Sons, Inc.